P9-DBN-386

Grzimek's
Animal Life Encyclopedia

Second Edition

●●●●

Grzimek's
Animal Life Encyclopedia

Second Edition

●●●●

Volume 12
Mammals I

Devra G. Kleiman, Advisory Editor
Valerius Geist, Advisory Editor
Melissa C. McDade, Project Editor

Joseph E. Trumpey, Chief Scientific Illustrator

Michael Hutchins, Series Editor
In association with the American Zoo and Aquarium Association

GALE®

THOMSON
GALE

Detroit • New York • San Diego • San Francisco • Cleveland • New Haven, Conn. • Waterville, Maine • London • Munich

Grzimek's Animal Life Encyclopedia, Second Edition
Volume 12: Mammals I

Project Editor
Melissa C. McDade

Editorial
Stacey Blachford, Deirdre S. Blanchfield, Madeline Harris, Christine Jeryan, Kate Kretschmann, Mark Springer, Ryan Thomason

Indexing Services
Synapse, the Knowledge Link Corporation

Permissions
Margaret Chamberlain

Imaging and Multimedia
Randy Bassett, Mary K. Grimes, Lezlie Light, Christine O'Bryan, Barbara Yarrow, Robyn V. Young

Product Design
Tracey Rowens, Jennifer Wahi

Manufacturing
Wendy Blurton, Dorothy Maki, Evi Seoud, Mary Beth Trimper

© 2004 by Gale. Gale is an imprint of The Gale Group, Inc., a division of Thomson Learning Inc.

Gale and Design® and Thomson Learning™ are trademarks used herein under license.

For more information contact
The Gale Group, Inc.
27500 Drake Rd.
Farmington Hills, MI 48331-3535
Or you can visit our Internet site at
http://www.gale.com

ALL RIGHTS RESERVED
No part of this work covered by the copyright hereon may be reproduced or used in any form or by any means—graphic, electronic, or mechanical, including photocopying, recording, taping, Web distribution, or information storage retrieval systems—without the written permission of the publisher.

For permission to use material from this product, submit your request via Web at http://www.gale-edit.com/permissions, or you may download our Permissions Request form and submit your request by fax or mail to: The Gale Group, Inc., Permissions Department, 27500 Drake Road, Farmington Hills, MI, 48331-3535, Permissions hotline: 248-699-8074 or 800-877-4253, ext. 8006, Fax: 248-699-8074 or 800-762-4058.

Cover photo of numbat (*Myrmecobius fasciatus*) by Frans Lanting/Minden Pictures. Back cover photos of sea anemone by AP/Wide World Photos/University of Wisconsin-Superior; land snail, lionfish, golden frog, and green python by JLM Visuals; red-legged locust © 2001 Susan Sam; hornbill by Margaret F. Kinnaird; and tiger by Jeff Lepore/Photo Researchers. All reproduced by permission.

While every effort has been made to ensure the reliability of the information presented in this publication, The Gale Group, Inc. does not guarantee the accuracy of the data contained herein. The Gale Group, Inc. accepts no payment for listing; and inclusion in the publication of any organization, agency, institution, publication, service, or individual does not imply endorsement of the editors and publisher. Errors brought to the attention of the publisher and verified to the satisfaction of the publisher will be corrected in future editions.

ISBN   0-7876-5362-4 (vols. 1–17 set)
0-7876-6573-8 (vols. 12–16 set)
0-7876-5788-3 (vol. 12)
0-7876-5789-1 (vol. 13)
0-7876-5790-5 (vol. 14)
0-7876-5791-3 (vol. 15)
0-7876-5792-1 (vol. 16)

This title is also available as an e-book.
ISBN 0-7876-7750-7 (17-vol set)
Contact your Gale sales representative for ordering information.

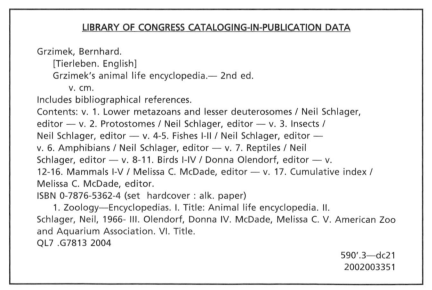

LIBRARY OF CONGRESS CATALOGING-IN-PUBLICATION DATA

Grzimek, Bernhard.
  [Tierleben. English]
  Grzimek's animal life encyclopedia.— 2nd ed.
    v. cm.
Includes bibliographical references.
Contents: v. 1. Lower metazoans and lesser deuterosomes / Neil Schlager, editor — v. 2. Protostomes / Neil Schlager, editor — v. 3. Insects / Neil Schlager, editor — v. 4-5. Fishes I-II / Neil Schlager, editor — v. 6. Amphibians / Neil Schlager, editor — v. 7. Reptiles / Neil Schlager, editor — v. 8-11. Birds I-IV / Donna Olendorf, editor — v. 12-16. Mammals I-V / Melissa C. McDade, editor — v. 17. Cumulative index / Melissa C. McDade, editor.
ISBN 0-7876-5362-4 (set hardcover : alk. paper)
  1. Zoology—Encyclopedias. I. Title: Animal life encyclopedia. II. Schlager, Neil, 1966- III. Olendorf, Donna IV. McDade, Melissa C. V. American Zoo and Aquarium Association. VI. Title.
QL7 .G7813 2004

590'.3—dc21
2002003351

Printed in Canada
10 9 8 7 6 5 4 3 2 1

Recommended citation: *Grzimek's Animal Life Encyclopedia,* 2nd edition. Volumes 12–16, *Mammals I–V,* edited by Michael Hutchins, Devra G. Kleiman, Valerius Geist, and Melissa C. McDade. Farmington Hills, MI: Gale Group, 2003.

Contents

Contents

Contents

Foreword

Earth is teeming with life. No one knows exactly how many distinct organisms inhabit our planet, but more than 5 million different species of animals and plants could exist, ranging from microscopic algae and bacteria to gigantic elephants, redwood trees and blue whales. Yet, throughout this wonderful tapestry of living creatures, there runs a single thread: Deoxyribonucleic acid or DNA. The existence of DNA, an elegant, twisted organic molecule that is the building block of all life, is perhaps the best evidence that all living organisms on this planet share a common ancestry. Our ancient connection to the living world may drive our curiosity, and perhaps also explain our seemingly insatiable desire for information about animals and nature. Noted zoologist, E. O. Wilson, recently coined the term "biophilia" to describe this phenomenon. The term is derived from the Greek *bios* meaning "life" and *philos* meaning "love." Wilson argues that we are human because of our innate affinity to and interest in the other organisms with which we share our planet. They are, as he says, "the matrix in which the human mind originated and is permanently rooted." To put it simply and metaphorically, our love for nature flows in our blood and is deeply engrained in both our psyche and cultural traditions.

Our own personal awakenings to the natural world are as diverse as humanity itself. I spent my early childhood in rural Iowa where nature was an integral part of my life. My father and I spent many hours collecting, identifying and studying local insects, amphibians and reptiles. These experiences had a significant impact on my early intellectual and even spiritual development. One event I can recall most vividly. I had collected a cocoon in a field near my home in early spring. The large, silky capsule was attached to a stick. I brought the cocoon back to my room and placed it in a jar on top of my dresser. I remember waking one morning and, there, perched on the tip of the stick was a large moth, slowly moving its delicate, light green wings in the early morning sunlight. It took my breath away. To my inexperienced eyes, it was one of the most beautiful things I had ever seen. I knew it was a moth, but did not know which species. Upon closer examination, I noticed two moon-like markings on the wings and also noted that the wings had long "tails", much like the ubiquitous tiger swallow-tail butterflies that visited the lilac bush in our backyard. Not wanting to suffer my ignorance any longer, I reached immediately for my *Golden Guide to North*

American Insects and searched through the section on moths and butterflies. It was a luna moth! My heart was pounding with the excitement of new knowledge as I ran to share the discovery with my parents.

I consider myself very fortunate to have made a living as a professional biologist and conservationist for the past 20 years. I've traveled to over 30 countries and six continents to study and photograph wildlife or to attend related conferences and meetings. Yet, each time I encounter a new and unusual animal or habitat my heart still races with the same excitement of my youth. If this is biophilia, then I certainly possess it, and it is my hope that others will experience it too. I am therefore extremely proud to have served as the series editor for the Gale Group's rewrite of *Grzimek's Animal Life Encyclopedia*, one of the best known and widely used reference works on the animal world. *Grzimek's* is a celebration of animals, a snapshot of our current knowledge of the Earth's incredible range of biological diversity. Although many other animal encyclopedias exist, *Grzimek's Animal Life Encyclopedia* remains unparalleled in its size and in the breadth of topics and organisms it covers.

The revision of these volumes could not come at a more opportune time. In fact, there is a desperate need for a deeper understanding and appreciation of our natural world. Many species are classified as threatened or endangered, and the situation is expected to get much worse before it gets better. Species extinction has always been part of the evolutionary history of life; some organisms adapt to changing circumstances and some do not. However, the current rate of species loss is now estimated to be 1,000–10,000 times the normal "background" rate of extinction since life began on Earth some 4 billion years ago. The primary factor responsible for this decline in biological diversity is the exponential growth of human populations, combined with peoples' unsustainable appetite for natural resources, such as land, water, minerals, oil, and timber. The world's human population now exceeds 6 billion, and even though the average birth rate has begun to decline, most demographers believe that the global human population will reach 8–10 billion in the next 50 years. Much of this projected growth will occur in developing countries in Central and South America, Asia and Africa—regions that are rich in unique biological diversity.

Finding solutions to conservation challenges will not be easy in today's human-dominated world. A growing number of people live in urban settings and are becoming increasingly isolated from nature. They "hunt" in supermarkets and malls, live in apartments and houses, spend their time watching television and searching the World Wide Web. Children and adults must be taught to value biological diversity and the habitats that support it. Education is of prime importance now while we still have time to respond to the impending crisis. There still exist in many parts of the world large numbers of biological "hotspots"—places that are relatively unaffected by humans and which still contain a rich store of their original animal and plant life. These living repositories, along with selected populations of animals and plants held in professionally managed zoos, aquariums and botanical gardens, could provide the basis for restoring the planet's biological wealth and ecological health. This encyclopedia and the collective knowledge it represents can assist in educating people about animals and their ecological and cultural significance. Perhaps it will also assist others in making deeper connections to nature and spreading biophilia. Information on the conservation status, threats and efforts to preserve various species have been integrated into this revision. We have also included information on the cultural significance of animals, including their roles in art and religion.

It was over 30 years ago that Dr. Bernhard Grzimek, then director of the Frankfurt Zoo in Frankfurt, Germany, edited the first edition of *Grzimek's Animal Life Encyclopedia*. Dr. Grzimek was among the world's best known zoo directors and conservationists. He was a prolific author, publishing nine books. Among his contributions were: *Serengeti Shall Not Die*, *Rhinos Belong to Everybody* and *He and I and the Elephants*. Dr. Grzimek's career was remarkable. He was one of the first modern zoo or aquarium directors to understand the importance of zoo involvement in *in situ* conservation, that is, of their role in preserving wildlife in nature. During his tenure, Frankfurt Zoo became one of the leading western advocates and supporters of wildlife conservation in East Africa. Dr. Grzimek served as a Trustee of the National Parks Board of Uganda and Tanzania and assisted in the development of several protected areas. The film he made with his son Michael, *Serengeti Shall Not Die*, won the 1959 Oscar for best documentary.

Professor Grzimek has recently been criticized by some for his failure to consider the human element in wildlife conservation. He once wrote: "A national park must remain a primordial wilderness to be effective. No men, not even native ones, should live inside its borders." Such ideas, although considered politically incorrect by many, may in retrospect actually prove to be true. Human populations throughout Africa continue to grow exponentially, forcing wildlife into small islands of natural habitat surrounded by a sea of humanity. The illegal commercial bushmeat trade—the hunting of endangered wild animals for large scale human consumption—is pushing many species, including our closest relatives, the gorillas, bonobos and chimpanzees, to the brink of extinction. The trade is driven by widespread poverty and lack of economic alternatives. In order for some species to survive it will be necessary, as Grzimek suggested, to establish and enforce a system of protected areas where wildlife can roam free from exploitation of any kind.

While it is clear that modern conservation must take the needs of both wildlife and people into consideration, what will the quality of human life be if the collective impact of short-term economic decisions is allowed to drive wildlife populations into irreversible extinction? Many rural populations living in areas of high biodiversity are dependent on wild animals as their major source of protein. In addition, wildlife tourism is the primary source of foreign currency in many developing countries and is critical to their financial and social stability. When this source of protein and income is gone, what will become of the local people? The loss of species is not only a conservation disaster; it also has the potential to be a human tragedy of immense proportions. Protected areas, such as national parks, and regulated hunting in areas outside of parks are the only solutions. What critics do not realize is that the fate of wildlife and people in developing countries is closely intertwined. Forests and savannas emptied of wildlife will result in hungry, desperate people, and will, in the long-term lead to extreme poverty and social instability. Dr. Grzimek's early contributions to conservation should be recognized, not only as benefiting wildlife, but as benefiting local people as well.

Dr. Grzimek's hope in publishing his *Animal Life Encyclopedia* was that it would "...disseminate knowledge of the animals and love for them", so that future generations would "...have an opportunity to live together with the great diversity of these magnificent creatures." As stated above, our goals in producing this updated and revised edition are similar. However, our challenges in producing this encyclopedia were more formidable. The volume of knowledge to be summarized is certainly much greater in the twenty-first century than it was in the 1970's and 80's. Scientists, both professional and amateur, have learned and published a great deal about the animal kingdom in the past three decades, and our understanding of biological and ecological theory has also progressed. Perhaps our greatest hurdle in producing this revision was to include the new information, while at the same time retaining some of the characteristics that have made *Grzimek's Animal Life Encyclopedia* so popular. We have therefore strived to retain the series' narrative style, while giving the information more organizational structure. Unlike the original *Grzimek's*, this updated version organizes information under specific topic areas, such as reproduction, behavior, ecology and so forth. In addition, the basic organizational structure is generally consistent from one volume to the next, regardless of the animal groups covered. This should make it easier for users to locate information more quickly and efficiently. Like the original Grzimek's, we have done our best to avoid any overly technical language that would make the work difficult to understand by non-biologists. When certain technical expressions were necessary, we have included explanations or clarifications.

Considering the vast array of knowledge that such a work represents, it would be impossible for any one zoologist to have completed these volumes. We have therefore sought specialists from various disciplines to write the sections with which they are most familiar. As with the original *Grzimek's*, we have engaged the best scholars available to serve as topic editors, writers, and consultants. There were some complaints about inaccuracies in the original English version that may have been due to mistakes or misinterpretation during the complicated translation process. However, unlike the original *Grzimek's*, which was translated from German, this revision has been completely re-written by English-speaking scientists. This work was truly a cooperative endeavor, and I thank all of those dedicated individuals who have written, edited, consulted, drawn, photographed, or contributed to its production in any way. The names of the topic editors, authors, and illustrators are presented in the list of contributors in each individual volume.

The overall structure of this reference work is based on the classification of animals into naturally related groups, a discipline known as taxonomy or biosystematics. Taxonomy is the science through which various organisms are discovered, identified, described, named, classified and catalogued. It should be noted that in preparing this volume we adopted what might be termed a conservative approach, relying primarily on traditional animal classification schemes. Taxonomy has always been a volatile field, with frequent arguments over the naming of or evolutionary relationships between various organisms. The advent of DNA fingerprinting and other advanced biochemical techniques has revolutionized the field and, not unexpectedly, has produced both advances and confusion. In producing these volumes, we have consulted with specialists to obtain the most up-to-date information possible, but knowing that new findings may result in changes at any time. When scientific controversy over the classification of a particular animal or group of animals existed, we did our best to point this out in the text.

Readers should note that it was impossible to include as much detail on some animal groups as was provided on others. For example, the marine and freshwater fish, with vast numbers of orders, families, and species, did not receive as detailed a treatment as did the birds and mammals. Due to practical and financial considerations, the publishers could provide only so much space for each animal group. In such cases, it was impossible to provide more than a broad overview and to feature a few selected examples for the purposes of illustration. To help compensate, we have provided a few key bibliographic references in each section to aid those interested in learning more. This is a common limitation in all reference works, but *Grzimek's Encyclopedia of Animal Life* is still the most comprehensive work of its kind.

I am indebted to the Gale Group, Inc. and Senior Editor Donna Olendorf for selecting me as Series Editor for this project. It was an honor to follow in the footsteps of Dr. Grzimek and to play a key role in the revision that still bears his name. *Grzimek's Animal Life Encyclopedia* is being published by the Gale Group, Inc. in affiliation with my employer, the American Zoo and Aquarium Association (AZA), and I would like to thank AZA Executive Director, Sydney J. Butler; AZA Past-President Ted Beattie (John G. Shedd Aquarium, Chicago, IL); and current AZA President, John Lewis (John Ball Zoological Garden, Grand Rapids, MI), for approving my participation. I would also like to thank AZA Conservation and Science Department Program Assistant, Michael Souza, for his assistance during the project. The AZA is a professional membership association, representing 215 accredited zoological parks and aquariums in North America. As Director/William Conway Chair, AZA Department of Conservation and Science, I feel that I am a philosophical descendant of Dr. Grzimek, whose many works I have collected and read. The zoo and aquarium profession has come a long way since the 1970s, due, in part, to innovative thinkers such as Dr. Grzimek. I hope this latest revision of his work will continue his extraordinary legacy.

Silver Spring, Maryland, 2001
Michael Hutchins
Series Editor

How to use this book

Gzimek's Animal Life Encyclopedia is an internationally prominent scientific reference compilation, first published in German in the late 1960s, under the editorship of zoologist Bernhard Grzimek (1909-1987). In a cooperative effort between Gale and the American Zoo and Aquarium Association, the series is being completely revised and updated for the first time in over 30 years. Gale is expanding the series from 13 to 17 volumes, commissioning new color images, and updating the information while also making the set easier to use. The order of revisions is:

Vol 8–11: Birds I–IV
Vol 6: Amphibians
Vol 7: Reptiles
Vol 4–5: Fishes I–II
Vol 12–16: Mammals I–V
Vol 1: Lower Metazoans and Lesser Deuterostomes
Vol 2: Protostomes
Vol 3: Insects
Vol 17: Cumulative Index

Organized by taxonomy

The overall structure of this reference work is based on the classification of animals into naturally related groups, a discipline known as taxonomy—the science through which various organisms are discovered, identified, described, named, classified, and catalogued. Starting with the simplest life forms, the lower metazoans and lesser deuterostomes, in volume 1, the series progresses through the more complex animal classes, culminating with the mammals in volumes 12–16. Volume 17 is a stand-alone cumulative index.

Organization of chapters within each volume reinforces the taxonomic hierarchy. In the case of the Mammals volumes, introductory chapters describe general characteristics of all organisms in these groups, followed by taxonomic chapters dedicated to Order, Family, or Subfamily. Species accounts appear at the end of the Family and Subfamily chapters To help the reader grasp the scientific arrangement, each type of chapter has a distinctive color and symbol:

● =Order Chapter (blue background)

⬟ =Monotypic Order Chapter (green background)

▲ =Family Chapter (yellow background)

△ =Subfamily Chapter (yellow background)

Introductory chapters have a loose structure, reminiscent of the first edition. While not strictly formatted, Order chapters are carefully structured to cover basic information about member families. Monotypic orders, comprised of a single family, utilize family chapter organization. Family and subfamily chapters are most tightly structured, following a prescribed format of standard rubrics that make information easy to find and understand. Family chapters typically include:

Thumbnail introduction
  Common name
  Scientific name
  Class
  Order
  Suborder
  Family
  Thumbnail description
  Size
  Number of genera, species
  Habitat
  Conservation status
Main essay
  Evolution and systematics
  Physical characteristics
  Distribution
  Habitat
  Behavior
  Feeding ecology and diet
  Reproductive biology
  Conservation status
  Significance to humans
Species accounts
  Common name
  Scientific name
  Subfamily
  Taxonomy
  Other common names
  Physical characteristics
  Distribution
  Habitat
  Behavior

Feeding ecology and diet
Reproductive biology
Conservation status
Significance to humans
Resources
  Books
  Periodicals
  Organizations
  Other

Color graphics enhance understanding

Grzimek's features approximately 3,000 color photos, including approximately 1,560 in five Mammals volumes; 3,500 total color maps, including nearly 550 in the Mammals volumes; and approximately 5,500 total color illustrations, including approximately 930 in the Mammals volumes. Each featured species of animal is accompanied by both a distribution map and an illustration.

All maps in *Grzimek's* were created specifically for the project by XNR Productions. Distribution information was provided by expert contributors and, if necessary, further researched at the University of Michigan Zoological Museum library. Maps are intended to show broad distribution, not definitive ranges.

All the color illustrations in *Grzimek's* were created specifically for the project by Michigan Science Art. Expert contributors recommended the species to be illustrated and provided feedback to the artists, who supplemented this information with authoritative references and animal skins from University of Michgan Zoological Museum library. In addition to species illustrations, *Grzimek's* features conceptual drawings that illustrate charactcristic traits and behaviors.

About the contributors

The essays were written by scientists, professors, and other professionals. *Grzimek's* subject advisors reviewed the completed essays to insure consistency and accuracy.

Standards employed

In preparing these volumes, the editors adopted a conservative approach to taxonomy, relying on Wilson and Reeder's *Mammal Species of the World: a Taxonomic and Geographic Reference* (1993) as a guide. Systematics is a dynamic discipline in that new species are being discovered continuously, and new techniques (e.g., DNA sequencing) frequently result in changes in the hypothesized evolutionary relationships among various organisms. Consequently, controversy often exists regarding classification of a particular animal or group of animals; such differences are mentioned in the text.

Grzimek's has been designed with ready reference in mind and the editors have standardized information wherever feasible. For **Conservation status,** *Grzimek's* follows the IUCN Red List system, developed by its Species Survival Commission. The Red List provides the world's most comprehensive inventory of the global conservation status of plants and animals. Using a set of criteria to evaluate extinction risk, the IUCN recognizes the following categories: Extinct, Extinct in the Wild, Critically Endangered, Endangered, Vulnerable, Conservation Dependent, Near Threatened, Least Concern, and Data Deficient. For a complete explanation of each category, visit the IUCN web page at <http://www.iucn.org/>.

• • • • •

Advisory boards

Series advisor

Michael Hutchins, PhD
Director of Conservation and Science/William Conway Chair
American Zoo and Aquarium Association
Silver Spring, Maryland

Subject advisors

Volume 1: Lower Metazoans and Lesser Deuterostomes

Dennis A. Thoney, PhD
Director, Marine Laboratory & Facilities
Humboldt State University
Arcata, California

Volume 2: Protostomes

Sean F. Craig, PhD
Assistant Professor, Department of Biological Sciences
Humboldt State University
Arcata, California

Dennis A. Thoney, PhD
Director, Marine Laboratory & Facilities
Humboldt State University
Arcata, California

Volume 3: Insects

Arthur V. Evans, DSc
Research Associate, Department of Entomology
Smithsonian Institution
Washington, DC

Rosser W. Garrison, PhD
Research Associate, Department of Entomology
Natural History Museum
Los Angeles, California

Volumes 4–5: Fishes I– II

Paul V. Loiselle, PhD
Curator, Freshwater Fishes

New York Aquarium
Brooklyn, New York
Dennis A. Thoney, PhD
Director, Marine Laboratory & Facilities
Humboldt State University
Arcata, California

Volume 6: Amphibians

William E. Duellman, PhD
Curator of Herpetology Emeritus
Natural History Museum and Biodiversity Research Center
University of Kansas
Lawrence, Kansas

Volume 7: Reptiles

James B. Murphy, DSc
Smithsonian Research Associate
Department of Herpetology
National Zoological Park
Washington, DC

Volumes 8–11: Birds I–IV

Walter J. Bock, PhD
Permanent secretary, International Ornithological Congress
Professor of Evolutionary Biology
Department of Biological Sciences,
Columbia University
New York, New York
Jerome A. Jackson, PhD
Program Director, Whitaker Center for Science, Mathematics, and Technology Education
Florida Gulf Coast University
Ft. Myers, Florida

Volumes 12–16: Mammals I–V

Valerius Geist, PhD
Professor Emeritus of Environmental Science
University of Calgary
Calgary, Alberta
Canada

Devra G. Kleiman, PhD
Smithsonian Research Associate
National Zoological Park
Washington, DC

Library advisors

James Bobick
Head, Science & Technology Department
Carnegie Library of Pittsburgh
Pittsburgh, Pennsylvania

Linda L. Coates
Associate Director of Libraries
Zoological Society of San Diego Library
San Diego, California

Lloyd Davidson, PhD
Life Sciences bibliographer and head, Access Services
Seeley G. Mudd Library for Science and Engineering
Evanston, Illinois

Thane Johnson
Librarian
Oklahoma City Zoo
Oklahoma City, Oklahoma

Charles Jones
Library Media Specialist
Plymouth Salem High School
Plymouth, Michigan

Ken Kister
Reviewer/General Reference teacher
Tampa, Florida

Richard Nagler
Reference Librarian
Oakland Community College
Southfield Campus
Southfield, Michigan

Roland Person
Librarian, Science Division
Morris Library
Southern Illinois University
Carbondale, Illinois

Contributing writers

Mammals I–V

Clarence L. Abercrombie, PhD
Wofford College
Spartanburg, South Carolina

Cleber J. R. Alho, PhD
Departamento de Ecologia (retired)
Universidade de Brasília
Brasília, Brazil

Carlos Altuna, Lic
Sección Etología
Facultad de Ciencias
Universidad de la República Oriental
del Uruguay
Montevideo, Uruguay

Anders Angerbjörn, PhD
Department of Zoology
Stockholm University
Stockholm, Sweden

William Arthur Atkins
Atkins Research and Consulting
Normal, Illinois

Adrian A. Barnett, PhD
Centre for Research in Evolutionary
Anthropology
School of Life Sciences
University of Surrey Roehampton
West Will, London
United Kingdom

Leonid Baskin, PhD
Institute of Ecology and Evolution
Moscow, Russia

Paul J. J. Bates, PhD
Harrison Institute
Sevenoaks, Kent
United Kingdom

Amy-Jane Beer, PhD
Origin Natural Science
York, United Kingdom

Cynthia Berger, MS
National Association of Science Writers

Richard E. Bodmer, PhD
Durrell Institute of Conservation and
Ecology
University of Kent
Canterbury, Kent
United Kingdom

Daryl J. Boness, PhD
National Zoological Park
Smithsonian Institution
Washington, DC

Justin S. Brashares, PhD
Centre for Biodiversity Research
University of British Columbia
Vancouver, British Columbia
Canada

Hynek Burda, PhD
Department of General Zoology Fac-
ulty of Bio- and Geosciences
University of Essen
Essen, Germany

Susan Cachel, PhD
Department of Anthropology
Rutgers University
New Brunswick, New Jersey

Alena Cervená, PhD
Department of Zoology
National Museum Prague
Czech Republic

Jaroslav Cerveny, PhD
Institute of Vertebrate Biology
Czech Academy of Sciences
Brno, Czech Republic

David J. Chivers, MA, PhD, ScD
Head, Wildlife Research Group
Department of Anatomy

University of Cambridge
Cambridge, United Kingdom

Jasmin Chua, MS
Freelance Writer

Lee Curtis, MA
Director of Promotions
Far North Queensland Wildlife Res-
cue Association
Far North Queensland, Australia

Guillermo D'Elía, PhD
Departamento de Biología Animal
Facultad de Ciencias
Universidad de la República
Montevideo, Uruguay

Tanya Dewey
University of Michigan Museum of
Zoology
Ann Arbor, Michigan

Craig C. Downer, PhD
Andean Tapir Fund
Minden, Nevada

Amy E. Dunham
Department of Ecology and Evolution
State University of New York at Stony
Brook
Stony Brook, New York

Stewart K. Eltringham, PhD
Department of Zoology
University of Cambridge
Cambridge, United Kingdom.

Melville Brockett Fenton, PhD
Department of Biology
University of Western Ontario
London, Ontario
Canada

Kevin F. Fitzgerald, BS
Freelance Science Writer
South Windsor, Connecticut

Theodore H. Fleming, PhD
Department of Biology
University of Miami
Coral Gables, Florida

Gabriel Francescoli, PhD
Sección Etología
Facultad de Ciencias
Universidad de la República Oriental
del Uruguay
Montevideo, Uruguay

Udo Gansloßer, PhD
Department of Zoology
Lehrstuhl I
University of Erlangen-Nürnberg
Fürth, Germany

Valerius Geist, PhD
Professor Emeritus of Environmental
Science
University of Calgary
Calgary, Alberta
Canada

Roger Gentry, PhD
NOAA Fisheries
Marine Mammal Division
Silver Spring, Maryland

Kenneth C. Gold, PhD
Chicago, Illinois

Steve Goodman, PhD
Field Museum of Natural History
Chicago, Illinois and
WWF Madagascar
Programme Office
Antananarivo, Madagascar

Nicole L. Gottdenker
St. Louis Zoo
University of Missouri
St. Louis, Missouri and The Charles
Darwin Research Station
Galápagos Islands, Ecuador

Brian W. Grafton, PhD
Department of Biological Sciences
Kent State University
Kent, Ohio

Joel H. Grossman
Freelance Writer
Santa Monica, California

Mark S. Hafner, PhD
Lowery Professor and Curator of
Mammals
Museum of Natural Science and De-
partment of Biological Sciences
Louisiana State University
Baton Rouge, Louisiana

Alton S. Harestad, PhD
Faculty of Science
Simon Fraser University Burnaby
Vancouver, British Columbia
Canada

Robin L. Hayes
Bat Conservation of Michigan

Kristofer M. Helgen
School of Earth and Environmental
Sciences
University of Adelaide
Adelaide, Australia

Eckhard W. Heymann, PhD
Department of Ethology and Ecology
German Primate Center
Göttingen, Germany

Hannah Hoag, MS
Science Journalist

Hendrik Hoeck, PhD
Max-Planck- Institut für Verhal-
tensphysiologie
Seewiesen, Germany

David Holzman, BA
Freelance Writer
Journal Highlights Editor
American Society for Microbiology

Rodney L. Honeycutt, PhD
Departments of Wildlife and Fisheries
Sciences and Biology and Faculty of
Genetics
Texas A&M University
College Station, Texas

Ivan Horácek, Prof. RNDr, PhD
Head of Vertebrate Zoology
Charles University Prague
Praha, Czech Republic

Brian Douglas Hoyle, PhD
President, Square Rainbow Limited
Bedford, Nova Scotia
Canada

Graciela Izquierdo, PhD
Sección Etología
Facultad de Ciencias
Universidad de la República Oriental
del Uruguay
Montevideo, Uruguay

Jennifer U. M. Jarvis, PhD
Zoology Department
University of Cape Town
Rondebosch, South Africa

Christopher Johnson, PhD
Department of Zoology and Tropical
Ecology
James Cook University
Townsville, Queensland
Australia

Menna Jones, PhD
University of Tasmania School of Zo-
ology
Hobart, Tasmania
Australia

Mike J. R. Jordan, PhD
Curator of Higher Vertebrates
North of England Zoological Society
Chester Zoo
Upton, Chester
United Kingdom

Corliss Karasov
Science Writer
Madison, Wisconsin

Tim Karels, PhD
Department of Biological Sciences
Auburn University
Auburn, Alabama

Serge Larivière, PhD
Delta Waterfowl Foundation
Manitoba, Canada

Adrian Lister
University College London
London, United Kingdom

W. J. Loughry, PhD
Department of Biology
Valdosta State University
Valdosta, Georgia

Geoff Lundie-Jenkins, PhD
Queensland Parks and Wildlife Service
Queensland, Australia

Peter W. W. Lurz, PhD
Centre for Life Sciences Modelling
School of Biology
University of Newcastle
Newcastle upon Tyne, United King-
dom

Colin D. MacLeod, PhD
School of Biological Sciences (Zool-
ogy)
University of Aberdeen
Aberdeen, United Kingdom

James Malcolm, PhD
Department of Biology
University of Redlands
Redlands, California

Contributing writers

David P. Mallon, PhD
Glossop
Derbyshire, United Kingdom

Robert D. Martin, BA (Hons), DPhil, DSc
Provost and Vice President
Academic Affairs
The Field Museum
Chicago, Illinois

Gary F. McCracken, PhD
Department of Ecology and Evolutionary Biology
University of Tennessee
Knoxville, Tennessee

Colleen M. McDonough, PhD
Department of Biology
Valdosta State University
Valdosta, Georgia

William J. McShea, PhD
Department of Conservation Biology
Conservation and Research Center
Smithsonian National Zoological Park
Washington, DC

Rodrigo A. Medellín, PhD
Instituto de Ecología
Universidad Nacional Autónoma de México
Mexico City, Mexico

Leslie Ann Mertz, PhD
Fish Lake Biological Program
Wayne State University
Detroit, Michigan

Gus Mills, PhD
SAN Parks/Head
Carnivore Conservation Group, EWT
Skukuza, South Africa

Patricia D. Moehlman, PhD
IUCN Equid Specialist Group

Paula Moreno, MS
Texas A&M University at Galveston
Marine Mammal Research Program
Galveston, Texas

Virginia L. Naples, PhD
Department of Biological Sciences
Northern Illinois University
DeKalb, Illinois

Ken B. Naugher, BS
Conservation and Enrichment Programs Manager
Montgomery Zoo
Montgomery, Alabama

Derek William Niemann, BA
Royal Society for the Protection of Birds
Sandy, Bedfordshire
United Kingdom

Carsten Niemitz, PhD
Professor of Human Biology
Department of Human Biology and Anthropology
Freie Universität Berlin
Berlin, Germany

Daniel K. Odell, PhD
Senior Research Biologist
Hubbs-SeaWorld Research Institute
Orlando, Florida

Bart O'Gara, PhD
University of Montana (adjunct retired professor)
Director, Conservation Force

Norman Owen-Smith, PhD
Research Professor in African Ecology
School of Animal, Plant and Environmental Sciences
University of the Witwatersrand
Johannesburg, South Africa

Malcolm Pearch, PhD
Harrison Institute
Sevenoaks, Kent
United Kingdom

Kimberley A. Phillips, PhD
Hiram College
Hiram, Ohio

David M. Powell, PhD
Research Associate
Department of Conservation Biology
Conservation and Research Center
Smithsonian National Zoological Park
Washington, DC

Jan A. Randall, PhD
Department of Biology
San Francisco State University
San Francisco, California

Randall Reeves, PhD
Okapi Wildlife Associates
Hudson, Quebec
Canada

Peggy Rismiller, PhD
Visiting Research Fellow
Department of Anatomical Sciences
University of Adelaide
Adelaide, Australia

Konstantin A. Rogovin, PhD
A.N. Severtsov Institute of Ecology and Evolution RAS
Moscow, Russia

Randolph W. Rose, PhD
School of Zoology
University of Tasmania
Hobart, Tasmania
Australia

Frank Rosell
Telemark University College
Telemark, Norway

Gretel H. Schueller
Science and Environmental Writer
Burlington, Vermont

Bruce A. Schulte, PhD
Department of Biology
Georgia Southern University
Statesboro, Georgia

John H. Seebeck, BSc, MSc, FAMS
Australia

Melody Serena, PhD
Conservation Biologist
Australian Platypus Conservancy
Whittlesea, Australia

David M. Shackleton, PhD
Faculty of Agricultural of Sciences
University of British Columbia
Vancouver, British Columbia
Canada

Robert W. Shumaker, PhD
Iowa Primate Learning Sanctuary
Des Moines, Iowa and Krasnow Institute at George Mason University
Fairfax, Virginia

Andrew T. Smith, PhD
School of Life Sciences
Arizona State University
Phoenix, Arizona

Karen B. Strier, PhD
Department of Anthropology
University of Wisconsin
Madison, Wisconsin

Karyl B. Swartz, PhD
Department of Psychology
Lehman College of The City University of New York
Bronx, New York

Bettina Tassino, MSc
Sección Etología

Facultad de Ciencias
Universidad de la República Oriental
del Uruguay
Montevideo, Uruguay

Barry Taylor, PhD
University of Natal
Pietermaritzburg, South Africa

Jeanette Thomas, PhD
Department of Biological Sciences
Western Illinois University-Quad
Cities
Moline, Illinois

Ann Toon
Arnside, Cumbria
United Kingdom

Stephen B. Toon
Arnside, Cumbria
United Kingdom

Hernán Torres, PhD
Santiago, Chile

Rudi van Aarde, BSc (Hons), MSc,
PhD
Director and Chair of Conservation
Ecology Research Unit
University of Pretoria
Pretoria, South Africa

Mac van der Merwe, PhD
Mammal Research Institute
University of Pretoria
Pretoria, South Africa

Christian C. Voigt, PhD
Research Group Evolutionary Ecology
Leibniz-Institute for Zoo and Wildlife
Research
Berlin, Germany

Sue Wallace
Freelance Writer
Santa Rosa, California

Lindy Weilgart, PhD
Department of Biology
Dalhousie University
Halifax, Nova Scotia
Canada

Randall S. Wells, PhD
Chicago Zoological Society
Mote Marine Laboratory
Sarasota, Florida

Nathan S. Welton
Freelance Science Writer
Santa Barbara, California

Patricia Wright, PhD
State University of New York at Stony
Brook
Stony Brook, New York

Marcus Young Owl, PhD
Department of Anthropology and
Department of Biological Sciences
California State University
Long Beach, California

Jan Zima, PhD
Institute of Vertebrate Biology
Academy of Sciences of the Czech
Republic
Brno, Czech Republic

• • • • •

Contributing illustrators

Drawings by Michigan Science Art

Joseph E. Trumpey, Director, AB, MFA
Science Illustration, School of Art and Design, University of Michigan

Wendy Baker, ADN, BFA

Ryan Burkhalter, BFA, MFA

Brian Cressman, BFA, MFA

Emily S. Damstra, BFA, MFA

Maggie Dongvillo, BFA

Barbara Duperron, BFA, MFA

Jarrod Erdody, BA, MFA

Dan Erickson, BA, MS

Patricia Ferrer, AB, BFA, MFA

George Starr Hammond, BA, MS, PhD

Gillian Harris, BA

Jonathan Higgins, BFA, MFA

Amanda Humphrey, BFA

Emilia Kwiatkowski, BS, BFA

Jacqueline Mahannah, BFA, MFA

John Megahan, BA, BS, MS

Michelle L. Meneghini, BFA, MFA

Katie Nealis, BFA

Laura E. Pabst, BFA

Amanda Smith, BFA, MFA

Christina St.Clair, BFA

Bruce D. Worden, BFA

Kristen Workman, BFA, MFA

Thanks are due to the University of Michigan, Museum of Zoology, which provided specimens that served as models for the images.

Maps by XNR Productions

Paul Exner, Chief cartographer
XNR Productions, Madison, WI

Tanya Buckingham

Jon Daugherity

Laura Exner

Andy Grosvold

Cory Johnson

Paula Robbins

• • • • •

Topic overviews

What is a mammal?

Ice Age giants

Contributions of molecular genetics to phylogenetics

Structure and function

Adaptations for flight

Adaptations for aquatic life

Adaptations for subterranean life

Sensory systems

Life history and reproduction

Mammalian reproductive processes

Ecology

Nutritional adaptations of mammals

Distribution and biogeography

Behavior

Cognition and intelligence

Migration

Mammals and humans: Domestication and commensals

Mammals and humans: Mammalian invasives and pests

Mammals and humans: Field techniques for studying mammals

Mammals and humans: Mammals in zoos

Conservation

• • • • •

What is a mammal?

At first sight, this is not a difficult question. Every child is able to identify an animal as a mammal. Since its earliest age it can identify what is a cat, dog, rabbit, bear, fox, wolf, monkey, deer, mouse, or pig and soon experiences that with anyone who lacks such a knowledge there would be little chance to communicate about other things as well. To identify an animal as a mammal is indeed easy. But by which characteristics? The child would perhaps explain: *Mammals are hairy four-legged animals with faces.*

A child answers: A hairy four-legged animal with a face

Against expectation, the three characteristics reported by this naive description express almost everything that is most essential about mammals.

Hair, or fur, probably the most obvious mammalian feature, is a structure unique to that group, and unlike the feathers of birds is not related to the dermal scales of reptiles. A mammal has several types of hairs that comprise the pelage. Specialized hairs, called vibrissae, mostly concentrated in the facial region of the head, perform a tactile function. Pelage is seasonally replaced in most mammals, usually once or twice a year by the process called molting. In some mammals, such as ermines, the brown summer camouflage can be changed to a white coat in winter. In others, such as humans, elephants, rhinoceroses, naked mole rats, and aardvarks, and in particular the aquatic mammals such as walruses, hippopotami, sirenia, or cetaceans, the hair coat is secondarily reduced (though only in the latter group is it absent completely, including vibrissae). In the aquatic mammals (but not only in them), the role of the pelage is performed by a thick layer of subcutaneous adipose tissue by which the surface of body is almost completely isolated from its warm core and the effect of a cold ambient environment is substantially reduced. Thanks to this tissue, some mammals can forage even in cold arctic waters and, as a seal does, rest on ice without risk of freezing to it. In short, the essential role of the subcutaneous adipose layer and pelage is in thermal isolation, in preventing loss of body heat. Mammals, like birds, are *endotherms* (heat is generated from inside of the body by continuous metabolic processes) and *homeotherms* (the body temperature is maintained within a narrow constant range).

The body temperature of mammals, about 98.6°F (37°C), is optimal for most enzymatic reactions. A broad variety of functions are, therefore, kept ready for an immediate triggering or ad hoc mutual coupling. All this also increases the versatility of various complex functions such as locomotion, defensive reactions, and sensory performances or neural processing of sensory information and its association analysis. The constant body temperature permits, among other things, a high level of activity at night and year-round colonization of the low temperature regions and habitats that are not accessible to the ectothermic vertebrates. In short, endothermy has a number of both advantages and problems. Endothermy is very expensive and the high metabolic rate of mammals requires quite a large energetic intake. In response, mammals developed a large number of very effective feeding adaptations and foraging strategies, enabling them to exploit an extreme variety of food resources from insects and small vertebrates (a basic diet for many groups) to green plants (a widely accessible but indigestible substance for most non-mammals). At the same time, mammals have also developed diverse ways to efficiently control energy expenditure.

Besides structural adaptations such as hair, mammals have also developed diverse physiological and behavioral means to prevent heat and water loss, such as burrowing into underground dens; seasonal migrations or heterothermy; and the controlled drop of body temperature and metabolic expenditure during part of the day, or even the year (hibernation in temperate bats, bears, and rodents as well as summer estivation in some desert mammals). So, considerable adaptive effort in both directions increases foraging efficiency and energy expenditure control. When integrated with morphological, physiological, behavioral, and social aspects, it is an essential feature of mammalian evolution and has contributed to the appearance of the mammalian character in many respects.

Four legs, each with five toes, are common not only to many mammals, but to all terrestrial vertebrates (amphibians, reptiles, birds, and mammals), a clade called Tetrapoda. Nevertheless, in the arrangement of limbs and the modes of locomotion that it promotes, mammals differ extensively from the remaining groups. The difference is so clear that it allows us to identify a moving animal in a distance as a mammal even in one blink of an eye. In contrast to the "splayed" reptilian

Red kangaroos (*Macropus rufus*) on the move. (Photo by Animals Animals ©Gerard Lacz. Reproduced by permission.)

groups. The presence of sophisticated mechanisms of social integration and an enlarged role in interindividual discrimination and social signaling are broadly characteristic of mammals. Nevertheless, each isolated component contributing to the complex image of the mammalian face says something important regarding the nature of the mammalian constitution, and, moreover, they are actually unique characters of the group. This is particularly valid for fleshy cheeks and lips, the muscular belt surrounding the opening of a mouth. The lips and the spacious pocket behind them between the cheeks and teeth (the *vestibulum oris*) are closely related to feeding, and not only in that they enlarge the versatility of food processing in an adult mammal. The lips, cheeks and *vestibulum oris* are completely developed at the time of birth and since that time have engaged in the first behavioral skill performed by a mammal. Synergetic contraction of lip and cheek muscles producing a low pressure in the *vestibulum oris* is the key component of the suckling reflex, the elementary feeding adaptation of a newborn mammal. All mammals, without exception, nourish their young with milk and all female mammals have large paired apocrine glands specialized for this role—the mammary glands, or mammae. Nevertheless, not

stance (i.e. horizontal from the body and parallel to the ground), the limbs of mammals are held directly beneath the body and move in a plane parallel to the long axis of the body. In contrast to reptiles, whose locomotion is mostly restricted to the lateral undulation of the trunk, mammals flex their vertebrate column vertically during locomotion. This arrangement enables a powered directional movement, such as sustained running or galloping, very effective for escaping from a predator, chasing mobile prey, or exploring spatially dispersed food resources. The respective rearrangements also bring another effect. By strengthening the vertebral column against lateral movement, the thoracic cavity can be considerably enlarged and the thoracic muscles released from a locomotory engagement, promoting changes to the effective volume of the thoracic cavity. With a synergetic support from another strictly mammalian structure, a muscular diaphragm separating the thoracic and visceral cavity, the volume of the thoracic cavity can change during a breathing cycle much more than with any other vertebrates. With the alveolar lungs, typical for mammals, that are designed to respond to volume changes, breathing performance enormously increases. This enables a mammal to not only keep its basal metabolic rate at a very high level (a prerequisite for endothermy) but, in particular, to increase it considerably during locomotion. In this connection, it should be stressed that the biomechanics of mammalian locomotion not only allow a perfect synchronization of limb movements and breathing cycles but, with the vertical flex of the vertebral column, are synergetic to the breathing movements and support it directly. As a result, the instantly high locomotory activity that characterizes a mammal increases metabolic requirements but at the same time helps to respond to them.

The face is the essential source of intra-group social information not only for humans but for many other mammal

A spotted hyena (*Crocuta crocuta*) stands on its meal of a baby elephant. (Photo by Harald Schütz. Reproduced by permission.)

Some mammals, such as this goat, have rather dramatic antlers or horns. (Photo by Animals Animals ©Robert Maier. Reproduced by permission.)

clades, in comparison to other vertebrates (excepting elasmobranchians and birds), the mammals are clearly the *K*-strategists (producing few; but well-cared for, offspring) in general.

The other components of the mammalian face provide correspondingly significant information on the nature of these animals. The vivid eyes with movable eyelids, external auricles, nose, and last but not least long whiskers (vibrissae, the hairs specialized for tactile functions), show that a mammal is a sensory animal. Most extant mammals are noctural or crepuscular and this was almost certainly also the case with their ancestors. In contrast to other tetrapods, which are mostly diurnal and perceive almost all spatial information from vision, mammals were forced to build up a sensory image of the world from a combination of different sources, in particular olfaction and hearing. Nevertheless, vision is well developed in most mammals and is capable of very fine structural and color discrimination, and some mammals are secondarily just optical animals. For example, primates exhibit a greatly enlarged capability for stereoscopic vision. In any case, all mammals have structurally complete eyes, though the eyes may be cov-

all mammalian newborns actually suck the milk. In the egg-lying monotremes (the Australian duck-billed platypus and spiny anteaters), mammary glands lack the common milk ducts and nipples, so young do not suck but instead lick the milk using their tongue. All other mammals, both marsupials and eutherians, together denoted as Theria, bear a distinctive structure supporting suckling—the paired mammary nipples. The nipples originate independently from mammary glands, they are present both in males and females, and their number and position is an important character of individual clades. The therian mammals are all viviparous. For the most vulnerable period of their lives they are protected first by the intrauterine development with placental attachment of the embryo and then by prolonged postnatal parental care. A milk diet during the latter stage postpones the strict functional control on jaws and dentition and enables postnatal growth, the essential factor for the feeding efficiency of an adult mammal. At the same time this provides extra time for development of other advanced and often greatly specialized mammalian characteristics: an evolving brain and the refinement of motor capacities and behavioral skills. Thanks to the extended parental investment that mammalian offspring have at the beginning of their independent life, they enjoy a much higher chance for post-weaning survival than the offspring of most other vertebrates. The enormous cost of the parental investment places, of course, a significant limit upon the number of offspring that can be produced. Despite the great variation in reproductive strategies among individual mammalian

A baby gray bat (*Myotis grisescens*). (Photo by Merlin D. Tuttle/Bat Conservation International/Photo Researchers, Inc. Reproduced by permission.)

Near Kilimanjaro, a giraffe (*Giraffa camelopardalis*) pauses to survey for predators. Giraffes are the tallest extant mammals, males reaching 18 ft (5.5 m) in height. (Photo by Harald Schütz. Reproduced by permission.)

ered by skin in some fossorial mammals (such as blind mole rats, or marsupial moles) or their performance may be reduced in some respect. In comparison with other vertebrates, the performance of vision is particularly high under low light intensities, and the eyes are quite mobile. The latter character may compensate for a reduced ability of head rotation in mammals due to the bicondylous occipital joint contrasting to a monocondylous joint in birds or reptiles. The eyes are covered by movable eyelids (not appearing in reptiles), significant both in protecting the eyes and in social signaling. The remaining two structures—nose and auricles—are particularly unique for mammals and are related to the senses that are especially important for mammals: olfaction and hearing. Not only the nose and auricles themselves, but also the other structures associated with the senses of smell and hearing feature many traits unique to mammals.

Mammals construct much of their spatial information with the sole aid of olfactory, acoustic, or tactile stimuli combined with information from low-intensity vision. This task necessitated not only a considerable increase in the capacity and sensory versatility of the respective organs, but also the refinement of the semantic analysis of the information they provide. As a result, the brain structures responsible for these tasks are greatly enlarged in mammals. The tectum mesen-

cephali, a center for semantic analysis of optical information, bi-lobed in other vertebrates, is supplemented by a distinct center of acoustic analysis by which the tectum of mammals becomes a four-lobed structure, the corpora quadrigemina. The forebrain or telencephalon, a structure related to olfactory analysis, is by far the largest part of the mammalian brain. Its enlargement is particularly due to the enlarging of the neocortex, a multi-layered surface structure of the brain, which further channels inputs from other brain structures and plays the role of a superposed integrative center for all sensory, sensory-motor, and social information.

A zoologist answers: A highly derived amniote

Many of the characters common to mammals do not appear in other animals. Some of them, of course, can be observed also in birds—a very high (in respect to both maximum and mean values) metabolic rate and activity level or complexity of particular adaptations such as advanced parental care and social life, increased sensory capacities, and new pathways of processing sensory information or enormous ecological versatility. Fine differences between birds and mammals suggest that the respective adaptations are homoplasies—that is, they evolved in both groups independently.

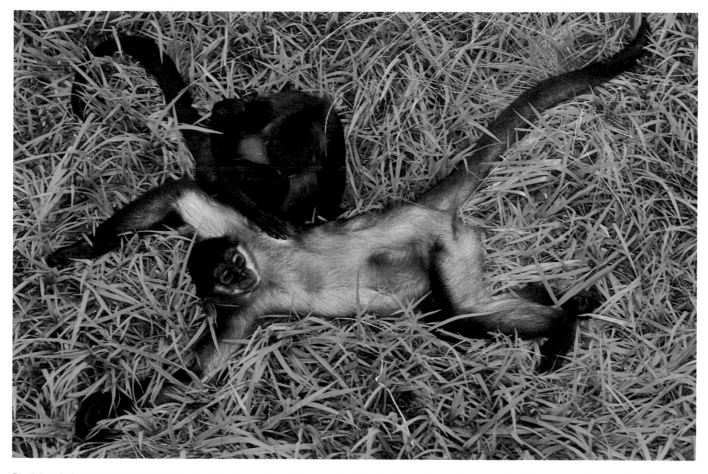

Black-handed spider monkeys (*Ateles geoffroyi*) grooming. (Photo by Gail M. Shumway. Bruce Coleman, Inc. Reproduced by permission.)

Other mammalian characteristics are synapomorphies of Amniota, the characteristics shared because of common ancestry. The amniotes, a group including reptiles, birds, and mammals, are the terrestrial vertebrates in which embryonic development takes place under the protection of fetal membranes (amnion, chorion, allantois). As in other amniotes, mammals are further characterized by an increased role of parental investment, internal fertilization, keratinized skin derivatives, an advanced type of kidney (metanephros) with a specific ureter, an advanced type of lung respiration, and the decisive role of dermal bones in skull morphology. Of course, at the same time, mammals share a large number of characteristics with all other vertebrates, including the general body plan, solid inner skeleton, the design of homeostatic mechanisms (including pathways of neural and humoral regulation), and functional integration of particular developmental modules. Mammals also share with other vertebrates the patterns of segmentation of trunk skeleton and muscles and the specific arrangements of the homeobox genes organizing the body segmentation as well as a lack of their expression in the head region, etc. These characters are synapomorphies of vertebrates, which are at least partly retained not only in some amniotes but throughout all other vertebrate clades. With respect to mammals, these are symplesiomorphies, the primitive characters that do not reveal closer relations of the class but on its broadest phylogenetic context.

Mammals also exhibit a large number of qualities that are fully unique to them, the autapomorphies. The autapomorphies are the characteristics by which a taxon can be clearly distinguished and diagnosed. Thus, though many characteristics of mammals are not specific just to them, answering the question "what is a mammal?" means first demonstrating the autapomorphies of that group. A simplified list of them includes:

(1) *The young are nourished with milk* produced by (2) *mammary glands*. These glands appear in all female mammals, and are the structure from which the class Mammalia got its name. (3) *Obligatory vivipary* (in Theria, i.e., marsupials and placentals) is the reproductive mode with a specialized organ interconnecting the embryo and maternal tissues, the *chorioallantoic placenta* (in Eutheria, i.e., placentals). (4) *Hairs*, covering the body, grow from deep invaginations of the germinal layer of epidermis called *follicles*. Similar to other amniotes, the hair is composed of keratin and pigments, but its structure is unique for mammals. (5) *Skin* is rich in various glands. Most mammals have sweat glands (contributing to water balance and cooling the body surface), scent glands, and sebaceous glands. (6) The *specific integumental derivatives*, characteristic of particular groups of mammals, are composed either exclusively of keratin (such as claws, nails, and hoofs, which protect the terminal phalanx of the digits and adapt them to a

A silverback jackal (*Canis mesomelas*) and an African elephant (*Loxodonta africana*) at a watering hole in Chobe National Park, Botswana. (Photo by © Theo Allofs/Corbis. Reproduced by permission.)

specific way of locomotion or foraging) or of keratin in combination with dermal bone structures (horns of bovids and antlers of cervid artiodactyls, which play a considerable role in social signaling). A large variety of integumental derivatives are included in defensive adaptations: dermal armors of armadillos or keratinized scales of pangolins, spines modified from hairs in echidnas, hedgehogs, tenrecs, porcupines, or spiny mice, or the accumulations of hairlike fibers keratinized into a horn structure in rhinoceroses. (7) *Limb position and function* are modified to support specific locomotory modes of mammals such as jumping, galloping, or sustained running and can be specifically rearranged. The extreme rearrangements are seen in bats, which fly using a forelimb wing, and in specialized marine mammals, pinnipedian carnivores, cetaceans, and sirenia, whose forelimbs take the shape of a fin (the external hind limbs are absent in the latter two groups). (8) *Pectoral girdle is simplified* in comparison to the non-mammalian state: coracoid, precoracoid and interclavicle bones are lost (except for monotremes, which retain them) or partly included in the scapula. Also the clavicle, the last skeletal element that fixes the limb to the axial and thoracic skeleton, is

lost in many groups. With these rearrangements the forelimbs get new locomotory qualities (such as extensive protraction), supporting abilities such as climbing and fine limb movements and providing a new spectrum of manipulative functions from cleaning hair to a variety of prey manipulations. (9) *The bones of the pelvic girdle are fused into a single bone*, with enlarged and horizontally prolonged ilium.

(10) A great degree of *regional differentiation of the vertebral column*. All mammals (except some edentates and manatees) have seven cervical vertebrae with the first two (atlas and axis) specifically rearranged to support powered head movements. (11) *The vertebral column is strengthened against lateral movements* but is greatly disposed to the vertical flexion. This is seen first of all in the lumbar section, whose vertebrae, in contrast to the non-mammalian ancestors, lack ribs. (12) The mammalian *skull is bicondylous* (the first vertebra, atlas, joints the skull via paired occipital condyles located on the lateral sides of the large occipital foramen), with (13) *an enlarged braincase*, (14) *massive zygomatic arches* (formed by the jugale and squamosum bones), and (15) a spacious *nasal cavity with a labyrith of nasal turbinalia* covered by vascularized tissue important both for olfaction (ethmoidal turbinalia) and/or heat and water exchange during breathing (maxillary turbinalia). (16) The nostrils open at a common structure called the *nose*, obviously the most prominent point of the head. The ancestral form of the nose, the *rhinarium*, is a hairless field of densely circular-patterned skin surrounding the nostril openings. The rhinarium is particularly large in macrosmatic (highly developed sense of smell) mammals (such as carnivores or artiodactyls), in lagomorphs, some rodents, and bats. In strepsirhine primates it is incised by a central groove, the phlitrum, while in some other groups such as in macroscelids or in elephants, the nose is prolonged and attains a number of supplementary functions. In contrast, all these structures are absent in cetaceans in which the nasal cavity is reduced and the nostrils (or a single nostril opening in Odontoceti) appear at the top of the head and their function is restricted to respiration. (17) Left and right maxillary and palatal bones are fused in early development and form the *secondary bony palate*, which is further extended by a fleshy soft palate. These structures provide a complete separation of the respiratory and alimentary tracts. The early appearance of such a separation is one of the essential prerequisites for suckling milk by a newborn and, hence, it seems probable that the secondary palate first appeared simply as an adaptation for this. (18) The *heart* is a large *four-chambered organ* (as in birds) with the *left aorta persistent* (not the right one, as in birds). (19) *Erythrocytes, the red blood cells, are biconcave and lack nuclei*. Thrombocytes are transformed to nonnucleated blood platelets.

(20) *Lungs have an alveolar structure*, ventilated by volume changes performed by the counteraction of two independent muscular systems, and a (21) *muscular diaphragm*, unique for mammals. (22) The *voice organ in the larynx*, with several pairs of membranous muscles, is unique for mammals. It is capable of very specialized functions such as the production of various communicative signals or high-frequency echolocation calls in bats and cetaceans. (23) There are *three ossicles in the middle ear* (malleus, incus, stapes). The former two are unique to mammals and are derived from the elements of the pri-

mary mandibular joint—articulare and quadratum—which still retain their original function in the immediate mammalian ancestors. The third bone of the primary mandibular joint, the angulare, changes in mammals into the tympanic bone, which fixes the tympanic membrane and finally enlarges into a bony cover of the middle ear—the bulae tympani. (24) The *sound receptor (Corti´s organ of the inner ear) is quite long and spirally coiled* in mammals (except for monotremes) and surrounded by *petrosum*, a very compact bone created by a fusion of several elements. (25) With an enlarged braincase, the middle ear and tympanic membranc arc thus located deeper in the head and open to the external environment by a *long auditory meatus* terminating with (26) a *large movable external auricle*. Auricles (pinnae) are specifically shaped in particular clades and contribute to the lateral discrimination of the auditory stimuli and directionality of hearing. They may be absent in some aquatic mammals (cetaceans, sirenia, walruses), while they are extremely pronounced and diversified in other groups such as bats, for which the acoustic stimuli (echoes of the ultrasonic calls they emit) are by far the most important source of spatial information. (27) In contrast to other amniotes, *the lower jaw, or mandible, is composed of a single bone*, dentary or dentale, which directly articulates with the temporal bone of the skull at the (28) *dentary-squamosal joint*. This arrangement not only fastens the jaw joint to resist the forces exerted during strong biting but also simplifies the functional rearrangements of jaw morphology responding to different demands of particular feeding specializations. (29) In all mammals, the posterior part of the mandible extends dorsally into the *ramus mandibulae*, which provides an area of attachment for the massive temporal muscles responsible for the powered adduction of the mandible.

(30) Essentially, all mammals have *large teeth* despite considerable variation in number, shape, and function in particular groups and/or the fact that some mammals secondarily lack any teeth at all (anteaters of different groups, and the platypus). Teeth are deep-rooted in bony sockets called alveoles. Only three bones host the teeth in mammals: the premaxilla and maxilla in the upper jaw and the dentary in the lower jaw. (31) Mammalian *dentition is generally heterodont (of different size, shape, etc.)*. Besides the conical or unicuspidate teeth (incisors and a single pair of canines in each jaw) mammals also have large complex multicuspidate molars (three in placentals, four in marsupials, in each jaw quadrant) and premolars situated between canines and molars whose shape and number varies considerably among particular groups. The latter two teeth types are sometimes called "postcanines" or "cheek teeth." (32) *The molars* are unique to mammals. The basic molar type ancestral to all particular groups of mammals is called tribosphenic. It consists of three sharp cones connected with sharp blades. In combination with the deep compression chambers between blades, such an arrangement provides an excellent tool both for shearing soft tissues and crushing insect exoskeletons. This type of molar is retained in all groups feeding on insects, such as many marsupials, tenrecs, macroscelids, true insectivores such as moles, shrews or hedgehogs, bats, tree shrews, and prosimian primates, but the design of the molar teeth is often extensively rearranged in other groups. The multicuspidate structure of molars bears enormous potential for morphogenetic and functional re-

Many young mammals practice skills needed for survival. These lion cubs practice hunting in the grass. (Photo by K. Ammann. Bruce Coleman, Inc. Reproduced by permission.)

arrangements, one of the prerequisites of the large diversity of feeding adaptations in mammals. (33) Mammalian dentition is *diphyodont*. This means that there are two generations at each tooth position (except for molars): the milk or deciduous teeth of the young and the permanent teeth of an adult mammal. Diphyodonty solves a functional-morphological dilemma: the size of teeth, an essential factor in feeding efficiency, is limited by the size of the jaws. While the jaws can grow extensively, the posteruption size of the teeth cannot be changed due to the rigidity of their enamel cover, which is the essential quality of a tooth. With diphyodonty, the size of the late erupting permanent teeth can be maximized and adapted to adult jaw size while the deciduous dentition provides a corresponding solution for the postweaning period. Dental morphology and the patterns of tooth replacement are specifically modified in some clades. In marsupials, only one milk tooth—the last premolar—comes in eruption, while the others are resorbed prior to eruption. Dolphins, aardvarks, and armadillos have a homodont dentition without any tooth replacement. No tooth replacement occurs in small and short-living mammals with greatly specialized dentition, such as shrews or muroid rodents (deciduous teeth are resorbed instead of eruption), while in some large herbivores tooth replacement can become a continuous process by which the tooth row enlarges gradually by subsequent eruption of still larger molar teeth in the posterior part of the jaws. In elephants and manatees, this process includes a horizontal shift of the erupting tooth, which thus replaces the preceding cheek tooth. All these processes are well synchronized with the growth of jaws, the course of tooth wear, and subsequent prolonging of time available for tooth development. (34) A general *enlargement of the brain* related perhaps not only to an increase in the amount of sensory information and/or a need to integrate sensory information from different sources, but also to more locomotory activity, high versatility in locomo-

A cheetah (*Acinonyx jubatus*) chases a Thomson's gazelle (*Gazella thomsonii*). The cheetah is the fastest land animal and can reach speeds of 70 mph (113 kph). (Photo by Tom Brakefield. Bruce Coleman, Inc. Reproduced by permission.)

tory functions, a greatly diversified social life, and a considerably expanded role for social and individual learning. (38) The extended spectrum of behavioral reactions and their interconnections with an increased capacity of social and individual learning and interindividual discrimination should also be mentioned. In fact, this characteristic is very significant for mammals, as are the following two: (39) *Growth is terminated* both by hormonal control and structural factors. The most influential structural aspect of body growth is the appearance of *cartilaginous epiphyseal discs separating diaphyses and epiphyses* of long bones. With completed ossification, the discs disappear and growth is finished. Corresponding mechanisms determine the size of the skull (except in cetaceans, which have a telescoped skull in which the posterior bones of the cranium overlap each other). (40) Sex is determined by chromosomal constitution (*XY system, heterogametic sex is a male*).

Almost all of these (and other) characteristics undergo significant variations and their modifications are often largely specific for particular clades of mammals. What is common for all is perhaps that in mammals all the characters are more densely interrelated than in other groups (except for birds). The morphological adaptations related to locomotion or feeding are often also integrated for social signaling, physiological regulation, or reproductive strategy, and often are controlled by quite distant and non-apparent factors. Thus, the excessive structures of ruminant artiodactyls, such as the horns of bovids and antlers of deer, are undoubtedly significant in social signaling, in courtship and display behavior, and frequently are discussed as excessive products of sexual selection. However, the proximate factor of these structures, the hereditary disposition for excessive production of mineralized bone tissue, can actually be selected rather by its much less

obvious effect in a female: her ability to produce a large, extremely precocial newborn with highly mineralized long bones that enable it to walk immediately after parturition. The female preference for the excessive state of the correlated characters in a male, his large body size and display qualities, possibly supported by social learning, supplement the mechanisms of the selection in quite a non-trivial way. Such a multi-layered arrangement of different factors included in a particular adaptation is indeed something very mammalian.

A paleontologist answers: The product of the earliest divergence of amniotes and index fossils of the Cenozoic

Mammals are the only extant descendants of the synapsids—the first well-established group of amniotes, named after a rounded temporal opening behind the orbit bordered by the jugale and squamosum bones. Since the beginning of amniotes, evolution of synapsids proceeded separately from the other amniotes, which later diversified in particular reptile lineages including dinosaurs and birds. The first amniotes recorded from the middle Carboniferous (320 million years ago) were just synapsids and just this clade predominated in the fossil record of the terrestrial vertebrates until the early Triassic. A large number of taxa appearing among early synapsids represented at least two different clades: Eupelycosauria and Caseasauria. The former included large carnivorous forms and the latter were generalized small- or medium-sized omnivores. Since the middle Permian (260 mya), another group of synapsids called Therapsida dominated the terrestrial record. In comparison with pelycosaurs, therapsids had much larger temporal openings, a single pair of large canines, and clear functional and shape differences between the anterior and the posterior teeth. Two lineages of that group, Dicynodontia and Cynodontia, survived the mass extinction at the Permian/Triassic boundary (248 mya).

Immediate ancestors of mammals are found among the cynodonts. Mammals are closely related to cynodont groups called tritylodontids and trithelodontids, which first appeared during the late Triassic. All three groups, including mammals, had additional cusps on posterior teeth, a well-developed ramus mandibulae, and a complete secondary palate. In some of them (*Diarthrognathus*), the jaw joint was formed both by the original articulation (articulare-quadratum) and by the mammal-like process (dentary-squamosal). In the oldest true mammals, the former jaw articulation is abandoned and removed in the middle ear. These characters are the index diagnostic features of a mammal in the fossil record (no. 23, 26, 27 of the above list).

The oldest mammals, *Sinoconodon*, *Adelobasileus*, *Kuehneotherium*, or *Morganucodon* (about 200–225 million years old), were all very small, with long heterodont dentition and a triangular arrangement of molar cusps designed for shearing. They were most probably quite agile night creatures resembling today's insectivores. The relative brain volume in the earliest mammals was close to that found in extant insectivores and about three times higher than in cynodonts. Of course, they still differed from the modern mammals in many respects. The derived characters of modern mammals (as reviewed in the preceding text) did not evolve together but were subsequently accumulated during the long history of synapsid evolution.

In contrast to the medium- to large-sized diurnal dinosaurs, birds, and other reptiles that had dominated the terrestrial habitats, the early mammals were quite small, nocturnal creatures. Nevertheless, since the Jurassic period they grew in greatly diversified groups and at least four lineages of that radiation survived the mass extinction at the Cretaceous/Tertiary boundary (65 mya). Three of these groups, monotremes, marsupials, and placentals, are extant; the fourth group, multituberculates, survived until the end of Oligocene. Multituberculates resembled rodents in design of dentition (two pairs of prominent incisors separated from a series of cheek teeth by a toothless diastema), but their cheek teeth and skull morphology were quite different from those in any other groups of mammals.

The major radiation of mammals appeared at the beginning of Tertiary, in the Paleocene. That radiation produced many groups that are now extinct (including nine extinct orders) as well as almost all the orders of modern mammals. During the Paleocene and Eocene, other groups occupied the niches of current mammalian groups. In Eurasia and North America it was Dinocerata, Taeniodonta, and Tillodontia as herbivores and Pantodonta and Creodonta as their predators. All these are extinct lineages not related to any of the recent orders. The most isolated situation was in Australia, which had been cut-off from the other continents since the Cretaceous and was not influenced by the intervention of the eutherian mammals. The mammalian evolution in South America after its separation from Africa at the early Paleocene was equally isolated. Besides the marsupials (clade of Ameridelphia) and edentates with giant glyptodonts, mylodonts, and megalonychids, whose relatives survived until recently, a great variety of strange eutherians appeared here during the Paleocene and Eocene. This includes the large herbivores of the orders Notoungulata, Astrapotheria, Litopterna, and Xenungulata, as well as the Pyrotheria (resembling proboscideans) and their giant marsupial predators, such as *Thylacosmilus*, resembling the large saber-toothed cats. The mammalian fauna of South America was further supplemented by special clades of hystricognathe rodents, haplorhinc primates, and several clades of bats, particularly the leaf-nosed bats. These groups probably entered South America during the Paleocene or Eocene by rafting from Africa. The evolution in splendid isolation of South America terminated with the appearance of a land bridge with North America some 3 mya, which heavily impacted the fauna of both continents. The impact of African and Asian fauna on the European mammalian evolution by the end of Eocene was of a similar significance.

It is important to remember that the fossil record of mammals, including detailed pathways of evolutionary divergences and/or the stories of particular clades, is much more complete and rich in information than in any other group of vertebrates. This is due to the fact that the massive bones of mammals, and in particular their teeth, which provide most information on both the relationship and feeding adaptation of a taxon, are particularly well suited to be preserved in fossil deposits. Due to this factor, the fossil record of mammals is perhaps the most complete among the vertebrates. Also, during the late Cenozoic, Neogene, and Quaternary, the fossil record of some mammalian groups (such as rodents, insectivores, and ungulates) is so rich that the phylogeny of many clades can be traced in surprisingly great detail by the respective fossil record. For the same reason, some of these fossils (e.g., voles in the Quaternary period) are the most important terrestrial index fossils and are of key significance not only for local biostratigraphies and precise dating of the late Cenozoic deposits, but also for large-scale paleobiogeography and even for intercontinental correlations. The late Cenozoic period is characterized by gradually increasing effects of climatic oscillations, including repeated periods of cold and dry climate—glacials—followed by the evolution of grass and the treeless grassland country. Many clades of mammals responded to these changes and produced the extreme specialists in food resources of the glacial habitats, such as mammoths, woolly rhinos, lemmings, cave bears, and cave lions.

The most diversified animals

There are about 4,600 species of mammals. This is a relatively small number compared to the 9,600 species of birds or 35,000 fish species and almost nothing in comparison to about 100,000 species of mollusks or some 10,000,000 species of crustaceans and insects. Even such groups as extant reptiles (with 6,000 species) and frogs (with about 5,200 species) are more diversified at the species level. Nevertheless, in diversity of body sizes, locomotory types, habitat adaptations, or feeding strategies, the mammals greatly exceed all that is common in other classes.

Only birds and arthropods may approach such variety. However, at least in diversity of body size, the mammals clearly surpass even them. The body mass of the largest extant terrestrial mammal—the African elephant *Loxodonta africana*— with shoulder height of 11.5 ft (3.5 m), reaches to 6.6 tons (6,000 kg). The extinct rhinocerotid *Baluchitherium* was about 18 ft (5.5 m) and 20 tons (18,000 kg), respectively. The largest animal to ever appear—the blue whale (*Balaenoptera musculus*)—with up to 98 ft (30 m) in length, reaches 220 tons (200,000 kg). In contrast to dinosaurs or elesmobranchians, which also produced quite large forms, the average mammal is a small animal the size of a rat, and the smallest mammals such as a pygmy white-toothed shrew (*Suncus etruscus)* or Kitti's hog-nosed bat (*Craseonycteris thonglongyai*) have a body length of just 1.2–1.6 in (3–4 cm) and weigh only 0.05-0.07 oz (1.5–2 g).

Mammals colonized almost all habitats and regions on the Earth. They now feed on flying insects hundreds of meters above the ground; jump through foliage in the canopy of a tropical forest; graze in lowland savannas and high mountain alpine meadows; hunt for fish under the ice cover of arctic seas; burrow the underground labyrinths to feed on diverse plant roots, bulbs, or insects; cruise the world's oceans, or dive there to depths of 1.8 mi (3 km) in the hunt for giant squid. Some even sit by a computer and write articles like this.

About 4,600 species of mammals are arranged in approximately 1,300 genera, 135 families, and 25 orders. Rodents with 1,820 species, 426 genera and 29 families are far the largest order, while in contrast, 8 orders include less than 10 species, and four of them are even monotypic (Microbiotheria, Notoryctemorphia, Tubulidentata, Dermoptera). Although interrelationship among individual orders is still the subject of a vivid debate, three major clades of mammals are quite clear: monotremes (2 families, 3 genera, 3 species), marsupials (7 orders, 16 families, 78 genera and 280 spp.), and eutherian or placentals (17 orders, 117 families, 1,220 genera, 4,300 spp.), the latter two clades are together denoted as Theria.

The essential differences among the three major clades of mammals are in mode of their reproduction and patterns of embryonic development. Monotremes (platypus and echidnas), restricted to the Australian region, show only little difference from their ancestral amniote conditions. They deliver eggs rich in yolk, and incubate them for 10 to 11 days. Young hatch from the egg in a manner similar to birds. Monotremes also retain the reptile conditions in the morphology of the reproductive system: the ovary is large and short oviducts come via paired uteri to a broad vagina, which opens with the urinary bladder and rectum into a common cloaca. Except for monotremes, all mammals are viviparous with intrauterine embryonic development and have quite small eggs, poor in yolk (particularly in eutherians).

There are essential differences between marsupials and eutherians in the earliest stages of embryonic development, as well as in many other characteristics. The reproductive tract in a female marsupial is bifurcated (with two vaginas), and also the tip of the penis in a male marsupial is bifurcated. Many marsupials have a marsupium, the abdominal pouch for rearing young, supported with the marsupial epipubic bones that are present in both sexes. The marsupial intrauterine development is very short and the embryo is attached to the uterine endometrium by the choriovitelline (yolk) placenta that lacks the villi penetrating deeper in the wall of uterus (except in bandicoots). The marsupial newborns are very small and little developed, and birth is nontraumatic. In contrast, the lactation period is much longer than in eutherians (only bats and some primates have proportionally long lactation periods). Nevertheless, the mother's total investment by the time of weaning young is roughly equal in both clades, but its distribution is different. The marsupial strategy is much less stressful for a mother and allows an extensive variation in tactics of reproduction. For instance, in the kangaroo, a mother can have three generations of young at one time: the young baby returning to drink low-protein but high-fat milk, the embryo-like young attached to a nipple nourished with high-protein but low-fat milk, and an embryo in the uterus for which development is delayed until the second-stage young is released.

A key agent of eutherian reproduction is the highly specialized organ supporting a prolonged embryonic development—the chorioallantoic placenta. Eutherian newborns are large and despite considerable variation over particular clades, are potentially capable of an independent life soon after birth. Large herbivores such as elephants, perissodactyls, and artiodactyls, as well as cetaceans, sirenians, hyraxes, and some primates, deliver single, fully developed newborns with open eyes, ears, and even the ability to walk immediately after birth. Such a newborn is called precocial in contrast to the altricial newborns of insectivores, bats, rodents, or carnivores, which are hairless, blind, and fully dependent on intensive mother's care. Both developmental strategies may, of course, appear within one clade as in lagomorphs (large litters and altricial young in a rabbit versus small litters and precocial young in a hare). Variations in reproductive strategies are closely interconnected with numerous behavioral adaptations and adaptations in social organization and population dynamics, all of which contribute significantly to mammalian diversity.

Recent molecular data strongly support the essential role of geographic factors in phylogenetic history and in taxonomic diversity of mammals. Thus, there is very strong support for the African clade Afrotheria, which is composed of the tenrecid and potamogalid insectivores, golden moles, macroscelids, aardvark, hyraxes, proboscideans, and sirenia. Also, the extensive covergences between Australian marsupials and particular eutherian clades and/or the paleontological data on mammalian evolution on particular continents suggest that on each continent, the adaptive radiation produced quite similar life forms: small to medium sized insectivores, rodent-like herbivores, large herbivores, and their predators. The niche of large herbivores seems to be particularly attractive (at least 18 different clades attained it) but at the same time, it is perhaps the most dangerous (13 of them are extinct).

Nearly one fourth of all mammals fly. This is pertinent to a number of species, the number of genera, and perhaps for the number of individuals as well. Bats, with more than 1,000 species in 265 genera, are the most common mammals in

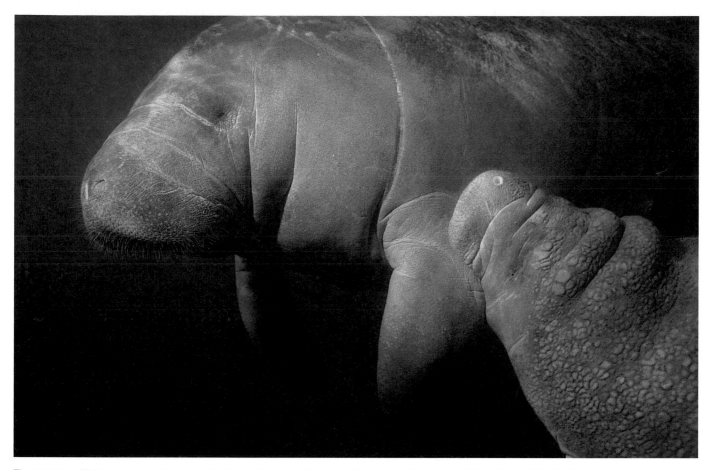

The manatee (*Trichechus manatus*) is primarily herbivorous. Here a mother nurses her young. (Photo by Jeff Foott. Bruce Coleman, Inc. Reproduced by permission.)

many tropical and subtropical habitats. Mostly active at night, bats hunt for various kinds of aerial prey (a basic strategy of the clade) or feed on fruit, nectar, or pollen. Some bats feed on frogs, reptiles, or other bats, and in the tropics of South America, the total biomass of bats exceeds that of all other mammals. Several Old World bats, such as false vampires, feed on small vertebrates, while others feed on fish plucked from the water surface. Frugivorous and nectarivorous bats are the essential agents for pollination and seed dispersal of many tropical plants, including banana and mango. Bats are often very social and form large colonies, including the largest assemblies known in mammals, such as the maternity colony of about 36 million Mexican free-tailed bats in Bracken Cave in Texas.

However, most of the extant mammals (nearly a half of all genera) maintain the basic mammalian niche. They are terrestrial, mostly nocturnal or crepuscular, and forage for different food resources that are available on the ground. In a tropical forest this may be seeds and fruits falling down from the canopy and the invertebrate or vertebrate animals feeding on them. In the subtropics and temperate regions, the significance of this habitat increases as the soil surface becomes the most significant crossroads of ecosystem metabolism. In a temperate ecosystem, the soil is the major conveyer of the energetic flow and an important source of free

energy that is available in a variety of food resources. It is no wonder that in the temperate regions terrestrial mammals form more than half of the local mammalian taxa (while it is one third or less in the tropics) and that their densities exceed those of all remaining mammalian species. Among them we find the groups that are the most progressive and most rapidly diversifying clades of the extant mammals (such as shrews or muroid rodents). Terrestrial mammals are, as a rule, quite small animals, and are often the r-strategists. They have short life spans, large litter sizes, several litters per year, and rapidly attain sexual maturity, sometimes even a few weeks after birth. Most of the small ground mammals dig underground burrows for resting. This reduces not only the risk of predation, but due to stable microclimatic conditions of the underground habitat, it also reduces metabolic stress by ambient temperature or by daytime changes in other weather conditions. Many mammals also tend to spend a considerable part of their active life underground, including food gathering. Those that combine it with terrestrial foraging are called semifossorial—most of the 57 genera of semifossorial mammals are rodents. Those that are entirely adapted to an underground way of life and often do not come above ground at all are called fossorial. The fossorial adaptations, which make them all quite similar in general appearance, are seen in 35 genera of 13 different clades and evolved convergently in all major geographic regions (Australian marsupial mole,

Holoarctic true moles, the African golden moles, and 10 groups of rodents in Holoarctic, Ethiopian, and Neotropical regions). Compared to their relatives, the fossorial mammals are all the K-strategists, some with pronounced tendencies to complex organization (mole rats).

The mammals also evolved another way to inhabit terrestrial habitats. It is called scansorial adaptation and is typical of large herbivores with an enormous locomotory capacity, enabling them to exploit distant patches of optimal resources and react actively to seasonal changes in them. In many instances these are social animals living in large nomadic herds. Kangaroos, the large macropodid marsupials of Australia, exhibit this scansorial adaptation. They move rapidly around their terrestrial habitat by hopping bipedally on their long, powerful hind legs, using their long tails for balance.

Locomotory modes are entirely different in the 156 genera of mammals that forage in arboreal habitats. Essentially arboricolous are primates, dermopterans, and tree shrews, as well as many marsupials, rodents, bats, and some edentates and carnivores. Typical for most of them are long forelimbs and a long tail, often prehensile. Other arboricolous mammals have a haired membrane between their legs, enabling them to glide between tree trunks. The mammals equipped for such gliding flight include flying lemurs (Dermoptera), several groups of rodents (flying squirrels, African anomalurids), and three genera of marsupials.

Roughly 107 genera and 170 species are aquatic or semiaquatic and mostly fish-eating. Three grades can be distinguished here: (1) terrestrial animals that enter aquatic habitats only temporarily for feeding only (African otter shrews, Old World water shrews, desmans, water opossum, more clades of rodents, including large rodents such as beaver and capybara, and several clades of carnivores, particularly otters); (2) marine mammals that spend most of their life in aquatic habitats but come to shore for breeding (all pinnipedian carnivores, such as seals, sea lions and walruses, and sea otters); and (3) the exclusively aquatic mammals incapable of surviving outside of the aquatic environment—sirenians and cetaceans. The latter group is quite diversified, and includes 78 species in 41 genera that can be subdivided into two major clades: Mysticeti, whales that filter marine plankton with baleen plates hanging from roof of the mouth cavity, and Odontoceti, dolphins and toothed whales, which echolocate and feed on fish or squid (including the giant deep-sea architeuthids as in the sperm whale). Cetaceans evolved various sophisticated adapatations for prolonged diving into deep oceanic waters, such very economic ways of gas exchange that include a reduced heart rate during diving and more oxygen-binding hemoglobin and myoglobin in blood than in other mammals. Cetaceans, though closely related to non-ruminant artiodactyls and recently included together with them in a common order, Cetartiodactyla, diverge from the common picture of "what is a mammal?" perhaps most of all.

The extreme diversity in feeding adaptations is among the most prominent characteristics of mammals. Feeding specializations such as grazing grass or herbal foliage, palynovory (eating pollen of plants), myrmecophagy (specialized feeding on ants and termites), and sanguivory (feeding on blood of birds and mammals, in five species of true vampires) are not known from any other vertebrates. At the same time, all the feeding adaptations occurring in other vertebrate clades occur also among mammals.

In all mammals, the efficiency of a feeding specialization depends upon the appropriate morphological, physiological, and behavioral adaptations. First, it concerns the design of the teeth and dentition. The generalized heterodont dentition and the tribosphenic molar teeth designed for an insectivorous diet (as retained in various marsupials, insectivores, tree shrews, prosimian primates, and bats) can be easily modified to the carnivorous diet. A carnivorous diet further demands enlarging the size of the canines and arrangements that increase the shearing effect of cheek teeth. A lower position of the jaw joint increases the powered action of temporal muscles at the anterior part of dentition, and in extremely specialized carnivores such as cats, the dentition is then considerably shortened and reduced except for canines and the carnasial cheek teeth (the last upper premolar and the first lower molar, generally the largest teeth of carnivores). There is no problem with digesting the tissues of vertebrates and thus no special arrangements of the alimentary tract are needed.

In contrast, herbivores, especially those specialized in feeding on green plant mass, require a modified jaw design. This kind of food is everywhere and easily accessible as a rule, but it is extremely difficult to digest for several reasons. One is that this diet is very poor in nutritive content and must be consumed in very large volumes; it must also be broken down mechanically into small particles. Hence, the dentition is overburdened by wear of occluding teeth and their abrasion with hard plant tissue. Efficiency of feeding depends directly on the design of the tooth crown, on the size of total area for effective occlusion, and the efficiency of masticatory action. Large teeth with flat surfaces and high crowns resistant to intensive wear are particularly required.

The major problem with a diet of plants is that mammals (as well as other animals) do not produce enzymes that break down cellulose. They must rely on symbiotic microorganisms residing in their alimentary tract, evolve an appropriate housing for them, and ensure a sufficient time for proper food fermentation. The mammals evolved several ways to fulfill these requirements. One is the foregut fermentation (digastric digestion system) characteristic of ruminant artiodactyls (bovids, cervids), kangaroos, and colobus monkeys. The fermentation chambers are situated in spacious folds of the stomach; from these fermentation chambers the partially fermented food can be regurgitated and chewed during a rest period, which also prolongs the movement of food through the gut. The microorganisms detoxify alkaloids by which growing plants defend against herbivores prior to digestion, but are very sensitive to tanins contained in the dry plant tissues. The foregut fermenters avoid dry plants but feed on growing parts of plants, selectively cut with the tongue and lips (ruminants even lack the upper incisors).

Perissodactyls, rodents, lagomorphs, hyraxes, and elephants evolved hindgut fermentation (monogastric digestion system), where fermenting microorganisms are housed in the caecum and large intestine. Food is not regurgitated and all mechanical disintegration of food must be performed at one mastication event. Except for caeca, the passage of food through the gut is almost twice as fast as in the foregut fermenters. Hindgut fermenters can survive on a very low-quality food, if it is available in large quantity. They can effectively separate the tanins and dry plant mass, both of which decrease the efficiency of the foregut fermenting. Correspondingly, the foregut and hindgut fermenters prefer different parts of plants and can both forage in the same habitats without any actual competition. The latter are, of course, under more intense pressure to evolve further adaptations to compensate for the energetic disadvantages of their digestion. One of them is extreme enlargement of caeca (as in rodents); another is considerable increase in the height of cheek teeth (maximized in several clades of lagomorphs and rodents, in which cheek teeth are hypselodont, or permanently growing). The third way is an increase in body size. This enlarges the length of the alimentary tract and prolongs the passage of food through it, while at the same time it reduces the rate of metabolism. The behavioral reduction of metabolic rate by a general decrease of activity level as in foliovore (leaf-eating) sloths or the koala produces the same results.

The gradual increase in body size is a feature of mammalian evolutionary dynamics, as it was repeatedly demonstrated by the fossil record of many clades. This is seen in most eutherians (not only in the herbivorous clades), but is much less apparent in marsupials. It seems that in addition to the common factors promoting a larger body size (a reduced basal metabolic rate, smaller ratio of surface area to body mass, and smaller heat transfer with ambient environment), something else comes into play, something which has to do with the essential differences of both the clades. This is the enormous stress of the eutherian way of reproduction. While intrauterine development is short and a litter weight is less than 1% of the mother body mass in a marsupial, the eutherian female must endure a very long pregnancy and the traumatic birth of a litter that in small eutherians such as insectivores, rodents, or bats, may weigh 50% of the mother's body mass.

With enlarging body size, the stress of pregnancy and parturition is reduced as the size of a newborn is relatively smaller (compared with 3-5% of a mother's mass in large mammals and 10-20% in smaller mammals). With a reduction of litter size, it further provides a chance to refine the female investment and deliver fully developed precocial young, as in ungulates or cetaceans. This aspect of mammalian adaptation and diversity should remind us that perhaps the ways in which a female does manage the stress of eutherian reproduction (the factor that magnified the strength of selection pressure) became the most influential source of viability of our clade.

Neighbors, competitors, and friends

Mammals and humans have been the closest relatives and nearest neighbors throughout the entire history of humankind. Mammals contribute essentially to our diet and we keep billions of domesticated mammals solely for that purpose. Hunting mammals for protein-rich meat became an essential background factor in human evolution several million years ago. More recently, the discovery of how to get such animal protein in another way started the Neolithic revolution some 10,000 years ago. The symbiotic coexistence with herds of large herbivores—which included taking part in their reproduction and consuming their milk and offspring—ensured the energetic base for a considerable increase in the human population of that time and became one of the most important developments in human history. Moreover, the other essential component of the Neolithic revolution may be related to mammals. Feeding on seeds of grass and storing them in the form of a seasonal food reserve could hardly have been discovered without inspiration from the steppe harvesting mouse (Mus spicilegus) and its huge corn stores or kurgans, containing up to 110 lb (50 kg) of corn. The theory that humans borrowed the idea of grain storage from a mouse is supported by the fact that the storage pits of Neolithic people were exact copies of the mouse kurgans. Mammals have even been engaged in the industrial and technological revolutions. Prior to the steam engine and for a long time in parallel with it, draft animals such as oxen, donkeys, and horses were a predominant source of power not only for agriculture, transport, and trade, but also for mining and early industry. Indeed, our civilization arose on the backs of an endless row of draft mammals.

At the same time, many wild mammals have been considered dangerous enemies of humans: predators, sources of epizootic infections, or competitors for the prey monopolized by humans. Many mammals were killed for these reasons, while some were killed merely because we could kill them. As a result, many species of wild mammal were drastically reduced in numbers leading to their local or global extinctions. The case of the giant sea cow (Hydrodamalis stelleri) is particularly illustrative here, but the situation with many other large mammals, including whales, is not much different. The introduction of cats, rats, rabbits, and other commensal species to regions colonized by humans has badly impacted the native fauna many times, and the industrial pollution and other impacts of recent economic activity act in a similar way on a global scale. About 20% of extant mammalian species may be endangered by extinction, mostly due to the destruction of tropical forest.

However, since the Paleolithic, humans also have kept mammals as pets and companions. Even now, the small carnivores or rodents that share our houses bring us a great deal of pleasure from physical and mental contact with something that, despite its apparent differences, can communicate with us and provide what often is not available from our human neighbors—spontaneous interest and heartfelt love. Contact with a pet mammal may remind us of something that is almost forgotten in the modern age: that humans are not the exclusive inhabitants of this planet, and that learning from the animals may teach us something essential about the true nature of the world and the deep nature of human beings as well.

Resources

Books

Anderson, S., and J. K. Jones, Jr., eds. *Orders and Families of Recent Mammals of the World.* New York: John Wiley and Sons, 1984.

Austin, C. R., and R. V. Short, eds. *Reproduction in Mammals.* Vols. 1, 2, 3 and 4. Cambridge, UK: Cambridge University Press, 1972.

Chivers, R. E., and P. Lange. *The Digestive System in Mammals: Food, Form and Function.* New York: Cambridge University Press, 1994.

Eisenberg, J. F. *The Mammalian Radiations, an Analysis of Trends in Evolution, Adaptation, and Behavior.* Chicago: University of Chicago Press, 1981.

Feldhammer, G. A., L. C. Drickamer, A. H. Vessey, and J. F. Merritt. *Mammalogy. Adaptations, Diversity, and Ecology.* Boston: McGraw Hill, 1999.

Griffith, M. *The Biology of Monotremes.* New York: Academic Press, 1978.

Kardong, K. V. *Vertebrates. Comparative Anatomy, Function, Evolution.* Dubuque, Iowa: William C. Brown Publishers, 1995.

Kowalski, K. *Mammals: an Outline of Theriology.* Warsaw, Poland: PWN, 1976.

Kunz, T. H., ed. *Ecology of Bats.* New York: Plenum Press, 1982.

Lillegraven, J. A., Z. Kielan-Jaworowska, and W. A. Clemens, eds. *Mesozoic Mammals: The First Two-Thirds of Mammalian History.* Berkeley: University of California Press, 1979.

Macdonald, D., ed. *The Encyclopedia of Mammals.* New York: Facts on File Publications, 1984.

Neuweiler, G. *Biologie der Fledermaeuse.* Stuttgart-New York: Georg Thieme Verlag, 1993.

Nowak, R. M. *Walker's Mammals of the World.* 5th ed. Baltimore and London: Johns Hopkins University Press, 1991.

Pivetau, J., ed. *Traité de paléontologie, Tome VII Mammiferes.* Paris: Masson et Cie, 1958.

Pough, F. H., J. B. Heiser, and W. N. McFarland. *Vertebrate Life.* 4th ed. London: Prentice Hall Int., 1996.

Ridgway, S. H., and R. Harrison, eds. *Handbook of Marine Mammals.* New York: Academic Press, 1985.

Savage, R. J. G., and M. R. Long. *Mammal Evolution, an Illustrated Guide.* New York: Facts on File Publications, 1986.

Starck, D. *Lehrbuch der Speziellen Zoologie. Band II: Wirbeltiere. 5. Teil: Säugetiere.* Jena-Stuttgart-New York: Gustav Fischer Verlag, 1995.

Szalay, F. S., M. J. Novacek, and M. C. McKenna, eds. *Mammalian Phylogeny.* New York: Springer-Verlag, 1992.

Thenius, E. *Phylogenie der Mammalia. Stammesgeschichte der Säugetiere (Einschliesslich der Hominiden).* Berlin: Walter de Gruyter and Co, 1969.

Vaughan, T. A., J. M. Ryan, and N. Czaplewski. *Mammalogy.* 4th ed. Belmont, CA: Brooks Cole, 1999.

Wilson, D. E., and D. M. Reeder, eds. *Mammal Species of the World: a Taxonomic and Geographic Reference.* 2nd ed. Washington, D.C.: Smithsonian Institution Press, 1993.

Young, J. Z. *The Life of Mammals.* 2nd ed. Oxford: Claredon Press, 1975.

Other

University of Michigan Museum of Zoology. Animal Diversity Web. <http://animaldiversity.ummz.umich.edu>

Animal Info—Information on Rare, Threatened and Endangered Mammals. <http://www.animalinfo.org>

BIOSIS. <http://www.biosis.org.uk>

Links of Interest in Mammalogy. <http://www.il-st-acad-sci.org/mamalink.html>

The American Society of Mammalogists. <http://www.mammalsociety.org>

Smithsonian National Museum of Natural History. <http://www.nmnh.si.edu/vert/mammals>

World Wildlife Fund. <http://www.worldwildlife.org>

Ivan Horácek, PhD

Ice Age giants

During the latter half of the Ice Ages, the Pleistocene, in response to the slow pulsation of continental glaciers, there evolved unique large mammals—man included. In their biology and appearance they diverged from anything seen earlier in the long Tertiary, the Age of Mammals. They did not merely adapt to the increasingly seasonal climates and greater extremes in temperature and moisture. Rather, in the sheer exuberance and breadth of their adaptations, they reflected both the new ecological riches and soil fertility generated by glacial actions as well as the long successions of biomes they evolved in prior to life in the face of glaciers. Their novelty resides in novel opportunities and seasonal resource abundance in the environments shaped by these glaciers. It is this which gave rise to their oddness in shape and biological eccentricity, which shaped many into giants, and which ushered in the Age of Man.

The Pleistocene epoch is the latter half of the Ice Ages and is characterized by major continental glaciations, which began about two million years ago. There have been about 20 of these. Minor glacial events building up to the major glacial periods characterized the latter part of the Pliocene epoch. The evolutionary journey of mammalian families that succeeded in adapting to the cold north began in moist tropical forests. It proceeded stepwise into tropical savanna, dry grasslands at low latitudes, and then either into the deserts or into temperate zones at higher latitudes and altitudes. From there it continued into the cold, but fertile environments formed through glacial action and on into the most inhospitable of cold environments: the tundra, the alpine, and the polar deserts. Such extreme environments developed with the great continental glaciations that cycled at roughly 100,000 year intervals between cold glacial and warm interglacial phases. There was massive ice buildup in the Northern Hemisphere during the former with a concomitant shrinkage of oceans and severe drop in ocean shorelines. During inter-glacials there was glacial melt-off, followed by a re-flooding of the ocean to roughly the current level. We live today towards the end of an interglacial period. The well-differentiated latitudinal climatic zones we take for granted are a characteristic of the Ice Ages we live in; during the preceding Tertiary period there were tropical forests in what are today polar deserts. Consequently, adaptations to the extreme environments of the Ice Ages are relatively new.

Species adapted to cold and glacial conditions are new because the environments generated by huge continental glaciers became extensive only in the Pleistocene. That was new. Habitats formed by small mountain glacier are, of course, old, but large glaciations allowed the spread of what once were rare ecosystems. Also new is a sharp climatic gradient between equator and poles, generating latitudinal successions of biomes with increasing seasonality, terminating in landscapes of glaciers and snow.

Glaciers are "rock-eaters" that grind rock into fine powder. This ground rock is spewed out by the glacier with melt water and flows away from glacial margins as silt. When the seasonal glacial melting declines and the freshly deposited silt dries under the sun's rays, it turns to fine dust which the winds blowing off the glaciers carry far, far away. Glacial times are dusty times. In the ice cores from Greenland glaciers, the glacial periods are characterized by their dust deposits. This wind-born dust is called by the German term "loess." The ecological significance of glacial dust lies first and foremost in its fertility. Loess has high pH levels. Where it falls after day it forms into the fertile loess steppe. Silt and loess are deposited in lakes and deltas. After the lakes drain, there remain fertile deep-soils deposits. These Pleistocene loess and silt deposits in Eurasia and North America, as well as the ongoing deposition of glacier-ground silt along major rivers such as the Nile, Mekong, or Yellow River, are not merely today's grain baskets, but the very foundations of great civilizations. The silt and loess deposits form rich virgin soils, unleached and undepleted of their soluble mineral wealth. These young, fertile soils foster rich plant growth wherever there is sunshine and moisture.

Glaciers generate their own climates. They foster katabatic, that is, warm winds blowing away from the glacier. On the melt-off edges they foster clear skies and sunshine. We still see such climates along the ice fronts of the large mountain glaciations in the western Yukon and Alaska, along with abundant, diverse, and productive flora and fauna. Glaciers are not hostile to life.

Along these ice fronts we also see that when melt-water retreats, lenses of alkali mineral salts form in the silt depressions due to evaporation. These "saltlicks," composed largely of sulfate salts, are avidly visited by large herbivores and

A skeletal comparison of a mastodon (left), modern elephant (center), and a woolly mammoth (right). (Photo by © David Worbel/Visuals Unlimited, Inc. Reproduced by permission.)

carnivores. The inorganic sulfur is converted in the gut by bacteria into sulfur bearing amino acids, cysteine and methionine, the primary amino acids for the growth of connective tissues, body hair, hooves, claws, horns, and antlers. Salt licks are avidly visited by lactating females and by all during the shedding and re-growth of a new coat of hair. They are essential for the growth of luxurious hair patterns and huge horns and antlers. These "evaporite lenses" are covered by more and more loess, becoming part of deep loess deposits. When water cuts through such deposits forming steep loess cliffs, these evaporates attract big game which gradually dig deep holes into loess cliffs.

The Pleistocene loess steppe is a haven for large grazers due to its fertility. It has been called the "mammoth steppe" based on remains of woolly mammoth associated with it, as well as an "Artemisia steppe" based on the fact that many species of sage thrive here. This fertile steppe was also home to wild horses, long-horned bison, camels, reindeer, saiga antelopes, giant deer, and wapiti, as well as wolves, hyenas, lions, saber-toothed cats of two species, and several species of bears. We may also call it the "periglacial" environment. It was extensive during glaciations. During the interglacial warm periods, without the fertilizing effect of glacial silt and loess, the acid tundra, alpine, polar deserts, and boreal forest were prevalent. Thus the development of diverse cold environments, some greatly affected by glacial actions and seasonally quite productive, invited the colonization by new types of mammals able to cope with the biological riches and the climatic hardships.

The evolutionary progression towards Ice Age giants begins in the tropical forests with old, primitive parent species that are, invariably, defenders of resource territories. They are recognizable as such by their weapons, which are specialized for injurious combat: long, sharp canines or dagger-like, short horns. Property defense is based on expelling intruders by inflicting painful injuries that also expose the intruder to greater risk of predation. Both males and females may be armed and aggressive. They escape predators by taking advantage of the vegetation for hiding or climbing and are excellent jumpers that can cross high hurdles.

In the subsequent savanna species the "selfish herd" becomes prominent as a primary security adaptation against predation. This is associated with a dramatic switch in weapon systems and mode of combat. That is, as individuals become gregarious, they fight mainly via wrestling or head-butting, and minimize cuts to the body that could attract predators. They also evolve "sporting" modes of combat, sparring matches, in which there are no winners or losers. This is a novelty permitted by the new mode of combat. Moreover, relative brain size increases, probably as a response to more complex social life. With adaptation to greater seasonality the species evolves the capacity to store surpluses from seasons of abundance into seasons of scarcity. That is, individuals develop the capacity to store significant amounts of body fat. Their reproduction tracks the seasonal growth of plants, whether triggered by rain falls or seasonal temperatures. Their mode of locomotion changes to deal with the predators of the open plains. They may evolve fast running, without giving up their ancestral ability to jump and hide in thickets.

As evolution progresses to the wide-open, grassy steppe, the plains-adapted species evolves capabilities to deal with low temperatures. Seasonal hair coats evolve. Because of the rich seasonal growth of forage, individuals experienced a "vacation" from want and from competition for food, evolving ornate hair coats and luxurious secondary sexual organs. This tended to go along with an increase in body size. Consequently, by the time species evolve in the cold environments close to continental glaciers, they may be giants of their respective families as well as their most ornate, brainy, and fat members. We may call these new Ice Age species "grotesque giants." They are exemplified by woolly mammoth and woolly rhino, the giant stag or Irish elk, the moose, caribou or reindeer, Przewalski's horse, Bactrian camels, the extinct cave bear and giant short-faced bear, and extant Kodiak and polar bear, and, of course, our own species, *Homo sapiens*. Compared to other species within our family or tribe, we are indeed a grotesque Ice Age giant. Indeed, two human species adapted to the glacial environments—the extinct Neanderthals and ourselves. Note: every Ice Age giant is the product of successful adaptations to a succession of climates and environments from tropical to arctic. Thus, they have a wide range of abilities built into their genomes.

The progression of species from primitive tropical forms to highly evolved arctic ones is well illustrated in the deer family, as is the varied nature of gigantism. Moreover, in the deer family both subfamilies of deer follow the very same evolutionary pattern. In the Old World deer it begins with the muntjacs, small tropical deer from southern Asia with one or two pronged antlers and long upper combat canines. They are largely solitary territory-defenders that escape predators by rapid bounding (saltatorial running) followed by hiding in dense cover. They are a very old group dating back to the mid-Tertiary.

The second step in the evolutionary progression is represented by species of tropical three-pronged deer. These are adapted to savanna, open wetlands and dry forest. They include the highly gregarious axis deer, hog deer, rusa and sambar, as well as the swamp-adapted Eld's deer and barasingha. These deer too are largely saltatorial runners and hiders, although they favor some open spaces. All have gregarious phases. All have antlers evolved for locking heads in wrestling matches. The upper canines are reduced or absent in adults. There is a split into more gregarious, showy meadow-species and more solitary forest-edge species. Although these species

A life-sized woolly mammoth (*Mammuthus primigenius*) model. (Photo by Stephen J. Krasemann/Photo Researchers, Inc. Reproduced by permission.)

differ in external appearance, nevertheless the identity of their body plan is readily apparent. The most gregarious forms have prominent visual and vocal rutting displays.

The third step in the progression is represented by the four-pronged deer. These are adapted to temperate climates with a short, mild winter. Only two species are alive today, the fallow deer and the sika deer. Besides the increased complexity of antlers, there is a stronger differentiation and showiness of the rear pole. While the three-pronged deer have a showy tail, the four-pronged deer have a rump patch in addition. That of the sika deer consists of erectable hair that may be flared during alarm and flight. These are highly gregarious deer with very showy vocal and visual displays.

The fourth step is represented by the five-pronged deer, all of which are primitive Asiatic subspecies of the red deer. They are found in regions with a distinctly harsher, colder, and more seasonal climate than the preceding four-pronged deer, including in high mountain areas of central Asia. These deer have progressed still further in the differentiation of the antlers, body markings, and rump patch and tail configurations. They are also much larger in body size. An evolutionarily advanced branch of red deer of some antiquity is the European red deer. These feature complex five-pronged antlers, a neck mane, and larger and more colorful rump patches.

The fifth step is represented by the six-pronged deer—the advanced wapiti-like red deer of northeastern Asia and North America. These are the ornate giants among Old World deer. They are much more cold-adapted and extend on both continents beyond 60°N. They occupy periglacial and cold montane, sub-alpine habitats, are more adapted to grazing than other red deer, and have a body structure similar to plains runners. They have the largest rump patch and the shortest tail, the greatest sexual body color dimorphism, and the most complex rutting vocalizations.

The Moreno glacier rises 197 ft (60 m) above Lago Argentino's water level in the National Park of Los Glaciares in Patagonia, Argentina. (Photo by Andre Jenny/Alamy Images. Reproduced by permission.)

Of the same evolutionary rank as the six-pronged wapiti was the now extinct giant stag or Irish elk. It grew the largest antlers ever and was also the most highly evolved runner among the deer. Besides the enormous antlers, it had a hump over the shoulders, a tiny tail, and probably had prominent body markings judging from cave paintings. It was a resident of the fertile glacial loess steppe and proglacial lakes. Of the same evolutionary rank among the New World deer are the moose and the reindeer, both found in extremely cold climates. In South America, in the cold southern pampas formed from loess, cold and plains adapted deer also evolved enormous reindeer-like antlers. They are now extinct.

Among the primates only the hominids leading to Neanderthals and modern humans have gone through a similar mode of evolution. Humans are the only primate that has been able to penetrate the severe ecological barriers posed by the dry, treeless steppe. That is what allowed them to spread into and adapt to northern landscapes. To conquer the treeless steppe humans had to be able to escape predation at night on the ground. They had to provide continually high quality food to gestating and lactating females irrespective of the season. Besides evolving the capacity to store very large quantities of body fat, which became a prerequisite of reproduction, they developed means to access the subterranean vegetation food stores encased in hard soils during the dry season. They successfully exploited the rich food resources of the inter-tidal zones and estuaries. Through hunting, they tapped into the rich protein and fat stores of the large mammals on the steppe. As the capacity to kill large mammals evolved, weapons developed that could stun opponents rendering them unable to retaliate, and cultural controls over killing augmented ancient biological inhibitions. This is a profound adaptation, and is thus biologically unique and not found among other mammals. The distinction between doing what is right and wrong must thus go back to the roots of tool and weapon use about two million years ago.

Here there is the familiar, step-wise progression from a tropical, forest-adapted, resource-defending ancestor similar to a chimp; to the savanna-adapted australopithecines who greatly reduced the canines—ancestral weapons of territorial defense; to the steppe-adapted *Homo erectus*, our parent species. *Homo erectus* appeared at the beginning of the major glaciations almost two million years ago and spread into cold-temperate zones in Eurasia. Unlike the deer family, however, which skipped past deserts and went directly into periglacial, arctic, and alpine environments, human evolution did not bypass deserts. It appears that with the massive Penultimate Glaciation beginning about 225,000 years ago,

Mountain goats (*Oreamnos americanus*) at a salt lick at Jasper National Park, Alberta, Canada. (Photo by © Raymond Gehman/Corbis. Reproduced by permission.)

which must have led to a maximum spread of deserts in Africa, *Homo sapiens* arose out of *Homo erectus* by adapting to deserts. Two branches survived to invade and thrive in the periglacial zones of Eurasia, the enigmatic Neanderthals and the modern *Homo sapiens* species. We can thus trace the rise of a "grotesque giant" primate—ourselves. Man is large in body, ornate in hair pattern and secondary sexual organs, and evolved a very large brain. Our reproductive biology depends on large stores of body fat, and we evolved highly sophisticated displays based on vocal and visual mimicry. Finally, we developed an insatiable urge to artistically modify everything we were able to modify, which led to culture. We are thus part and parcel of a greater evolutionary phenomenon, that of the Ice Age giants.

However, the tropics too produce giants, represented among primates by the larger of the great apes, foremost by the gorilla and orangutan. Tropical giants built on primitive body plans are, invariably, "coarse food" giants. Small-bodied mammals have a high metabolic rate per unit of mass compared to large-bodied mammals. This is related to the fact that to keep a constant core-temperature of about 98.6°F (37°C), which is essential for optimum enzyme functioning, small mammals must burn more fuel per unit of mass than do large mammals. Small mammals, because of the very large surface to mass ratio, lose heat rapidly compared to large mammals with their low surface to mass ratio. Consequently, a mouse must metabolize per unit of mass much more food than an elephant. In order to maintain the high metabolic rate required the mouse needs to digest its food very rapidly, compared to an elephant, and must consequently select only rich, highly digestible food. Elephants, by comparison, can feed on very coarse, fibrous food that may remain for some time in their huge digestive tracts. The same principle applies to tiny and gigantic tropical primates. The former feed

on buds, flowers, fruit, insects, etc., while the gorilla feeds on fibrous, much more difficult to digest vegetation. The chimpanzee, which stands so close to our ancestral origins, is somewhere in between large and small, and its omnivorous food habits reflect that fact.

Ice Age giants reflect totally different conditions. Their size depends, in part, on the large seasonal surpluses of high quality food during spring and summer. Large size, however, is also an option in insuring minimum predation. That is, a high diversity and density of predators, such as those that characterized North America's Pleistocene, generates gigantic herbivores with highly specialized anti-predator adaptations. Conversely, herbivores stranded on a predator-free oceanic island decline rapidly in size and loose their security adaptations. They become highly vulnerable "island dwarfs". Elephants for instance have shrunk to 3 ft (0.9 m) in shoulder height on islands. Oddly enough, large body size is not related to ambient temperatures in winter, despite the fact that the surface to mass ratio declines with body size, favoring heat conservation. This is the principle behind the famous, but invalid Bergmann's Rule. Contrary to its predictions, body size in the same species does not increase steadily with latitude. Rather, body size increases only to about 60-63°N and then reverses rapidly. That is, individuals of a species beyond 63°N become rapidly smaller with latitude, some, such as caribou and musk oxen reaching dwarf proportions closest to the North Pole. Lowering the surface to mass ratio as an adaptation to cold is so inefficient that the absolute metabolic costs of maintaining ballooning bodies outstrips whatever metabolic savings might be gained by the reduction in surface relative to mass. Bergmann's Rule has thus neither empirical nor theoretical validity. That predation plays a role in driving up body size is not only indicated by North America's Pleistocene fauna of gigantic predators and prey or the biology of island dwarfs, but also by the fact that the largest deer, the Irish elk, was also the most highly evolved runner among deer. For humans adapting to the dry steppe, hunting must have played a role in increasing body size, while periods of low food abundance favored a reduction in body size.

A woolly mammoth (*Mammuthus primigenius*) skeleton. (Photo by John Cancalosi/OKAPIA/Photo Researchers, Inc. Reproduced by permission.)

A man stands next to a life-sized woolly mammoth model in the Royal British Columbia Museum. (Photo by © Jonathan Blair/Corbis. Reproduced by permission.)

However, by far the most striking attributes of the grotesque Ice Age giants are their showy, luxurious hair coats, secondary sexual organs and weapons, and their showy social displays. The enormous tusks of mammoths and their long hair coats; the huge antlers of Irish elk, moose, and caribou as well as their striking hair coats; the enormous horn-curls of giant sheep and bighorns; the beards, pantaloons, and hair-mops of bison and mountain goats; and the sharply discontinuous hair patterns and large fat-filled breast and buttocks in our species all stand in sharp contrast to comparable organs in tropical relatives. The great surpluses of food in summer do permit the very costly storage of fat as well as horn, tusk, and antler growth. However, seasonally abundant food is only a necessary condition for "luxury organs" to evolve, but not a sufficient one. The rise of animal behavior as a science has informed us about these luxury organs. Their size, structure, and distribution over the body, as well as the manner in which they are displayed during social interactions indicate that they are

signaling structures evolved under sexual selection. Predation lurks in the background in some lineages, as illustrated by the way in which the gigantic antlers, horns, and tusks of northern plains-dwelling herbivores have evolved.

Envision a deer moving from tree and bush-studded savanna to open grasslands void of cover. The more open the landscape, the more difficult it is to hide a newborn adequately, particularly in already large-bodied species. Also, hiding becomes increasingly more risky, as visits by the female to suckle and clean her young are now quite readily observed as they are out in the open. Predators can thus find newborns in the open terrain. The way out of this dilemma is to bear young that can quickly get to their legs and follow their mothers at high speed. This must be followed by nursing the young with milk exceptionally rich in fat and protein. Then the young are able to grow rapidly to "survivable size," at which endurance as well as speed can match that of adults. This, however, places a great burden on the female. In order to be

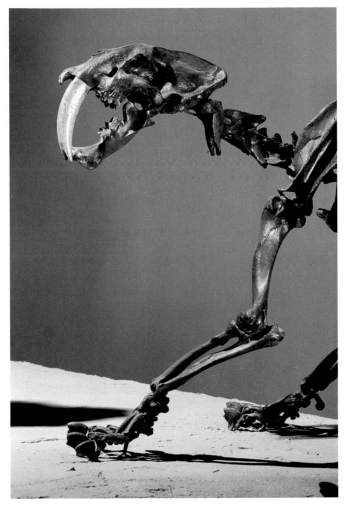

A saber-toothed cat (*Smilodon californicus*) skeleton taken from the Rancho La Brea tar pits. (Photo by Tom McHugh/Photo Researchers, Inc. Reproduced by permission.)

the female should choose as a father for her daughters a male with very large, symmetrical antlers. If so, then the larger the antlers, the more males ought to advertise with their antlers during courtship. Therefore, the larger the antlers of a species, the larger and more advanced the newborn, the richer the milk, the better the parents are at high-speed running, and the more males flout their antlers in courtship. All of these expected correlations are found.

What antlers do for deer, horns or tusks and elaborate hair coats do for other species that are associated with wide-open, cold, but productive Ice Age landscapes. These luxury organs evolve in relation to the security of newborns. Therefore, severe predation pressures ought to enhance the evolution of these luxury organs in plains dwellers. It is not surprising, therefore, that in Pleistocene North America, the large number of specialized predator species are associated not only with very large-bodied prey, but also with body structures in prey that enable speedy running, as well as immense tusks in mammoth, enormous horns in long-horned bison, and huge, complex antlers in the stag-moose and caribou.

In the hominid line leading to our species, the highly developed luxury organs are also related to reproduction and sexual selection. Not only need females have an exceptionally high body fat content for mammals before conception and pregnancy is possible, but the female's "hour glass" distribu-

The grizzly bear (*Ursus arctos*), standing in a meadow at Strawberry Valley, Utah, USA, evolved a thick coat to deal with its cold environment close to continental glaciers. (Photo by © Galen Rowell/Corbis. Reproduced by permission.)

well developed and quick to rise and run, a newborn needs to be born large in body and advanced in development. That, and the subsequent production of rich milk requires that the female not only have a high food intake, but take risks feeding in dangerous areas just to maintain a high food intake. She must also spare nutrients and energy from her body growth in favor of that of her child. What father must such a female choose so that her daughters may have the highest genetic potential to grow large babies and rich milk?

Since the fathers contribute only their genes to the welfare of their children, they must—somehow—advertise their success at foraging and escaping predators. In the case of deer, any energy and nutrients ingested above the need of the body for maintenance and growth are automatically diverted into antler growth. Therefore, the better a male is at finding high quality food, the better its antler growth and the larger its antlers. However, high quality, uncontested food may be in insecure areas favoring predators. Therefore, the larger the antlers the more daring the male and the more able it is at detecting and escaping predators. The symmetry of the antlers is also a factor. Symmetry is a proxy for health. Therefore,

A recreation of how *Megantereon,* a species of saber-toothed cat, may have looked. (Photo by Tom McHugh/Photo Researchers, Inc. Reproduced by permission.)

tion of fat over her body signals her fertility and health. In addition to breast and buttocks, which are shaped largely through triglyceride (fat) deposits, the density and quality of hair acts as sexual signals. Hair is formed from sulfur-bearing amino acids such as cysteine and methionine, which are rare and cannot be readily synthesized in our bodies in adequate amounts. Luxurious hair growth, formed from rare amino acids signals, therefore, biological success as it is based on the intake of high quality food. Our notion of beauty and grace rests in the first instance on symmetry of body features and function. Symmetry is a function of good nutrition. Our brain consists largely of fat. Its superior function depends on a rich and balanced intake of essential fatty acids. Even mother's milk reflects this demand as, compared to other species, it is exceptionally rich in brain-building omega-3 and omega-6 fatty acids. For superior growth and development, humans thus require omnivorous food sources fairly rich in fats that contain these essential fatty acids, such as the fat of wild game, fish, and seafood. In the Upper Paleolithic of Europe humans reach exceptional physical development and brain sizes; the archeological record indicates that their food was primarily reindeer and salmon. Neanderthal, the human super-predator, reached earlier comparable brain sizes living off the largest of periglacial herbivores, woolly mammoth, woolly rhino, giant deer, horses, and steppe bison. The cerebral cortex is a tissue

of low growth priority. That is, it grows to maximum size only under exceptionally favorable food intake, provided of course that it is also well stimulated daily through an active social life and adventures.

There were many Ice Age giants in the periglacial environment: huge beavers; enormous bears such as the carnivorous short-faced bear of North America or the cave bear of Europe or the current Kodiak brown bear and the polar bear. Northern cave-lions and cave-hyenas reached body sizes twice the mass of their African relatives, as did the American Pleistocene cheetah compared to the African or Near East forms. And, of course, humans reached maximum body and brain dimensions in the periglacial landscapes. Why then did these large-bodied forms not grow larger still? We get a glimpse of the answer for our own species by looking at what Neanderthals did.

Neanderthals, exceptionally robust and powerful, with huge joints in arms and legs and massive bones, approached us in height and were probably equal in body mass. As super-carnivores they specialized in the largest and hairiest herbivores, which were brought down apparently by one hunter skillfully tackling and attaching himself to the hairy exterior of his prey, while a second hunter jumped in to disable or kill the distracted beast. The tackled beast, however, must have gone through some violent rodeo-like bucking and jumping to dislodge the attached first hunter. Consequently, dislodged hunters would have been catapulted through the air followed by harsh landings, often on frozen ground. Not too surprisingly, Neanderthal skeletons reveal frequent bone breaks and the pattern of bone breaks resembles that of rodeo cowboys. Neanderthal hunters did not enjoy a long life expectancy, unlike the modern humans in the Upper Paleolithic that followed them. A body size larger than that attained would have placed the hunters at a disadvantage as it would have been more difficult to hold on, while falling off would have generated harder impacts and generated even more bone breaks. That is, as body size increases, it reduces mobility and acceleration, but increases muscular strength and impact force during a fall. If agility and acceleration are compromised at large body size, then the ability of hunters to evade attacks by large prey they antagonized or wounded is compromised. Furthermore, should there be periods of food shortage that strike during the growth and development of an individual, then the smaller-bodied individual has a better chance of growing brain tissue than a large individual. Small body size, while excelling at agility and acceleration, suffers from lower strength, maximum running speed, and endurance. Body mass is thus based on a compromise of factors that allow optimum performance in the tasks demanded by adaptation.

The evolutionary road to periglacial environments is virtually a one-way road to herbivores and omnivores. There is practically no return of Ice Age species to habitats at lower elevation and in more benign climates, with a few exceptions. Ice Age herbivores must be generalists that can deal with a great diversity of seasonal temperatures and foraging conditions. They cannot compete against the food-specialists at lower elevations. These evolved to deal with the myriad of chemical and structural defenses of plants against being eaten by herbivores. Cold climate vegetation, by comparison, is much less defended in large part because much of it may be hidden under a long-lasting snow-blanket and unavailable for grazing. Moreover, Ice Age species are exposed to fewer parasites and pathogens than are species in warmer climates. Consequently, the diseases and parasites of more primitive warm-climate species can be devastating to Ice Age mammals. The tropics in particular are characterized by an abundance of microorganisms and parasites against which northern species have little or no immunity. Nevertheless, some species have succeeded in re-invading warm climates, humans in particular. That humans from high latitudes do have significant problems in the tropics is attested to by medical history.

With the rise of humans to ecological dominance their fellow Ice Age giants did not fare well. Most went extinct, and though their passing is shrouded in controversy, the weight of opinion points to humans as the cause of their extinction. Human capabilities improved sharply between 60,000 to 40,000 years ago and the largest of the Ice Age giants began to fade then. Massive extinctions of the megafauns followed the last retreat of the glaciers. However, that was also a time of great hardships for humans, followed by the rise of agriculture, which brought some relief. Most of the Ice Age giants must have become victims of very hungry and very able human hunters.

The Ice Ages generated new, increasingly seasonal and climatically harsh environments from the equator to the poles. Each Ice Age giant is thus the product of successive adaptations to ever more challenging landscapes. Each has the genome of ancestors that were highly successful in the tropics, in the savanna, in the steppe, in deserts, in cold-temperate zones as well as in the climatically extreme periglacial, arctic, and alpine habitats. They were thus constructed genetically with more and more abilities to cope with more and more challenges. That may be the reason why in the Pleistocene, and not earlier, there evolved the most uniquely gifted Ice Age giant of all: the modern human.

Resources

Books

Bonnichsen, Robson, and Karen L. Turnmire, eds. *Ice Age People of North America.* Corvallis: Oregon State University Press. 1999.

Geist, Valerius. *Life Strategies, Human Evolution, Environmental Design.* New York: Springer Verlag, 1978.

Geist, Valerius. *Deer of the World.* Mechanicsburg, PA: Stackpole Books. 1998.

Guthrie, Dale R. *Frozen Fauna of the Mammoth Steppe.* Chicago: University of Chicago Press. 1990.

Martin, Paul S., and Richard G. Klein. *Quaternary Extinctions.* Tucson: University of Arizona Press, 1984.

Nilsson, Tage. *The Pleistocene.* Stuttgart: Ferdinand Emke Verlag, 1983.

Valerius Geist, PhD

Contributions of molecular genetics to phylogenetics

Introduction

Traditional studies of evolutionary relationships among living organisms (phylogenetics) relied predominantly on comparisons of morphological characters. However, phylogenetic studies have increasingly benefited from additional inputs from molecular genetics. Reconstruction of evolutionary relationships among mammals provides a prime example of such benefits, and the broad outlines of the phylogenetic tree of mammals have now been convincingly established. For instance, it has proved possible to identify four major clusters (superorders) of placental mammals: (1) Afrotheria—elephants, manatees, hyraxes, tenrecs, golden moles, elephant shrews, and the aardvark; (2) Xenarthra—sloths, anteaters, and armadillos; (3) Euarchontoglires—rodents, lagomorphs, primates, dermopterans, and tree shrews; (4) Laurasiatheria—artiodactyls (including cetaceans), perissodactyls, carnivores, pangolins, bats, and most insectivores ("eulipotyphlans": hedgehogs, shrews, and moles).

All phylogenetic reconstructions based on molecular data depend on studies of the genetic material DNA or of proteins, whose synthesis is governed by individual DNA sequences. The primary components of the double-stranded DNA molecule are nucleotide bases, sugar groups, and phosphate groups. There are four nucleotide bases (adenine, cytosine, guanine, and thymine), and specific chemical bonds between pairs of these (adenine with thymine; cytosine with guanine) provide the backbone for DNA's double helix structure. These specific bonds between pairs of bases also ensure that, if one strand is separated, the missing strand will be faithfully replicated. Because the basic unit in the double-stranded DNA molecule is hence a bonded pair of bases (one base in each strand), the length of a DNA sequence is measured in base pairs (bp). The sequence of nucleotide bases in the DNA molecule provides the basis for protein synthesis through the genetic code, with a group of three bases ("triplet") in the DNA sequence corresponding to one amino acid in the protein sequence. Protein sequences hence depend directly upon DNA sequences, and a score of different amino acids are combined into chains of specific composition through translation of sequences of nucleotide bases in DNA, assisted by two kinds of RNA (messenger RNA and transfer RNAs). However, the original simple concept of "one gene, one protein" has needed modification. One major reason for this is that a

DNA sequence corresponding to a particular protein sequence often contains non-coding regions (introns) between the coding regions (exons). Only the exons are ultimately reflected in the amino acid sequence of the corresponding protein. Furthermore, the products of individual DNA sequences can be spliced together to produce a protein.

Both DNA sequences and protein sequences are particularly suitable for phylogenetic reconstruction because they consist of relatively simple components arranged in linear series (nucleotide bases and amino acids, respectively) that can be easily compared between species. Initially, comparisons between species were based on laborious step-by-step determination of the amino acid sequences of proteins, as there was no straightforward technique for studying DNA sequences themselves. The first phylogenetic trees derived from molecular genetics were therefore based on amino acid sequences of proteins, and relatively few species were included in comparisons because of the time-consuming procedure involved in protein sequencing. At first, it was also technically very difficult to determine DNA sequences. However, a major breakthrough came with development of the capacity for generating large quantities of individual DNA sequences through amplification using the polymerase chain reaction (PCR). This opened the way to relatively straightforward and rapid direct determination of DNA sequences, and heralded the transition from studies of gene products (proteins) to studies of the genes themselves (DNA sequences). In fact, because it became easier and faster to determine DNA sequences directly, a protein sequence is now commonly inferred from the DNA sequence of the corresponding gene rather than from sequencing of the protein.

It is important to note that there are two different sets of genetic material (genomes) in mammalian cells, as in animals generally. The primary genome is contained in the chromosomes in the nucleus (nuclear DNA), but each mitochondrion in the cell cytoplasm also contains a number of copies of a separate small genome (mitochondrial DNA). As several mitochondria are present in each cell, there are numerous copies of the mitochondrial genome, whereas there is only one nuclear genome per cell. However, the basic structure of DNA is the same for nuclear DNA (nDNA) and mitochondrial DNA (mtDNA), with chains of nucleotide bases, although mtDNA is organized in a ring whereas nDNA exists as linear sequences within chromosomes. The mitochondrion is a respiratory power plant found in all organisms with a cell nucleus (eukaryotes). It is, in fact, derived from a free-living bacterium that took up residence in the cell cytoplasm in an ancestral eukaryote more than a billion years ago, in an arrangement that was of mutual benefit (symbiosis). Originally, the mitochondrial genome contained many more genes than are now present in mammals, whose mitichondria retain only a small number of protein-coding genes that are all connected with respiration.

Reconstruction of phylogenetic trees

Regardless of whether morphological or molecular data are analyzed, reconstruction of phylogenetic relationships between species depends on interpretation of shared similarities. In principle, it is relatively easy to survey similarities between species for individual characters. This task is particularly straightforward with molecular data because the individual components at defined positions in sequences of DNA (nucleotide bases) or proteins (amino acids) are relatively simple and directly comparable. Furthermore, because the primary process underlying evolutionary change is point mutation (random replacement of one nucleotide base by another in a DNA sequence), comparison of DNA sequences directly reveals basic evolutionary steps. Most changes in DNA sequences lead to changes in corresponding protein sequences, but there is some degree of redundancy in the genetic code, because up to six different triplet sequences of nucleotides can correspond to a single amino acid. For this reason, about 25% of point mutations in DNA are "silent" and do not lead to a change in protein sequences. Such redundancy applies particularly to the third base position in DNA triplets.

Analysis of similarities between species to construct phylogenetic trees is more difficult than it seems at first sight. In the first place, similarities can arise independently through convergent evolution at any time after the separation between two lineages. For instance, rodent-like incisor teeth have developed several times independently during the evolution of mammals. But reconstruction of the relationships between species depends on exclusion of convergent similarities and identification of homologous similarities that have been inherited through descent from a common ancestor. In the case of morphological characters, it is often possible to identify convergent similarities directly because development of similar characters is typically driven by similar functional requirements. For example, rodent-like incisors develop in response to selection pressure for gnawing behavior. For complex morphological characters, convergent similarity is typically only superficial because it merely needs to meet a particular functional requirement. Hence, detailed examination of such characters commonly reveals fundamental differences. With incisors, for instance, a rodent-like pattern can develop without altering the structure of enamel that characterizes a particular group of mammals. With molecular characters, by contrast, each type of nucleotide base or amino acid shows complete chemical identity, so it is impossible to determine from direct examination whether convergent evolution has occurred. Instead, convergence in molecular

Due to DNA evidence, cetaceans like the bottlenosed dolphin (*Tursiops truncatus*) have been linked to hippopotamuses (*Hippopotamus amphibius*). (Photo by © Tom Brakefield/Corbis. Reproduced by permission.)

evolution is recognizable only from the phylogenetic tree after it has been generated on the assumption that the tree requiring the smallest total amount of change (the most parsimonious solution) is the correct one. Because there are so few possibilities for evolutionary change at the molecular level (4 nucleotide bases; 20 amino acids), convergent evolution is very common. As a rule, about half of the similarities between species recorded in any tree that is generated must have arisen independently through convergent evolution. Convergence is therefore a major problem with any tree derived from molecular data, particularly because functional aspects of changes in nucleotide bases and amino acids are rarely considered (thus excluding any possibility of identifying functional convergence). Moreover, precisely because there are so few possibilities for change in DNA base sequences, repeated point mutation at a given site will mask previous changes and can easily lead to chance return to the original condition. Although it is now standard practice to make a global correction for repeated mutation at a given site in molecular trees, it is virtually impossible to reconstruct the mutational history of individual sites if repeated change has occurred.

In fact, there is a further problem in interpreting similarities for the reconstruction of phylogenetic trees. Even if it is possible to exclude certain cases of convergent evolution, as is often true with complex morphological similarities, an important distinction remains with respect to inherited homologous similarities. For any group of species considered, a particular set of features will be present in the initial common ancestor. If such a primitive feature is retained as a homologous similarity in any descendants, it reveals nothing about

The red panda (*Ailurus fulgens*) is very hard to classify, but was placed with the other Ursidae species due to similar DNA. (Photo by Tim Davis/Photo Researchers, Inc. Reproduced by permission.)

branching relationships within the tree. The only homologous features that provide information about branching within a tree are novel features that arise at some point and are subsequently retained by descendants as shared derived similarities. This crucial distinction between primitive and derived homologous similarities is particularly relevant if there are marked differences between lineages in the rate of evolutionary change. For instance, members of two slowly evolving lineages can retain many primitive similarities and would thus be grouped together on grounds of overall homologous similarity if no special attempt were made to identify derived similarities. It was once believed that rates of change are reasonably constant at the molecular level, thus reducing the need to distinguish between primitive and derived homologous similarities, but the availability of large molecular data sets has revealed that there can be major differences in rates between lineages.

In conclusion, the increasing availability of molecular data has provided a major benefit for the reconstruction of phylogenetic trees. The large numbers of directly comparable characters included in molecular data sets provide a highly informative basis for quantitative comparisons. On the other hand, because the methods used do not explicitly tackle the crucial distinction between convergent, primitive, and derived

similarities, the results are subject to error. Accordingly, if there is a conflict between a tree based on molecular data and one based on morphological data, it should not be automatically assumed that the latter is necessarily incorrect. After all, there is quite often a similar conflict between trees based on two different molecular data sets. The safest procedure is therefore to take a balanced approach that gives due consideration to both morphological and molecular evidence. Combined studies that do precisely this with comprehensive data sets are becoming increasingly common.

Mitochondrial DNA

The ring-shaped, double-stranded mtDNA molecule has the same basic structure in all mammals. It is approximately 16,500 bp in length and contains coding sequences for 13 genes, 2 ribosomal RNA molecules (12S and 16S), and 22 transfer RNA molecules, together with a non-coding control region (D-loop). In contrast to nuclear genes, there are no introns in mtDNA. Furthermore, mtDNA differs from nDNA in another crucial respect that simplifies analysis of its evolution. In mammals, mtDNA is exclusively or almost exclusively inherited maternally (i.e., from the mother), and there is no recombination of genes when the mitochondrion

The giant anteater (*Myrmecophaga tridactyla*) belongs to the Xenarthra cluster of placental mammals. (Photo by © Tom Brakefield/Corbis. Reproduced by permission.)

divides. Phylogenetic reconstructions may be based on part of mtDNA (e.g. using an individual gene, such as cytochrome *b*) or on the entire molecule, and many complete mtDNA sequences are now available for analysis. Overall, mtDNA tends to accumulate changes more rapidly than nDNA (about five times faster overall), and for this reason it is more suitable for analyses of relatively recent changes in the evolutionary tree of mammals. Because rapidly evolving DNA sequences become saturated with changes at an earlier stage, they are unsuitable for probing early parts of the tree. However, there are differences in rate of evolution between individual parts of the mtDNA molecule, so it is possible to select regions that are suitable for particular stages of mammalian evolution. Mitochondrial DNA sequences can be crudely divided into those that evolve relatively rapidly, hence being useful for comparisons of quite closely related species (e.g. control region, ATPase gene) and those that evolve relatively slowly, thus being useful for comparisons of more distantly related species (e.g. ribosomal genes, tRNA genes, cytochrome *b* gene). For example, golden moles are of presumed African origin. This implies that there was an extensive African radiation from a single common ancestor that gave rise to ecologically divergent adaptive types. DNA studies suggest that the base of this radiation occurred during Africa's isolation in the Cretaceous period before land connections were developed with Europe in the early Cenozoic era. In another study, scientists examined the mtDNA of 654 domestic dogs, looking for variations. They were trying to determine whether dogs were domesticated in one or several places, and then attempting to identify the place and time that such domestication occurred. Their results show that our common domestic dog population originated from at least five female wolf lines. They went on to speculate that while the archaeological record cannot define the number of geographical origins or their locations, their own data indicate a single origin of domestic dogs in East Asia some 15,000 to 40,000 years ago.

Nuclear DNA

Surprisingly, only a small fraction of the nuclear DNA (nDNA) contained in chromosomes consists of gene sequences that code for production of proteins. It is estimated that less than 5% of human DNA consists of genes that code for approximately 30,000 different proteins. Much of the rest (95%) has no well-established function and is often labeled "junk DNA". A large part of this DNA consists of repetitive sequences that in some cases are present as many thousands of copies. Such DNA sequences that have been inserted into the genome are known as "retroposons", but their function remains essentially unknown.

Reconstruction of phylogenetic relationships using DNA sequences that code for protein sequences (or using the protein sequences themselves) hence involves only a small part of the nuclear genome. Nevertheless, there are many different nuclear genes available for analysis, and the sequence data set for mammals is increasing rapidly. Certainly, the potential total sequence information that can be obtained from the 30,000 protein-coding genes in the nuclear genome is vastly greater than that provided by the 13 protein-coding genes in the mitochondrial genome. As a general rule, the reliability of phylogenetic trees generated with molecular data increases both with the number of species included in comparisons and with the number of DNA sequences analyzed. However, there are some unresolved problems with the methods currently employed for reconstruction of trees using molecular data. Furthermore, there are practical limits to the quantity of data that can be effectively analyzed, so various short-cuts are necessary.

In addition to protein-coding DNA sequences, some categories of retroposons (inserted sequences) are becoming increasingly useful as tools for reconstructing phylogenetic relationships. This is particularly true of inserted sequences known as short interspersed nuclear elements (SINEs) and long interspersed nuclear elements (LINEs), respectively.

The African elephant (*Loxodonta africana*) has been grouped into the Afrotheria cluster of placental mammals. (Photo by © Craig Lovell/Corbis. Reproduced by permission.)

Four phylogenetic mammal trees have been established. The branch Laurasiatheria includes carnivores such as the gray wolf (*Canis lupus*). (Photo by Tom Brakefield/Bruce Coleman, Inc. Reproduced by permission.)

Because SINEs and LINEs apparently arise at random and occur widely throughout the genome, they are almost ideal derived characters. Given the vast array of DNA sequences in the nuclear genome, the probability of convergent evolution in the insertion of a SINE or LINE is exceedingly small. It is highly improbable that one of these sequences will be inserted at exactly the same site in the genome in two separate lineages. Secondly, because each insertion is a unique event that is unlikely to be reversed, SINEs and LINEs provide excellent markers for the recognition of groups of related organisms descended from ancestors possessing specific insertions. A very good example of the use of such evidence comes from discussion of the relationships between cetaceans (dolphins and whales) and artiodactyls (even-toed hoofed mammals). It has been accepted for some time that cetaceans are in some way related to artiodactyls. However, accumulating evidence from DNA sequences (both mtDNA and nDNA) indicated that cetaceans are, in fact, specifically related to hippopotamuses and thus nested within the artiodactyl group. This interpretation conflicts with the standard interpretation of the morphological evidence, according to which cetaceans constitute the sister-group to all artiodactyls. Certain fossil forms (mesonychians) that were regarded as relatives of whales and dolphins lacked a characteristic double-pulley adaptation of the ankle joint that is found in all artiodactyls (and was most probably present in their common ancestor). It had therefore been concluded that cetaceans branched away before the emergence of ancestral artiodactyls. This conflict of evidence was convincingly resolved by the discovery that cetaceans and hippopotamuses share a number of SINEs that are not found in any other mammals. Subsequently, early whale fossils possessing the typical artiodactyl ankle joint were discovered. Hence, it is now well established that cetaceans and hippopotamuses are sister-groups and that mesonychians are not direct relatives of the cetaceans after all.

Gene duplication

Studies of the evolution of DNA and protein sequences have generally concentrated on changes arising through point mutations of individual nucleotide bases. Indeed, molecular evolution has often been portrayed essentially as the progressive accumulation of point mutations in genes. However, evolution of DNA can also take place in other ways. One of the most important changes that can occur is tandem duplication of genes. This arises through slippage during the replication of DNA during cell division. Once a gene has been duplicated, the way is open for divergent evolution of the original and its copy. Indeed, over long periods of evolutionary time, gene duplication can occur repeatedly, such that quite large families of genes can result. A prime example is provided by globin genes, which are thought to have arisen from an original single gene through repeated duplication. The hemoglobin molecule, which plays a vital role in respiration, consists of four globin chains. In the blood of adult humans, the hemoglobin molecule contains two alpha-chains and two beta-chains. Sequence comparisons indicate that the beta-chain arose from the alpha-chain through an ancient duplication. There are also special hemoglobins that are temporarily present during the embryonic and fetal stages. Embryonic hemoglobin contains two epsilon-chains, while fetal hemoglobin contains two gamma chains. Both the epsilon-chains and the gamma chains also arose from the beta-chain through relatively recent duplications that took place during the evolutionary radiation of the placental mammals. This illustrates how gene duplication can provide an alternative route for the evolution of new functional properties of genes.

During the long history of evolution of living organisms with a cell nucleus containing chromosomes (eukaryotes), there have also been cases where the entire set of chromosomes has been multiplied (ploidy), for example through doubling of their number. Once sex chromosomes became established, as is the case with all living mammals (XX for females and XY for males), such doubling of the entire set of chromosomes became virtually impossible. Doubling of a male set of chromosomes (to XXYY) would result in the presence of two X chromosomes, thus disrupting the normal process of sex determination in which males have only a single X chromosome. However, at an earlier stage of evolution,

A mounted specimen of the now-extinct thylacine, or Tasmanian wolf (*Thylacinus cynocephalus*). (Photo by Tom McHugh/Photo Researchers, Inc. Reproduced by permission.)

The molecular clock

In addition to permitting reconstruction of relationships among species to generate a phylogenetic tree, molecular genetics can also yield valuable information with respect to the timescale for that tree. With the very first reconstructions conducted using molecular data, it was observed that the rate of change in amino acid sequences of particular proteins (and hence in the DNA sequences of the genes responsible) seemed to be relatively constant along different lineages. This led on to the notion of the "molecular clock", according to which the degree of difference between DNA sequences or amino acid sequences in any two species can provide an indication of the time elapsed since their separation. However, it should be noted that accumulating evidence has indicated that the rate of molecular change is in fact quite variable. In the first place, it was obvious from the outset that some genes evolve faster than others, and it was then shown that the overall rate of change in mtDNA is considerably greater than that in nDNA. Moreover, it also became clear that rates of change differ markedly even within individual genes. Some of this variation in rate of change within genes is to be expected. For

prior to the development of typical mammalian sex chromosomes, multiplication of chromosome sets would still have been possible. Although the evidence is controversial, there is a strong possibility that two successive duplications of the entire chromosome set took place during vertebrate evolution leading up to the emergence of the mammals. For example, two successive doublings of an original set of 12 chromosomes could have let to a set of 48 chromosomes, which is the modal condition found in placental mammals. Although duplications of an entire chromosome set can, of course, be subsequently masked by secondary modifications of individual chromosomes, quadrupling of the chromosomes prior to the emergence of the ancestral placental mammals should still be reflected in the presence of four copies of many individual genes. This does, indeed, seem to be the case for many sets of genes, such as the homeobox genes that play an important part in development.

Duplication of individual genes or entire chromosomes in fact poses an additional problem for reconstruction of phylogenetic trees using molecular data sets. When a tree is based on nucleotide sequences for any individual gene, care must be taken to ensure that it is really the same gene that is being compared between species. If there are multiple copies of a particular gene in the genome, there is always the danger that comparisons between species might involve different copies. A striking example of this danger is provided by mitochondrial genes. Although gene duplication has never been recorded within the mitochondrial genome, individual mitochondrial genes have been repeatedly copied into the nuclear genome, where they generally remain functionless. Inadvertent inclusion of such redundant nuclear copies in comparisons of mitochondrial genes between species has led to serious errors in interpretation. For instance, a supposed mitochondrial gene sequence reported for a dinosaur turned out to be an aberrant nuclear copy of that sequence in human DNA.

The Euarchontoglires grouping of placental mammals is represented by mammals such as the woodchuck (*Marmota monax*). (Photo by © John Conrad/Corbis. Reproduced by permission.)

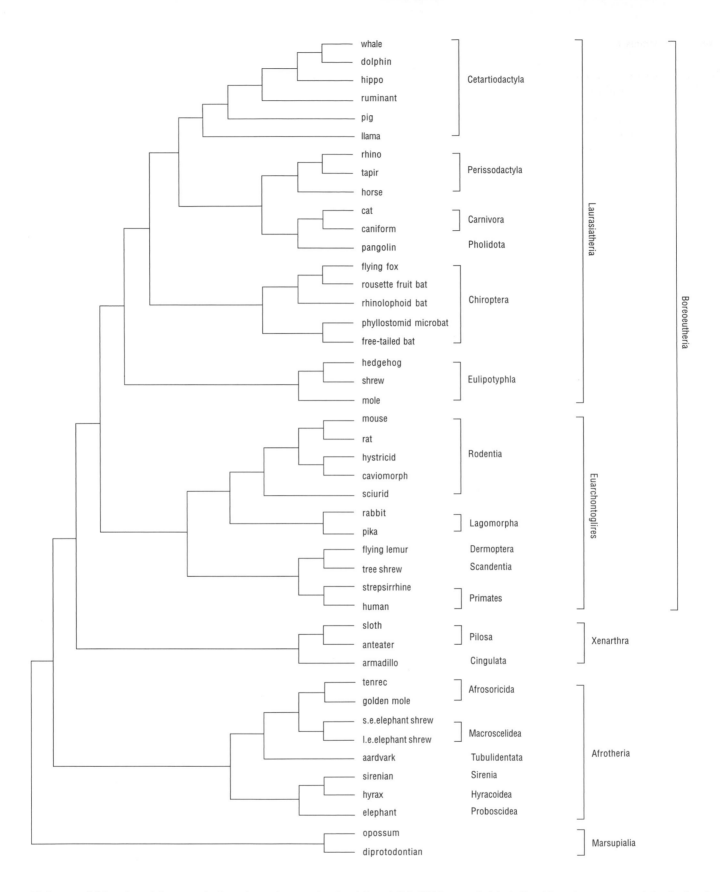

Phylogeny of living placental mammals, based on a large molecular data set (16,397 base pairs) including 19 nuclear genes and 3 mitochondrial genes. Based on Murphy et al. 2001. (Illustration by GGS. Courtesy of Gale.)

Resources

Ursing, B. M., and U. Arnason. "Analyses of mitochondrial genomes strongly support a hippopotamus-whale clade." *Proceedings of the Royal Society, London B* 265 (1998): 2251–2255.

Waddell, P. J., Y. Cao, M. Hasegawa, and D. P. Mindell. "Assessing the Cretaceous superordinal divergence times within birds and placental mammals by using whole mitochondrial protein sequences and an extended statistical framework." *Systemic Biology* 48 (1999): 119–137.

Robert D. Martin, PhD

Structure and function

Introduction

Mammalian evolutionary history goes back 230 million years. The earliest mammals occupied a nocturnal niche and developed a suite of traits that allowed them to adapt to the cooler temperatures of the night. Mammals are endothermic (able to produce their own heat) and most are homeothermic (able to maintain their body temperature within a particular range). Many mammalian structures and their functions are involved with maintaining body temperature, which requires efficient generation and conservation of heat.

Modern mammals evolved from early mammal-like reptiles called therapsids, which had a mosaic of reptile and mammalian traits. In defining what a mammal is, we must utilize characteristics that are preserved in the fossil record, which are largely skeletal. When a transitional fossil species has a mammal trait it is usually accompanied by a reptile trait. Consequently, there is controversy as to when the actual mammal-reptile division occurred.

The key identifier in mammalian fossils is found in the dentary-squamosal articulation because the mammalian lower jaw (the mandible) is unique. It consists of two bones, the dentaries, which articulate directly with the cranium. This is the key criterion that defines mammals. And there is a relationship between the repitian jaw and the mammalian ear. The reptilian jaw consists of two bones, the articular and the quadrate. In mammals, these were modified to become the malleus and the incus bones of the ear, which, along with the stapes (called the columella in reptiles), form the auditory ossicles in the mammalian skull. Thus, mammals have three bones in their middle ears (malleus, incus, and stapes), and reptiles have only one (the columella).

Living mammals also have many soft anatomy traits that further define them as mammals. For example, at some stage of their lives, all mammals have hair. (However, it has been suggested that the reptilian flying pterosaurs had fur, so we must be cautious about saying that hair is exclusive to mammals). On the other hand, mammary glands *are* unique to mammals among the living vertebrates. Another structure unique to mammals is the respiratory diaphragm that separates the thoracic and abdominal cavities. Only mammals possess this structure as a muscle, whereas other vertebrates have either a membranous diaphragm or no diaphragm at all. The mature red blood cells of mammals are enucleated (without a nucleus), whereas the red

blood cells of other vertebrates contain a nucleus. The mammalian heart differs from other vertebrates in that only the left aortic arch is developed in adult mammals. In mammal brains, the neopallium (neocortex) is elaborated and expanded compared to reptile brains. Each of these unique mammal structures are discussed in context in the rest of this entry.

Integumentary form and function

The integumentary system is composed of the skin and its accessory organs. The mammalian integument has many of the characteristics that we consider mammalian. Generally mammalian skin is thicker than the skin of other vertebrates because of its function in retarding heat and water loss. The integument consists of two major regions, the epidermis and dermis. Squamous cells are produced by a basal (or germinative) layer on the border of the epidermis and dermis. As cells are produced at the basal layer they push the cells above them toward the surface of the epidermis. As they move toward the surface the squamous cells fill with the protein keratin and produce the corneum, a tough waterproof layer of dead cells on the outermost layer of the epidermis. Epidermal cells are continuously shed and replaced as they serve as mechanical protection against environmental insults.

The dermis is mainly a supportive layer for the epidermis and binds it to underlying tissues. Blood vessels in the dermis pass near the basal layer of the epidermis and provide the cells of the avascular epidermis with nutrients. The dermis also contains muscle fibers, associated with hair follicles, and nervous tissue that provides assessment of the environment. A subcutaneous layer lies below the dermis and is a site of adipose (fat) deposition, which serves as both insulation and energy storage.

Mammals have a number of skin glands that are found in no other vertebrate. Mammals have two types of coiled, tubular sweat glands, apocrine (or sudoriferous) and eccrine. Apocrine sweat glands are usually associated with a hair follicle, and secrete the odorous component of sweat. Eccrine sweat glands secrete sweat onto the surface of the skin to remove heat through evaporative cooling. Most mammals have both these glands in the foot pads. They are more widely distributed on a few mammals, including humans. Those species with a limited distribution often use a supplementary method for cooling such as panting by dogs or immersion into cool mud or water by members of the pig family. Some small mammals such as insectivores

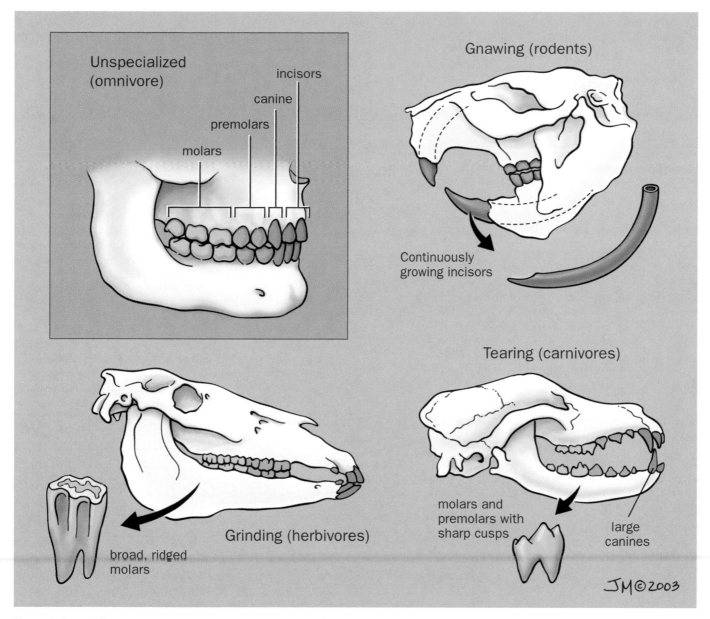

Mammals have different tooth shapes for different functions. Herbivores typically have large, flattened teeth for chewing plants. Rodents' ever-growing incisors are used for gnawing. Carnivores have teeth for holding and efficiently dismembering their prey. (Illustration by Jacqueline Mahannah)

and small rodents, bats, and aquatic mammals, do not experience heat loading and therefore do not have sweat glands.

Most mammals also have sebaceous glands distributed widely throughout the integument. Sebaceous glands consist of specialized groups of epithelial cells that produce an oily substance, sebum, that keeps hair and skin pliable, waterproof, and soft. These glands are usually associated with hair follicles. In some marine mammals such as otters and sea lions sebum is especially important in waterproofing the pelage and keeping cold water from contacting the skin, thereby preventing heat loss.

Scent glands are odoriferous glands used for social interactions, territory marking, and defense. One type of secretion is a pheromone, which elicits a behavioral or physiological effect on a conspecific (member of the same species). During

the breeding season it may advertise the sexual receptivity of the individual. Pheromones in the urine of some rodent species is even believed to induce estrus. Scent glands are used to delineate territory (i.e., marking). The distribution of scent glands is highly variable; they may be located on the wrists (carpal glands), throat region, muzzle, the chest (sternal glands), on the head, or the back, but most commonly scent glands are found in the urogenital area (anal glands). Among the mustelids (weasel family including skunks and minks), modified anal glands are able to squirt a smelly irritant several feet (or meters) when the animal is threatened.

Ceruminous glands are the wax-producing glands located in the skin of the ear canal. They help to prevent the tympanic membrane from drying out and losing its flexibility. Ceruminous glands are modified apocrine sweat glands.

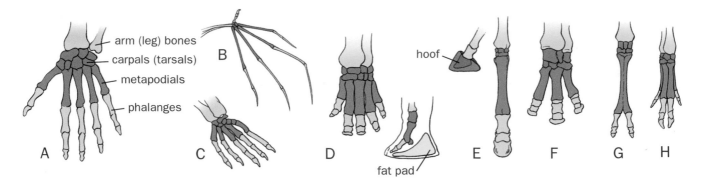

Mammal foot diversity: A. Hominid; B. Bat; C. Pinniped; D. Elephant; E. Equid; F. Odd-toed ungulate; G. Two-toed ungulate; H. Four-toed ungulate. (Illustration by Patricia Ferrer)

Mammary glands (mammae) are also generally believed to be modified apocrine sweat glands, although it has been suggested that they could have been derived from sebaceous glands. Mammae secrete milk that is used to feed the newborn mammal. The mammary glands in most mammals consist of a system of ducts that culminate in a nipple or teat. The one exception is found in the monotremes (egg-laying mammals). In monotremes, the mammary glands secrete milk onto hair associated with the glands and the hatchlings suck milk from these hairs. The number and location of mammae is variable among mammal species and is related to the normal size of the litter. The fewest number of mammae is two, but up to 27 are found in some marsupials. Mammae are usually on the ventral surface of placental mammals; in the marsupials they may be located in the pouch.

Hair is often described as a unique mammalian characteristic that has no structural homologue in any other vertebrate. Its distribution varies from heavy, thick pelages (fur coats) on many mammals to just a few sensory bristles (e.g., on the snout of whales or seacows). Mammalian hair originates in the epidermis, although it grows out of a tubular follicle that protrudes into the dermis. Growth occurs by rapid replication of cells in the follicle. As the shaft pushes toward the surface the cells fill with keratin and die. Each hair is composed of an outer scaly layer called the cuticle, a middle layer of dense cells called the cortex, and (in most hair shafts) an inner layer of cuboidal cells called the medulla. Each hair is associated with a sebaceous gland and a muscle (called the arrector pili) that raises the hair. Raising hair serves as a threat signal in social interactions but also increases insulation properties. The evolution of hair is part of the suite of adaptations that enabled mammals to be active at night. It served to retard heat loss by insulating the body. There are two layers of hair that form the pelage. The dense and soft underfur functions primarily as insulation by trapping a layer of air. The coarse and longer guard hair serves to shelter the underfur, keeping it dry in aquatic mammals, and to provide coloration.

Although the primary purpose of hair is insulation it has assumed other roles in living mammals. Color in hair comes from the pigments melanin and phaedomelanin. The main function of coloration appears to be camouflage, which helps the animal blend in with its surroundings. Mammals tend to have pelage colors that match their environment. One example of this is countershading. The pelage tends to be darker on the top and sides of the animal and lighter below and underneath, which under normal lighting conditions functions to obscure the form of the animal. In addition, there are various patterns on the pelage. Patterns on predators such as a tiger's stripes (Panthera tigris) help to conceal the predator. Stripes found on prey tend to confuse predators. Eye spots located above the eyes (e.g., Masoala fork-marked lemurs [Phaner furcifer] or four-eyed possums) may divert attention from the eye, confusing predators. Such patterns are called disruptive coloration. Another functional pattern is the white rump patch of the tail in mule deer (Odocioleus hemionus), which may serve as a silent alarm signal to conspecifics. But, when the tail is lowered, a predator whose eyes are fixed on the white patch might lose sight of the deer. Coloration may also identify conspecifics in visually oriented species. Blue monkeys (Cercopithecus mitis) and red-tailed monkeys (Cercopithecus ascanius) are sympatric (live in the same geographic location) and closely related, but they normally do not mate. It is believed that their distingushing facial patterns are the reason. Color patterns may also differ within a species. Often sexual dimorphism is expressed in color differences between males and females. Infants and juveniles may have different pelage colors or patterns from adults. In monkeys—one of the most visually oriented species—pelage patterns on the face, rump, or tail are used to communicate with one another. Another function of pelage patterns is that of warning potential enemies, e.g., the white stripe found on skunks might be a signal that it is well armed and can defend itself. Color changes can occur over a mammal's lifetime because hair, like skin, is replaced over time. Most mammals have two annual molts, usually correlated with the seasons. Yet others, such as humans, have hairs that grow to a particular length and then are shed. Some populations of snowshoe hares undergo three seaonal molts: a brownish-gray summer coat, a gray autumn coat, and a white winter coat.

Hair has undergone other modifications in addition to color. The guard hairs of some species have been modified for specific functions. For example, spines (or quills) are enlarged stiff hairs that are used for defense. In North American porcupines these quills have barbs that work their way into the flesh of an attacker. Vibrissae (or whiskers) are another modification of hair. These are supplied with nerves to provide tactile (touch) sensory information. These hairs are commonly found on the muzzles of many mammals such as

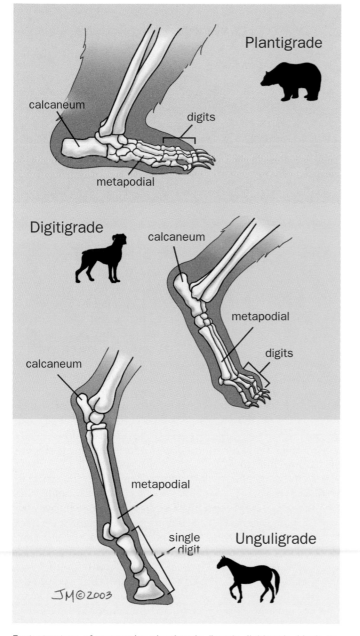

Foot structure of mammals: plantigrade (bear); digitigrade (dog); unguligrade (horse). (Illustration by Jacqueline Mahannah)

very similar to the claws found in other vertebrates. A claw consists of a dorsal plate called the unguis and a ventral plate called the subunguis. The unguis is curved both in length and width and encloses the subunguis, which is connected to the digital pad at the distal end of the digit. In addition to protection, claws assist predator species, such as lions and tigers, in holding their prey. They provide traction for some arboreal species (e.g., squirrels) when scampering on branches. Sloths have long curved claws that serve as hooks for hanging. Digging mammals, such as anteaters and moles, have long claws that help them dig.

Nails are modified claws found on the first digit of some arboreal mammals and on all the digits of some primates. Nails cover only the dorsal part of digits. The unguis (called a nail plate in human anatomy) is broad and flat, and the subunguis is vestigial. It has been suggested that nails evolved in primates to prevent rolling and provide flat support for the large pad of tactile sensory tissue found on the underside of the digit. Thus nails allow both increased tactile perception and enhanced manipulative abilities. The Callitrichidae (small monkeys found in South and Central America) have secondarily evolved claws, which are not true claws because they are derived from the laterally compressed nails of their ancestors. Nails and claws may be found on the same mammal (e.g., hyraxes).

cats and mice, but they can be found in other body locations as well. For example, tactile hairs may be located on the wrists, above the eyes, or on the back of the neck. These hairs allow mammals to sense objects around them when low-light conditions do not allow them to see well.

Claws, nails, and hooves

The distal ends of mammal digits possess keratinized sheaths or plates that are epidermal derivatives forming claws, nails, or hooves. Only the members of the whale and sirenia (seacows) families lack these structures. Claws are usually sharp, curved, and pointed. In many cases mammal claws are

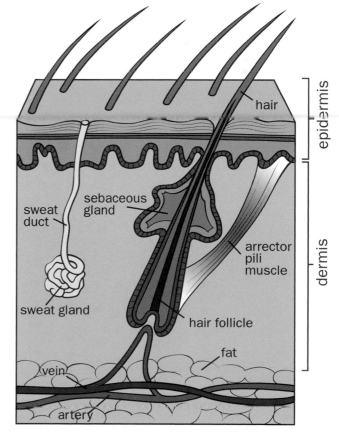

Cross section of mammal skin. (Illustration by Patricia Ferrer)

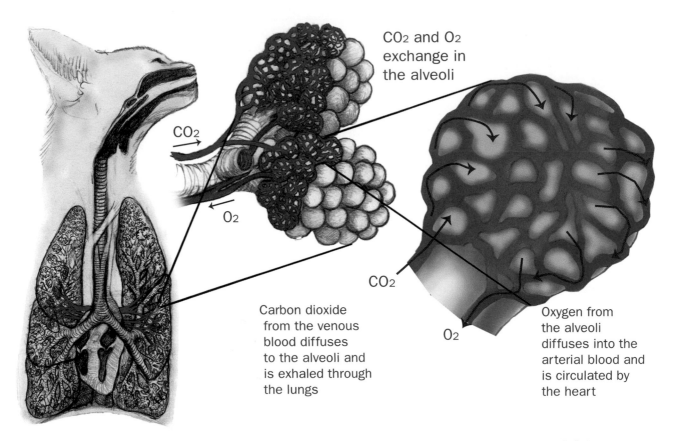

CO₂ and O₂ exchange in the alveoli

CO₂

O₂

Carbon dioxide from the venous blood diffuses to the alveoli and is exhaled through the lungs

CO₂

O₂

Oxygen from the alveoli diffuses into the arterial blood and is circulated by the heart

Gas exchange between the air and bloodstream takes place in the capillaries of the lungs' alveoli. (Illustration by Wendy Baker)

Hooves are constructed of a prominent unguis that curves around the digit and encloses the subunguis. The well-developed unguis has lent its name to the group that have hooves: the ungulates (although this is not a true taxonomic group). For ungulates, the unguis is much harder than the subunguis and does not wear away as quickly, thus developing a sharp edge.

Horns and antlers

Horns and antlers are found in the order Artiodactyla (cattle, sheep, deer, giraffes, and their relatives). Several other types of mammals have similar head structures, but true horns, originating from the frontal bone of the skull and found only among the Bovidae (cattle, antelopes, buffalo), consist of a bony core enclosed by a tough keratinized epidermal covering or sheath. True horns are not branched, although they may be curved. Horns grow throughout the life of the animal and are used for defense, display, and intraspecies combat (e.g., contests between males for mates). A variation of the true horn is the pronghorn, found on the North American pronghorn (*Antilocapra americana*) in the artiodactyl family Antilocapridae. A pronghorn branches and its epidermal sheath is shed on an annual basis; the sheath on a true horn is not shed. Antlers are found among the Cervidae (deer, caribou, moose, and their relatives). Mature antlers are entirely made of bone and are branched. They develop from buds covered by in-

tegument that is richly innervated and vascularized, called velvet. As the antlers grow the velvet dies and the animal usually rubs it off on tree trunks. Antlers are used for combat between males for mates. After the breeding season they are shed and replaced by a larger pair the next year; this continues until they reach their full growth. The small bony horns of giraffes (*Giraffa camelopardalis*) originate from the anterior portion of the parietal bones. Because they do not arise from the frontal bones they are not considered true horns. Giraffe horns are covered by furred skin and persist throughout life. Another type of horn is found on the rhinoceroses of the order Perissodactyla, the only living mammals outside the artiodactyls with a horn. The rhinoceros horn is centered over elongated nasal bones, but it lacks a bony core. It is a solid mass composed of dermal cells interspersed with tough epidermal cells.

Body design and skeletal system

As endotherms, mammals require more energy than ectothermic animals. Consequently, many mammal traits evolved to conserve energy. This is particularly true of the mammal skeleton. Mammals differ as a group from other living quadrupedal vertebrates in that their limbs are positioned directly below the body, allowing more energy-efficient locomotion. The lateral placement of the limbs on reptiles and amphibians requires them to spend considerable energy keeping their bodies lifted off of the ground while they un-

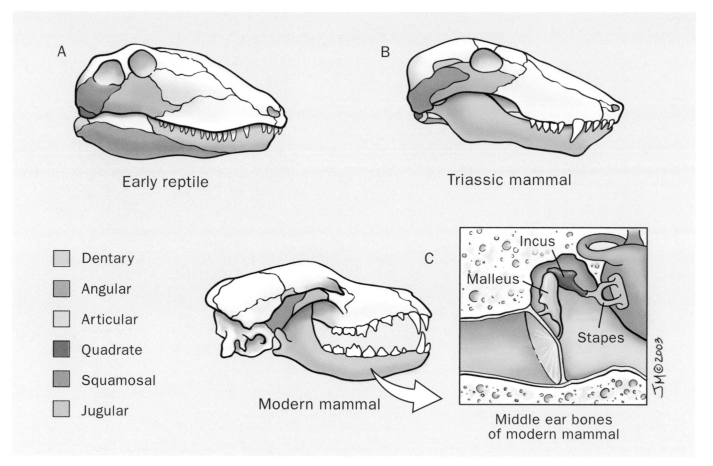

A. Early reptile
B. Triassic mammal
C. Modern mammal

Dentary
Angular
Articular
Quadrate
Squamosal
Jugular

Incus
Malleus
Stapes

Middle ear bones
of modern mammal

Jaw and ear structure in mammals. Mammals have only one bone in the lower jaw, not several as reptiles do. However, the bones of the former reptilian jaw are now part of the mammalian inner ear structure. (Illustration by Jacqueline Mahannah)

dulate laterally (rather than moving straight forward as mammals do). The vertical limb placement in mammals also allows removal of size constraints so that mammals may become much larger than amphibians or reptiles. (The large Mesozoic reptiles actually had limbs placed under the body also.) Another difference in the mammal skeleton is that it ceases growth in the adult, saving metabolic energy. Mammal long bones grow from bands of cartilage positioned between the diaphysis (shaft) and epiphyses (the ends). This allows permanent articulations between bones and forms well-established joints. The mammal skeleton has been simplified in that many bones have fused, decreasing the number of growth surfaces overall, and saving metabolic energy that would be used for maintenance. An example is the mammal skull, which is more ossified and simpler than those of other vertebrates. Bones that are fused in mammal skulls are separated by cartilage in reptiles. Ossification provides more surface area on bones with larger sites available for muscle attachments. Exceptions to many of these mammalian characteristics are found among the living monotremes. Some mammalogists actually believe that monotremes should be classified as therapsid reptiles because they have laterally placed forelimbs and their skeletons contain separated bones that are fused in other mammals, and, like reptiles, they retain cervical ribs. Ribs in most mammals are attached to the vertebral column only in the thoracic region; reptiles have

ribs attached to cervical, thoracic, and lumbar vertebrae. This rib arrangement allows mammals to lie on their sides for resting or to suckle their young.

The bones that make up the limbs in mammals are part of the appendicular skeleton. The forelimbs of mammals are attached to the pectoral girdle, consisting of a scapula and a clavicle (collar bone). The scapula is held in place by musculature, which provides the high mobility important in many locomotor modifications. The pectoral girdle articulates with the axial skeleton (skull, vertebral column, thoracic cage) only through sternal bones, which allows a large range of motion in the shoulder. The pelvic girdle supports the hind limbs and consists of a coxal bone (the fusion of three different elements from the reptile condition) that attaches to the axial skeleton at the sacrum.

Locomotor adaptations

The musculoskeletal design of the mammalian body has accomodated many diverse means of locomotion, not only in terrestrial environments but also in aquatic and aerial niches. Many mammal species are capable of using several different means of locomotion, but much of the body configuration is determined by the dominant mode of locomotion used by a particular species.

Mammal tail diversity reflects different functions. 1. A jerboa's tail is used as a counterweight and balance; 2. A spider monkey's tail is prehensile and used in locomotion; 3. A narwhal's tail propels it through the water; 4. A mule deer's tail can communicate an alarm; 5. A red kangaroo uses its tail as a support while upright; 6. A northern flying squirrel uses its tail as a rudder. (Illustration by Gillian Harris)

Most mammals that are ambulatory (walking) do so on all four limbs, i.e., they are quadrupedal. Most ambulators are pentadactyl (possessing five digits) and plantigrade (walking on the soles, or plantar surface, of their feet). Pentadactyly is the primitive condition in mammals, although many lineages have reduced this number. Ambulators include bears and baboons. Some ambulators are large. As they approach a ton (0.9 tonne) in weight, adaptations for their large size are a necessity. Such animals are said to be graviportal. They have a rigid backbone and their limbs take on the appearance of a column with each element directly above the one below it. They retain all five digits in a pad that provides cushioning. Elephants are an example of a graviportal mammal. Elephants are not able to run, instead they trot, increasing their speed by walking quickly.

Cursorial locomotion (running) is accomplished in diverse ways. Among mammals there is a range of adaptations and abilities for this way of moving. Many cursors are digitigrade, i.e., their metacarpals and metatarsals are permanently raised above the substrate with only the phalanges in contact with the ground. Often the metacarpals and metatarsals are elongated and the number of digits reduced. For example, in

equids (horses), the leg is supported on a single central digit of their mesaxonic foot. Other mammals, such as deer and hyenas, have legs with a paraxonic foot with two toes contacting the ground. Some mammals have one set of limbs that are paraxonic and another set that are mesaxonic.

A number of characteristics allow the generation of high speed in cursory locomotion. Reduction or loss of the clavicle that would impede forelimb movement is one adaptation. In addition, most of the musculature has shifted to the upper limb, and the lower limb has become thinner and elongated. In many cases it is the metacarpals and metatarsals that are the most elongated. In hoofed mammals the number of digits is reduced. The horse is the most extreme example with only a single digit. The elongation of the leg relative to body length produces a longer stride, thereby increasing the speed, which is equal to the length of the stride multiplied by stride rate. Another trait that increases speed is a pliant vertebral column that enables the mammal to place the hind feet in front of the fore feet when running at full clip. At very high speed, all four feet may be simultaneously off the ground. This is seen in horses, greyhounds, gazelles, and cheetahs. Cheetahs can cover a fixed distance faster than any other mammal. They are able to sprint at over 60 mph (97 kph).

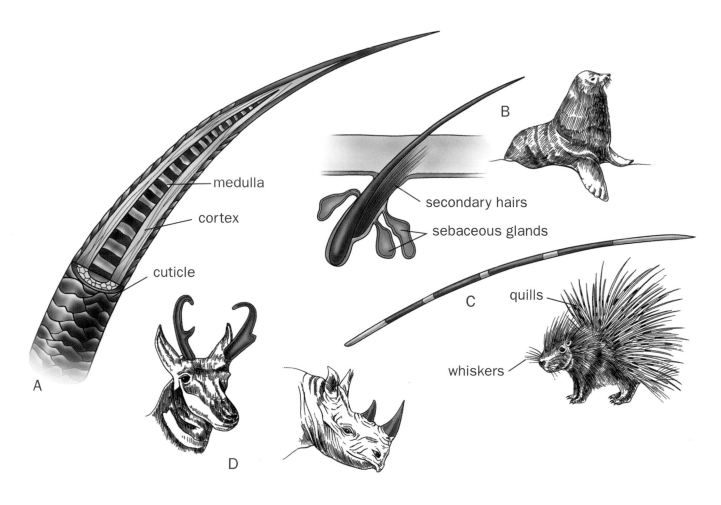

A. Cross section of a hair. B. Hairs may provide insulation and waterproofing. Specialized hair includes quills, whiskers (C), and horns (D). (Illustration by Patricia Ferrer)

Saltatory mammals use hopping as a means of locomotion. Some are quadrupedal, such as hares, using all four limbs to make their leaps. Others use another form of hopping called ricochet saltation. This is a bipedal locomotion in which only the two hind legs are involved in propulsion. Ricochet saltators have a long tail (often tufted at the end) that is used for counterbalance. In the case of the macropods (kangaroos) it is also used as a support during rest. Other ricochet saltators include kangaroo rats, jerboas, springhares, and tarsiers. In most cases ricochet saltators have reduced forelegs, long hind legs, and elongated feet. Generally it is the metatarsals have that been the most elongated. One exception is the tarsier, a small primate that retains longer forelimbs. It is the tarsal bones in the hind foot that are elongated, rather than the metatarsals. This leaves the tarsier with grasping ability in its hind paws. It is a ricochet saltator on a horizontal substrate, but in the shrub layer it employs vertical clinging and leaping, i.e., it pushes off vertical supports with its powerful hind legs and grasps a branch with its forelimbs and the upper part of its hind legs.

Arboreal mammals have many adaptations to life in the trees. One is stereoscopic vision (depth perception) in leapers and gliders. Grasping paws and opposable thumbs are often found, but they are not required. Squirrels scamper about on branches using sharp claws that provide traction. Sloths have elongated curved claws from which they hang under branches. Gibbons (small apes) have long fingers that serve as hooks, a reduced thumb, and a more dorsal scapula that allow them to swing underneath branches in the manner that children do on "monkey bars" at playgrounds. Prehensile tails are found in many arboreal mammals ranging from marsupials to primates. These tails are capable of wrapping around branches and serving as a "fifth limb." An arboreal adaptation among gliders (e.g., flying squirrels and colugos) is a ventral membrane, called the patagium, that can be spread out to generate lift for gliding. Except in the colugo, the tail is free of the patagium and is used for maneuvering. Gliders cannot ascend in flight, so they must climb before they launch. Despite that limitation, gliding can be very efficient. Colugos are able to cover over 400 ft (122 m) with a loss of only 40 ft (12 m) in altitude.

Among mammals, powered flight has evolved only in the bats, which are nocturnal fliers. In the order Chiroptera (literally "handwing") the mammalian forelimb has been modi-

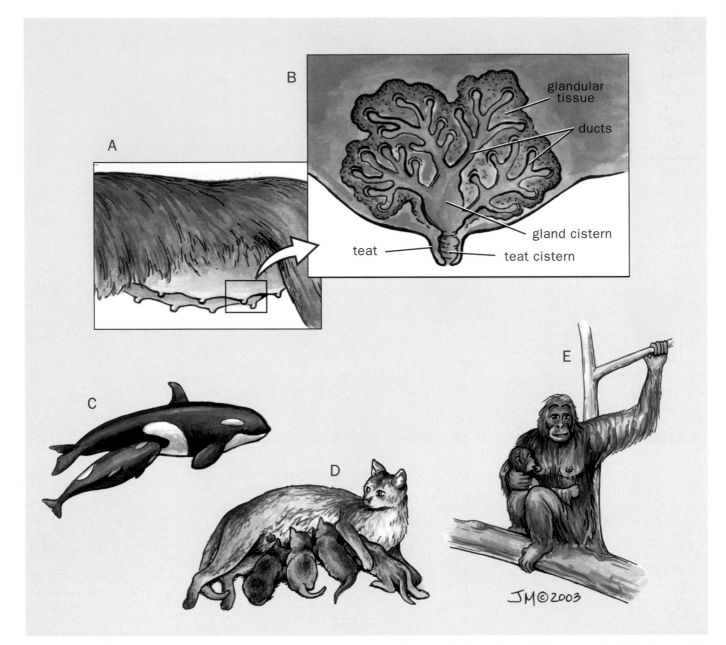

A. Mammary glands are commonly on the ventral surface of the abdomen; B. Cross section through a typical mammary gland. Numbers and location of teats may be due to litter size and life style of the mammal, as shown by C. Killer whales; D. Domestic cat; E. Orangutan. (Illustration by Jacqueline Mahannah)

fied to form a wing with the main portion composed of the elongated bones of the hand. Bat hind limbs have been modified to help control the rear portion of the flight membrane. In addition, the feet have curved claws that enable these animals to hang in an upside down position. Flight requires a whole suite of characteristics including circulatory and thermoregulatory adaptations, and echolocation for flight in a dark environment. Most of the Old World fruit bats do not have echolocation and do not fly in complete darkness.

Some mammals have successfully adapted to aquatic environments. Semiaquatic mammals spend some time on land (e.g., yapoks [or water opossums], beavers, otters, and duck-

billed platypuses). Their locomotor adaptations include a body approaching a fusiform (torpedo) shape, webbed feet (at least on the hind leg), and valvular openings to the nose and ear that can be closed to keep out water. Seals, sea lions, and walruses are more aquatic and have evolved flippers that provide efficient propulsion in water, although they still retain some locomotive ability on land. Seals propel themselves mainly by undulation of the flexible vertebral column assisted by the hind limbs. Completely aquatic mammals include the Cetacea (whales, dolphins, and porpoises) and the seacows (dugongs and manatees). The body is more fusiform than other marine mammals. The skeletal elements of the hind limbs have been lost, having been replaced by soft tissue forming a fluke that

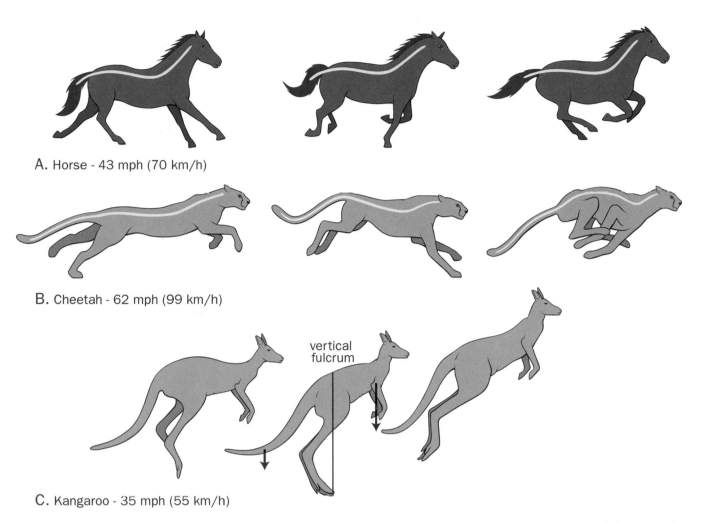

A. Horse - 43 mph (70 km/h)

B. Cheetah - 62 mph (99 km/h)

vertical fulcrum

C. Kangaroo - 35 mph (55 km/h)

A. The horse has a stiffer spine, making it better suited for endurance. B. The cheetah is the fastest land mammal; flexibility in its spine allows for longer strides. C. The kangaroo has more upward movement with each leap, and does not move forward as quickly as the horse or cheetah. (Illustration by Patricia Ferrer)

propels the animal in concert with undulation of the posterior vertebral column. The forelimbs are flippers used for maneuvering. The cervical vertebrae of completely aquatic mammals are short, and they have completely lost their pinnae.

Cardiovascular and respiratory systems

In order to distribute nutrients and oxygen needed for metabolism, mammals need a highly efficient circulatory system. The main differences in circulatory structure between mammals and most other vertebrates are in the heart and in the red blood cells. The mammalian heart has four chambers (as do birds and crocodilian reptiles) compared to the three chambers found in the reptiles (except the crocodilians). The additional chamber is the result of a muscular wall (or septum) that divides the ventricle (lower half of the heart) into two chambers. In reptiles there is a single ventricle in which deoxygenated blood from the right atrium mixes with oxygenated blood from the left atrium. In mammals deoxygenated blood enters the right ventricle only and is then pumped to the lungs. Oxygenated blood returns from the

lungs to the left atrium and then enters the left ventricle from which it is sent to the systemic circuit and the rest of the body. Thus, mammals have separated the pulmonary and systemic circuits with the result that mammalian blood is more fully oxygenated than the blood of terrestrial vertebrates with three-chambered hearts. Additionally, mature mammalian red blood cells (erythrocytes) are enucleated, i.e., they lack a nucleus, and are concave in shape. Space saved by the lack of a nucleus leaves room for additional hemoglobin molecules, the oxygen-binding molecule. The concave shape also increases surface area and places the membrane surface closer to the hemoglobin molecules facilitating gas diffusion. Thus, mammalian blood is capable of carrying more oxygen than reptilian blood.

The mammal heart is large, as are the lungs, and together these organs occupy most of the thoracic cavity. In certain taxa, these organs may be even bigger. Bats, for example, have a heart that is three times larger than the average terrestrial mammal of the same size. The mammal lung is sponge-like and consists of branched airways that terminate in microscopic sacs called

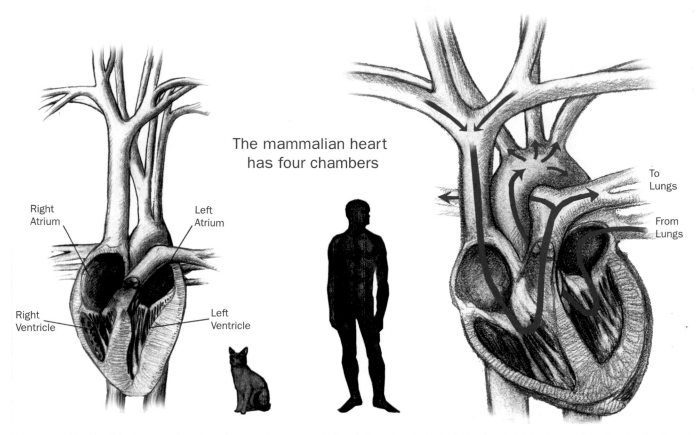

The mammalian heart has four chambers

Right Atrium

Left Atrium

Right Ventricle

Left Ventricle

To Lungs

From Lungs

The mammalian heart is four-chambered, and separates oxygenated and deoxygenated blood. Oxygenated blood flows from the lungs, through the heart, to the body. Deoxygenated blood flows from the body, through the heart, to the lungs, where it is reoxygenated. The smaller heart on the left belongs to a cat. (Illustration by Wendy Baker)

alveoli. The alveoli walls consist of epithelial tissue through which gas exchange occurs. This structural arrangement of smaller and smaller tubes and saccules increases the surface area available for respiration. For example, the respiratory membrane of human lungs is between 750 and 860 ft^2 (70–80 m^2), which is about 40 times the surface area of the skin.

Mammals have a layer of muscle that separates the thoracic cavity from the abdominal cavity. This is called the respiratory (or muscular) diaphragm. When this muscle is relaxed its domed shape forms the floor of the thoracic cavity. When it contracts it moves towards the abdominal cavity, increasing the volume of the thoracic cavity, which draws air in from the external environment.

Digestive system

To fuel endothermy, mammals require more calories per ounce (or gram) of tissue than do ectothermic vertebrates such as reptiles. This is accomplished by more efficient digestion of food stuffs and more efficient absorption of nutrients. This efficiency begins with specialization of the teeth. Mammals have four different kinds of teeth (heterodonty) that are ideally shaped to cut, slice, grind, and crush food. An exception is the toothed whales in which all the teeth are similar (homodonty). The four types of teeth are incisors for slicing, canines for pierc-

ing, premolars for crushing or slicing, and molars for crushing. They are commonly represented by a notation called a dental formula, e.g., I2/2 C1/1 P3/3 M2/3, the dental formula for the Egyptian fruit bat (*Rousettus aegyptiacus*). The first in each group of two numbers represents the teeth in the upper jaw and the second is the number of teeth in the lower jaw. Multiplying the dental formula by 2 gives the total number of teeth, 34. The Egyptian fruit bat's dental formula indicates that the full set of teeth for the upper jaw is four upper incisors, two upper canines, six upper premolars, and four molars. There is a great deal of variation in the number and type of teeth present. For example, the prosimian primate, the aye-aye (*Daubentonia madagascariensis*), has a dental formula of I1/1 C0/0 P1/0 M3/3, which illustrates a reduction in number of some teeth and the complete loss of others. In the case of some herbivorous mammals the upper incisors are either reduced in number or completely replaced by a hard dental or gummy pad that functions as a cutting board for the lower incisors. In some gnawing mammals, such as rodents and rabbits, the upper and lower incisors grow throughout the entire life span and the canines have been lost. Modification of teeth may be extreme, such as complete loss in most anteaters, or the formation of large tusks, derived from the second upper incisors in elephants, or from the canines in walruses (*Odobenus rosmarus*).

The relationship between dental structure and function is so precise that the diets of long-extinct mammals can be de-

A longtail weasel (*Mustela frenata*) in winter coat. (Photo by Bob and Clara Calhoun. Bruce Coleman, Inc. Reproduced by permission.)

A longtail weasel (*Mustela frenata*) in its summer coat. (Photo by Tom Brakefield. Bruce Coleman, Inc. Reproduced by permission.)

duced from their teeth. The teeth are often the only fossil remains recovered from paleontological sites. Teeth perform mechanical (or physical) digestion by breaking down a food morsel into smaller pieces, providing additional surface area for action by digestive enzymes. Premolars in herbivorous mammals usually have ridges for grinding. In some carnivores such as wolves, the last upper premolar has a blade that shears against the first lower molar. Fruit-eating mammals such as flying foxes often have flattened premolars and molars.

Other modifications for efficient digestion occur in the stomach, a portion of the gastrointestinal tract. The stomach serves as a storage receptacle in most mammals and as a site of protein breakdown. A simple stomach is found in most mammal species, including some that consume fibrous plant material. In other mammals that consume a high fiber diet the stomach has become enlarged and modified to handle more difficult digestion. These modifications comprise a foregut digestive strategy, for which the stomach contains compartments where symbiotic microbes break down cellulose and produce volatile fatty acids (VFA) that can be utilized by the mammal. Foregut fermentation has been developed to the greatest degree among the mammal order Artiodactyla, which includes pigs, peccaries, camels, llamas, giraffes, deer, cattle, goats, and sheep. Rumination, reprocessing of partially digested food, is accomplished by the four-compartment stomachs of giraffes, deer, cattle, and sheep. Less complex tubular and sacculated stomachs are found in kangaroos, colobus monkeys, and sloths. Stomachs in foregut fermenting species are neutral or only slightly acidic, around pH 6.7, to provide a favorable environment for symbionts.

Food moves from the stomach to the intestines; which consist of the small intestine, where most digestion and absorption occurs, and the large intestine. The wall of the small

intestine contains epithelial tissue with small finger-like projections called villi. In turn, each villus has smaller extensions called microvilli. The villi secrete enzymes for further carbohydrate and protein digestion. The microvilli absorb the digested nutrients. The presence of the villi and microvilli in the mammal small intestine increases the absorptive surface area by at least 600 times that of a straight smooth tube. The

Black rhinoceros (*Diceros bicornis*) profile showing its horns used for protection and fighting for supremacy and social heirarchy. (Photo by Leonard Lee Rue III. Bruce Coleman, Inc. Reproduced by permission.)

An elephant can reach a long way up for a meal using its trunk. (Photo by K & K Ammann. Bruce Coleman, Inc. Reproduced by permission.)

villi of the human small intestine, for example, provide 3,230 ft² (300 m²) of surface area whereas the surface area of a smooth tube of the same size as the small intestine is about 5.4 ft² (0.5 m²). Nutrient absorption occurs through the membranes of the microvilli of each intestinal epithelial cell. Also distributed throughout the small intestine are glands that secrete special enzymes for further digestion of proteins, carbohydrates, and lipids.

For mammals, diet and the length of the small intestine are closely correlated. Mammals that consume a diet that is either digested in the stomach (such as animal protein consumed by faunivores) or easily absorbed (such as nectar consumed by nectarivores) have a shorter small intestine than

A free-ranging yak (*Bos grunniens*) exhibits the rare golden color phase. (Photo by Harald Schütz. Reproduced by permission.)

An aardvark (*Orycteropus afer*) unearths the subterrranean nests of ants and termites through active digging using powerful and well-adapted claws. (Photo by Rudi van Aarde. Reproduced by permission.)

other mammals. Herbivores that eat very fibrous plant matter tend to have the longest small intestine. The small intestines of fruit-eaters tend to be intermediate in length.

The foregut fermentation strategy of herbivores requires a medium or large body size to accommodate the necessarily large stomach. A strategy generally used by smaller herbivores is hindgut fermentation (although there are large hindgut fermenters such as horses, elephants, and howler monkeys). The hindgut, also called the large intestine, consists of the cecum and the colon. The cecum is a blind pouch that serves as the principal fermentation chamber in the hindgut strategy. As in the stomach of foregut herbivores, colonies of symbionts in the cecum of hindgut fermenters break down cellulose and excrete products advantageous to the mammal. Nutrients appear to be absorbed through the wall of both the cecum and colon, especially in the larger mammals.

Small hindgut fermenters, such as many rodents and rabbits, have the problem that food can only be retained in the gut for a short time. As it leaves the hindgut, digestion is incomplete and many valuable nutrients may be left unabsorbed. This problem is solved by a behavioral adaptation: a soft pellet is produced in the cecum, defecated, and immediately

heat production. Mammals also regulate their body temperature within a stable range, generally between 87 and 103°F (30–39° C). This is called homeothermy. Having a constant temperature allows mammals to maintain warm muscles, which gives them the ability to react quickly, either to secure food or to escape predation. They can also maintain the optimum operating temperature for many enzymes, providing a more effective physiology. Some mammals are heterothermic (able to alter their body temperature voluntarily). Many insectivorous bats are heterothermic. When in torpor they lower their body temperature to the ambient temperature, conserving calories that would otherwise be used for heat production.

To regulate body temperature, mammals must also have the means to retain a certain amount of the heat they produce. Small mammals lose heat more rapidly than larger mammals because they have a greater proportion of surface area to volume (or, equivalently, to their body mass). Heat is lost through surface area. The higher the surface area–to-mass ratio, the greater the rate of heat loss. Fur helps to insulate a small mammal to some degree, but often it is not enough to prevent the high rate of heat loss. Small mammals often compensate by obtaining more calories per unit of time by continuously eating foods that are quickly digested and absorbed. Larger mammals' surface area–to-mass ratio decreases as their mass increases, and they lose heat at a lower rate.

Reproductive system

There are three different modes of reproduction used by mammals. The monotremes, whose extant members are the echidnas and duck-billed platypuses, lay eggs. The therians (marsupial and placental mammals) give birth to live young. Marsupial newborns are undeveloped (some mammalogists call them embryos). After only a short gestation period they must make their way to a teat outside the mother's body (a teat that may be in a pouch in species that have pouches) to finish development. The embryos of placental mammals remain in the uterus during development, and they have a nutritive connection with the mother through the placenta. The young of placental mammals are born more mature than the young of the other two groups.

The female reproductive tract in monotremes is very much like a reptile's. A cloaca (also found in amphibians, reptiles, and birds) is a common chamber for the digestive, urinary, and reproductive system. The eggs are conveyed from the ovaries through the oviducts where fertilization occurs. After fertilization the eggs are covered with albumen and a leathery shell produced by the shell gland. In therian females the reproductive organs are separate from the urinary and digestive systems. The marsupial female has two uteri, each with its own vagina. Eutherian females may have either a single uterus or paired uteri, but always a single vagina. The placental embryo implants and develops in the uterine wall.

In all therians, the male urinary and reproductive systems share a common tract, the urethra. A problem for endothermic mammals is that their body temperature may be too high to sustain viable sperm. This is not a problem for monotreme males because their body temperature is lower than that of therians, and their testes are contained in the abdominal cavity. The testes of therian males are typically contained in a scrotum, a sac-like structure that lies outside the body cavity. The testes may descend into the scrotum from the abdominal cavity only during breeding season or they may be permanently descended. The penis differs in the three main groups of mammals. The monotreme penis is attached to the ventral wall of the cloaca. The marsupial penis is directed posteriorly, contained in a sheath, and the glan penis (tip) is bifid, which accommodates the two vaginas in the marsupial females. The eutherian penis is directed forward. It may hang freely or be contained in an external sheath. In many species, including most primates, a bone called the baculum supports the penis.

Mammary glands (see also the discussion under integument) provide nourishment for the young mammal. While milk requires energy to produce, it also conserves energy for the mother: Mammals do not have to make numerous trips to find food and return with it to feed their offspring. Observations of bird parents making trip after trip in order to feed insatiable hungry mouths at the nest illustrate this point. A mammal mother obtains her food, returns to the nest or den, and can feed her young in comparative safety.

Resources

Books

Feldhamer, G. A., ed. *Mammalogy: Adaptation, Diversity, and Ecology.* San Francisco: McGraw-Hill, 2003.

Hildebrand, M. *Analysis of Vertebrate Structure.* 4th ed. New York: John Wiley & Sons, 1994.

MacDonald, D., ed. *The Encyclopedia of Mammals.* New York: Facts on File, 2001.

Martin, R. E., R. H. Pine, and A. F. DeBlase. *A Manual of Mammalogy.* 3rd ed. San Francisco: McGraw-Hill, 2001.

Neuweiler, G. *The Biology of Bats.* New York: Oxford University Press, 2000.

Novak, R. M. *Walker's Mammals of the World.* 6th ed. Baltimore: John Hopkins University Press, 1999.

Pough, F. H., C. M. Janis, and J. B. Heiser. *Vertebrate Life.* 6th ed. Upper Saddle River, NJ: Prentice Hall, 2001.

Romer, A. S. and T. S. Parson. *The Vertebrate Body.* 6th ed. San Francisco: Saunders College Publishing, 1985.

Welty, J. C., L. Baptista, and C. Welty. *The Life of Birds.* 5th ed. New York: Saunders College Publishing, 1997.

Vauhan, T. A., James M. Ryan, and Nicholas Czaplewski. *Mammalogy.* 5th ed. Philadelphia: Saunders College Publishing, 1999.

Periodicals

Young Owl, M., and G. O. Batzli. "The Integrated Processing Response of Voles to Fibre Content of Natural Diets." *Functional Ecology* 11 (1998): 4–13.

Marcus Young Owl, PhD

• • • • •

Adaptations for flight

Adaptation for flight in bats

Bats, and an uneasy creeping in one's scalp As the bats swoop overhead! Flying madly. Pipistrello! Black piper on an infinitesimal pipe. Little lumps that fly in air and have voices indefinite, wildly vindictive; Wings like bits of umbrella. Bats!

The poet, D. H. Lawrence, seemed to find bats disgusting, but these creatures of the night are the only mammals to have evolved powered flight. Occupying the nocturnal flier niche has been extremely successful—so successful that one out of every four mammal species is a bat.

Three vertebrate taxa have evolved lineages capable of powered flight: the pterosaurs (Reptilia), birds (Aves), and bats (Mammalia). In all three cases, the forelimbs of these vertebrates were modified over time to form wings. This is an example of convergence, the independent evolution of a common structure that performs a similar function among unrelated species. The pterosaurs, the only reptiles to evolve true flight, were the first vertebrates to develop powered flight. Pterosaurs (order Pterosauria) appeared about 225 million years ago and lasted about 130 million years until they became extinct at the end of the Mesozoic era. The most diverse lineage of flying vertebrates is the birds (class Aves), which underwent extremely rapid evolution during the Cretaceous period, approximately 150 million years ago (mya). Bats (order Chiroptera) appear to be the most recent flying lineage among vertebrates, although precisely how recent is uncertain because only a few examples are represented in the fossil record. The oldest unquestioned fossil bat dates back to the early Eocene (about 50 mya) and is already a well-developed bat. Fossils from the early Paleocene (65–60 mya) attributable to bats consist mainly of teeth and jaws. They are often disputed as belonging to the order Primates.

Advantages to flight

Flight in a vertebrate provides several advantages. First, the flying animal has access to food sources unavailable to terrestrial species. This includes insects flying above the ground level that cannot be reached by earthbound animals as well as fruits and flowers on the terminal ends of thin branches. Second, the flier has a ready means of escape from non-flying (or non-volant) predators and can rest in places that are not accessible to earthbound predators. Third, flight gives a species great mobility and the ability to cover large expanses rapidly and cheaply. Although the amount of energy required to initiate flight is great, once the animal is airborne, flying is the most economical form of locomotion per distance traveled in a terrestrial environment. In addition to daily foraging advantages, flight provides the means to compensate for seasonal changes in climate and food availability. A fourth advantage is at the evolutionary level. Fliers can overcome geographic barriers such as large bodies of water and, consequently, can disperse to locations not easily traversed by non-volant terrestrial animals. For example, bats are the only mammals native to New Zealand, to many remote Pacific Islands, and to the Azores in the Atlantic. Before humans arrived on Australia with dogs, bats and a few rodents (apparently arriving from New Guinea) were the only eutherian mammals among all the terrestrial fauna on the continent.

Nocturnal flight adaptations

The focus of this entry will be the adaptation for flight among bats, the only mammals to evolve structures for powered flight. Bats are not just fliers, they are mammalian, nocturnal fliers. Consequently, their adaptation to flight involves more than just the evolution of wings, but also requires solutions to nocturnal navigation, thermoregulatory problems, and energy considerations.

Over a span of 65 million years of evolutionary history, natural selection acted to balance several physical considerations to accommodate demands of flight: body mass and shape, wing morphology, flying style (i.e., control of wing shape, orientation, and motion), and physiology (to meet the energy requirements for flight). To understand flight adaptation, it is useful to gain an understanding of the forces exerted on the animal in powered flight. Adaptation for flight of bats is guided by the need to generate and withstand, or minimize, these forces during flight. However, before looking at flight it is imperative to look at a prerequisite for nocturnal flight: some way to navigate in darkened space. Before flight could evolve in bats, a bat ancestor must have devel-

52

Grzimek's Animal Life Encyclopedia

The bulldog bat (*Noctilio albiventris*) uses its claws to swoop in and grab prey while in flight. (Photo by T. E. Lee Jr./Mammal Images Library of the American Society of Mammalogists.)

oped echolocation. There is more to being a flying bat than just having wings.

Echolocation

Before there could be nighttime fliers, there had to a way to navigate in the dark. Bats are active at night and they often inhabit darkened areas such as caves or the inside of hollow trees.

Echolocation is an adaptation for navigating in visually limited environments. There are other mammals that employ echolocation, including various marine mammals and possibly members of the order Insectivora such as shrews. There is also some suggestive evidence that the colugo, a nocturnal glider, has some form of echolocation. Marine mammals such as whales and dolphins move through a medium that transmits light very poorly. The water they swim in is often obscured by murkiness from plankton and other suspended particles. At depths greater than 656 ft (200 m), a routine diving depth for marine mammals, the surroundings become completely dark. The shrew is a terrestrial mammal with tiny eyes and presumably poor vision. They are active at night. Shrews are fossorial, i.e., they burrow, dig, and forage in the leaf litter of wooded areas. Consequently, they also occupy a

visually limited environment. It should also be pointed out that only two bird species are known to echolocate. Oilbirds are nocturnal and inhabit caves. The other, the Asiatic cave swiftlets, frequently fly in dark caves.

Birds have been part of the terrestrial fauna for at least 150 million years. They fill the available diurnal (daytime) flying niches. By the time bats appeared in the Eocene, birds were completely developed and no latecomer mammal would have been able to out-compete them in the daytime. Bats most likely descended from small nocturnal insectivorous mammals. Therefore, protobats were already in the nocturnal niche when there was an opening for a nocturnal flier. However, flying in the dark can be dangerous. In addition to the open nighttime environment, most bats roost in caves or inside hollow trees, which are darker environments than outside. There is also the danger of mid-air collisions with other bats. Consequently, before nocturnal flying could be feasible, some way of avoiding obstacles had to evolve. Of course, there is no way to know when echolocation actually evolved in bats. However, it had to be very early in their development. As mentioned previously, shrews have a crude form of echolocation, and shrews and other insectivores are often cited as a mammalian rootstock. If so, it is not unreasonable to suggest that echolocation developed sometime before flight.

The flying fox (*Pteropus* sp.) can have a wingspan of up to 6.6 ft (2 m). (Photo by © W. Perry Conway/Corbis. Reproduced by permission.)

It was the Italian physiologist Lazzaro Spallanzani who first experimented with obstacle avoidance in bats and owls in the eighteenth century. He discovered that owls would not fly in complete darkness, but this did not deter bats. He hung wires from his ceiling with small bells attached. Bats could fly throughout his study and never jingle the bells. When he blinded the bats, again they did not touch the wires. He finally inserted brass tubes into their ear canals. This was observed to impair the bats ability to avoid the wires. Spallanzani was still baffled. No sound came from the wires while they were simply hanging. Nevertheless, he attributed the bats' ability to avoid the wires to keen hearing. Of course, he was not able to hear the high-pitched sounds that the bats were actually emitting.

In the 1930s, the first microphone capable of detecting ultrasound (beyond the hearing of humans) was produced. American zoology and comparative psychology student, Donald Griffin, prominent in the 1980s for his work on animal cognition, found that placing one of these microphones in the middle of a group of quiet bats suddenly changed these relatively quiet animals into loud chatterboxes. At about the same time, the Dutch zoologist, Sven Dijkgraf, who had very keen hearing, discovered that he could hear sounds coming from Geoffroy's bat. When he placed muzzles over their jaws, preventing the emission of the sounds, these bats became dis-

oriented and crashed into objects. From these discoveries, early researchers were able to gain some understanding of the mechanisms of echolocation. However, to date, the details of detection and interpretation of these signals by the bat are still a very active area of research (e.g., the bat project at the Auditory Neuroethology Lab at the University of Maryland, College Park).

Echolocation in bats results from the production of a high-pitched sound by the larynx and emitted through either the mouth or the nostrils. Often, the nose has been modified into a nose leaf, a fleshy process on the upper snout, which helps direct these sounds. Sound waves travel until hitting an object and bouncing back. The pinnae (external ears) of bats are large, highly modified structures designed to receive the returning signal of the bounced sound. The tragus is a small flap located in front of the ear canal. It acts as an antenna and allows the bat to discern the direction from which the sound is coming. Different species of bats utilize different frequencies. Individuals of the same species will alter their frequencies slightly to prevent confusion of signals that could lead to mid-air collisions.

Echolocation is also used for foraging. In fact, echolocation may have originally developed in a bat ancestor that was foraging in the forest litter. Bats can catch insects "on the fly,"

Bats have adapted to flight by developing extremely light and slender bones. (Photo by Barbara Strnadova/Photo Researchers, Inc. Reproduced by permission.)

ing their tongue. This is different from microchiropteran echolocation. Although echolocation is not necessary for flight in itself, it is a required adaptation for fliers who travel in pitch-black darkness.

The physics of powered flight

Once a means for detecting and avoiding obstacles was developed in a bat ancestor, the lineage was free to expand into the nocturnal flier niche. Powered flight allows access to flying insects. Because gliders do not have the maneuverability to pursue flying insects, this feeding niche was wide open during early bat evolution. The difference between powered flight and other modes of traveling through the air is maneuverability. Gliders such as the colugo have extra skin at the body's sides, which can both stretch out and change angle during flight to control both the rate and the angle of descent. Therefore, gliding has both a downward and a horizontal component of motion. However, the starting point is always higher than the final position of the animal. This is because gravitational potential (the energy determined by a body's position in a gravity field) is the only source of kinetic energy (energy of motion) in this mode of traveling through the air. To obtain a greater height above the starting position, gliders must utilize other means (e.g., tree climbing). Power flyers can oppose the force of gravity and increase their height above the ground by using wings and the power generated by their own muscles. They are also capable of controlling the magnitude and direction of their forward speed without depending on gravity or air currents.

part of what makes them successful as nocturnal fliers. However, an evolutionary arms race exists because some moths have developed a defense against bat echolocation. They possess sound sensors on their thorax that enable them to detect the ultrasonic pulses being aimed at them by the bats. They then engage in erratic flight patterns in an attempt to evade the foraging bats. Some moths have even developed countermeasures. They produce sounds directed at the bats that seem to deter them. It is possible that these sounds are jamming the bats' echolocation in the way that aluminum foil was used to jam radar signals during World War II.

One group of bats is notable for not having echolocation. These are the large flying foxes and fruit bats (Megachiroptera). These bats depend on vision during activity under low-light conditions at dusk and dawn, a cycle referred to as a crepuscular activity cycle. They are also active during all moon phases, except the new moon when there is no moonlight. Megachiroptera lack the large pinnae and elaborate nose leafs found on the echolocating insectivorous bats (Microchiroptera). There is one exception: rousette bats that roost in dark caves (which is unusual for a megachiropteran) use a form of echolocation in which they produce sounds by slowly click-

The southern flying squirrel (*Glaucomys volans*) uses a thin membrane that extends from its hands to its feet to glide up to 80 yd (73 m). (Photo by © Joe McDonald/Corbis. Reproduced by permission.)

These little red flying foxes (*Pteropus scapulatus*) hang upside down and as soon as they drop, they are in flight. (Photo by B. G. Thomson/Photo Researchers, Inc. Reproduced by permission.)

Powered flight is possible because air is a fluid. In everyday usage, the word "fluid" brings to mind a liquid such as water or gasoline. But technically, a fluid obeys the law that the faster an object moves through it, the greater the force exerted on the object. In the terminology of fluid mechanics, the force exerted on an object in a direction perpendicular to the direction in which the object moves through a fluid is called dynamic lift, which is generated when an object moving through a fluid changes the direction of the fluid flow.

Another fluid force exerted on an object is dependent on the shape of the object. This is called Bernoulli lift, which may be involved in natural selection pressure for the wing and body shape of the bat. The Bernoulli principle in fluid mechanics states that the faster a fluid flows over a surface, the lower the pressure on that surface perpendicular to the fluid flow. Therefore, the pressure is lower on the top than it is on the bottom. This pressure difference results in Bernoulli lift upward. Experimentally it has been determined that Bernoulli lift alone is not sufficient for power flying, but most likely provides a selection pressure favoring a particular wing shape, body streamlining, and flight style.

In summary, the forces that must be overcome in powered flight are inertia (the resistance to change in motion that is a property of all masses), weight (the force exerted on the mass by gravity), and drag (the fluid force exerted by air on any object moving through it). To change the height from the ground and the speed and direction of forward motion, the bat has to use its wings to manipulate the airflow to generate the forces of lift and thrust. The wing structures themselves must also be able to withstand the stresses of moving through the air.

Bat wing morphology and its role in powered flight

Chief among the many adaptations of the bat for powered flight is the bat wing, and the flapping flight style that uses muscle power to generate lift and thrust. The bat wing evolved from the forelimbs of a terrestrial mammalian ancestor. The mammalian forelimb is exceedingly mobile because the shoulder joint between the scapula (shoulder bone) and the humerus (upper forelimb bone) is loosely held together with muscles. This allows for actual rotation of the arm around the shoulder joint in many species. Primates have this mobility, and so do bats.

The taxonomic name of the bats, order Chiroptera (meaning, "hand-wing"), perfectly describes the morphology of the bat wing. The skeletal structure of the bat wing consists of the humerus, a well-developed radius, and a much-reduced ulna. In humans, the ulna is a major bone of the forearm and forms a hinge joint in the elbow region with the humerus. The highly elongated hand (metacarpal) and finger (phalanges) bones form the rest of the bat wing skeleton. Only the pollex (or thumb) retains the claw of the mammal ancestor, although on fruit bats and flying foxes the second digit also retains a claw. The bones of the wing provide a segmented skeletal frame for support and control of the flight membrane.

The flight membrane (called a patagium) is a flexible double-layered structure consisting of skin, muscle, and connective tissue. It is richly supplied with blood vessels. The region of the patagium that stretches from the sides of the body and the hind limbs to the arm and the fifth digit is called the plagiopatagium. Other portions of membrane extend from the shoulder to the pollex (first digit) along the anterior portion of the wing (propatagium), between the fingers (the chiropatagium), and from the hind limbs to the tail (the uropatagium, also called the interfemoral membrane). The wing operates on an airfoil design, with the flexible membrane segments changing shape to produce variable pressure gradients along the wing surface that results in variable amounts of lift and thrust. The bats' fine control of the shape of the patagium gives them a maneuverability that cannot be matched by birds.

Bat flight is controlled by seventeen different pairs of muscles. Three different muscles provide power for the downstroke. Another three muscles execute the upstroke. This is very unlike birds, where two pairs of muscles provide the power for the depression and elevation of the wings. The sternum (breastbone) in bats is not particularly well developed, while in birds it is very prominent with a well-developed keel. The pectoralis muscle that originates from the sternum is the largest bat flight muscle and it has the richest supply of blood vessels known for any mammal. Other muscles that originate from along the vertebral column (backbone) and the scapula help to provide tension to the membrane and adjust the position of the wing. Muscles fibers embedded in the membrane assist in regulation of the tension of the patagium. Many muscles that exist in terrestrial mammals have been slightly repositioned, while others unique to bats assist in keeping the patagium taut. The wing operates on an airfoil design, with the flexible patagium segments changing shape to provide variable amounts of lift and thrust.

The hind limb possesses a bony spur unique to bats called a calcar that projects inwardly from the tibia. This bone attaches to the uropatagium and functions to keeps the tail portion from flapping during flight. The legs can also form a pouch out of the uropatagium used for catching insects. In most bats, the hind limbs have rotated 90° outwardly and assumed a reptilian-like position. The legs are used to control the uropatagium during flight. Another important adaptation of the hind legs is as a hook, an adaptation for hanging upside down. Bats are able to hook the claws of their hind paws onto horizontal supports or rough edges on walls or on ceilings of caves. The claws have developed a locking mechanism that allows them to hold without any muscular involvement. Hanging upside down allows bats to occupy areas unavailable to birds and allows a bat to use gravity to initiate flight by dropping. It is often believed that bats are completely helpless on the ground because of the arrangement of their legs; this is not true. Some species hop while others move quadrupedally. If a bat falls in water, it can swim to land. However, they do not use these forms of locomotion habitually. The arrangement of the bat hind limbs has probably constrained the bat lineage to being flyers. There are no flightless bats nor are there swimming bats comparable to those found among the birds (e.g., ostriches and penguins, respectively).

Bat flight

The superior aerobatic ability of the bat in flight is due to the wing segmentation provided by the skeletal frame, the flexibility of the membrane segments, and the very fine controls provided by the wing musculature. To date, the best analytical theory of animal flight is the vortex theory first introduced by Ellington in 1978 and further developed by Rayner in 1979. According to vortex theory, bats fly by generating volumes of circulating air (called vortices, singular vortex) that create pressure differences on different parts of the bat's wing. The resulting fluid forces push the animal in the direction it wants to go, at the speed it wants to go. The bats' flight motions are similar to the motions of a human swimmer doing the butterfly stroke. During the downstroke, the wing is fully extended. It envelops the maximum possible volume of air and pushes it down, generating a region of high pressure beneath the wing and low pressure above the wing. The pressure differences add up to a resultant force that has two components: a thrust component that opposes the drag exerted on the animal by its motion through the air, and a lift component perpendicular to the drag that opposes the action of gravity on the mass of the animal (the animal's weight). The numerical value of each component depends on the an-

The spectacled flying fox (*Pteropus conspicillatus*) shows the thin membrane where blood vessels, nerves, and tendons are held. (Photo by David Hosking/FLPA–Images of Nature. Reproduced by permission.)

gle of attack. The steeper the angle, the higher the lift and lower the thrust. During steep-angle ascent, the bat increases the curvature of the propatagium to prevent stalling. To maximize lift, the uropatagium is curved downward. During the upstroke, the bat flexes the wing and extends the legs to decrease drag by decreasing the surface area perpendicular to the airflow. Each wing segment contributes different relative amounts of lift and thrust. The wing segments closest to the sides of the body, the plagiopatagium, generate mostly lift, and the distal wing segments (the chiropatagium) provide most of the thrust. The exact pattern of airflow over different wing segments is not yet known, however, computer simulations of the aerodynamics of the bat in flight are an area of very vigorous research. For example, a project to simulate the airflow around the changing geometry of the bat wing in flight is under way at Brown University. Preliminary results were published by Watts in 2001.

Wing form and flying strategies

The wing forms of bats are highly variable from species to species. A particular form (e.g., either long and narrow or short and broad) may reveal a relationship between flight style and foraging habits because it is likely that selection pressure favors the evolution of the best wing form for a particular feeding style. The two primary quantities used for comparing wing morphology to flight style are wing loading (WL) and aspect ratio (AR). WL is the ratio of body weight to the surface area of the wing, which demonstrates the size of the wing relative to the size of the bat. In general, the higher the WL, the faster the bat has to fly to generate sufficient lift with relatively small wings. One calculates AR by squaring the wingspan and dividing that number by the wing's surface area. AR measures the broadness of the wing. The higher the AR, the narrower and more aerodynamically efficient (lower drag)

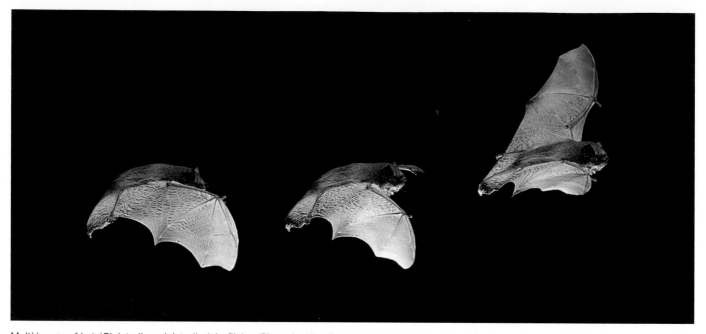

Multi-image of bat (*Pipistrellus pipistrellus*) in flight. (Photo by Kim Taylor. Bruce Coleman, Inc. Reproduced by permission.)

is the wing. Bats with high-AR wing morphology are faster flyers, but lack the agility of bats with low AR. The surface areas of the uropatagium and the plagiopatagium are large in slow, agile flyers because these areas provide most of the lift during flight. The propatagium alters the leading edge curvature of the wing, and prevents stalling during steep-angle flying. If the surface areas of these regions are large compared to the wingspan, giving a low AR, the agility of the bat is very high. Examining the wing form can provide clues about the bat's specialization in foraging. There are no exact correlations because bats are very adaptable and highly flexible in their foraging habits. Also, the wing form suitable for a certain foraging style may be a disadvantage in other aspects of bat behavior. Generalizations must be made with caution. With that in mind, observers have noted that some tendencies do emerge. In general, bats with wings having high WL and high AR are bats that fly fast and forage in open air above vegetation. These bats regularly fly long distances in a short amount of time, feeding on insects while in flight. Bats with wings having low WL and low AR are able to fly slowly without stalling and can make tight maneuvers. They are gleaners and hoverers, able to navigate in heavy vegetation and to take off from the ground while carrying heavy prey. Fruit-eating bats that forage among vegetation and carnivorous bats that catch prey from the ground both fit in this category. High-WL and low-AR wings tend to belong to bats that fly fast, but have short, broad wings and are capable of maneuvering in cluttered spaces. They tend to be expert hoverers, and their flight speed allows them to visit among separated patches of vegetation in a minimum amount of time. They also tend to specialize in nectar or pollen feeding. Low-WL and high-AR wings are found among fishing bats that fly slowly over open water with very little tight maneuvering required. The low body weight allow these fishers to carry off the day's catch for later consumption.

Body design

To understand bat body adaptations for flight, it may be instructive to examine bird bodies. Bird bodies are designed for mass reduction. They do this in a number of ways. They have lost teeth and the accompanying heavy jaws and jaw musculature over evolutionary time. They have thin, hollow, and strong bones. Many bones are fused or reduced in size. The long bony tail of their ancestors has been greatly reduced to the small vestigial pygostyle. Birds have a series of air sacs in the body that serve to reduce weight. They do not have a urinary bladder to store urine nor do they have a urethra. The kidneys excrete uric acid into the cloaca where it is mixed with intestinal contents to produce the white guano associated with birds. Birds have lost one ovary, and lay eggs so they do not have to carry a fetus. The most distinctive feature of birds is their feathers, which provide lift, insulate them against heat or cold, streamline the body, and reduce mass.

Bats, as mammals, must address these weight reduction issues differently. In general, bats are much smaller in size than birds. Most bats belong to the suborder Microchiroptera (the insectivorous bats or microbats, also called the "true bats") and range from 0.07 oz (2 g) (Kitti's hog-nosed bat, perhaps the smallest mammal) to 8.1 oz (230 g), but fewer than 50 species weigh more than 1.8 oz (50 g). The larger flying foxes (Megachiroptera) may reach 56.4 oz (1,600 g) with wingspans of 6.5 ft (2 m), but they are never as large as the largest birds. Bat bones are thinner and lighter than those of most mammals, but not as light as bird bones. Bat bones have marrow in the shafts, whereas bird bones are hollow. Several bones in the bat skeleton (ulna, caudal [or tail] vertebrae) have been reduced, while several have been lost altogether (fibula, caudal vertebrae in fruit bats). The distal phalanges have less mineralization and a flatter cross-section than normally found in mammal bones, which provides more flexibility in the wing frame. Birds, on the other hand, have more mineralized bones that are somewhat more brittle. If present in the bat wing, these could actually break under the stresses on the wing frame during flight. Bats have not lost any organs as birds have. Bats still retain teeth. To compensate for the extra skull mass, they have a short neck that helps to keep the center of gravity in the middle of the torso. The bat body as a whole has been shortened and some of the vertebrae have fused, making for a stiff backbone. The diets of bats are high-energy foods, such as insects, fruit, or nectar, that pass through the gut quickly so as not to load the animal down with bulky fiber. This high-energy diet also meets the energy requirements for flight. Bats, because they are mammals, have fur instead of feathers. Fur has some limited lifting properties, produces rough surfaces that change airflow, and has some malleability for streamlining, but it is inferior in those properties to feathers. Fur does insulate, but not as efficiently as feathers.

The most important difference between bats and birds is that birds are daytime flyers and bats are nighttime flyers. As nocturnal flyers, bats face problems not faced by birds. The first problem they had to solve is navigation in a visually limited environment. Other problems bats must solve are getting sufficient oxygen and nutrients to tissues and thermoregulation. Bats have dealt with these problems very successfully.

Energy

Powered flight has enormous energy costs. Flight is energetically cheaper than walking or running once the bat is up in the air. However, it takes a considerable amount of calories to get airborne. Flying is very demanding on bat physiology. In some species, the heart rate may rise to approximately 1,000 beats per minute in order to supply oxygen to the tissues during flight. Because of these demands, the heart and lungs are larger in bats than in comparably sized mammals.

Bats do not consume fibrous plant material. Such a diet simply would not supply enough calories. Also, the gut passage time and the gut modifications needed to digest high-fiber material would increase the weight of the animal. Bats consume easily digestible, high-calorie items such as insects, fruit, or nectar. Some species also eat small vertebrates like fish, frogs, mice, or even other smaller bats.

Thermoregulation

Associated with the metabolic costs of flight is thermoregulation. Bats have unusual problems to solve in this regard. Bats probably have the most complex thermoregulatory problems to solve of any mammal. Most bats are small. Small mammals must overcome heat loss problems because they have a greater proportion of surface area in relation to their volume (or equivalently, their body mass). Heat is lost through surface area. The higher the surface area–to–mass ratio, the greater is the rate of heat loss. For this reason, small mammals have higher rates of heat loss than larger mammals. Fur helps to insulate a small mammal to some degree, but often it is not enough to prevent the high rate of heat loss. Small

Gould's wattled bat (*Chalinolobus gouldii*) echolocating from a branch on Mt. Isa, Queensland, Australia. (Photo by B. G. Thomson/Photo Researchers, Inc. Reproduced by permission.)

mammals need to obtain more calories per unit time to produce heat (via muscles) by continuously eating foods that are quickly digested and absorbed. This is quite the opposite of larger mammals. As mammals get larger, their surface area–to–mass ratio decreases as the mass increases. Elephants have a heat load problem, not a heat conservation problem. Besides being a small mammal, bats also have additional surface area from their wing membranes. Therefore, this flight adaptation results in about six times greater surface area than that present in non-volant mammals of comparable size, which increases the heat loss rate many fold.

Bats solve these extreme heat loss problems through heterothermy, a temporary reduction in body temperature to conserve calories. Mammals are endothermic ("warm-blooded"), i.e., they generate internal heat. Most mammals are also homeothermic, which means that they regulate their body temperature within a particular range (generally, 95–102°F [35–39°C]). Bats, however, can reduce their body temperature to conserve energy. This strategy is called heterothermy and results in torpor. Bats can lower their body temperature to the environmental (or ambient) temperature and therefore do not have to devote calories to produce heat, much of which would be lost to the environment. Additionally, bats are able to reduce blood flow to the extremities and to the wing membrane that reduces heat loss through these surface areas.

During flight, the bat's thermoregulatory problems are reversed. The problem becomes how to dissipate the heat generated from the flight muscles. Bat wings have a rich supply of blood vessels. Heat is transferred from the blood to the wing membrane and is radiated off the surface. Bats do not have sweat glands, but a small amount of water vapor passes through the skin onto the membrane surface. As water evaporates off the surface of the wing, it also carries away some heat. Another area where the echolocating bats lose heat is from the blood vessels of the large external ear. Breathing also helps remove heat. Water vapor is one of the byproducts of respiration and, when the animal exhales, more heat is dissipated.

Some species can build up a heat load while they are resting during the day. This is more likely to occur among the larger bats, but some smaller bats that roost in sunny locations face this problem as well. In these situations, temperature can be regulated through behavior by moving to a shadier location. Some bats will also use their wings to fan themselves. Sometimes, they also lick themselves to promote evaporative cooling from the saliva.

Cardiovascular and respiratory adaptations

Like birds, bats have hearts that are about three times bigger than those of comparably sized mammals. The heart muscle fibers (or cells) in bats possess higher concentrations of ATP (the molecule that is utilized for energy by cells) than observed in any other mammal. These adaptations enables bats to pump more blood during a flight, a period of peak demand for oxygen. Resting bats may have heart rates as low as 20 beats per minute. Within minutes of initiating flight, the heart rate may rise to between 400 and 1,000 beats per minute. Bats also have relatively larger lungs than most mammals, providing a larger respiratory membrane for gas exchange. This is in response to the demands for oxygen required for muscle metabolism during flying.

Bats have highly vascularized wings (i.e., rich in blood vessels) that supply the wing membrane with oxygen and other nutrients. Because of this circulation, damage to the wing membrane can heal very quickly. An unusual feature of the bat wing circulation is sphincters (muscular valves) that can close off blood flow to the capillaries and shunt blood directly from the arteries to the veins. It is not exactly known when and why this is done. Some biologists believe that the sphincters are closed and blood flows through the shunts during flight. The sphincters may open during rest to allow blood to flow into the capillaries and nourish the wing membrane. A problem that exists for wing circulation is that the flapping of the wings creates a centrifugal force that impedes the flow of blood back to the heart, causing pooling in the extreme ends of the wings. To compensate for this, the veins of the wings have regions in between venous valves that contract rhythmically. These have been referred to as "venous hearts." When venous hearts contract, the vein is constricted and pushes venous blood back towards the heart. The valves in mammalian veins prevent back flow, ensuring that blood will only travel in one direction. Bat blood is capable of carrying more oxygen per fl oz (ml) than other mammals. In fact, it carries more oxygen than bird blood. It appears that this is accomplished by increasing the concentration of red blood cells (RBC), which contain the iron pigment heme that binds to oxygen. Bat blood has smaller individual RBCs than normally found in mammals and a larger number of RBCs within the same circulating blood volume. These smaller cells also provide a relatively larger surface area for gas exchange to occur. The actual mechanism of bat circulation is still not completely understood. For budding bat biologists, the bat circulatory system offers many research possibilities because little experimentation has been done on many aspects of this system.

Bat lungs are larger than the lungs of terrestrial mammals, but they do not contain the respiratory volume found in birds. The alveoli, the tiny sacs that help form the respiratory membrane, are smaller in the bat lungs than in the lungs of other mammals. The smaller the alveoli are, the greater the functional surface area for gas exchange. In addition, the alveoli are richly endowed with capillaries that bring a rich flow of blood for gas exchange. Bats are superior to other mammals at extracting oxygen from the environment, approaching the capability of birds. Bats do not have the lung volume of birds, but they have high respiratory rates that facilitate aeration. The high respiratory rates are also believed to be associated with heat removal via water vapor.

Bats own the night sky

Bats are extremely successful nocturnal mammal fliers. Their anatomy, physiology, and ecology are a finely tuned integration of many different body organs and organ systems that enable these animals to dominate the night sky. Bat adaptations for flight include more than just wings. The diet consists of high-calorie food that is easy to digest, assimilate, and pass quickly through the gut. They have solved thermoregulatory problems ranging from the heat loss due to small size to the heat load of flight metabolism. The cardiovascular and respiratory systems are highly adapted for efficient distribution. All of these adaptations work together efficiently to make the bat a well-integrated nighttime flying machine.

Resources

Books

Altringham, J. D. *Bats: Biology and Behaviour*. New York: Oxford University Press, 2001.

Anderson, D. F., and S. Eberhardt. *Understanding Flight*. New York: McGraw-Hill, 2001.

Fenton, M. *Bats*. New York: Checkmark Books, 2001.

Fenton, M. *Just Bats*. Toronto: University of Toronto Press, 1983.

Hall, L., and G. Richards. *Flying Foxes: Fruit and Blossom Bats of Australia*. Malabar, FL: Krieger Publishing Company, 2000.

Hildebrand, M. *Analysis of Vertebrate Structure*, 4th edition. New York: John Wiley & Sons, Inc., 1995.

Hill, J. E., and J. D. Smith. *Bats: A Natural History*, 1st edition. Austin: University of Texas Press, 1984.

Neuweiler, G. *The Biology of Bats*. New York: Oxford University Press, 2002.

Norberg, U. "Flying, Gliding and Soaring." In *Functional Vertebrate Morphology*, edited by M. Hildebrand, D. Bramble, K. Liem, and D. Wake. Cambridge, MA: Belknap Press, 1985.

Norberg, U. "Wing Form and Flight Mode in Bats." In *Recent Advances in the Study of Bats*, edited by M. B. Fenton, P. Racey, and J. Rayner. London: Cambridge University Press, 1987.

Vaughan, T., J. Ryan, and N. Czaplewski. *Mammalogy*, 5th edition. Philadelphia: Saunders College Publishing, 1999.

Welty, J., L. Baptista, and C. Welty. *The Life of Birds*, 5th edition. New York: Saunders College Publishing, 1997.

Periodicals

Carpenter, R. E. "Flight Physiology of Australian Flying Foxes, *Pteropus poliocephalus*." *Journal of Experimental Biology*, 114 (1984): 619–647.

Maina, J. N. "What It Takes to Fly: The Structural and Functional Respiratory Refinements in Birds and Bats." *Journal of Experimental Biology*, 203 (2000): 3045–3064.

Morris, S., A. Curtin, and M. Thompson. "Heterothermy, Torpor, Respiratory Gas Exchange, Water Balance and the Effect of Feeding in Gould's Long-eared Bat (*Nyctophilus gouldi*)." *Journal of Experimental Biology*, 197 (1994): 309–335.

Padian, K. "The Origins and Aerodynamics of Flight in Extinct Vertebrates." *Palaeontology* 28 (1985): 4132–433.

van Aardt, J., G. Bronner, and M. de Necker. "Oxygen Dissociation Curves of Whole Blood from the Egyptian Free-tailed Bat, *Tadarida aegyptiaca* E. Geoffroy, Using a Thin-layer Optical Cell." *African Zoology*, 37, no. 1 (April 2002): 109–113.

Watts, P., E. Mitchell, and S. Swartz. "A Computational Model for Estimating the Mechanics of Horizontal Flapping Flight in Bats: Model Description and Validation." *Journal of Experimental Biology*, 204 (2001): 2873–2898.

Other

Auditory Neuroethology Lab Webpage. University of Maryland, College Park. 2002. <http://www.bsos.umd.edu/psych/batlab>.

Swartz Lab Webpage. Brown University, Providence, RI. 2003. <http://www.brown.edu/Departments/EEB/EML/about/about.html>.

Vertebrate Flight Exhibit Webpage. University of California, Berkeley. January 11, 1996. <http://www.ucmp.berkeley.edu/vertebrates/flight/enter.html>.

Weinstein, R. *Simulation and Visualization of Airflow around Chiroptera Wings during Flight*. Brown University, Providence, RI. May 1, 2002. <http://www.cs.brown.edu/rlweinst/bat.pdf>.

Marcus Young Owl, PhD

Adaptations for aquatic life

Life in water

In the beginning, all life on Earth was aquatic. Although water covers over two-thirds of our planet, precisely how life in the oceans came to be is one of our unanswered questions. Many of these animals have been around for millions of years. Over time, they have adapted in such a way that allows them to live and reproduce in water. One unusual example of long-term ocean survival is that of the coelacanth. Fossils of this armored fish dating back more than 75 million years have been discovered, and it was thought to have been extinct. In 1938, however, one was caught off the coast of South Africa. Since then, more than 100 of these prehistoric, deep-dwelling fish have been examined. They have no scales or eyelids, as do "modern" fish, and have quietly kept to themselves in the deepest areas of the ocean.

The hippopotamus (*Hippopotamus amphibius*) has its nostrils on the top of its snout allowing it to spend most of its time in the water. (Photo by Alan Root/Okapia/Photo Researchers, Inc. Reproduced by permission.)

The Pacific white-sided dolphin's (*Lagenorhynchus obliquidens*) coloring helps to act as camouflage underwater. (Photo by Phillip Colla/OceanLight.com. Reproduced by permission.)

For the most part, aquatic creatures spend their entire lives submerged. However, a few aquatic animals—those that are descended from land animals—come all or part of the way out of the water for one reason or another: sea turtles, pinnipeds, and penguins come ashore to breed, for example. Mammals, such as whales and dolphins, have also acquired some handy adaptive techniques for life in the water, coming to the surface only to breathe.

The smallest of the marine mammals is the sea otter (*Enhydra lutris*), at 5 ft (1.5 m) long, including the tail, and up to 70 lb (32 kg). The largest is the blue whale (*Balaenoptera musculus*)—the largest animal alive—which can be 110 ft (33.5 m) long and weigh 300,000 lb (136,000 kg). To varying degrees, these mammals that have returned to the water have retained vestiges of their terrestrial forms, including hair, which only mammals have. Sea otters, seals, and sea lions are thickly furred; manatees and dugongs have a sparse pelage, but they have many whiskers around their mouths. Dolphins and whales are hairless, but in some species hairs are present at birth (they are soon lost). Sea otters have hand-like paws on their front legs, but their hind feet have become webbed, so that they're almost flippers. The four legs of pinnipeds have become flippers, and the sirenians have front flippers (some of them have fingernails), but no hind legs, and a flattened tail for propul-sion. Whales and dolphins have no hind legs, flippers instead of forelegs, and a horizontal tail (fluke) for propulsion.

Evolution of aquatic animals

Marine fossils paint an idyllic scene of aquatic animal life in its infancy some 670 million years ago (mya): soft coral fronds arch from the ocean floor, jellyfishes undulate in the currents, and marine worms plow through the ooze. But a geologically brief 100 million years later, at the dawn of the Cambrian period, the picture suddenly changes. Animals abruptly appear cloaked in scales and spines, tubes and shells. Seemingly out of nowhere, and in bewildering abundance and variety, the animal skeleton emerges.

For more than a century, paleontologists have tried to explain why life turned hard. Hypotheses abound, some linking the skeletal genesis to changing chemistries of the seas and skies. Yet a recent analysis of old fossil quarries in Canada and new ones in Greenland is providing evidence supporting the notion that the skeletal revolution was more than a chemical reaction—it was an arms race.

High in Canadian Rockies of British Columbia, in an extraordinary 540-million-year-old fossil deposit called the

The Galápagos sea lion (*Zalophus californianus wollebaecki*) has well developed flippers to provide locomotion both in the water and on land. (Photo by Tui De Roy/Bruce Coleman, Inc. Reproduced by permission.)

predators strongly influenced the elaborate new skeletal designs of the mid-Cambrian.

What sort of creature could gouge such wounds in a tough trilobite? One likely culprit is *Anomalocaris*, the largest of Cambrian predators. This half-meter-long creature glided through the seas with ray-like fins and chomped with a ring of spiked plates that dispatched trilobite shells like a nutcracker.

From the treacherous maw of *Anomalocaris* to the healed wounds of *Wiwaxia*, much of the support for the arms race argument hinges on the Burgess shale collection. But what about the small shelly fauna that emerged 30 million years earlier? For an arms race hypothesis to be complete, predators must have roamed then, too.

New finds strengthen the case for an early Cambrian arms race. From an extraordinary fossil bed discovered in 1984 in north Greenland, predating the Burgess shale by perhaps as much as 15 million years, comes a jigsaw puzzle already assembled: a suspiciously familiar, slug-like beast sheathed in chain-mail armor, proposed to be the long-sought ancestor of the armored slug *Wiwaxia*.

The creature sports a disproportionately large, saucer-like shell at each end of its elongated body. From another fossil discovery at a quarry in south China, which appears even older than the Greenland site, emerges the bizarre *Microdictyon*. Unveiled in 1989 by Chinese paleontologists, *Microdictyon* is a wormish creature with a row of pointed appendages and a body studded with oval phosphate plates. About 30 quarries

Burgess shale, a mid-Cambrian marine community comes to life. Like many less exceptional deposits, the Burgess harbors mollusks, trilobites (the ubiquitous, armored "cockroaches" of the Cambrian seas), and clam-like brachiopods. But other imprints in the smooth black shale dispel any image of a peaceful prehistoric aquarium. In these waters lurked a lethal cast of predators, eyeing little shells with bad intent: *Sidneyia*, a flattened, ram-headed arthropod with a penchant for munching on trilobites, brachiopods, and cone-shelled hyolithids; *Ottoia*, a chunky burrowing worm that preferred its hyolithids whole, reaching out and swallowing them with a muscular, toothed proboscis; and even some trilobites with predatory tastes. These findings have helped resurrect the arms race hypothesis: the 80-year-old idea that skeletons evolved primarily as fortresses against an incoming wave of predators.

Take *Wiwaxia*, a small, slug-like beast sheathed in a chain-mail-like armor. With two rows of spikes running along its back, *Wiwaxia* was the mid-Cambrian analogue to a marine porcupine. The chinks in its armor are telling. Some of *Wiwaxia's* spines appear to have broken and healed. The healed wounds of trilobite and *Wiwaxia* specimens suggest that

The sea otter's (*Enhydra lutris*) extremely dense fur enables it to trap air bubbles which help to keep the otter afloat while sleeping. (Photo by Richard R. Hansen/Photo Researchers, Inc. Reproduced by permission.)

than two seconds, half the time humans take, even though the whale breathes in 3,000 times more air. Exhaling and inhaling takes about 0.3 seconds in bottlenosed dolphins (*Tursiops truncatus*). When swimming quickly, many pinnipeds and dolphins jump clear out of the water to take a breath. Cetaceans have the advantage of having a blowhole on top of the head. This allows them to breathe even though most of the body is underwater. It also means that cetaceans can eat and swallow without drowning.

The long, deep dives of aquatic mammals require several crucial adaptations. For one thing, they must be able to go a long time without breathing. This involves more than just holding their breath, for they must keep their vital organs supplied with oxygen. To get as much oxygen as possible before dives, pinnipeds and cetaceans hold their breath for 15 to 30 seconds, then rapidly exhale and take a new breath. As much as 90% of the oxygen contained in the lungs is exchanged during each breath, in contrast to 20% in humans. Not only do diving mammals breathe more air faster than other mammals, they are also better at absorbing and storing the oxygen in the air. They have relatively more blood than nondiving mammals. Their blood also contains a higher concentration of red blood cells, and these cells carry more hemoglobin. Furthermore, their muscles are extra rich in myoglobin, which means the muscles themselves can store a lot of oxygen. To aid in diving, marine mammals also increase buoyancy through bone reduction and the presence of a layer of lipids (fats or oils).

Aquatic mammals have adaptations that reduce oxygen consumption in addition to increasing supply. When they dive, their heart rate slows dramatically. In the northern elephant seal, for example, the heart rate decreases from about 85 beats per minute to about 12. A bottlenose dolphin's average respiratory rate is about two to three breaths per minute.

The bottlenosed dolphin (*Tursiops truncatus*), with its streamlined shape and powerful tail, can swim at speeds of up to 33.5 mph (54 kph). (Photo by François Gohier/Photo Researchers, Inc. Reproduced by permission.)

The air-breathers

Some water birds, such as cormorants and pelicans, simply hold their breath until completely out of the water. However, it is not appropriate for all air breathers to leave the water to breathe, especially if only a small portion of them can do it. This also has two evolutionary advantages: it reduces the amount of time at the surface of the water so they can spend more time feeding, and it reduces the amount of wave drag they encounter. The external nares of aquatic mammals, such as beavers, hippopotamuses, and dolphins, are always dorsal in position, and the owner seems always to know when they are barely out of water. A ridge deflects water from the blowhole of many whales. When underwater, the nares are automatically tightly closed. Sphincter muscles usually accomplish this, but baleen whales use a large valve-like plug, and toothed whales add an intricate system of pneumatic sacs so that great pressure can be resisted in each direction.

To avoid inhaling water, aquatic mammals take very quick breaths. Fin whales can empty and refill their lungs in less

A manatee's (*Trichechus manatus*) heartbeat slows while diving, enabling it to stay underwater for a longer period of time. (Photo by Phillip Colla/Bruce Coleman, Inc. Reproduced by permission.)

Blood flow to nonessential parts of the body, like the extremities and the gut, is reduced, but it is maintained to vital organs like the brain and heart. In other words, oxygen is made available where it is needed most.

Another potential problem faced by divers results from the presence of large amounts of nitrogen in the air. Nitrogen dissolves much better at high pressures, such as those experienced at great depths. When nitrogen bubbles form in the blood after diving, they can lodge in the joints or block the flow of blood to the brain and other organs. Aquatic mammals have adaptations that prevent nitrogen from dissolving in the blood, whereas human lungs basically work the same underwater as on land. When aquatic mammals dive, their lungs actually collapse. They have a flexible rib cage that is pushed in by the pressure of the water. This squeezes the air in the lungs out of the places where it can dissolve into the blood. Air is moved instead into central places, where little nitrogen is absorbed. Some pinnipeds actually exhale before they dive, further reducing the amount of air, and therefore nitrogen, in the lungs.

Buoyancy

Unlike fishes, secondary swimmers (terrestrial animals that returned to an aquatic environment) have no such specific adaptations to the buoyancy problem. They all rely on simple density adaptations to help them. For example, the bones of diving birds are less pneumatic, and their air sacs are reduced (loons, penguins). Mammals that dive deep may hyperventilate before submerging, but they do not fill their lungs. Indeed, they may exhale before diving. Deep-diving whales have relatively small lungs. Sirenians, which may feed while resting on the bottom or standing on their tails, have unusually heavy skeletons; their ribs are swollen and solid. Likewise, the skeleton of the hippopotamus is also unusually heavy. The presence of blubber in marine mammals also contributes to their overall density, and walruses (Odobenidae) have two large air pouches extending from the pharynx, which can be inflated to act like a life preserver to keep the animals' head above water while sleeping.

Convergence

The largest group of marine mammals, the cetaceans, is also the group that has made the most complete transition to aquatic life. While most other marine mammals return to land at least part of the time, cetaceans spend their entire lives in the water. Their bodies are streamlined and look remarkably fish-like. Interestingly, even though all marine mammals have evolved from very different evolutionary groups, there are certain similarities in lifestyle and morphology, and they are considered good examples of the principle of convergence. Convergent evolution is the process by which creatures unrelated by evolution develop similar or even identical solutions to a particular problem; in this case, life in water.

Resources

Books
Ellis, Richard. *Aquagenesis*. New York: Viking, 2001.

Periodicals
Alexander, R. McNeill. "Size, Speed, and Buoyancy Adaptations in Aquatic Animals." *American Zoologist* 30 (Spring 1990): 189–196.

Butler, Patrick J., and David R. Jones. "Physiology of Diving Birds and Mammals." *Physiological Reviews* 77 (1997): 837–894.

Fish, Frank E. "Transitions from Drag-based to Life-based Propulsion in Mammalian Swimming." *American Zoologist* 36 (December 1996): 628–641.

Graham, Jeffrey B. "Ecological, Evolutionary, and Physical Factors Influencing Aquatic Animal Respiration." *American Zoologist* 30 (Spring 1990): 137–146.

Thompson, David, and Michael Fedak. "Cardiac Responses of Grey Seals During Diving at Sea." *Journal of Experimental Biology* 174 (1993): 139–164.

Webb, P. M., D. E. Crocker, S. B. Blackwell, D. P. Costa, et al. "Effects of Buoyancy on the Diving Behavior of Northern Elephant Seals." *Journal of Experimental Biology* 201 (August 1998): 2349–2358.

Gretel H. Schueller

Adaptations for subterranean life

Subterranean mammals

Across the globe, some 300 (7%) of the extant species of mammals belonging to 54 (5%) genera and representing 10 (7.5%) families of four mammalian orders spend most of their lives in moist and dark, climatically stable, oxygen-poor and carbon dioxide-rich, self-constructed underground burrows, deprived of most sensory cues available above ground. The subterranean ecotope is safe from predators, but relatively unproductive and foraging is rather inefficient. These mammals are fully specialized for their unique way of life in which all the foraging, mating, and breeding takes place underground. These animals are called "subterranean" ("sub" means under, and "terra" means earth or soil), whereas animals that construct extensive burrow systems for shelter but search for their food (also) above ground are denoted "fossorial" ("fossor" means digger). Of course, there is a continuum from fossorial through facultative subterranean to strictly subterranean lifestyles.

The subterranean niche opened to mammals in the upper Eocene (45–35 million years ago [mya]) and then extended into upper Tertiary (Oligocene and Miocene, i.e., 33.7–5.3 mya) and Quaternary (some two mya) when in the course of global cooling and aridization, steppes, savannas, semideserts, and deserts expanded. In seasonally dry habitats, numerous plants (the so-called geophytes) produce underground storage organs (bulbs and tubers) that can be a substantial source of food for herbivorous animals. (Underground storage organs of some plants such as potatoes, sweet potatoes, yams, cassava, etc. are among the most important human staple foods.) There have been several waves of adaptive radiation when, independently in space and time, mammals in different phylogenetic lineages occupied the underground niche either to feed on geophytes (in the case of rodents) or to feed on invertebrates, which themselves find food and shelter underground (in the cases of insectivores, armadillos, and marsupial moles). Thus, two morphological and ecological subtypes of subterranean mammals have evolved. Nevertheless, they all have been subjected to similar environmental stresses and, as a consequence, have much in common. Although the subterranean ecotope is relatively simple, monotonous, stable, and predictable in many aspects, it is very specialized and stressful in others. Consequently, the adaptive evolution of subterranean mammals involves structural and functional changes, which are both regressive (degener-

ative) and progressive (compensatory) in nature. The mosaic convergent global evolution of subterranean mammals due to similar constraints and stresses is a superb example of evidence for evolution through natural selection, evidence obtained through comparative methods.

Who cares about subterranean mammals?

Although at least some of the underground dwellers have been known for many years, their biology has remained unstudied. This may be explained by the cryptic way of life of subterranean animals, and technical problems related with keeping, breeding, and observing them. The fact is, scientists were always more fascinated by animals coping with complicated environments and solving seemingly difficult and complex problems than by those encountered by mammals underground (sensitive vision versus blindness; echolocation in a high-frequency range versus hearing in a human auditory range; navigating across hundreds or thousands of miles versus maze orientation across tens of feet; thermoregulation in cold environments versus life in a thermally buffered burrow, etc.). Interestingly, although many preserved specimens of moles (i.e., insectivorous subterranean mammals) and mole-rats (subterranean rodents) have been collected and deposited in museums, not even the study of morphological digging specializations has received the attention it has deserved. Textbooks of biology in general and evolutionary biology in particular have brought diverse examples for convergent evolution, yet one of the most remarkable examples—convergent evolution of subterranean mammals—has rarely been mentioned.

It may be of interest to examine the literature dealing with ecology, evolution, morphological, physiological, and behavioral adaptations of subterranean mammals. Although there are some relevant scientific papers published as early as at the beginning of the nineteenth century, the real exponential growth of the research and publishing activity referring to subterranean mammals started in the 1940s. Since then, the number of publications has doubled about every 10 years. Thus, 56% of about 1,300 scientific papers addressing at least partly adaptations of subterranean mammals and published to date (March 2003) appeared after 1990, a further 25% are dated 1981 to 1990, and another 11% appeared between 1971 and 1980. The interest in adaptations of subterranean mammals

The common mole-rat (*Cryptomys hottentotus*) has a oval-shaped body with short legs to enable it to move through small tunnels. (Photo by © Peter Johnson/Corbis. Reproduced by permission.)

cles have been authored and co-authored by very few persons or research teams. Thus, 31% (410 out of 1,300) of the scientific papers bear a signature of one or more of just five authors (Bennett, Burda, Heth, Jarvis, and Nevo), while the most influential of them, Eviatar Nevo, has authored or co-authored 225 of them. As a consequence—since every scientist observes the world through her/his own eyes and is constrained by her/his own research possibilities (professional training, knowledge and experience, interests, affiliation, geography, available funding, etc.)—the research has many biases. Although the validity of the published data and findings is not questioned, their interpretation may be influenced by science's limited knowledge and/or ideology molded by the philosophy of the author and the world she/he lives in. However, this is a general problem of scientific research.

The taxonomic treatment is uneven in that almost 85% of all the papers on subterranean mammals and their adaptations deal with just a few species belonging to 10 (out of 54) genera (*Arvicola, Cryptomys, Ctenomys, Geomys, Heterocephalus, Spalacopus, Spalax, Geomys bursarius, Talpa,* and *Tachyoryctes*). Some of the species, like the mole (*Talpa europaea*), the northern water

has been triggered particularly by two seminal papers on the blind mole-rat, *Spalax*, published in 1969, both authored or co-authored by Eviatar Nevo of the University of Haifa. In 1979, a review article (which has since become a citation classic) by Nevo stimulated considerable research into the physiology, sensory biology, communication, temporal and spatial orientation, ecology, taxonomy, and phylogeny of burrowing rodents. A second stimulus triggering the interest in subterranean mole-rats, particularly in their social behavior, came in 1981 with the pioneering studies of Jennifer U. M. Jarvis, when she reported on eusociality in the naked mole-rat, and in 1991, when a book was published on the evolution and behavioral ecology of naked mole-rats and related bathyergids. Since then, several international symposia on subterranean mammals have been convened and four books in English were published by renowned publishing houses within just two years (1999–2001).

Biased knowledge

Still, general knowledge on the subject is rather incomplete and heavily biased in several aspects. Most of the arti-

The star-nosed mole (*Condylura cristata*) uses its snout and tentacles as touch receptors. (Photo by Dwight Kuhn/Bruce Coleman, Inc. Reproduced by permission.)

The Columbian ground squirrel (*Spermophilus columbianus*) spends the majority of its time hibernating in underground burrows. (Photo by François Gohier/Photo Researchers, Inc. Reproduced by permission.)

vole (*Arvicola terrestris*), and pocket gophers (Geomyidae), have been studied to a much greater extent than considered here; yet most of the earlier studies on these animals have not specifically addressed subterranean adaptations. Usually, few species within (mostly) speciose genera have been studied in only a few localities within a broader geographic and ecological range of distribution. Studies are biased also as to which aspect of biology (and from which point of view) has been investigated. Thus, the (about 150) articles on the naked mole-rat (*Heterocephalus glaber*) primarily concentrate on the spectacular social behavior of this species, and some authors are prone to think of the naked mole-rats as if they were the only subterranean mammals. In spite of the fact that the naked mole-rat has attracted such intense interest by sociobiologists, only some 7% of all studies published on this species involved field ecological research and/or investigation of wild-captured animals. Most information on (social) behavior of the naked mole-rat has been based on the study of captive colonies. Yet, as S. Braude has demonstrated recently, long-term intensive and extensive field research may lead, at least in some aspects, to different results than the short-term study of captive animals. Interestingly, the problem of why the naked mole-rat is hairless (on both proximate and ultimate levels) has received very little attention from scientists. Similarly, although the question of whether subterranean mammals are blind or not (and if so, why do they still have miniscule eyes?) has been pertinent since Aristotle, the answer for most species is not yet known.

How to get through?

The most significant challenge is a mechanical one: soil is a dense, more or less hard and compact medium that cannot

be penetrated easily. Movement through soil is energetically very costly. Vleck (1979) has estimated that a 5.3-oz (150-g) pocket gopher burrowing 3.3 ft (1 m) may expend 300–3,400 times more energy than moving the same distance on the surface. To keep the energy costs of burrowing at the minimum, the tunnel should have a diameter as small as possible. To achieve this, subterranean mammals have a cylindrical body with short limbs and no protruding appendages. Even testes of most underground dwellers are seasonally or permanently abdominal. Subterranean mammals are mostly small-sized animals weighing 3.5–7 oz (100–200 g), but ranging from 1 oz (30 g) (Namib golden mole, naked mole-rat, and mole-vole) to 8.8 lb (4 kg) (bamboo rat). In order to penetrate the mechanically resistant medium, subterranean mammals need efficient digging machinery. Subterranean rodents dig (loosen soil) primarily with their procumbent, ever-growing incisors, or use teeth and claws, whereas subterranean insectivores, armadillos, and the marsupial mole use only robust, heavily muscled and large-clawed forelimbs. In teeth-diggers, the whole skull is subservient to incisors and well-developed chewing muscles. Interestingly, subterranean rodents belonging to the suborder Hystricognathi (Bathyergidae, Octodontidae) transport loosened soil backwards by pushing or kicking the soil with hind limbs, whereas representatives of the suborder Sciurognatha (Muridae, Geomyidae) turn in the burrow and push out the loosened soil with the head. Desert golden moles as well as the marsupial mole do not dig permanent tunnels (except for their nest burrow), but "swim" through the sand. Although sand-swimming requires less than a tenth of the energy required by mammals that dig permanent tunnels through compact soil, it is still much more expensive than running on the surface. Sand-swimming at a mean velocity of 25–97 ft/h (7.6—29.6 m/h) (as recently estimated for the Australian marsupial mole [*Notoryctes typhlops*]

The eastern mole (*Scalopus aquaticus*) has no external ears and a thin layer of skin protects its eyes. (Photo by Kenneth H. Thomas/Photo Researchers, Inc. Reproduced by permission.)

The naked mole-rat (*Heterocephalus glaber*) has hair between its toes that sweeps the dirt out of the pathways in its burrow. (Photo by J. Visser/Mammal Images Library of the American Society of Mammalogists.)

and Namib desert golden mole [*Eremitalpa granti*], respectively) is also substantially slower than walking or running above ground (about 1,476 ft/h [450 m/h]). It would apparently be energetically impossible for these mammals to obtain enough food by foraging only underground. Indeed, in one study of free-living Namib desert golden moles, the mean daily track length was 0.87 mi (1.4 km), but only 52.5 ft (16 m) of it was below the surface.

Subterranean mammals can move backwards with the same ease as forwards. The skin is usually somewhat slack, and the fur tends to be short and upright, brushing in either direction. These all may be burrowing adaptations to match frictional resistance, to facilitate moving and turning in tunnels. The extremes such as the total alopecia (hairlessness) of the naked mole-rat or the long hairs and thick pelage of the silvery mole-rat (*Heliophobius argenteocinereus*) are exceptions to the rule and should not be considered burrowing adaptations per se. Reduction or even absence of auricles (pinnae) may be beneficial for digging and moving in tunnels because of the reduced friction. The popular assumption that auricles are reduced or missing because, otherwise, they would have to act as shovels collecting all the dirt cannot withstand critical comparative analysis. Many burrowing rodents have rather prominent auricles and are apparently not handicapped. Probably more im-

portant than whether auricles are an advantage or disadvantage for burrowing is whether they are required for sound localization. If not needed for hearing, only then would they be reduced. The tail tends to be shortened in subterranean and fossorial mammals, yet there is no clear explanation as to the adaptive value of this feature. For instance, African mole-rats of two related genera, *Heterocephalus* and *Cryptomys*, differ in this trait markedly. Similarly, fossorial-subterranean octodontids have medium-sized tails, whereas related surface dwelling cavies have reduced tails. Vibrissae in subterranean mammals are also shorter and less protruding than in many surface dwellers. In sand-swimming golden and marsupial moles, they are inconspicuous, sometimes even missing.

How to acquire oxygen

Subterranean mammals also have to cope with problems from the burrow atmosphere. The oxygen concentration may be as low as 6%, compared to 21% prevailing in the ambient atmosphere at sea level altitude. This means that the oxygen concentration even a few inches underground may be lower than that on Mount Everest. The carbon dioxide concentration in burrows ranges between 0.5–13.5% (compared to 0.03% in the above ground atmosphere). Surprisingly, some recent measurements in foraging tunnels of three species of African mole-rats have revealed, however, that concentrations of oxygen and carbon dioxide did not differ greatly from ambient values above ground. Nevertheless, gas concentrations may change rapidly after rains, they may differ in different soil types and depths, and, above all, they must change dramatically in the immediate vicinity of the nose of a burrowing animal. Normally, there are no air currents in underground burrows. One can imagine that an animal moving through the narrow tunnel acts like a piston securing the ventilation of burrows, much like a train in a subway tunnel. As suggested by Arieli, who significantly contributed to knowledge of respiratory physiology of mole-rats, enhanced burrowing activity following rains may serve as a means of replenishing the burrow atmosphere when the hypoxic-hypercapnic situation becomes aggravated. Of course, working under such conditions becomes even more difficult.

Whereas adaptations to extreme atmospheric conditions have been extensively studied in diving mammals and in mammals living at high altitudes, much less is known in this regard about subterranean mammals. The combination of extreme hypercapnia and hypoxia normally encountered and tolerated by subterranean animals is unique. Interestingly, whereas surface dwellers respond to hypercapnic conditions by increasing breathing frequency, subterranean mammals display lower ventilation rates than would be expected. In fact, ventilation is not effective for releasing carbon dioxide from blood when there is a high concentration of it in the inspired gas. The lungs of subterranean mammals do not show any specific morphological specializations—on the contrary they seem (at least in the few species studied so far) to be rather simplified and juvenile-like. Although the oxygen transport properties of blood vary markedly among subterranean mammals, and no generalizations can be made, relatively high hemoglobin affinity for oxygen has been reported consistently for several species of subterranean rodents. Because of the higher amount of carbon dioxide inhaled, a higher concentration of this gas in the blood and, thus, also higher blood acidity can be expected. It has been shown in the blind mole-rat that urine contains high values of bicarbonates and may serve as a pathway to bind and void carbon dioxide. Further adaptations to the extreme burrow atmosphere may involve higher capillary density in muscles (including the heart), higher volume of muscle mitochondria (found in the blind mole-rat), and, particularly, low metabolic rates and relaxed thermoregulation.

How to regulate temperature

The microclimate of the subterranean ecotope is rather stable. Particularly in the nest chamber, which in giant Zambian mole-rats (*Cryptomys mechowi*) is usually 23.6 in (60 cm) (in some cases, even 6.6 ft [2 m]) below ground, there are minimal daily or seasonal fluctuations in temperature and humidity. This constant temperature enables a lower basal rate of metabolism. In the thermally buffered environment of the underground "incubator," it is possible to abandon complex and complicated morphological and physiological mechanisms of thermoregulation. Indeed, subterranean mammals tend to hypothermia (lowering the body temperature—on average 89.6–96.8°F [32–36°C]). Body temperature is partly de-

The European mole (*Talpa europaea*) uses its long claws to dig its burrows. (Photo by Hans Reinhard/OKAPIA/Photo Researchers, Inc. Reproduced by permission.)

pendent upon the ambient temperature. This relaxed thermoregulation (heterothermia) is most pronounced in the smallest (and the only hairless representative) among the subterranean mammals, the naked mole-rat. High relative humidity (about 93%) in underground burrows results in a relatively low vapor pressure gradient and low rate of water loss through exhalation or through the skin. This is beneficial for water balance as these animals do not drink free water, but instead obtain water from the food they consume.

However, high humidity and relatively high temperatures, which can occur on sunny days in shallow foraging burrows, may result in thermoregulatory problems. In the absence of evaporative and convective cooling, overheating and thermal stress would seem to be inevitable, since burrowing is energetically demanding and most mammals can tolerate dry, warm climate better than humid, warm climate. Subterranean mammals living in warmer environments have high thermal conductance, which means that the animals may exchange heat (cool or warm themselves) relatively easily through direct physical contact between themselves and the soil. As in poikilothermic reptiles, behavioral thermoregulation is of particular importance in heterothermic mammals. Thus, the animals can adapt timing and duration of their working activity to ambient temperatures in shallow burrows. Comparative and experimental physiological research of thermoregulation and energetics has a long tradition since McNab in 1966 first compared the metabolic rate of five subterranean rodent species and emphasized their shared adaptive convergence syndrome: low resting metabolic rate (involving also lower

The lips of the valley pocket gopher (*Thomomys bottae*) close behind its teeth to keep dirt out. (Photo by Tom McHugh/Photo Researchers, Inc. Reproduced by permission.)

ventilation and heart rates than would be expected on the basis of body size), low body temperature, and high thermal conductance. Since then, additional physiological data have been obtained on diverse species of subterranean mammals supporting the earlier conclusions by McNab.

How to avoid rickets

The underground ecotope is dark. Apart from the consequences for sensory orientation and communication, absence of light also influences some physiological functions. One of them is mineral metabolism. On the one hand, subterranean rodents have an especially high requirement for calcium because their large teeth are constantly worn down during digging. In the African mole-rat, *Cryptomys*, the visible part of the incisors regenerates completely every week. Also, calcium may be excreted in high concentrations as calcium carbonate through urine, a mechanism to deplete tissues of carbon dioxide. On the other hand, it is common knowledge that vitamin D (and principally D_3, cholecalciferol), which is needed for effective absorption of calcium from the gut (its deficiency causes rickets), is synthesized in the skin by the action of sunlight. Rochelle Buffenstein and colleagues have demonstrated that several species of African mole-rats, although in the perpetual state of vitamin D_3 deficiency due to their lightless environment, have evolved other physiological mechanisms to absorb calcium effectively (indeed, up to 91% of minerals can be extracted) from their diet.

How to tell time

Light-dark rhythm (photoperiod) is known to regulate production of the hormone melatonin that, in turn, regulates circadian (meaning around a day, as in a 24-hour period) rhythms by a feedback mechanism. In surface-dwelling vertebrates (including human), melatonin is produced during dark hours. It can therefore be expected that subterranean mammals living in constant darkness would display high melatonin levels. This does seem to be the case, yet the role of melatonin in regulating activity rhythms in subterranean animals remains obscure.

Permanent darkness in subterranean burrows makes sight and eyes rather useless and, apart from the fact that it precludes visual orientation in space, it also makes orientation in time problematic. Virtually all surface-dwelling mammals exhibit more or less pronounced circadian cycles of activity and diverse functions. These cycles are maintained by an endogenous pacemaker synchronized (entrained) by the so-called *zeitgeber* (time-giver). The most universal *zeitgeber* is the light-dark cycle perceived by photosensors. Considering the stability of the environment, constant availability of plant food, darkness underground, and poor sight or even blindness, one might expect that herbivorous subterranean rodents would not exhibit distinct diurnal activity/sleeping patterns. Field and laboratory studies on American pocket gophers (*Geomys bursarius* and *Thomomys bottae*) and African mole-rats (*Heliophobius argenteocinereus* and *Cryptomys hottentotus*) have revealed dispersed activity occurring throughout the 24 hours, for instance, arhythmicity and lack of distinct sleep-wake cycles. Interestingly, there are some other species of subterranean rodents that exhibit endogenous circadian activity rhythms that are free running in constant darkness and synchronized by light-dark cycles. These animals have been found to be either mostly diurnal such as the east Mediterranean blind mole-rat (*Spalax ehrenbergi* species complex), the East African mole rat (*Tachyoryctes splendens*), and the Kalahari mole-rat (*Cryptomys damarensis*); or predominantly nocturnal such as the African mole-rats (*Georychus capensis* and *Heterocephalus glaber*) and the Chilean coruro (*Spalacopus cyanus*).

There is no apparent correlation between the circadian activity pattern, on the one hand, and the degree of confinement to subterranean ecotope, development of eyes, seasonality of breeding, or social and mating systems, on the other. It should be noted, however, that findings in the same species have frequently been inconsistent. Available data have been obtained by different examination methods and may not be fully comparable. Moreover, there may be differences between individuals, sexes, and populations, between seasons of the year and habitats, as well as between the laboratory and the field. The methodological problem can be demonstrated in the example of the naked mole-rat, which had been considered to be arrhythmic by previous authors, yet was shown to display clear circadian rhythms and ability to synchronize them by the light-dark cycle if given an opportunity to work on a running wheel in the laboratory. There is a similar finding in *Cryptomys anselli*. Further studies of activity patterns in other species of mammals are clearly needed. A possible *zeitgeber* determining digging activity can be also temperature and humidity, which may fluctuate in shallow tunnels (although certainly not in deep nest chambers), as well as consequent changes in activity of invertebrates, which may affect foraging activity of moles.

The blind mole-rat, which is visually blind and has degenerated eyes, still has a hypertrophied retina and a large harderian gland in which the so-called circadian genes as well as the recently discovered photopigment melanopsin are expressed in high concentrations, and these apparently contribute to regulation of photoperiodicity. In other words, the double function of any vertebrate eye changed: instead of sight and circadian functions, only a circadian eye remains.

The ability to perceive and recognize the length of the photoperiod is important for seasonal structuring of reproductive behavior. It has also been suggested that melatonin may suppress production of gonadotrophins, hormones which, in turn, control activity of gonads and, thus, the sexual behavior and reproductive biology. Nothing is known about this aspect in subterranean mammals. Also unclear is how circannual (approximately one year) cycles are synchronized in subterranean mammals. These cycles are associated particularly with seasonal breeding in solitary territorial animals. It is assumed that the length of the photoperiod may play a role in seasonal breeders from temperate zones. Nevertheless, factors triggering breeding in mammals with long gestation from the tropics (where there is minimal variability in the daylight throughout the year) remain unknown and enigmatic in many cases. This is also the case for the eastern African silvery mole-rat. Alternating rains and drought represent the main periodic environmental factor. However, as shown recently, mating takes place at the end of the rainy and beginning of the dry season so that there is a substantial lag between the onset of rains (and subsequent softening of the soil and change of vegetation that could provide a triggering signal) and onset of breeding behavior.

How to find the way

Subterranean mammals construct, occupy, and maintain very long and extensive burrow systems. An average subterranean mammal (single, weighing 5.3 oz [150 g]) controls about 203 ft (62 m) of burrows. Of course, there are species-specific, habitat, and seasonal differences. This also implies that an average mole-rat living in a group consisting of 10 family members has to be familiar with at least 2,034 ft (620 m) of burrows. The longest burrow systems were found in *Cryptomys* mole-rats and coruros. Yet, the burrow system is a complicated, complex, three-dimensional network. Although there is evidence that subterranean mammals have an extraordinary spatial memory based on well-developed kinesthetic sense (controlled, in part, by sensitive vestibular organs), this fact does not explain how subterranean mammals can steer the course of their digging and what, in absence of visual landmarks, is the nature of external reference cues for the kinesthetic sense. In 1990, the first evidence that Zambian mole-rats (*Cryptomys anselli*) show directional orientation based upon the magnetic compass sense was published. In a laboratory experiment, mole-rats collected nest materials and built a nest in a circular arena. They showed a spontaneous tendency to position their nests consistently in the southeast sector of the arena. When magnetic north was shifted (by means of Helmholtz coils), the animals shifted the position of the nest accordingly. This laboratory experiment on mole-rats has become the first unambiguous evidence for magnetic compass orientation in a mammal and a paradigm for further tests of magnetic compass orientation in small mammals.

The black-tailed prairie dog (*Cynomys ludovicianus*) is well known for its large burrows. (Photo by Tim Davis/Photo Researchers, Inc. Reproduced by permission.)

Convergent spontaneous directional magnetic-based preference for location of nests in the laboratory was demonstrated also in taxonomically unrelated blind mole-rats from Israel. In 2001, Nemec and associates observed, for the first time in a mammal, structures in a brain (populations of neurons in colliculus superior), which are involved in magnetoreception in *Cryptomys anselli*.

The problem of orientation underground was addressed as recently as 2003, when it was reported that blind mole-rats could avoid obstacles by digging accurate and energy-conserving bypass tunnels. Apparently, the animals must possess both the means to evaluate the size of the obstacle as well as the ability to perceive its exact position relative to the original tunnel that it will rejoin. At present, information about potential sensory mechanisms can be only speculated.

How to find food

Subterranean mammals are animals that live and forage underground. However, the underground ecotope is low in productivity, burrowing is energetically demanding, and, in addition to these costs, foraging seems to be inefficient. It is widely assumed that subterranean rodents must forage blindly without using sensory cues available to and employed by surface dwellers. Indeed, vision is ineffective underground, there are no air currents to transmit airborne odorants over longer distances, high frequency sounds are damped by the soil, and low frequencies cannot be localized easily; touch and taste are only useful on contact. Carnivorous and/or insectivorous subterranean mammals such as moles can dig a stable foraging tunnel system into which prey may be trapped. Moles run-

The thumb of the alpine marmot (*Marmota marmota*) has a nail instead of a claw to aid in digging. (Photo by St. Meyers/Okapia/Photo Researchers, Inc. Reproduced by permission.)

ning along existing burrows can locate prey by hearing their movement in the tunnel system. Prey animals may also leave scent trails that the insectivorous predator can follow. In most cases, the food is detected at encounter or in the immediate vicinity through touch. The most spectacular example for this type of foraging is the star-nosed mole (*Condylura*), with its Eimer-organ-invested rostral tentacles. Mason and Narins have shown that the Namib golden mole may use low-frequency vibrations produced by isolated hummocks of dune grass and orient its movement toward the hummocks and the invertebrate prey occurring there. However, in contrast to the prey of moles, tubers, bulbs, and roots are stationary and silent.

It has been demonstrated that subterranean rodents are able to dig in relatively straight lines until they encounter a food-rich area and then make branches to their tunnels to harvest as much as possible from this area. Tunneling in relatively straight burrows conserves energy because the animals do not search in the same area twice. Because geophytes are generally distributed in clumps and patches, extensive burrowing around one geophyte upon encountering it increases the chances of encountering another. Although this dual strategy has been described and its functional meaning recognized in different species of subterranean rodents, sensory mecha-

nisms that may underlie the switch from linear to reticulate digging have not been addressed until a recent study. It has been shown that subterranean rodents can smell odorous substances leaking from growing plants and diffusing around the plant through the soil. Thus, herbivorous mammals are able to identify the presence of the plants and possibly even to identify particular types of plants specifically. They may be able to orient their digging toward areas that are more likely to provide food sources.

How to find a partner

In order to reproduce, mammals have to find and recognize an appropriate mate (belonging to the same species, opposite sex, adult, in breeding mood, sexually appealing). Monogamous mammals undergo this search once in life; solitary mammals have to seek mates each year. Subterranean mammals do not differ in this respect from their surface-dwelling counterparts. In 1987, two research teams reported, simultaneously and independently, the discovery of a new, previously unconsidered, mode of communication in blind mole-rats: vibrational (seismic) signaling. The animals can put themselves into efficient contact through vibrational signals produced by head drumming upon the ceiling of the tunnel. Communicative drumming by hind feet was reported for solitary African mole-rats (*Georychus* and *Bathyergus*). This behavior, however, could not be found in another solitary African mole-rat, the silvery mole-rat (*Heliophobius*). It can be speculated that seismic signaling evolved in those solitary species that disperse and look for potential mates underground. Subterranean mammals that usually occur at lower population densities and whose burrow systems are far apart from each other have to cover larger distances (which would be impossible to do by digging) in order to find a partner. They dare to carry out their courting above ground. These mammals such as the silvery mole-rat, the naked mole-rat, or the European mole have not invented seismic communication. Moles wandering at the surface in hopes of finding a burrow of a female probably are led by olfactory signals.

Seeing, or not seeing

Sensory perception plays a pivotal role not only in spatial and temporal orientation, foraging, and recognition of food, but also in communication with conspecifics. Like their surface dwelling counterparts, subterranean animals also must find and recognize a mate, recognize kin or intruders, and be warned of danger. This all is very difficult in a monotonous, dark world where transmission of most signals and cues is very limited. Some senses such as sight are apparently useless, whereas others have to compensate for their loss. One of the most prominent features of subterranean mammals is, no doubt, reduction of eyes and apparent blindness. The question of whether and what subterranean mammals see has been studied by Aristotle, Buffon, Geoffroy, Cuvier, and Darwin, among others. Still, today, no general unambiguous reply can be given. Thus, for instance, in comparing two of the most specialized subterranean rodents, the east Mediterranean blind mole-rat (*Spalax*) and the African mole-rat (*Cryptomys*), both are strictly confined to the underground and are of similar appearance, externally. They both have very small eyes and they both are behaviorally blind, yet whereas the eyes of the blind mole-rat are structurally degenerated and lie under the skin, the eyes of the African mole-rat are prominent, only miniaturized but in no respect degenerate; on the contrary, they are fully normally developed. As has been shown by several research groups, they also use different kinds of cone-pigments. Whereas the degenerated subcutaneous eye of the blind mole-rat has apparently adapted to a function in circadian photoreception, the function of a normally developed eye of the African mole-rat remains enigmatic.

Interestingly, whereas spalacines and bathyergids in the Old World lost their sight and have become completely subterranean, their New World counterparts, geomyids and octodontids, converged to similar habitats and habits retaining their eyes and sight. Unfortunately, the eyes and the sight of most subterranean mammals have not yet been studied.

As do blind people, blind subterranean mammals compensate for loss of sight by well-developed somatosensory perception, which was shown in the blind mole-rat, in the star-nosed mole, and in the naked mole. This somatosensory perception is over-represented also in the brain cortex where it occupies areas dominated in seeing mammals by visual projections.

How to hear underground

For communication and orientation in darkness, acoustic signals seem to be predestined, as exemplified by bats and dolphins. However, high sound frequencies (characterized by short wavelength), which can be well localized by small mammals and which are employed in echolocation, are quickly dampened underground. Low frequency sound, on the other hand, is characterized by long waves and cannot be localized easily. Indeed, it was demonstrated that, in a burrow of the blind mole-rat, sounds with a frequency of about 500 Hz were more efficiently transmitted than sounds of low and higher frequencies. Although nothing is known so far about the effect of the tunnel diameter, soil characteristics, temperature, and humidity, it is assumed that acoustic features of all burrows are rather similar. Consistent with the results of these measurements, the vocalization of all nine species (representing six genera) of subterranean rodents studied so far was also tuned to a lower frequency range. Corresponding with acoustics of the environment and with vocalization characteristics, the hearing of five species (of five genera) of subterranean mammals studied exhibited its highest sensitivity in the lower frequency range (0.5–2 kHz). This is quite unusual among small mammals, because hearing and vocalization in related surface dwellers of a comparable body size are usually much higher: about 8–16 kHz, or higher. The hearing range of subterranean mammals is very narrow, and frequencies of about 16 kHz and higher cannot be perceived (similar to humans). Frequency of 500 Hz is characterized by a wavelength of 27 in (68.6 cm). To be able to localize this frequency (to which hearing is tuned), an animal would need to have a head of a corresponding width; this is not possible. However, the transmission of airborne sounds in a tunnel is unidirectional, anyway. Consequently, animals that are confined to their un-

The European badger (*Meles meles*) builds onto underground tunnels, or setts, inherited from previous generations which results in setts that can be centuries old. (Photo by Roger Wilmshurst/Photo Researchers, Inc. Reproduced by permission.)

derground burrows do not need auricles for directional hearing. Some scholars would tend to label this restricted hearing as degenerated. However, looking at the morphology of the middle and inner ear, many progressive structural specializations enabling tuning of hearing to the given lower frequency range are observed. Indeed, several papers have described diverse morphological features of the middle and inner ear, as well as the expansion of auditory brain centers, which can be found consistently in non-related species of underground mammals and provide an example of convergent evolution. However, while the ear is clearly a low-frequency-tuned receiver, apparently the evolution has not fully utilized all the possibilities, as demonstrated in ears of desert animals, to enhance the sensitivity. On the contrary, some features of the external ear canal, eardrum, and middle ear ossicular chain indicate that sensitivity has been secondarily and actively reduced. Too little is known about the acoustics of burrows, and also the suggestion of Quilliam, 40 years ago, that sound in burrows can be amplified as in an ear-trumpet, has not yet been tested. Should there really be such a stethoscope effect, reduction of sensitivity (in order to avoid deafening) would have to be considered an adaptation in the same way as its increase is in other species.

How to avoid tetanus

Humid, thermally stable soil swarms with many microorganisms (bacteria, fungi, protozoans), eggs, and larval stages of diverse helminths and arthropods, many of which are pathogens. A renowned representative is *Clostridium tetani*, a bacterium that is particularly prevalent in soil contaminated with animal droppings and that, after entering wounds, causes a serious disease, tetanus. Subterranean mammals are parasitologically understudied, and the results are inconsistent and do not allow any generalization. Many more factors probably influence whether an animal will be infected or infested by parasites. Higher lung infection by adiaspores of *Emmonsia parva* was reported in burrowing voles, as compared to more surface-dwelling mice. Yet, the preliminary examination of *Spalax* and *Cryptomys* provided variable, inconsistent results. The ectoparasites in African mole-rats are, however, very rare, and infestation with endoparasitic helminths is unusual. Surely, the examination of the immune system of subterranean mammals may be of great interest and importance for human medicine. At least, the few existing studies of *Spalax* do suggest the opening of new research horizons.

Living in a safe, predictable world

The underground ecotope is not only challenging, it is also quite safe from predators. Certainly, naked mole-rats would not be able to survive in any other ecotope than in the safe, humid, and ultraviolet-light-protected underground burrows. No wonder that convergent patterns can be seen also in life histories of subterranean mammals. They all show tendencies to K-strategy, breed rather slowly, have slow and long prenatal and postnatal development, slow rates of growth, and

unusually long lifespans (an age of over 25 years has been recorded in the naked mole-rat in captivity, and there is an African mole-rat, *Cryptomys anselli*, that is at least 22 years old and is still breeding). On the proximate level, slow growth is surely correlated with low metabolic rates. However, the effect of phylogeny is very strong and phylogenetic relationships can best explain the length of pregnancy and many other parameters of life histories, such as mating behavior and mating system, as well as social systems.

In spite of all the similarities of the subterranean ecotope, there are differences in many of its biotic and abiotic parameters in different geographic regions and different habitats, which in turn also influence underground dwellers. Thus, subterranean mammals may serve not only as an example for convergent evolution but they provide cases to study adaptive divergence as well.

Last, but not least

Last, but not least, it should be mentioned that 40 million years of evolution underground, including hypoxia tolerance, light absence, etc., may prove important to biomedical research and human gene therapy. Subterranean mammals may become unique laboratory and model animals of the next generation.

Resources

Books

Bennett, Nigel C., and Chris G. Faulkes. *African Mole-rats. Ecology and Eusociality.* Cambridge, UK: Cambridge University Press, 2000.

Lacey, Eileen A., James L. Patton, and Guy N. Cameron, eds. *Life Underground. The Biology of Subterranean Rodents.* Chicago: University of Chicago Press, 2000.

Nevo, Eviatar. *Mosaic Evolution of Subterranean Mammals: Regression, Progression and Global Convergence.* Oxford: Oxford University Press, 1999.

Nevo, Eviatar, and Osvaldo A. Reig, eds. *Evolution of Subterranean Mammals at the Organismal and Molecular Levels.* New York: Wiley-Liss, 1990.

Nevo, Eviatar, Elena Ivanitskaya, and Avigdor Beiles. *Adaptive Radiation of Blind Subterranean Mole Rats.* Leiden: Backhuys Publishers, 2001.

Sherman, Paul W., Jennifer U. M. Jarvis, and Richard D. Alexander. *The Biology of the Naked Mole-rat.* Princeton, New Jersey: Princeton University Press, 1991.

Hynek Burda, PhD

• • • • •

Sensory systems

Introduction

Mammals, like other animals, can be expected to use whatever information is available to them when making decisions about activities such as foraging, mating, navigating, selecting shelter, or locating habitats. The range of information actually used by any one species can be predicted from its sensory apparatus—the stimuli they can perceive. Lifestyle plays an important role here, so that moles and other fossorial (also known as subterranean) mammals, including golden moles, some rodents, and at least one species of marsupial, can be expected to use vision less than species that are active aboveground, including the lion (*Panthera leo*), vervet monkey (*Cercopithecus aethiops*), or moose (*Alces alces*). Everyone who has walked a dog (*Canis familiaris*) or experienced the spraying of a male housecat (*Felis cattus*) knows the importance of odor in the lives of these mammals. The important role that sound plays in the lives of mammals becomes obvious when listening to the echolocation calls of a bat attacking an insect or to the bugling of a male elk (*Cervus elaphus*) during the rut.

The media

Vision

Most of the 5,000 or so living species of mammals have eyes and, in many, the keenness of their vision (visual acuity) is at least equivalent to that of humans. A few mammals have very limited vision, such as river dolphins (Platanistidae, Lipotidae, Pontoporiidae, and Iniidae) that live in extremely murky water or moles (Talpidae) that live in total darkness; indeed, in some moles, the optic nerve has actually degenerated. In mammals' eyes, a lens focuses light on the retina, a layer of light-sensitive cells in the back of the eye. Different chemicals (photopigments) in the cells of the retina convert optical information to electrical signals that are transmitted via the optic nerve to the brain. The retina has two main types of photosensitive cells: rods (that respond to black and white) and cones (that respond to color, which are different wavelengths of light). Color vision in mammals is uncommon, being present mainly in primates, some rodents, and some carnivores. In nocturnal mammals such as any microchiropteran bats, rodents (Muridae), and shrews (Soricidae), rods are often prevalent,

while cones may be absent. To these mammals, the world is black, white, or shades of gray. The eyes of some diurnal mammals (for example, primates in the families Lorisidae and Leumuridae, or rodents in the Sciuridae) have both rods and cones, and these mammals can see color. Other mammals such as some cats (Felidae) have color vision, but only perceive a few colors.

Mammals show a range of overlap between the field of view of left and right eyes—this is the degree of binocularity. The position of eyes in the face and the size and shape of the muzzle influence the degree of binocularity. Humans, with eyes side-by-side and no muzzle to speak of, have a high degree of binocular overlap, which means they have stereoscopic vision. Stereoscopic vision allows mammals (and other animals) to locate objects in space with accuracy. This is the ability to perceive depth, which plays an important role in hand-eye coordination. The distance between the eyes also affects binocularity. For example, African elephants (*Loxodonta africana*) or blue whales (*Balaenoptera musculus*), with eyes situated on the sides of huge faces, have almost no binocular overlap. In animals such as California leaf-nosed bats (*Macrotus californicus*), the degree of binocularity depends upon the direction in which the bat is looking. There is minimal binocular overlap when the bat looks down its muzzle, and a high degree of overlap when it looks across the top of its muzzle.

Arboreal animals such as many species of primates (lemurs, galagos, and lorises) tend to have higher degrees of binocularity than more terrestrial species (horses, cows, and pigs, in the orders Perrisodactyla and Artiodactyla, respectively). Finally, in some cases, the significance of binocularity in the animal's life is not known (for example, in the case of the wrinkle-faced bat, *Centurio senex*, of South and Central America).

It is common for nocturnal mammals to have a tapetum lucidum behind the retina. The tapetum lucidum is a layer of cells on the back of the eye that reflects light back through the retina, amplifying the stimulation of retinal cells by ensuring one round of stimulation as the light goes through, and another as it is reflected back. Tapeta lucida account for the "eyeshine" when catching a house cat or raccoon (*Procyon lotor*) in a car's headlights or in the beam of a flashlight. Pinnipeds (Phocidae, Otariidae, and Odobenidae) and odontocetes (toothed whales and dolphins) also have tapeta lucida for

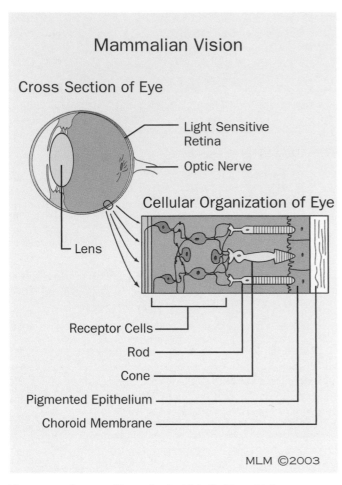

Mammalian Vision

Cross Section of Eye

Light Sensitive Retina

Optic Nerve

Lens

Cellular Organization of Eye

Receptor Cells

Rod

Cone

Pigmented Epithelium

Choroid Membrane

MLM ©2003

The mammalian eye. (Illustration by Michelle Meneghini)

vert chemical signals to electrical ones that are conveyed to the brain via the olfactory nerves. Many species of mammals also use Jacobson's organs (structures in the roof of the mouth) to obtain additional olfactory data through the "Flehman" response (the curling of its upper lip as a male horse [*Equus caballus*] or an impala [*Aepyceros melampus*] smells the urine of a female). One advantage of olfaction is that some odors are persistent and may continue to produce signals for long periods of time, unlike visual displays, which are immediate. Distinctive aromas signal the locations of the permanent dens of river otters (*Lontra canadensis* or *Lutra lutra*) or the burrows of shrews (Soricidae). Other olfactory materials such as mating pheromones in rodents are volatile and persist for only a short period of time. Pheromones can be quite potent, causing the "strange male (or "Bruce") effect" in some rodents (e.g., house mice, *Mus musculus*). With this effect, the mere presence of another male's urine can cause a female to miscarry a litter.

An individual's olfactory signature is often the product of the interaction of odors from different sources. Familiar examples include the aromas of sweat and breath and, in some situations, body products such as urine, feces, or oil from glands. An animal's scent can reveal a great deal about its condition and status, while yet more detailed information can be obtained from the aromas of its urine and/or feces. Bull elk during the rut rub urine on their chests, providing a conspicuous signal to females and other males of their condition. Male white-tailed deer (*Odocoileus virginianus*) leave urine and feces in specific locations in the woods to announce their presence to other deer. Male pronghorn antelopes (*Antilocapra americana*) mark the boundaries of territories with piles of feces, as do male white rhinos (*Ceratotherium simum*), which

helping gather available light at dark ocean depths, resulting in keen underwater vision. Visual displays from the tapetum lucidum are also common to the communication of diurnal mammals, but require that the individuals be close in proximity to each other.

Olfaction

Terrestrial mammals often have distinctive scents. In some societies, humans go to great lengths (and expense) to mask or alter olfactory information, as is reflected in the sales of deodorants and perfumes, respectively. Zookeepers recognize the importance of smell in mammals because, immediately after they have cleaned a cage, the animal often defecates, urinates, or otherwise marks its area again. Individual olfactory signatures may be less likely in mammals that spend most of their lives in water, which would at the least dilute, if not wash away body odors. Water does not allow permanent scent-marking locations, whereas land provides many places to position a long-term scent mark. In fact, whales and dolphins have completely lost the olfactory-sensing portions of their brains.

Mammals use their noses to collect information about odors. Specifically, olfactory epithelium (sensors on the mucosal surfaces of mesethmoid bones nose) in the nostrils con-

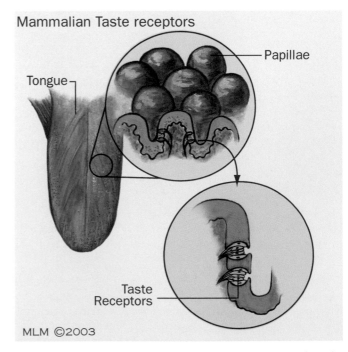

Mammalian Taste receptors

Papillae

Tongue

Taste Receptors

MLM ©2003

Mammals' taste receptors, the taste buds, are concentrated on the surface of the tongue. (Illustration by Michelle Meneghini)

The timber wolf (*Canis lupus*) uses its keen sense of smell to track its prey. (Photo by Wolfgang Baye. Bruce Coleman, Inc. Reproduced by permission.)

terminate in nuzzling. Grooming often involves touching, such as in two chimpanzees (*Pan troglodytes*) carefully stroking and picking at each other's fur. Primates have an especially well-developed sense of touch, having friction ridges (finger prints) on the tips of their digits used for careful investigation of objects. Although dolphins do not have limbs for grasping, their sleek, hairless skin is especially sensitive to touch at various locations on the body, specifically, the gape of the mouth, the gum, and tongue, and the insertion point of the flipper. Dolphins commonly swim close to each other, touching and rubbing their bodies together. The spectacular nasal appendages of the star-nosed mole (Talpidae) are extremely sensitive to touch and are used to locate and identify prey.

Vibration

Aye-ayes (*Daubentonia madagascarensis*) are among the mammals most obviously specialized to use vibrations. These Madagascar natives have long, slender third fingers. A foraging aye-aye taps branches with its elongated fingers and listens for reverberations that it uses to find hollows. The vibrations, combined with the noises made by insects moving through tunnels in wood or chewing to excavate tunnels, help aye-ayes find their prey.

Vibrations can also serve in communication. Vibrations tend to be low frequency, readily sensed by specialized hairs (whiskers) or other body parts. Nearly furless naked mole rats (*Heterocephalus glaber*) live in a burrow system and announce their presence to nearby conspecifics in other burrows by tapping their heads against the roofs of tunnels. Elephants are thought to use vibrations to sense danger or intruders over long distances. Recent studies of captive elephants showed that male elephants in mating condition (musth) moved their foreheads in and out, movements coinciding with the production of low-frequency sounds. Although African elephants can detect acoustic signals of about 115 Hz at distances of 1.5 mi (2.5 km), they need to be closer (0.6–0.9 mi [1–1.5 km]) to extract individual-specific information about the signaler(s). Researchers also believe that elephants sense very-low-frequency vibrations with their large, flat feet, enabling them to detect the movements and signals of other elephants from great distances. Foot drumming is a common way to generate vibrations that are used in communication by mammals such as lagomorphs (rabbits and hares) as well as kangaroo rats and subterranean mammals.

Infrared

Around their nose leaves, vampire bats (*Desmodus rotundus*) have sensors sensitive to infrared energy. The bat's sensors

Spectral bat (*Vampyrum spectrum*) locating prey. (Photo by Animals Animals ©Stephen Dalton. Reproduced by permission.)

lack a lens, so they provide poor spatial resolution of infrared sources. Vampire bats probably use infrared cues to locate places on a mammal or bird's body where blood flows close to the skin, ideal places to bite and obtain a blood meal. Among mammals, some felids have vision that extends into the infrared spectrum. Elsewhere among vertebrates, some pit vipers (rattlesnakes) use infrared sensors on the roofs of their mouths to locate and track warm-blooded prey in cool desert nights.

Chemoreception (taste)

Mammals detect a wide range of flavors as odors and tastes. Bottlenosed dolphins (*Tursiops truncatus*) readily detected different concentrations of bitter, sweet, and sour liquids presented to them. Unlike some terrestrial mammals, bottlenosed dolphins are not sensitive to subtle changes in salinity, suggesting that an animal living in salt water would not be averse to the taste of salt in its mouth.

Geomagnetic

The ability to orient to geomagnetic fields has been demonstrated in several species of migrating birds and in some rodents. In mammals, the geomagnetic sensing ability is correlated with the presence of magnetite in the brain. Some classic studies on trained rodents showed that the animals, when spun around 360°, could choose a particular orienta-

tion. Since the time of Aristotle, people have recognized that some odontocetes (toothed whales and dolphins) strand or beach themselves, often in large groups. Stranded animals may be completely out of the water and face certain death. Some locations where cetaceans often strand themselves are in areas with abnormal or unpredictable geomagnetic fields.

Role of sensory data

Mammals sense or gather information about their environment and use it to make decisions that affect their survival and reproduction. Sometimes, species initially respond to one type of cue. For example, female hammer-headed bats (*Hypsignathus monstrosus*) in Africa locate groups of males by listening to their distinctive calls. Picking a male to mate with, however, is a decision females appear to make only after visiting several in the line of displaying suitors. A female's actual choice may involve more than just her response to the males' calls.

Mammals typically use clues collected from several modalities. For example, vervet monkeys must cope with different predators. Social animals, vervets have keen vision and extensive vocal repertoires that include several types of warning calls, which indicate the presence of a predator. Using different warning calls, vervets can alert group members to specific threats. One alarm call is given in response to snakes, another to mammals such as leopards (*Panthera pardus*), and yet another to raptors such as eagles. These predators pose different kinds of threats. Vervets typically see the predators, but use sound to alert their group mates to the danger. Because each type of predator requires different defensive behavior, the vervets have specific acoustic signals to increase the precision of their communication.

The ability of recognizing other individuals in mammals begins at birth. Female mammals are expert at recognizing

The common tenrec (*Tenrec ecaudatus*) uses its whiskers to sense vibrations. (Photo by Albert Visage/FLPA–Images of Nature. Reproduced by permission.)

Spectral bat (*Vampyrum spectrum*) with a mouse caught at night using echolocation. (Photo by Animals Animals ©Stephen Dalton. Reproduced by permission.)

their own young. This is a valuable behavior because milk is expensive to produce and vital to the survival of young. The level of challenge to the mother varies with different mammals. Ewes recognize their lambs by smell, and she usually finds her own lamb quite easily. The lamb imprints on its mother within in a few days of birth. A female Brazilian free-tailed bat (*Tadarida brasiliensis*) faces a more difficult challenge. She typically leaves her single young in a creche with hundreds or thousands of others. When she returns from foraging and looks for her young, she initially relies on spatial memory to locate the general area where her young might be, then she uses the calls of her young to pinpoint their location, and finally ensures that she is feeding the right young by smelling its scent. When females depend upon odor to recognize their offspring, the distinctive smell could be something produced by the young, something in the milk she has fed it, or her own distinctive aroma.

In mammals, distinctive odors do more than mediate interactions between mothers and young. Young piglets (*Sus scrofa*) can recognize other piglets by the odor of their urine, which allows them to distinguish between familiar and strange individuals. In summer, Bechstein's bats (*Myotis bechsteinii*) live together in small groups (colonies). Individuals have specific odor signatures produced by a gland located between their ears. In other bats, such as big browns (*Eptesicus fuscus*), lesser-crested mastiff bats (*Chaerephon pumila*), or pipistrelles (*Pipistrellus pipistrellus*), odors allow bats to identify their home groups or roosts. It is typical for individual aromas to reflect a combination of odor sources, not just the products of a single gland.

Groups of mammals can also have distinctive vocalizations. Greater spear-nosed bats (*Phyllostomus hastatus*) use group-specific screech calls to locate other group members when they are feeding. In big brown bats and little brown bats (*Myotis lucifugus*), echolocation calls provide cues to group membership. Humpback whales (*Megaptera novaeangliae*) have distinctive songs that consist of phrases and arrangement of phrases in a music-like organization. Songs within a pod slowly change, so that last year's song is slightly different than the current year's song. Furthermore, phrases come and go in the overall repertoire. Songs also play a role in maintaining group cohesion. Male humpbacks sing from an inverted position and project their sounds over large ocean expanses. In the Antarctic, Weddell seals (*Leptonychotes weddellii*) have a large vocal repertoire of 34 underwater vocalizations, 10 of which are used only by males.

Echolocation

Animals, including some mammals and birds, use echoes of sounds they produce to locate objects in their surroundings. This is echolocation behavior. Microchiropteran bats produce echolocation signals by vibrating their vocal cords—exactly the same operation humans use in speaking. Echolocating mammals hear echoes through their auditory systems, just as humans hear sounds. While echolocating bats collect echoes of their own sounds by their pinnae (external ears), odontocetes appear to collect sounds via their lower jaws.

An echolocating animal uses the differences between the sound it produces and the reflected echoes to collect information about its surroundings. Echoes arrive sometime after the production of the outgoing signal, providing time cues to the echolocator. Echoes can differ in frequency composition from the outgoing signal, encoding information about the target's surface. Echolocating mammals, including some toothed whales and many bats, use echolocation to detect targets (food items) in their path, whether fish or insects, as well as objects such as trees or seamounts. They use echolocation to determine three factors about detected targets: identity; distance; and movement, whether toward or away from the echolocator. Some of this information is obtained by comparing information from

Beluga whales (*Delphinapterus leucas*) use echolocation to navigate, locate holes in the ice, and to find their deep dwelling prey. (Photo by Ed Degginger. Bruce Coleman, Inc. Reproduced by permission.)

The naked-mole rat (*Heterocephalus glaber*) has sensory whiskers on its face and tail. (Photo by Neil Bromhall/Naturepl.com. Reproduced by permission.)

sequences of calls and their echoes. Some microchiropteran bats and toothed whales also use echolocation to obtain fine details about objects. Echolocating dolphins can distinguish between echoes from same-sized spheres of aluminum and glass (down to accuracy of 0.001 in [0.0025 cm]), while some echolocating bats detect insects smaller than midges and can distinguish flying moths from flying beetles.

It is not obvious that other echolocating mammals (some species of shrews and tenrecs) collect and use such detailed information. Shrews and tenrecs appear to use echolocation while exploring, providing another medium for collecting general information about their surroundings rather than about specific targets. Most species of pteropodids, plant-visiting flying foxes of the Old World, do not echolocate. Furthermore, not all bats echolocate. Or, while Egyptian fruit bats (*Rousettus aegyptiacus*) and perhaps some other species in this genus (*Rousettus*) echolocate, they produce echolocation sounds by clicking their tongues. A further complication is that the role echolocation plays in the lives of some other bats is not known. Then, some phyllostomid, nectar-feeding bats, visit flowers that are specialized to deliver strong echoes of ultrasonic (echolocation) calls and thus guide the bats to the nectar they seek.

While toothed whales living in turbid waters may use echolocation to find prey, it is not clear how often these animals use echolocation to find food in clear waters. It is not known if any of the mysticete (baleen) whales use echolocation because none has been held in captivity to conduct the necessary perceptual studies.

The echolocation signals of toothed whales, shrews, and tenrecs are short, click-like sounds composed of a range of frequencies (broadband). The echolocation signals of microchiropteran bats are tonal, because they show structured changes in frequency over time. The echolocation signals of toothed whales and some microchiropteran bats are very intense, measured at over 110 decibels; in the toothed whales, it is measured at over 200 decibels. Other microchiropteran bats and shrews and tenrecs produce signals of low-intensity (<60 decibels). Many species of echolocating bats that hunt flying insects and other species that take prey from the surface of water change the details of their calls according to the situation. Longer calls, often consisting of a narrow range of frequencies (narrowband) and dominated by lower frequency sounds, are produced when the bats are searching for prey. Once prey has been detected, the bats often produce shorter, broadband signals. Over an attack sequence, the calls get progressively shorter as does the time between them. The end sequence (or terminal feeding buzz) has closely spaced signals for precisely timed capture of the prey. Odontocetes can adjust the frequency and amplitude of their echolocation signals, depending on the amount of environmental noise. In areas of noise from snapping shrimp, dolphins produce louder and higher frequency signals to avoid the masking sounds of the shrimps' snaps.

To ensure that they can hear faint returning echoes, echolocating mammals typically separate pulse and echo in time. In other words, most echolocating mammals cannot transmit signals and receive echoes at the same time because strong outgoing pulses deafen the echolocator to faint returning echoes, which means that collecting information about close objects requires signals that are short enough to be over before the echoes return. Differences in the density of water and air indicate that sound travels much faster in water, and cetacean echolocation signals are much shorter than those commonly used by bats.

Some microchiropteran bats, including species of horseshoe bats (Rhinolophidae), Old World leaf-nosed bats (Hipposideridae), and Parnell's moustached bat (*Pteronotus parnellii*; Mormoopidae), take a different approach to echolocation. They separate pulse and echo in frequency, meaning that they can transmit signals and receive echoes at the same time. These bats depend upon Doppler shifts in the frequencies of their echolocation sounds to collect information about their surroundings and targets.

Microchiropteran bats that separate pulses and echo in time produce short echolocation calls separated by long periods of silence. Separating pulse and echo in frequency produces much longer signals that are separated by shorter periods of silence. Some bats produce short signals that are called low-duty cycle, while others produce longer signals that are called high-duty cycle, reflecting the percentage of time that the signal is on (10% versus >50%, respectively). Anatomical evidence from Eocene fossil bats indicates that both high-duty cycle and low-duty cycle approaches to echolocation had evolved over 50 million years ago.

The distance over which an animal can use echolocation to collect information will depend upon the strength of the original signal and the sensitivity of the echolocator's auditory system. As a signal moves away from the source (an echolocator's mouth), it loses energy through spreading loss and, for higher frequency signals in air, by attenuation. The same rules apply to the echo returning from the target. For a big brown bat, this means that the effective range for detecting a spherical target 0.7 in (19 mm) in diameter is 16.4 ft (5 m). This assumes that the initial signal was 110 decibels and that the bat's hearing threshold is 0 decibels. For 0.7-in (19-mm) diameter spheres located 32.8 ft (10 m) in front of the bat, the original signal would reach the target, but the echo would not have sufficient energy to return to the bat.

The same general situation applies to echolocating porpoises and dolphins, although the distances are greater because of a combination of original signal strength and the sound-conducting properties of water. A false killer whale (*Pseudorca crassidens*), for example, can detect a 3-in (76-mm) diameter water-filled, aluminum sphere at 377 ft (115 m), a range that is far beyond visual detection. Similarly, an echolocating big brown bat would detect a 0.7-in (19-mm) long insect at 16.4 ft (5 m), but see it only at 3.2 ft (1 m). However, because of spreading loss and atmospheric attenuation over longer distances, the same bat would detect a tree-sized object at 49.2 ft (15 m), but would have been able to see it at over 328 ft (100 m).

Echolocation is an orientation system that allows animals to detect objects in front of them, and some species also use it to detect, track, and assess prey. The short operational range of echolocation in terrestrial mammals means that it is of much less value in navigation. But short operational range used in combination with local knowledge can be effective in a longer-range orientation. Greater spear-nosed bats deprived of vision and taken 31 mi (50 km) from their home caves could find their way back. Odontocetes use echolocation for navigation in murky, deep, dark waters and a varied underwater topography. Beluga whales (*Delphinapterus leucas*) use echolocation to survey the irregular undersurfaces of their ice-covered habitats.

Information leakage is an important disadvantage to echolocation. The signals one animal uses in echolocation are available to any other animals capable of hearing them. Many species of insects (e.g., some moths, beetles, mantids, crickets, and lacewings) have ears that allow them to detect the echolocation calls of bats. Moths with bat-detecting ears evade capture in 60% of attacks by echolocating bats, while deaf moths are most often caught. Some herring-like fish change their behavior when they hear the echolocation clicks of dolphins. Weddell seals dramatically reduce their underwater vocalization rate from 75 calls per minute to no calls when they hear sounds from their predators, killer whales (*Orcinus orca*) and leopard seals (*Hydrurga leptonyx*).

The most obvious eavesdroppers on echolocation calls are members of the same species. Bats often use the echolocation calls of other bats to detect patches of prey or vulnerable prey. Echolocation calls also can serve as communication signals promoting cohesion in groups of flying bats.

Echolocating animals also have repertoires of social calls. While echolocation signals can serve a communication func-

Bottlenosed dolphins (*Tursiops truncatus*) use echolocation for hunting and communication. (Photo by Jeff Rotman/Photo Researchers, Inc. Reproduced by permission.)

tion, many social calls are too long to be useful in echolocation. A long signal masks echoes that return as the call is being produced.

Navigation

Some species of mammals make long-distance migrations, typically repeated annual movements between summer and winter ranges. Gray whales (Eschrichtiidae), bowhead whales (*Balaena mysticetus*), right whales (*Eubalaena* species), and humpback whales make predictable, long-range migrations between summer feeding areas and winter breeding areas. Manatees (*Trichechus manatus*) make shorter seasonal migrations up and down the east coast of Florida. Other migrating mammals include some species of bats, caribou (*Rangifer tarandus*), and antelopes that regularly move between summer and winter ranges.

The navigational cues used by migrating mammals appear to vary. Over shorter distances, perhaps first traveled with their mothers or group members, young may learn the landmarks that guide them from one place to another. This appears to be true of manatees off the east coast of Florida. Over longer distances, geomagnetic cues may be more important. However, compared to the situation in migrating birds, relatively little is known about the mechanisms of navigation in mammals.

Resources

Books

Attenborough, D. *The Life of Mammals.* London: BBC, 2002.

Au, W. W. L. *The Sonar of Dolphins.* New York: Springer-Verlag, 1993.

Fenton, M. B. *Bats: Revised Edition.* New York: Facts On File Inc., 2001.

Fay, R. R., and A. N. Popper (editors). *Comparative Hearing: Mammals.* New York: Springer-Verlag, 1994.

Griffin, D. R. *Listening in the Dark. The Acoustic Orientation of Bats and Men.* Ithaca, NY: Cornell University Press, 1986.

Nowak, R. M. *Walker's Mammals of the World, Volume 1.* Baltimore: Johns Hopkins University Press, 1999.

Popper, A. N., and R. R. Fay (editors). *Hearing by Bats.* New York: Springer-Verlag, 1995.

Sebeok, T. A. *How Animals Communicate.* Bloomington, IN: Indiana University Press, 1977.

Smith, W. J. *The Behavior of Communicating.* Cambridge, MA: Harvard University Press, 1977.

Thomas, J. A., and R. A. Kastelein. *Sensory Abilities of Cetaceans: Laboratory and Field Evidence.* New York: Plenum Press, 1990.

Thomas, J. A., C. A. Moss, and M. A. Vater (editors). *Echolocation in Bats and Dolphins.* Chicago: University of Chicago Press, 2003.

Vaughan, T. A., J. M. Ryan, and N. J. Czaplewski. *Mammalogy, Fourth Edition.* Forth Worth, TX: Saunders College Publishing, 2000.

Walther, F. R. *Communication and Expression in Hoofed Mammals.* Bloomington, IN: Indiana University Press, 1984.

Periodicals

Deutsch, C. J., J. P. Reid, R. K. Bonde, D. E. Easton, H. I. Kochman, and T. J. O'Shea. "Seasonal Movements, Migratory Behavior, and Site Fidelity of West Indian Manatees along the Atlantic Coast of the United States." *Wildlife Monographs,* 151 (2003): 1–77.

Erickson, C. J. "Percussive Foraging in the Aye-aye, *Daubentonia madagascariensis.*" *Animal Behaviour,* 41 (1991): 793–801.

Scully, W. M. R., M. B. Fenton, and A. S. M. Saleuddin. "A Histological Examination of Holding Sacs and Scent Glandular Organs of Some Bats (Emballonuridae, Hipposideridae, Phyllostomidae, Vespertilionidae and Molossidae)." *Canadian Journal of Zoology,* 78 (2000): 613–623.

Voight, C. C. "Individual Variation in Perfume Blending in Male Greater Sac-winged Bats." *Animal Behaviour,* 63 (2002): 907–913.

Jeanette Thomas, PhD
M. B. Fenton, PhD

Life history and reproduction

Suckling as a defining feature

Patterns of reproduction are truly fundamental to mammal biology. This is at once apparent from the word mammal itself. In all species of the class Mammalia (monotremes, marsupials, and placentals), females suckle their offspring, and almost all of them have teats (mammae) to deliver the products gathered from the milk-generating glands. As defining features of the class Mammalia, mammary glands and milk production (lactation) are clearly central to mammalian evolution. Indeed, these features undoubtedly appeared at an earlier stage than the birth of live offspring (vivipary). Whereas marsupials and placentals give birth to live offspring after a period of development within the mother's body, monotremes (platypuses and echidnas) still lay large, yolk-rich eggs. Although the milk-generating glands of monotremes release their products through milk patches in the pouch rather than through a small number of teats, suckling of the offspring is clearly evident. Hence, suckling occurs in all extant mammals, and most species show a characteristic duration of this behavior (lactation period) as one of their reproductive hallmarks. Unfortunately for biologists concerned with the reconstruction of mammalian evolution, reproductive features are very rarely preserved in fossils. For this reason, the origin of mammals is defined for practical purposes by the emergence of a new jaw hinge between the dentary and the squamosal, replacing the original reptilian jaw hinge between the articular and the quadrate. Even in the dentition, however, there are features that reflect the mammalian pattern of reproduction. One defining dental feature of mammals is the presence of only two sets of teeth throughout life (diphyodonty), contrasting with the typical reptilian pattern of continuous, wave-like replacement of teeth (polyphyodonty). Young mammals usually have an initial set of deciduous teeth containing only incisors, canines, and premolars, which is replaced by a permanent set of teeth in which molars are also added. It is in itself revealing that the deciduous teeth of young mammals are referred to as "milk teeth", although the replacement of teeth may continue well after the end of the lactation period.

An intriguing question that arises is why lactation and suckling of offspring are consistently limited to female mammals. In principle, it should be possible for male mammals to produce milk as well and thus contribute directly to the survival of their offspring. This question is all the more appropriate because most male mammals (including the human male) have teats that serve no apparent function. As a rule, anatomical structures that have no function tend to disappear in the course of evolution, one striking example being the reduction and eventual loss of eyes in cave-living animals that live in the dark. Yet teats are so widespread among male mammals that it seems quite likely that male teats were present in the common ancestor of the marsupials and placentals. So why are they still present in most male mammals today, after some 150 million years of evolution? We are still awaiting a satisfactory answer to this enigma. All that can be said is that the initial structure that eventually gives rise to teats and (in females) to functional mammary glands appears very early in fetal development in both sexes, in the form of so-called milk lines, one on each side of the ventral surface of the body. Such early appearance in development is, in fact, a further indication of the basic importance of lactation and suckling in the evolution of mammals. But we still have no explanation for the fact that milk lines develop not only in the female fetus but also in the male.

The development of mammary glands is linked to another universal feature of mammals, namely the possession of hair and associated sweat glands, both developed in connection with the evolution of a relatively constant body temperature (homeothermy). It seems highly likely that mammary glands were derived from sweat glands through a secondary conversion that enabled them to produce a nutrient fluid instead of sweat.

Reproductive tract

Two key features in the initial development of the mammalian reproductive tract are crucial for understanding its evolution in both males and females. First, the system begins development as essentially separate left and right halves that are virtually mirror images of one another (bilateral symmetry). Second, there is a very close connection between the development of the reproductive tract and that of the kidneys and allied structures (renal system). According to species, these initial conditions are progressively reduced to varying extents in the course of development, but they provide im-

A sheep (*Ovis aries*) giving birth. (Photo by Animals Animals ©G. I. Bernard, OSF. Reproduced by permission.)

portant clues for distinguishing between primitive and advanced features.

The basic features of the female reproductive tract are common to all mammals. On each side of the body, there is an ovary that discharges the egg(s) into an oviduct, which leads to a uterus that is in turn connected with the vagina. Like other land-living vertebrates, all mammals have internal fertilization, which requires insertion of the male's erectile penis through the external opening (vulva) into the vagina. One variable feature of the vaginal region of the female reproductive tract in mammals concerns the relationship with the outlet of the urinary system (urethra) on either side of the body. In most mammals, the ureter enters the female tract at some distance from the vulva and there is a combined urinary and reproductive passage (urogenital sinus). This is the case for most small mammals, such as numerous marsupials, many insectivores, tree shrews, many rodents and carnivores. Perhaps the most spectacular case is the female elephant, which has a urogenital sinus that is close to 2 ft (61 cm) in length. In contrast, there is no urogenital sinus and the ureter has a separate opening adjacent to the vulval orifice in primates (including humans) and certain other mammals. Because a close connection between the urinary and reproductive systems is known to be primitive, the loss of the urogenital sinus is undoubtedly a secondary development. It should be noted, in-

cidentally, that in many mammals (e.g. monotremes, marsupials, and some placentals) the reproductive, urinary, and digestive tracts all open into a common structure known as the cloaca. Indeed, the word monotreme means "one-holed", referring to the fact that there is a single cloacal opening. In several groups of placental mammals, such as hoofed mammals and primates, the cloaca has been completely lost and there is a wide separation between the anus and the reproductive/urinary outlets.

The form of the uterus also shows substantial variation among mammals. In the widespread primitive condition, there are two separate uterine chambers (bicornuate uterus), reflecting the initial development of two completely separate reproductive tracts in the female. A bicornuate uterus is found in marsupials and most placentals, although there is variation in the degree of separation between the two uterine chambers. In contrast, some placental mammals have the two original uterine chambers completely fused to form a single midline structure (simplex uterus). This relatively unusual condition is found in simian primates (monkeys, apes, and humans), but not in prosimian primates (lemurs, lorises, and tarsiers), which have retained the primitive bicornuate condition. A single-chambered, simplex uterus is also found in some edentates (armadillos and sloths) and in a few bat species. Interestingly, several bats show conditions intermediate be-

A deer mouse (*Peromyscus maniculatus*) nursing its two-day-old young. (Photo by Joe & Carol McDonald. Bruce Coleman, Inc. Reproduced by permission.)

with seals, sea-cows, hippopotamus, whales, and dolphins) and heavily built pachyderms (elephant and rhinoceros). It has long been suspected that descent of the testes is in some way connected with avoidance of the relatively constant, elevated core body temperature that characterizes mammals. However, recent evidence suggests that it is not the actual production of sperm (spermatogenesis) that requires a lower temperature but rather sperm storage. For instance, in some mammal species that lack descent of the testis as such, the tail of the epididymis migrates to a position close to the ventral abdominal wall. This adaptation, which ensures a lower temperature for sperm storage but not for sperm production, is found, for example, in hyraxes and elephant shrews.

A further special feature of both male and female reproductive organs in many mammals is the presence of a baculum. This is commonly present both in the penis of the male (os penis) and in the clitoris of the female (os clitoridis), although the baculum is typically significantly larger in the male. Various mammals have secondarily lost the baculum. This is, for example, true of certain higher primates, including humans. The function of the baculum is still unclear, although there is some evidence that the os penis may play a role in stimulating the female during copulation. However, rather like teats in males, no function has been proposed for the os clitoridis in females.

Reproductive processes

The primary reproductive process in female mammals is the production of eggs (ova) from follicles in the ovary. In a non-pregnant female mammal, production of eggs is typically a cyclical process, although there are varying degrees of seasonal restriction such that some female mammals do not show repeated cycles. Seasonality of reproduction in mammals is mainly governed by annual variation in rainfall and vegetation, and hence becomes increasingly common at high latitudes. In many mammals, seasonality of reproduction is indirectly triggered by annual variation in day length, but in some mammals, it is a direct response to rainfall or food avail-

tween the typical bicornuate uterus and the advanced, single-chambered form, thus providing clues to the evolution of the simplex uterus. All placental mammals show some degree of midline fusion of the right and left female reproductive tracts in that there is a single vaginal passage (with or without a urogenital sinus). In marsupials, there are separate left and right vaginal passages and the penis is correspondingly bifid in males. This difference between marsupials and placental mammals may have arisen through a chance difference in development. In marsupials, the ureters pass between the paired vaginal tracts, thus preventing full midline fusion, whereas in placentals, the ureters are located lateral to the vagina and do not stand in the way of midline fusion.

In the male reproductive system, sperm are produced in the testis, stored in the epididymis and transported into the penis by the vas deferens. As a further consequence of the early close connection between the urinary and reproductive systems, the bladder and the vas deferens from each testis open into a common channel in the penis, which conveys both urine and seminal fluid to the outside world. An unusual phenomenon found in most marsupials and placental mammals (but not in monotremes) is descent of the testes into special scrotal sacs outside of the main abdominal cavity. Once again reflecting the close developmental connection between the urinary and reproductive systems, the testes initially develop close to the kidneys. In most mammals, they subsequently migrate into external scrotal sacs. In addition to monotremes, mammals that show no descent of the testes, or only partial migration within the abdominal cavity, notably include burrowing marsupials, insectivores and rodents, various aquatic mammals (certain marsupials, insectivores, and rodents, along

The African lion (*Panthera leo*) courtship ritual includes hitting, spitting, and roaring. (Photo by Jack Couffer. Bruce Coleman, Inc. Reproduced by permission.)

Galápagos fur seals (*Arctocephalus galapagoensis*) in their breeding territory in Ecuador. (Photo by Harald Schütz. Reproduced by permission.)

ability. The typical ovarian cycle of mammals begins when one or more follicles ripen to the point where the egg can be released (ovulation). Following ovulation, the residue of the follicle is converted into a corpus luteum (yellow body), which produces progesterone that maintains at least the early part of pregnancy. The basic stages of the ovarian cycle are common to all mammals, with a follicular phase preceding ovulation and a luteal phase afterwards. However, there is a fundamental difference between different mammal groups with respect to the occurrence of ovulation and changes in the ovary. In many mammals, ovulation occurs only if mating takes place (induced ovulation), and therefore formation of a corpus luteum also requires mating. Some species show a slightly different condition in which ovulation takes place without mating, but mating is necessary for formation of a corpus luteum (induced luteinization). In both cases, mating is required for a corpus luteum to form, such that without mating ovarian cycles are confined to follicular phases and are correspondingly short. By contrast, in other mammals both ovulation and formation of a corpus luteum occur regardless of whether mating takes place (spontaneous ovulation). In these species, cycles always include combined follicular and luteal phases and are correspondingly long. Induced ovulation and induced luteinization are found in mammals that breed relatively rapidly, as is the case with most insectivores, tree shrews, many rodents, and numerous carnivores. On the

other hand, spontaneous ovulation is typical of mammals characterized by slow breeding, such as hoofed mammals, cetaceans (whales and dolphins), hystricomorph rodents, and primates.

Following ovulation in marsupials and placentals, the egg travels down the oviduct, where fertilization will take place if the female has been inseminated. The fertilized egg (zygote) begins to divide as it completes its journey down the oviduct. By the time the zygote reaches the uterus, it has transformed into a hollow ball of cells (blastocyst). In placentals, the blastocyst is ready to implant in the wall of the uterus as the first stage in the development of placentation that will nourish the developing embryo/fetus. (By definition, a developing embryo becomes a fetus when recognizable organs are formed.) In both marsupials and placentals, development of the embryo/fetus within the uterus involves four embryonic membranes that play different roles. The chorion is the outermost membrane and remains intact throughout development right up to birth. Hence, any nutrients supplied by the mother to the developing offspring must first of all pass through the chorion. In all placental mammals, the chorion is in intimate contact with the wall of the uterus in the placenta. A second embryonic membrane, the amnion, surrounds the developing embryo/fetus throughout pregnancy and its fluid content (amniotic fluid) provides a protective hydrostatic cushion. The

A mother cheetah (*Acinonyx jubatus*) play fighting with cubs, in Kenya. (Photo by Joe McDonald. Bruce Coleman, Inc. Reproduced by permission.)

remaining two embryonic membranes, the yolk sac (vitelline sac) and the allantois, play a crucial role in transfer of nutrients from the mother and in transfer of waste products in the opposite direction across the placenta, to be disposed of by the mother. In the enclosed egg of reptiles, which has been retained by monotremes, the vitelline sac contains a nutrient-rich yolk that is absorbed by blood vessels running over the surface of the sac, while the allantois stores waste products deposited by similar superficial waste products. When a reptile or montreme emerges from the egg, the waste-filled allantois is shed. In the development of the embryo and fetus from the yolk-poor egg in marsupials and placentals, nutrients must be provided directly by the mother and waste products must be removed in some way. In a fascinating reversal of function, the superficial blood vessels of the yolk sac in marsupials and placentals absorb maternal nutrients arriving from outside rather than absorbing yolk from inside. In many cases, the superficial blood vessels of the allantois also absorb nutrients coming from the mother as a substitute for the original function of depositing waste products inside the sac. In accordance with the original functional adaptations, in marsupials and placentals the blood vessels of the yolk sac typically develop their exchange role first (chorio-vitelline placenta), while blood vessels of the allantois do so secondarily (chorio-allantoic placenta).

Because monotremes still lay eggs, they are commonly labeled Prototheria to distinguish them from the Theria (marsupials and placentals), which all have live births. Although the fertilized egg is retained within the mother's body for the initial phase of development in marsupials, the impression is often given that there is no placentation in marsupials. It is, indeed, true that in all marsupials a shell membrane is present over the chorion at least for the major part of pregnancy. Widespread use of the name "placental mammal" has unfortunately tended to reinforce the false impression that placentation is lacking in all marsupials. In fact, some form of placentation is developed in certain marsupials and a few of them, such as the bandicoot (*Perameles*), even develop a relatively advanced chorio-allantoic form of placentation. For this reason, many mammalogists prefer the terms metatherian for marsupials and eutherian for placentals, derived from the formal names Metatheria and Eutheria. The fact remains, how-

ever, that proper formation of a placenta is characteristic of all eutherians, whereas it has secondarily been developed only in some marsupials, so continued use of the easily understandable term "placental mammal" is surely acceptable.

In placental mammals, there is considerable variation in the form of the definitive chorio-allantoic placenta, although as a general rule each order, or at least suborder, of mammals tends to have a particular kind of placentation. Following Grosser (1909), a basic classification of types of placentation into three major categories, reflecting different degrees of invasiveness, can be made with respect to the relationship between the chorion and the inner wall of the uterus. In the least invasive type of placentation, the placenta is diffuse and the chorion is simply apposed to the inner epithelial lining of the uterus. It is labeled epitheliochorial placentation. In the other two kinds of placentation, invasion of the uterine wall occurs to some degree and the placenta is accordingly relatively localized (discoid). When moderate invasion occurs, the uterine wall is broken down in the region of the placenta and the chorion comes into contact with the walls of maternal blood vessels (endothelium). This type of placentation is called endotheliochorial. In the most invasive form of placentation, the walls of

Gray squirrel (*Sciurus carolinensis*) babies in their tree nest. (Photo by E. R. Degginger. Bruce Coleman, Inc. Reproduced by permission.)

Burchell's zebras (*Equus burchellii*) with their albino offspring. (Photo by Fritz Polking. Bruce Coleman, Inc. Reproduced by permission.)

resent divergent specializations away from a moderately invasive ancestral condition. It is noteworthy that during pregnancy, mammals with epitheliochorial placentation show great proliferation of uterine glands in the wall of the uterus. These uterine glands produce a nutrient secretion (so-called uterine milk) that is absorbed by special structures (chrionic vesicles) on the surface of the chorion. The selective advantages of the different basic types of placentation have yet to be identified. However, it is clear that the degree of invasiveness of the placentation has little to do with development of large-bodied offspring or of offspring with relatively large brains. Instead, it seems likely that the degree of invasiveness of the placenta reflects a trade-off between the advantages of an intimate placental connection between the mother and her developing offspring and the disadvantages of potential immunological conflict between the mother and her embryo/fetus.

The process of spermatogenesis typically takes place throughout the life span of male mammals, although it may be subject to periodic interruption in those species with a seasonal pattern of breeding. Spermatogenesis occurs as a wave-like process along the seminiferous tubules and completion of

the maternal blood vessels are themselves broken down in the region of the placenta, such that the chorion is directly bathed by maternal blood (haemochorial placentation). Epitheliochorial placentation is found in a few insectivores and it is uniformly characteristic of hoofed mammals, whales and dolphins, hyraxes, and strepsirrhine primates (lemurs and lorises). Endotheliochorial placentation is found in some insectivores, tree shrews, carnivores, sloths, anteaters, armadillos, elephants, and sea cows. Haemochorial placentation is found in many insectivores, rodents, bats, and haplorhine primates (tarsiers, monkeys, apes, and humans).

The evolutionary history of the three basic types of placentation is still subject to debate. It is often stated that the least invasive, epitheliochorial kind of placentation is the most primitive. This seems to be only logical, as the initial development of placentation must surely involve simple superficial contact between the chorion and the inner lining of the uterus. Because it is regarded as primitive, the epitheliochorial placenta is also often believed to be inefficient, notably with respect to development of the brain. By contrast, the highly invasive, haemochorial type of placentation is commonly thought to be very advanced and efficient. Human beings have the largest brain size (relative to body size) found among mammals and they also have highly invasive haemochorial placentation, so this is often seen as proof of the advanced nature of that very invasive type of placentation. However, many mammals with endotheliochorial or haemochorial placentation have relatively small brains, while dolphins, which have noninvasive epitheliochorial placentation, come a close second to humans with respect to relative brain size. In fact, there is much to be said for the alternative interpretation that ancestral placental mammals already had a moderately invasive type of placenta, following a long previous history of development. According to this view, noninvasive epitheliochorial and highly invasive haemochorial types of placentation rep-

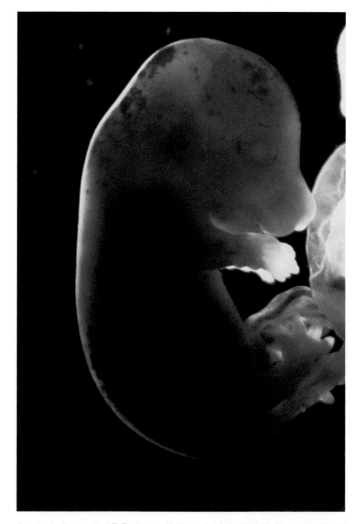

A rat embryo at 17.5 days. (Photo © Science Pictures Limited/ Corbis. Reproduced by permission.)

A gemsbok (*Oryx gazella*) with its nursing calf in Etosha National Park, Namibia. (Photo by Animals Animals ©Ana Laura Gonzalez. Reproduced by permission.)

sperm development in any one region takes between several days and a number of weeks. After transfer to the epididymis, the sperm are then stored until ejaculation takes place.

Gestation and neonate type

With only a few exceptions, each mammal species has a characteristic gestation period that shows remarkably little variation. In comparisons between species, gestation periods tend to increase as body size increases. However, effective comparisons of gestation periods among mammals must take into account a fundamental distinction in the state of the neonate at birth. As a general rule, it is possible to distinguish fairly clearly between mammals that give birth to several poorly developed (altricial) offspring and those that give birth to a few (usually just one) well-developed (precocial) offspring. Altricial offspring are largely hairless at birth and their eyes and ears are sealed with membranes. They are relatively helpless at birth and are typically deposited in a nest. By contrast, precocial offspring, which are usually born with a well-developed coat of hair and with their eyes and ears open, are typically able to move around quite actively at birth and are rarely kept in a nest. Other things being equal, it is obvious that the gestation period should be relatively longer for pre-

cocial offspring than for altricial offspring. In principle, it might be expected there would be a smooth continuum between altricial and precocial offspring. In practice, however, there is a fairly sharp division between them. When the relationship between gestation period and body size is examined for altricial and precocial mammals separately, it is found that there is a wide gap between them. At any given maternal body size, the gestation period for precocial offspring is about three times as long as that for altricial offspring. Furthermore, each main mammal group (order or suborder) is distinguished by the typical condition of the neonate and the relative length of the gestation period. Most insectivores, tree shrews, carnivores, and many rodents (myomorphs and sciuromorphs) give birth to altricial offspring after a relatively short gestation period, whereas hoofed mammals, hyraxes, elephants, cetaceans, pinnipeds, primates, and hystricomorph rodents give birth to precocial offspring after a relatively long gestation period. This is one of the few reproductive characters for which there is supporting evidence. Pregnant fossil horses from the Eocene and Miocene have consistently been found to have only one fetus, while an Eocene fossil bat has been found with twin fetuses. This shows that the small litter size of horses and bats, at least, have characterized those groups for at least 45 million years.

Opossum babies in their mother's pouch. (Photo by © Mary Ann McDonald/Corbis. Reproduced by permission.)

An inverse relationship between the average number of offspring produced at birth (litter size) and the gestation period is only to be expected. For a given uterus volume, there is clearly a trade-off between the number of developing offspring and the extent to which they can develop prior to birth. One corollary of this is that, for any given adult body size, altricial offspring must grow more after birth than precocial offspring.

Postnatal development

Across mammal species generally, there is a fairly consistent relationship between the average litter size and the typical number of teats possessed by the mother. As a rule, it can be said that there is one pair of teats for each offspring in the typical litter. However, suckling of the offspring is just one aspect of parental care in mammals. Maternal care, which can include nest building, grooming of the offspring, and infant carriage, is found in all mammals. Paternal care is relatively rare and is usually restricted to grooming and/or carriage. Predictably, paternal care in mammals is usually restricted to monogamous species in which there is a relatively high level of certainty of paternity.

Once the effects of body size have been taken into account, it emerges that the pattern of maternal care for any mammal species is quite closely reflected in milk composition. Three principal components of mammalian milk are carbohydrates, fats, and proteins. As a crude approximation, it can be said that the carbohydrate content of milk reflects immediate energy needs of the offspring, while the fat content indicates energy needs over a longer term. The protein content of milk provides a fairly good indication of requirements for growth. Milk composition also provides an indication of maternal behavior. Here, a major distinction can be drawn between mothers that feed on schedule and those that feed on demand. For offspring that are fed on schedule, it is the mother that determines the suckling frequency. Commonly, this applies to offspring that are left in a nest. These tend to be fast growing but relatively inactive altricial offspring that must maintain their body temperature unaided in the absence of the mother. As a result, the milk of such species tends to be high in protein and fat but relatively low in sugar. The most extreme example of suckling on schedule known for mammals is found in certain tree shrews species that keep their offspring in a separate nest and suckle them only once every 48 hours. The milk of these tree shrews is extremely concentrated, containing more than 20% fat and 10% protein. By contrast, in mammals that feed on demand, it is the offspring (usually a singleton) that determines suckling frequency. Usually, this requires close proximity between the offspring and its mother, so that it can signal its intention to suckle. Suckling on de-

ative to body size, than species with monogamous or polygynous systems. This indicates that males show increased levels of sperm production in cases where sperm competition is relatively intense.

Sexual dimorphism

Male and female mammals obviously differ in various features that are directly linked to reproduction, as is the case with the sex organs of both sexes and the mammary glands of females (primary sexual characteristics). However, males and females can also differ in a variety of features that are not directly associated with reproduction (secondary sexual characteristics). Such secondary differences between males and females, like the human facial beard, are collectively labeled sexual dimorphism. The simplest form of sexual dimorphism distinguishing male and female mammals involves overall adult body size. As a general rule in mammals, males tend to be bigger than females, but there are some cases in which females are bigger than males (reverse sexual dimorphism). Differences in body size between the sexes are often relatively mild, as is the case in humans, where adult males are about 20% heavier than females; but there are also some striking contrasts. In the most extreme case of dimorphism in body size found among mammals, namely in the elephant seal (*Mirounga*), adult males are about four times heavier than females (8,000 lb [3,629 kg] compared to 2,000 lb [907 kg]). Sexual dimorphism in mammals can also affect other features, notably involving differences in external appearance (e.g. coat coloration), the size of the canine teeth and special appendages such as the antlers of deer (Cervidae). Overall, there seems to be a general tendency for the degree of sexual dimorphism in size of the body or its appendages to increase with increasing body size (Rensch's Rule), although the validity of this generalization has been questioned. An additional generalization that can be made is that sexual dimorphism in body size, canine size, and the size of such appendages as horns is generally lacking from species with a monogamous pattern of social organization. However, this does not apply to sexual dimorphism in coat coloration, as is shown by striking differences in external appearance between males and females in certain species of monogamous gibbons and lemurs.

The baseline expectation for mammals is that males and females will be similar in size and other features unless some special selective factor intervenes. However, there is no real reason why males and females should be similar in the size of the body and its external appearance or appendages. Given the major inequality in contribution to reproduction that characterizes all mammals, because gestation and lactation are exclusive to females, the baseline expectation should surely be

that male and female strategies are quite likely to diverge. We should be more surprised by the numerous cases in which males and females are very similar in size and appearance than we are by sexual dimorphism. The standard explanation for sexual dimorphism in mammals is that selection acts on the male to increase the size of the body or its appendages because of competition for mating access to females (sexual selection). In elephant seals, for example, the big bull males fight one another to establish mating territories and maintain harems that may contain three dozen females. The large body size of males and their large canine teeth are therefore reasonably interpreted as features that increase male success in competition for females. A similar explanation is provided for the development of large antlers in male deer. Among primates, this interpretation is also applied to various species that show conspicuous sexual dimorphism. A prime example is the mandrill (*Mandrillus sphinx*), in which males are more than twice as heavy as females, vividly colored and equipped with very prominent canine teeth.

There has been a general tendency to overlook the potential part played by selection on females in the evolution of sexual dimorphism in mammals. In the first place, it is obvious that a single unitary explanation for sexual dimorphism, such as improved competitive ability of males, is inadequate because the different kinds of sexual dimorphism (e.g. body size, canine size, and coloration) can vary independently to a large degree. In primates, for example, it is possible to find marked dimorphism in coat coloration and mild variation in canine size without any matching difference in body size. Furthermore, most lemur species lack any kind of sexual dimorphism despite sometimes fierce competition among males for mating access to females. When males and females of a species differ in body size, it is commonly assumed that selection has operated to increase male body size, but it should be remembered that selection might also act to reduce (or increase) female body size. Because many reproductive features are scaled to body size (e.g. neonate size and age at sexual reproduction), one effect of reduction of female body size will be to decrease nutritional requirements for reproduction and increase the rate of reproductive turnover. In principle, sexual dimorphism between males and females in adult body size could be achieved by an increased rate of growth in males, by an extension of the growth period in males, or by some combination of these two possibilities. In practice, sexual dimorphism in mammalian body size is always associated with at least some delay in the attainment of sexual maturity in males relative to females, so there are consequences for reproductive dynamics in every case. Hence, it seems likely that sexual dimorphism reflects the effects of diverging selection pressures operating on both sexes.

Resources

Books

Austin, Colin R., and Roger V. Short. *Reproduction in Mammals. Book VI: The Evolution of Reproduction*. Cambridge: Cambridge University Press, 1976.

Austin, Colin R., and Roger V. Short. *Reproduction in Mammals. Book I: Germ Cells and Fertilization*. Cambridge: Cambridge University Press, 1982.

Bronson, Frank H. *Mammalian Reproductive Biology*. Chicago: University of Chicago Press, 1989.

Resources

Hayssen, V., and A. Van Tienhoven. *Asdell's Patterns of Mammalian Reproduction: A Compendium of Species-Specific Data.* Ithaca, New York: Comstock Publishing Associates, Cornell University Press, 1993.

Johnson, Martin H., and Barry J. Everitt. *Essential Reproduction.* Oxford: Blackwell Scientific Publications, 1980.

Martin, Robert D., Lesley A. Willner, and Andrea Dettling. "The evolution of sexual size dimorphism in primates." In *The Differences between the Sexes,* edited by Roger V. Short and Evan Balaban. Cambridge: Cambridge University Press, 1994, pp. 159–200.

Mossman, Harland W. *Vertebrate Fetal Membranes.* New Brunswick, New Jersey: Macmillan/Rutgers University Press, 1987.

Perry, John S. *The Ovarian Cycle of Mammals.* Edinburgh: Oliver & Boyd, 1971.

Stearns, Steven C. *The Evolution of Life Histories.* Oxford: Oxford University Press, 1992.

Steven, D. H. *Comparative Placentation: Essays in Structure and Function.* London: Academic Press, 1975.

Periodicals

Ben Shaul, D. M. "The composition of the milk of wild animals." *International Zoology Yearbook* 4 (1962): 333–342.

Charnov, E. L. "Evolution of life history variation among female mammals." *Proceedings of the National Academy of Sciences of the United States of America–Biological Sciences* 88 (1991): 1134–1137.

Cowles, R. B. "The evolutionary significance of the scrotum." *Evolution* 12 (1958): 417–418.

Daly, M. "Why don't male mammals lactate?" *Journal of Theoretical Biology* 78 (1979): 325–345.

Derrickson, E. M. "Comparative reproductive strategies of altricial and precocial eutherian mammals." *Functional Ecology* 6 (1992): 57–65.

Fenchel, T. "Intrinsic rate of natural increase: The relationship with body size." *Oecologia* 14 (1974): 317–326.

Harvey, Paul H., and Zammuto, R. M. "Patterns of mortality and age at first reproduction in natural populations of mammals." *Nature* (London) 315 (1985): 319–320.

Kenagy, G. J., and S. C. Trombulak. "Size and function of mammalian testes in relation to body size." *Journal of Mammalogy* 67 (1986): 1–22.

Kleiman, Devra G. "Monogamy in mammals." *Quarterly Review of Biology* 52 (1977): 39–69.

Long, C. A. "Two hypotheses on the origin of lactation." *American Naturalist* 106 (1972): 141–144.

Martin, Robert D. "Scaling effects and adaptive strategies in mammalian lactation." *Symposium of of the Zoological Society of London* 51 (1984): 87–117.

Martin, Robert D., and Ann M. MacLarnon. "Gestation period, neonatal size and maternal investment in placental mammals." *Nature* (London) 313 (1985): 220–223.

Møller, Anders P. "Ejaculate quality, testes size and sperm production in mammals." *Functional Ecology* 3 (1989): 91–96.

Oftedal, O. T. "Milk composition, milk yield, and energy output at peak lactation: a comparative review." *Symposium of the Zoological Society of London* 51 (1989): 33–85.

Pianka, E. R. "On r- and K-selection." *American Naturalist* 104 (1970): 592–597.

Pond, Caroline M. "The signficance of lactation in the evolution of mammals." *Evolution* 31 (1977): 177–199.

Ralls, K. "Mammals in which females are larger than males." *Quarterly Review of Biology* 51 (1976): 245–276.

Read, A. F., and Paul H. Harvey. "Life history differences among the eutherian radiations." *Journal of Zoology, London* 219 (1989): 329–353.

Rose, R. W., Claire M. Nevison, and Alan F. Dixson. "Testes weight, body weight and mating systems in marsupials and monotremes." *Journal of Zoology, London* 243 (1997): 523–531.

Sacher, George A., and E. Staffeldt. "Relation of gestation time to brain weight for placental mammals." *American Naturalist* 108 (1974): 593–615.

Schaffer, William M. "Optimal reproductive effort in fluctuating environments." *American Naturalist* 108 (1974): 783–798.

Western, David. "Size, life history and ecology in mammals." *African Journal of Ecology* 17 (1979): 185–204.

Robert D. Martin, PhD

Reproductive processes

Mammalian reproduction

Reproduction is pivotal to the continuation of life. From an evolutionary standpoint, there is no single factor that has more impact on the development of species. The impetus to reproduce shapes morphology, physiology, life history, and behavior of all animals, mammals included. From the egg-laying platypus (*Ornithorhynchus anatinus*) to the wildebeests (genus *Connochaetes*) that have neonates that can run mere seconds after birth, a wide variety of strategies have evolved to successfully bear offspring in a multitude of environments.

Fundamentals of mammalian reproduction

Mammals reproduce sexually, and both sexes must unite to conceive offspring. The physical contact of both sexes does not constitute reproduction, but instead, it is the union of the sexual cells or gametes produced by each sex that constitutes the first step to reproduction. In females, the gamete is the egg or ovum. In males, the gamete is sperm. Each gamete contains one copy of each of the parental chromosomes, and thus, when the gametes unite, they form a zygote, the first complete cell of a new animal. Division of this first cell will result in the development of a full-grown animal. But before the two gametes can be united, several events must occur, all contributing to the phenomenon of mammalian reproduction. First, animals must find and choose mates. This very basic step will allow for a variety of adaptations to conquer, convince, attract, or seduce mates of the other sex. Second, mammals must get their gametes together, and this will be discussed under copulation and fertilization. Third, offspring growth and development to adulthood will be discussed under ontogeny and development. Because several aspects of development will be affected by the role of the two sexes, the importance and implications of mating systems will be discussed as well as strategies of reproduction and associated life history. Finally, some of the peculiar reproductive strategies present in mammals, and their role in the evolution and development of mammalian reproductive processes will be explored.

Mate choice

For a mammal to reproduce it must unite its gametes with the gametes of a member of the other sex. Because each ga-

mete provides half of the genetic material of the offspring, the choice of mate has direct implications on the resulting genotype (genetic makeup) of the offspring. For this reason, animals do not mate randomly and instead choose mates. Mate choice is among the most important pressures affecting the evolution of species because failure to be able to select a "good" mate results in poor offspring quality (offspring that may be less adept to survive or reproduce), or worse yet, no offspring. Animals that fail to reproduce disappear from the gene pool, so mate choice is a critical factor.

In mammals, both sexes produce gametes of different size. Females produce relatively large eggs, and, typically, in limited number. In contrast, males produce tiny, cheap (from an energy standpoint), and extremely abundant sperm. Thus, from the outset, females adopt a "quality" strategy whereas males adopt a "quantity" strategy. This very basic difference in the size of gametes and parental investment in gamete production will trickle down and affect almost all subsequent reproductive processes. The larger initial investment of females will also result in females taking care of growing embryo(s). Production of offspring in mammals involves placental growth (monotremes and marsupials are exceptions—see below), and this is performed by females. Once born, young are nursed with milk, also produced by the mother. So although the genetic contribution of each sex in the offspring is equal, the investment of females in offspring is greater. Thus, females have more to lose from bad mating decisions. This asymmetry in investment between sexes will be reflected throughout most sexual adaptations, and will lead to females being almost invariably the most selective in mate choice.

Sexual selection and the evolution of species and their attributes

The pressures caused by females choosing males will lead to two types of evolutionary selection: inter-sexual selection (adaptations to win members of the other sex), and intra-sexual selection (adaptations to win access to mates over members of the same sex). Both vary in importance according to species and environments.

Inter-sexual selection leads to the development of adaptations, morphological, physiological, or behavioral, to seduce

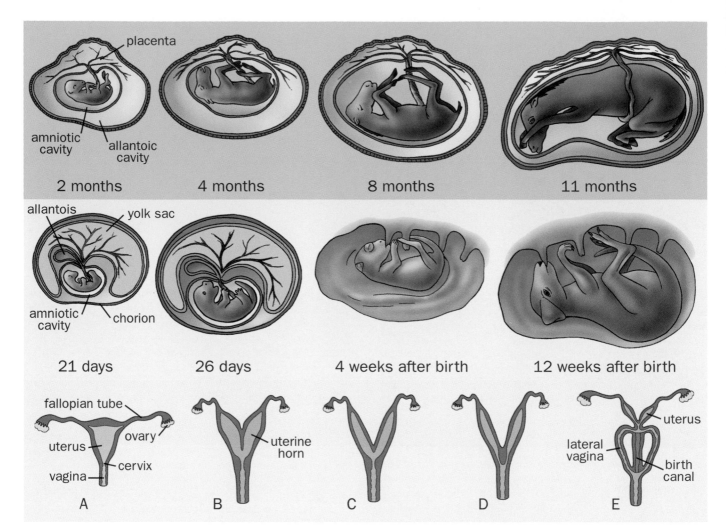

Top: Placental mammal development. Middle row: Marsupial mammal development. Types of uterus: A. Simplex; B. Bipartite; C. Bicornuate; D. Duplex; E. Marsupial. (Illustration by Patricia Ferrer)

mates. Typically, females are choosers, so most of the inter-sexual selection targets males. Adaptations resulting from inter-sexual selection are meant to reveal genetic quality, so males will harbor features, or perform behaviors that indicate quality. Examples of inter-sexual selection are numerous in mammals, and the best known morphological examples are probably the growth of horns in ungulates such as giraffe (*Giraffa camelopardalis*), or bighorn sheep (*Ovis canadensis*).

In contrast to inter-sexual selection, which arises from pressures imposed by the other sex, intra-sexual selection leads to adaptations as a result of pressures imposed by the same sex. Displays of strength or resources may serve to indicate dominance and establish hierarchy among males, allowing better individuals to access more mates. Best examples in the mammals are the sparring competitions of ungulates such as deer (genus *Odocoileus*), or the banging of heads in musk oxen (*Ovibos moschatus*). In these examples, some of the morphological attributes such as large antlers may be used for both inter-sexual selection (seduction of mates), intra-sexual selection (indication of dominance), and even individual selection (defense against predators). But multiple functions do

not lessen their attractiveness as large antlers indicate that the males that harbor them are able to find resources to grow and carry heavy antlers, and thus indicate that they are in good health and have good genes. The extinct Irish elk possessed the largest antlers ever. They were up to 6 ft (1.8 m) in length. However, the purpose of these was not to fight. It is unlikely they could be used for this purpose as they were too big. These massive antlers seem to have evolved because of runaway sexual selection. Females prefered males with larger and larger antlers, so the antlers got bigger and bigger.

Sperm and egg formation

Sperm cells are made in the testes of males. Through a process of cellular division called meiosis, sperm-producing cells with regular genetic material (diploid cells, meaning they posses two copies of each chromosome) undergo division with the end product being two cells each with only one copy of each chromosome (haploid, half of the parent cell). Because sperm production is optimal at temperatures slightly colder than average body temperature, testes are housed in a pouch

These giraffes illustrate flehmen behavior through which the male assesses the sexual receptivity of the female by sniffing her urine and allowing it to run over the Jacobson's organ in the roof of the mouth. (Photo by Rudi van Aarde. Reproduced by permission.)

of skin just outside the body, the scrotum. But not all species have scrotal testes year-round, and some species such as bats have testes that are kept internal for most of the year and become external prior to breeding. Other species such as aquatic cetaceans (whales and dolphins) or sirenians (dugong and manatees), as well as terrestrial armadillos, elephants, and sloths retain internal testes even during reproductive season; interestingly, sperm production readily occurs at normal body temperatures in these species.

Once formed, the sperm cells are stored in the epididymis where they remain until sperm is ejaculated. Not surprisingly, the mating system of species is often correlated with the size of testes, and many species with promiscuous mating systems will have larger testes simply because copulations are more frequent and males require more sperm, for example lions. Sperm production in humans is constant throughout reproductive life, but in seasonally breeding animals it only occurs immediately prior to the reproduction period as sperm made too far in advance degrades with age. Because millions of sperm are released with each ejaculate, the number of sperm produced by a male during a lifetime is astronomically high.

In females, the process is much different. At birth, females already possess in their ovaries all the eggs they will ever produce. Eggs are stored as follicles, and they will start developing into fully functional eggs once sexual maturity is reached. As follicles mature, they expand in size on the surface of the ovary until ovulation is triggered by release of luteinizing hormone. Peaks of this hormone occur either as part of the estrous cycle or are triggered by physical stimuli in induced ovulators such as cats (Felidae) and rabbits (lagomorphs).

Copulation and fertilization

Once eggs are released from the female ovaries, they migrate down into the uterus. Eggs are not self-propelled, and migration occurs passively by gliding over cells that have minuscule sweepers (cilia). In contrast, sperm cells each have a long flagellum that provides mobility. But for sperm to reach the egg or eggs, copulation must first occur.

Copulation in most terrestrial species occurs as the male straddles the female from behind. Typical examples of this type of copulation occur in deer, elephants, mice, and cats. Most animals remain in this position, but some, especially dogs (Canidae), may then turn 180 degrees and continue a prolonged copulation in a copulatory lock, where the penis points 180 degrees away from the head, and both animals face in the opposite direction. Another situation occurs in Cetaceans (whales and dolphins) where copulation occurs as

Two bison (*Bison bison*) bulls sparring, a common play-fight among young bulls in which there are neither winners nor losers. (Photo by © Wally Eberhart/Visuals Unlimited, Inc. Reproduced by permission.)

both animals lay or swim belly to belly. In primates, including humans, copulation positions are more flexible but most often consist of the ventral-dorsal mount or the ventral-ventral position.

The position assumed during copulation helps transfer the gametes from the male to the female tract, and physical constraints of size and position necessitate the use of a gamete-transfer organ, the penis. Depending on challenges faced by each species, characteristics of the penis such as size, length, or structure will vary. But all penes have the same function: to facilitate the transfer of the sperm cells from the male body to the female reproductive tract. Unlike fishes, which can release their sperm in water, mammalian sperm loses its mobility when exposed to air and internal copulation and fertilization maximizes the chances of successful transfer of viable sperm.

Both sexes have evolved adaptations to facilitate every step of reproduction, and copulation is no different. For example, females that come into heat will often produce thick vaginal secretions that not only attract males, but that also serve to lubricate the reproductive tract in preparation for copulation. Males also produce a lubricant via their bulbo-urethral glands to help facilitate copulation. In some species, including humans, the vaginal tract is highly acidic to serve

as a barrier against diseases or microbial infections that may occur in or on the reproductive organs of males. To neutralize the acidity of the female tract that could damage or affect the motility of sperm cells, males have a prostate gland that secretes an alkali buffer to bring the pH of the female tract closer to neutral (pH = 7) so that sperm mobility and survival is optimized.

Once sperm cells reach the egg, a single sperm cell fuses with the egg cell. An enzyme then allows penetration of the genetic material of the sperm into the egg cell. Once this is done, the membrane of the egg becomes sperm proof to prevent inclusion of additional genetic material that would unbalance the process and possibly render the zygote non-viable.

Although this process seems simple, males release millions of sperm in each ejaculation so obviously not all sperm reach eggs. Sperm mobility may be affected by external conditions (such as acidity in the female tract), but also by their age, and older sperm become less mobile as they age. Strength and mobility of sperm cells is crucial, especially if females copulate with numerous males, and in these cases, sperm of multiple males may be present at the same time, forcing males to compete again for the available eggs, this time via sperm wars within the female's reproductive tract.

After the koala (*Phascolarctos cinereus*) leaves its mother's pouch at seven months, it stays on her back for another four months. (Photo by Tom McHugh/Photo Researchers, Inc. Reproduced by permission.)

Sperm competition

Gametes evolved to different sizes. Females by definition are the sex that produces larger gametes. Once they've deposited their smaller gametes, males are in the advantageous position to limit energetic input (e.g., leave). The females are then left with the decision to raise offspring or not, and this decision has important energetic implications. But males also have one evolutionary uncertainty to overcome: the certainty of paternity. If males release millions of sperm and can father multiple litters within a single reproductive season (by breeding with several females), the certainty of paternity is seldom assured, and the uncertainty of paternity may well be the greatest challenge that mammalian males face when it comes to reproduction. Not surprisingly, a myriad of adaptations has evolved to overcome this uncertainty—from mate guarding and mate defense to strategies inside the reproductive tract such as sperm competition. Also, if there are fights to prevent other males from mating, these males will fight to overcome barriers put in place by previous males. One could argue that when males fight, females ultimately win because whichever male succeeds probably has better genes or the kind of genes

that will enable her male progeny (i.e., son) to produce more children (i.e., increased fitness).

Mate guarding and defense following mating is a simple way for a male to reduce the odds of another male copulating with a female. However, the trade-off is obvious: staying with one mate precludes males from courting others, and strategies that allow males to protect their paternity without being present would yield great advantages. Sperm competition is one such process that can be simply summarized as any event that leads to sperm of two or more males being present at once inside the reproductive tract of a female. Males that release seminal fluids with greatest number of sperm, and sperm with the greatest mobility are thus more likely to fertilize female eggs.

Other strategies also exist for males to ensure paternity. In some species of primates such as the Senegal bush baby (*Galago senegalensis*) or ring-tailed lemur (*Lemur catta*), males have a penis that is highly spinuous, and the function of these spines is to alter the reproductive tract of the mated female so that she become less receptive to subsequent mating from other males. In carnivores such as wolverines (*Gulo gulo*) or American mink (*Mustela vison*), the penis bone may also play a role in causing enough stimulation for females to abort the first set of fertilized eggs, thus allowing males with larger penis bones to father more offspring. This also would allow females to compare male quality via the size of males' penis bones inside the reproductive tract instead of by classic displays.

In many species of rodents such as brown rats (*Rattus norvegicus*), primates, or bats, some of the seminal fluids will form a copulatory plug. This plug is formed soon after copulation and it appears that its main function is to prevent leakage of sperm from the female reproductive tract following

A Dall's sheep ram (*Ovis dalli*) in the November rut season in a urine-testing pose, testing the female's urine for signs of estrus. (Photo by © Hugh Rose/Visuals Unlimited, Inc. Reproduced by permission.)

Florida manatee mother nursing her calf in Crystal River, Florida, USA. Photo by Animals Animals ©Franklin J. Viola. Reproduced by permission.)

copulation. The longer the sperm stays inside the female, the better the odds of fertilization by maximizing the amount of sperm and hence number of sperm cells active in the female tract. Interestingly, the tip of the penis (the glans) is used by males to remove sperm plugs deposited by other males.

Ontogeny and development

Mammalian ontogeny and development can, from a physiological standpoint, be separated into three general strategies. First, the monotremes such as platypus and echidnas are oviparous, meaning that they conceive young via copulation, but give birth to young inside an eggshell. After a short period of development in the egg, young hatch and then suckle the mother's milk as it leaks into the fur and not through a nipple as monotremes do not have nipples.

In contrast, marsupials and placental mammals are viviparous, meaning they give birth to live young. But their respective strategies differ, especially with regards to level of development of the young at birth. Marsupials give birth to live young, but they have a short gestation and produce young that are extremely altricial, i.e., very early in their development stage. Marsupial offspring are born blind and naked, and with underdeveloped organs except for a pair of extremely well-developed and strong front limbs. The young are also minuscule in size compared to adults. They spend most of their growth phase outside of the female reproductive tract but securely attached to the maternal nipple for nourishment, often, but not always, in a pouch. Kangaroos for example give birth to tiny young which then crawl to the pouch (with the strong front limbs) and find a nipple. Attachment to the nipple ensures that the blind and naked

young does not fall off the mother and always remains close to its source of nourishment.

Placental mammals constitute the largest group of mammals. In these species, which includes, cats, dogs, horses, bats, rats and humans, fertilized ova migrate to the uterus where they implant and fuse with the lining of the uterus called endometrium, which then leads to the creation of a placenta, a highly vascular membrane that acts as the exchange barrier between embryo(s) and mother. Young develop inside the female tract to varying degrees, but even the most altricial of placental mammals (polar bears *Ursus maritimus*, for example) still are more developed at birth than marsupials. Internal development can be extremely advanced and lead to birth of young that are able to stand and run almost immediately after birth. Wildebeests, elephants and guinea pigs all have precocial young (offspring born fully developed) in this category.

Milk and lactation

At birth, the young no longer can rely on the direct exchange of nutrients through the placenta (or in monotremes, through nutrient stored in the egg). Thus, nutrition of young requires an additional process, and milk is the nutrient that serves that purpose. Milk is unique to mammals, and all species of mammals are capable of producing milk. Milk production occurs in the mammary glands, which resemble sweat glands in form but become mature only following parturition or birth of young. The milk is delivered through nipples or teats (except in monotremes), and typically the number of teats is roughly twice the average litter size. Although males have fewer and smaller teats that are vestigial, only females produce milk and consequently they have larger teats. All rules have their exception and in mammals, there is one species in which males can also produce milk and nurse young, the Dyak fruit bat (*Dyacopterus spadiceus*).

A female cape fur seal (*Arctocephalus pusillus pusillus*) reacting to a male. (Photo by David Dennis/Nature Portfolio. Reproduced by permission.)

A short-beaked echidna (*Tachyglossus aculeatus*) hatching in the pouch of the mother. (Photo by McKelvey/Rismiller. Reproduced by permission.)

Sexual maturity

Sexual maturity in many species occurs when body size reaches adult size. However, there are some notable exceptions: male least weasels (*Mustela nivalis*) often seek maternity dens of females and will copulate the newly born females, as soon as 4 hours after birth. At that time, neonates still have their eyes and ears closed, are pink and hairless. This strategy enables females to have a first litter within weeks of birth (least weasels do not exhibit delayed implantation), and then again before the end of their first year. Another example of early sexual maturity is in musk shrews (*Suncus murinus*) in which mating and repeated ejaculations from males induce puberty and ovulation in virgin females.

Mating systems

Depending on the environment, mammals will adopt various mating strategies. In some species such as beaver (genus *Castor*), the maintenance of a pond, lodge, and dam to maintain water level and insure security of the offspring requires the efforts of both parents, and thus both sexes must combine efforts to raise young to adulthood successfully. In such species, the sexes not only combine gametes, but also efforts and they remain together throughout the mating season, or for life. The term monogamy describes such systems, where animals remain with one mate either annually or permanently. Only 3% of mammals are monogamous but monogamy is found within nearly all mammalian orders and predominates in some families, for example in foxes, wild dogs and gibbons. Some examples of seasonal monogamy would be in red foxes

(*Vulpes vulpes*), in which males provide parental care, but where couples break-up after rearing of young. Permanent monogamy would be best exemplified in North American beavers that mate for life (*Castor canadensis*). Generally speaking, monogamy occurs in species where the support from the males is instrumental to the rearing of young, and males gain more from limiting the number of offspring (by staying with one female) and investing instead in the growth of their offspring. However, many so-called monogamous males will opportunistically attempt breeding with other females, paired or not, to increase their genetic fitness, and true monogamy occurs rarely when lack of potential mates occurs, or under pressures from the environment.

Polygynous mating systems occur when males do not provide paternal care, and hence pursue matings with numerous females. Such are probably best exemplified in ungulates and pinnipeds, where males maintain access to several females simultaneously. In these harems, males try to control female breeding by asserting their dominance over other males, and if successful, winning males sire offspring from several females, whereas females only breed with a single male.

Not all females accept a single male, and in some species, both males and females will mate with numerous individuals. Promiscuity describes such a system, and is probably best exemplified in wide-ranging carnivores such as mink or wolverines. In these species, females cannot compare mates because they are spatially widely scattered, so females may mate with numerous individuals, and rely on other mechanisms to choose mates such as sperm competition. Promiscuous females also occur in certain social structures where uncertainty of paternity in males prevents them from killing the offspring of the female (infanticide). Such a social structure occurs in prides of lions (*Panthera leo*).

Finally, a mating system exists where male alliances may form to care for the offspring and allow females to spend most of their energy producing, and not caring for, offspring. This system, called polyandry, may occur in mammals under extremely biased sex-ratio in adulthood where males are extremely abundant, and females extremely rare, as for example with the African wild dog or the naked mole rat, with one "queen" and all else workers. This scenario is much less common but occurs in some human cultures.

Life history strategies

Just like gamete production, offspring production can follow the same two strategies: quantity or quality. The quality strategy is best exemplified in African elephants (*Loxodonta africana*), the largest living land animal. African elephants produce one young, rarely two, after a gestation of 22 months. Young are born precocial, and can stand up and follow the mother within 15–30 minutes. They are nursed for 2–3 years (sometimes up to 9 years), and reach sexual maturity at 8–13 years of age. In their lifetime of 55–60 years, female elephants average four calves, with a range of one to nine.

In contrast to the "quality" strategy of elephants, numerous rodents and lagomorphs have multiple litters each year, each with numerous young, and spend little time investing in

An African elephant (*Loxodonta africana*) bull in "musth," a state of sexual excitement characterized by aggressive posturing, continuous dribbling of urine, and secretions running from the temporal glands. (Photo by Rudi van Aarde. Reproduced by permission.)

offspring growth (quantity strategy). Obviously, a continuum exists between the two extremes and typically these life history strategies are influenced by the size and life span of the animal and the environment. For example, small rodents grow faster, and live shorter lives, and thus invest in "faster" reproduction such as earlier age at maturity, and smaller but more numerous neonates. In contrast, larger mammals such as ungulates, elephants, and whales have relatively larger but fewer neonates, and attain sexual maturity much later. The latter species are typically longer-lived, and thus spread their reproductive efforts over a longer time span than smaller mammals.

Reproduction in monotremes

Three species of mammals differ completely from the more common placental animals: the short-beaked echidna (*Tachyglossus aculeatus*), the long-beaked echidna (*Zaglossus bruijni*), and the duck-billed platypus (*Ornithorhynchus anatinus*). Monotremes differ from placental mammals mostly because development of the offspring occurs outside of the female re-

productive tract. Monotreme females conceive young via copulation, but fertilized ova then move to a cloaca (a common opening of urinary and reproductive tract) where they are coated with albumen and a shell, and eggs are laid after 12–20 days. Female echidnas carry the egg into a pouch whereas platypus lay the eggs into a nest of grass. At hatching, young echidnas remain in the pouch and subsist on the milk that drips from the fur (monotremes lack nipples). Once large enough, young are deposited in a nest where they are nursed until weaning. In contrast, female platypus lay their eggs in a grass nest inside a burrow, incubate them until hatching 10 days later, and then nurse the young in the den. Young emerge from the burrow when fully furred and approximately 12 in (30 cm) in length.

Reproduction in marsupials

In marsupials, ova are shed by both ovaries into a double-horned or bicornate uterus. The developing embryos remain in the uterus for 12–28 days, and most of the nourishment comes from an energy sac attached to the egg (yolk sac). There is no placenta (except for one groups of marsupials, the bandicoots, that have an interchange surface that resembles a true placenta). Gestation is thus short (less than one month), and much of the development of the young will occur outside of the female reproductive tract.

At birth, the offspring are extremely altricial (poorly developed). In many marsupials such as kangaroos, offspring will migrate to the nipples where they attach. This will ensure that they remain in contact with their nourishment sources, and also probably serve to secure the young and prevent them from falling off the mother at an age when they do not have the strength to hold on by themselves. In red kangaroo (*Macropus rufus*), the largest of all marsupials, young climb unaided to the pouch within a few minutes of birth, remain on the nipple for 70 days, protrude from the pouch at 150 days, emerge on occasion at 190 days, and permanently leave the pouch at 235 days, but is fully weaned only after a year.

Marsupials have the tiniest young in relation to adult size. In the eastern gray kangaroo (*Macropus giganteus*), young at birth weigh less than 0.0001% of the female mass. Put into context, this would be similar to a 150 lb (68 kg) human female giving birth to individual babies that would each weigh 0.0048 lbs (22 g), or 0.08 oz. But extreme altriciality is not a disadvantage in evolutionary terms. In fact, many scientists believe that this is instead a great advantage as the small investment in each neonate allows females to minimize investment in young and be more flexible and responsive to environmental conditions. Mechanistically, if environmental conditions become too tough to raise young successfully, starvation would terminate the production of milk and lead to rapid death of young, thereby saving energy lost (versus placental mammals that have a greater energy investment). This would give marsupial females a competitive advantage over animals with internal pregnancy (placental mammals) in unpredictable environments.

Some marsupials such as the eastern gray kangaroo also display other reproductive oddities. Pregnancy following copulation does not affect the cycle and the female can become pregnant again as her first young (litter size usually is one) move to the pouch. The second fertilized egg undergoes diapause (halt in development) until the first young either reaches adulthood or dies. In this case, the diapause is facultative or changes length depending on circumstances such as food availability or season. Then, the second egg immediately resumes development so that birth occurs as soon as the mother's pouch is available. So at any one time, the female can have three young: one in the placenta, one in the pouch, and one joey out of the pouch and still suckling.

Placental mammals

Placental mammals constitute the largest group of mammals. In placental mammals, fertilized eggs migrate to the uterus or to the uterine horns where they implant and begin to develop. In the process, a placenta is grown to act as the interface between mother and offspring. The highly vascular placenta then connects to growing embryos via the umbilical cord, and exchanges of nutrients and waste between mother and fetus occur in the placenta as fluids are not shared between mother and fetus in the womb.

Not all placental mammals have the same reproduction scheme. Most females have estrous cycles with well-defined "heat" periods and will only accept mates and breed during this period. The estrous period often is fairly conspicuous and recognized by mates through hormonal, pheromonal, or behavioral cues. Most mammals fit in this category, and best known examples may include deer, dogs, and otters.

Primates are recognized by many as the most advanced mammals, possibly because of advanced cognitive skills and a larger brain relative to body size. Many primates including humans do not have an estrous cycle, but instead a menstrual cycle. The menstrual cycle typically leads to more frequent ovulation (every 28 days on average in humans) and considerable bleeding associated with breakdown of the endometrial lining in the uterus (menstruations). Moreover, females of advanced primate species including humans do not show clear sign of ovulation and in many females, ovulation fits into a process called the menstrual cycle. In this cycle, the uterus is prepared prior to ovulation, the egg or eggs are released several days later, and if fertilization does not occur, the lining of the uterus degenerates and is shed during a period called the menstruations. The combination of regular but hidden ovulation in females probably allow primates to evolve promiscuous mating systems because males cannot assess when ovulation occurs, and thus mate guarding either occurs throughout the year (monogamy in many human societies), or uncertainty of paternity leads to child care in large promiscuous groups (chimpanzees) and lessens the risk of infanticide.

Peculiar mechanisms: Induced ovulation

Induced ovulation occurs when release of eggs in females is triggered by a stimulus, most often physical such as copulation, but also behavioral or pheromonal such as the vicinity of males. In contrast to spontaneous ovulators, or species where the release of eggs depends on the seasonal photope-

African lioness (*Panthera leo*) with her cubs. (Photo by Joe McDonald. Bruce Coleman, Inc. Reproduced by permission.)

riod signal, species with induced ovulation develop ova that are ready for release but require a stimulus for release.

Induced ovulation occurs in many species, but is best understood in mammalian carnivores. Examples of species with induced ovulation include cats (Felidae), bears (Ursidae), and numerous Mustelidae such as wolverine (*Gulo gulo*), striped skunk (*Mephitis mephitis*), and North American river otter (*Lontra canadensis*). For species with induced ovulation, it appears that a certain level of stimulation is required for eggs to be released, and thus it has been hypothesized that females may use the ability of a male to induce her ovulation as an indicator of male vigor, hence male quality. In these species, females may not be able to compare males simultaneously and because of the severity of the environment, may not be sure of her ability to find mates. In this case, the best strategy for the females would be to mate with all males encountered, and bear offspring from the male that induces the greatest stimulus. Evidence in black bears (*Ursus americanus*) of multiple paternity within single litters suggests that induced ovulation may be used by females as a mate choice strategy within the reproductive tract. For males, inducing ovulation may be a method of ascertaining paternity when pair bonds must be short to allow for encounters with other females. It is also possible that induced ovulation evolved as a strategy against sexual coercion in carnivores. In this case, females forced into copulation by males of lesser quality could abort eggs if they subsequently bred with a better quality male that provided a greater stimulus. Although the complexity of induced ovulation is still being investigated, it appears that benefits may exist for both males and females of species living in highly seasonal environments.

Peculiar processes: Delayed fertilization and delayed implantation

The opportunity to find a suitable mate is essential for reproduction, but because gestation is fixed in duration, timing of mating has direct implications on the timing of parturition. However, mammals have evolved two strategies to separate

A western barred bandicoot (*Perameles bougainville*) joey suckling in mother's pouch. (Photo by © Jiri Lochman/Lochman Transparencies. Reproduced by permission.)

in time mating and parturition: delayed fertilization and delayed implantation.

Delayed fertilization

Sperm storage occurs in bats inhabiting northern temperate regions such as the little brown bat (*Myotis lucifugus*), and also in many bats such as noctule (*Nyctalus noctula*). In the little brown bat, the testes become scrotal in the spring, and most sperm production is completed by September. The sperm are then stored until copulation commences months later. Females are inseminated in the fall and winter, while they are in hibernation. Sperm are then stored again, this time in the female reproductive tract, the uterus, where they remain motile for almost 200 days in the noctule. Females ovulate much later, and active development of the embryos starts in the spring. For bats, delayed fertilization allows males to copulate when they are in best condition in the fall, and parturition to occur just prior to emergence of insects. Because of the energy required for copulation, mating in the spring would be at the time of worst male condition. Delayed fertilization also allows females to give birth immediately after spring arrives, thus allowing more time for offspring growth before the next hibernation period. Thus, delayed fertilization is especially advantageous for species with long periods of dormancy, and allows females to compare breeding males via sperm competition.

Delayed implantation

Delayed implantation is a peculiar reproductive process in which fertilized eggs come to a halt in their development, usually at the blastocyst stage. After a period of time that varies among species and that is somewhat flexible, the blastocysts implant on the uterine wall and start developing for a duration called "true gestation." Delayed implantation occurs in

marsupials, rodents, roe deer, and bats, but probably is best known in carnivores. In the Carnivora, not all species exhibit delayed implantation, but the best examples probably are in the Mustelidae, Ursidae, and pinnipeds (Odobenidae, Phocidae, Otariidae). Examples of species that have delayed implantation include American marten (*Martes americana*), wolverine (*Gulo gulo*), black bear (*Ursus americanus*), giant panda (*Ailuropoda melanoleuca*), and northern fur seal (*Callorhinus ursinus*). In bears, mating occurs after den emergence in the spring, but birth of offspring does not occur until late winter of next year, a full 9-10 months later. Delayed implantation likely evolved to uncouple the tight relationship between mating and birth, and because it is most prevalent in northern species, would provide an advantage to allow parturition at the time of greatest food availability and possibly mating to occur at the time of greatest mate availability. Possibly, mating systems may influence delayed implantation and at least two species in North America show variable delay, the American mink (*Mustela vison*) and the striped skunk (*Mephitis mephitis*), where the delay varies from 0–14 days. In the pinnipeds, delayed implantation would allow females to mate when conditions are favorable to maximize male competition and availability, thus allowing females to breed at a time when seals are aggregated, thus facilitating mate choice.

Peculiar mechanisms: Inbreeding avoidance

Inbreeding is a word that describes breeding of one individual to another that is related, and in most animals, mammals included, is relatively rare. This is likely so because breeding with relatives has deleterious effects on the survival of the offspring, and often leads to reduced fertility. In evolution, inbreeding is rapidly selected against. Not surprisingly then, animals go to great lengths to avoid breeding with animals to whom they are related. To accomplish this, animals must be able to either recognize relatives (kin recognition) or, as an alternate strategy, recognize situations that could lead to breeding with relatives.

Kin recognition has been demonstrated in numerous mammals, and is probably most developed in humans. The alternate scenario of recognizing situations that lead to breeding with relatives is thought to explain sex differences in dispersal. In mammals, dispersal from natal areas is most common in males whereas females tend to stay closer to the home range or territory of their mother. Although several explanations for dispersal have been proposed, at the top of the list is that dispersal may reduce possibility of inbreeding. This would be most common in polygamous species where males copulate with numerous females. In this scenario, many of the resident females present during the next breeding season would be related to the males, and hence, males often disperse after a successful breeding season.

Peculiar mechanism: Reproduction in armadillos

Armadillos are placental mammals (Order Xenarthra) that occupy the southern United States, Central America, and the northeastern half of South America. They differ from other placental mammals in numerous ways. The uterus is simplex, just like that of humans, but there is no vagina and instead, a urogenital sinus serves as vagina and urethra. Males have internal testes and no scrotum, and have among the longer penes of mammals, reaching one third the length of the body in nine-banded armadillos (*Dasypus novemcinctus*). The oddities are not limited to morphology, but include the physiology as well.

Nine-banded armadillos have unusual delayed implantation. Armadillos breed in June, July or August, and the only fertilized egg becomes a blastocyst after 5-7 days at which point it enters the uterus. Development then ceases and the blastocyst remains free-floating in the uterus until November or December when it implants, and then the zygote divides twice to form four identical embryos. After 5 months of gestation, four identical quadruplets are born usually in May. Although identical twins are known to occur in humans, nine-banded armadillos regularly have identical offspring, most often four of them. But the mystery does not end there: a captive female held in solitary confinement gave birth to a litter of four females 24 months after her capture, or 32 months after she could have mated in the wild! Although the mechanism is not clearly understood, either her implantation delay lasted 23-24 months, or this particular female had produced two eggs, one of which would have remained dormant for at least 15 months.

Peculiar morphology: The mammalian penis bone

A peculiar bony structure exists in the penis of many mammalian species, and this bone, often referred to as baculum or os penis, is probably one of the most puzzling and least understood bones of the mammalian skeleton. Present in a variety of orders including Insectivora, Chiroptera, Primates, Rodentia, and Carnivora, this bone does not occur in all Orders in the Mammalia, but also does not occur in all species within each Order. Within a species the bone also varies in size, with older individuals typically possessing longer penis bones. In the mammals, the largest penis bone in absolute and relative size occurs in the walrus (*Odobenus rosmarus*), where the baculum may reach up to 22 in (54 cm) in length. In contrast, the bone is mostly vestigial in rodents such as North American beavers (*Castor canadensis*), and very small in all felids (cats), which have spines on the penis.

There are obvious costs to possessing a penis bone as evidenced from accounts of penis bone fractures. Historical hypotheses suggested that the bone may provide additional support to the penis for copulation, may protect the urethra from collapsing and blocking sperm passage in species that copulate for long periods, or else may help stimulate females into ovulation. However, all hypotheses have weaknesses and the most current hypothesis explaining the evolution of the mammalian penis bone in carnivores suggest that the largest penis bone evolved in species with promiscuous mating systems as a way for females to assess male quality during copulation. However, explanations may not be exclusive and possibly other functions may exist in other taxa. Undoubtedly, the full significance of this bone into the evolution of mammals has not yet been fully understood and remains an enigmatic puzzle to solve for mammalian scientists.

Resources

Books

Adams, C. E., ed. *Mammalian Egg Transfer*. Boca Raton: CRC Press, 1982.

Asdell, S. A. *Patterns of Mammalian Reproduction*. 2nd ed. Ithaca, NY: Cornell University Press, 1964.

Eberhard, W. G. *Sexual Selection and Animal Genitalia*. Cambridge, MA: Harvard University Press, 1985.

Kosco, M. *Mammalian Reproduction*. Eglin, PA: Allegheny Press, 2000.

Tomasi, T. E. *Mammalian Energetics: Interdisciplinary Views of Metabolism and Reproduction*. Ithaca, NY: Comstock Publishing Association, 1996.

Zaneveld, L. J. D., and R. T. Chatterton. *Biochemistry of Mammalian Reproduction*. New York: John Wiley & Sons, 1982.

Periodicals

Ferguson, S. H., J. A. Virgl, and S. Larivière. "Evolution of delayed implantation and associated grade shifts in life history traits of North American carnivores." *Écoscience* 3 (1996): 7–17.

Jaffe, K. "The dynamics of the evolution of sex: Why the sexes are, in fact, always two?" *Intersciencia* 21 (1996): 259–267.

Jia, Z., E. Duan, Z. Jiang, and Z. Wang. "Copulatory plugs in masked palm civets: Prevention of semen leakage, sperm storage, or chastity enhancement?" *Journal of Mammalogy* 83 (2002): 1035–1038.

Kenagy, G. J., and S. C. Trombulak. "Size and function of mammalian testes in relation to body size." *Journal of Mammalogy* 67 (1986): 1–22.

Larivière, S., and S. H. Ferguson. "On the evolution of the mammalian baculum: Vaginal friction, prolonged intromission or induced ovulation?" *Mammal Review* 32 (2002): 283–294.

Larivière, S., and S. H. Ferguson. "The evolution of induced ovulation in North American carnivores." *Journal of Mammalogy* 84 (2003): in press.

Laursen, L., and M. Bekoff. "*Loxodonta africana*." *Mammalian Species* 92 (1978): 1–8.

McBee, K., and R. J. Baker. "*Dasypus novemcinctus*." *Mammalian Species* 162 (1982): 1–9.

Miller, E. H., and L. E. Burton. "It's all relative: Allometry and variation in the baculum (os penis) of the harp seal, *Pagophilus groenlandicus* (Carnivora: Phocidae)." *Biological Journal of the Linnean Society* 72 (2001): 345–355.

Miller, E. H., A. Ponce de León, and R. L. DeLong. "Violent interspecific sexual behavior by male sea lions (Otariidae):

Resources

Evolutionary and phylogenetic implications." *Marine Mammal Science* 12 (1996): 468–476.

Moller, A. P. "Ejaculate quality, testes size and sperm production in mammals." *Functional Ecology* 3 (1989): 91–96.

Pasitschniak-Arts, M., and L. Marinelli. "*Ornithorhynchus anatinus.*" *Mammalian Species* 585 (1998): 1–9.

Patterson, B. D., and C. S. Thaeler Jr. "The mammalian baculum: Hypotheses on the nature of bacular variability." *Journal of Mammalogy* 63 (1982): 1–15.

Poole, W. E. "*Macropus giganteus.*" *Mammalian Species* 187 (1982): 1–8.

Rissman, E. F. "Mating induces puberty in the female musk shrew." *Biology of Reproduction* 47 (1992): 473–477.

Sharman, G. B. "Reproductive physiology of marsupials." *Science* 167 (1970): 1221–1228.

Sharman, G. B., and P. E. Pilton. "The life history and reproduction of the red kangaroo (*Megaleia rufa*)." *Proceedings of the Zoological Society of London* 142 (1964): 29–48.

Stockley, P. "Sperm competition risk and male genital anatomy: Comparative evidence for reduced duration of female sexual receptivity in primates with penile spines." *Evolutionary Ecology* 16 (2002): 123–137.

Storrs, E. E., H. P. Burchfield, and R. J. W. Rees. "Superdelayed parturition in armadillos: A new mammalian survival strategy." *Leprosy Review* 59 (1988): 11–15.

Serge Larivière, PhD

The black-footed ferret (*Mustela nigripes*) takes over abandoned prairie dog (*Cynomys* sp.) burrows. (Photo by © D. Robert & Lorri Franz/Corbis. Reproduced by permission.)

the species and ecosystems studied to answer it: population cycles of voles, lemmings, and snowshoe hares, mostly in a 3–5 year period, have long been suggested to be driven by specialist predators. However, controlled removal of weasels (*Mustela nivalis*) in one study did not prevent population crashes of field voles, nor did it influence population dynamics at any other stage of the cycle. Also, for several populations of snowshoe hares, both cyclic and non-cyclic ones, predators (weasels, mink, bobcats, lynx, coyotes, and several birds of prey) were the most frequent cause of death for radio-collared hares, and hare population cycles did heavily influence the reproduction, mortality, and movement of the predators. Nevertheless, at or near peak densities, predator activity appeared to have almost no influence on hare density. At lowest density, no influence was evident either, provided that enough cover was available for the hares to retreat into. Survival of hares was directly related to good cover and good feeding conditions. Juvenile and malnourished individuals were more at risk not only due to predators but also due to hard winters.

Caribou (*Rangifer tarandus*) eat a wide variety of vegetation, which enables them to survive through harsh winters. Much of their native land has been taken over for human use, but large areas have been protected by the government to preserve their habitiat. (Photo by © Annie Griffiths Belt/Corbis. Reproduced by permission.)

dominant species in these communities regularly excludes the others from the "best" temporal niche. Similarly, intra-specific competition often drives subordinate individuals into other temporal niches and is often the first step in excluding an individual from a group or litter, long before agonistic expulsion is to be seen.

In general, there is a clear relationship between body size and activity patterns of mammals: the smaller the species, the more likely it is to be nocturnal. This is confirmed for both herbivores and carnivores. Being nocturnal offers better protection from being detected by predators as well as competitors (which is a more important factor when belonging to a small species). In small carnivores, additionally, the effect of finding more prey during the night further enhances this preference. There are, however, exceptions to this rule. Microtine rodents are characterized by very short, ultradian activity patterns, which are very adaptive to the present ecological conditions. Insectivorous and gregarious small mongooses are diurnal, and tree squirrels are all diurnal.

Predators and prey—are they influencing each other's population biology? The answer to this question is as variable as

An adult hippopotamus (*Hippopotamus amphibius*) feeds on grass on the banks of the Chobe River in Botswana. Hippopotamuses usually feed at night, but on relatively cool mornings they may wander out of the water to feed on nearby grass. (Photo by Rudi van Aarde. Reproduced by permission.)

In a large, comparative, long-term study of many species of predators and prey (wolf and lynx, weasels and stoats, and raptors and owls, and prey from European bison and moose down to amphibians and shrews), it was found that the largest species were barely influenced by predators at all, that amphibians were mostly influenced in their population densities by weather, and that predators in general could neither influence prey densities nor fluctuations. There were, however, a few exceptions: lynx were able to limit roe deer densities below carrying capacity, and both wolves and lynx were obviously able to influence population densities of roe and red deer. The reason might be that, contrary to most other predator-prey systems, both predators are smaller and have a higher reproductive rate than their prey. In those cases, predators might be able to react (numerically, by means of litter size and survival) more rapidly to changing conditions than their prey does. In many ecosystems, both temperate and tropical, large species of prey migrate and thus leave the areas with highest predator activity. Both migrating gnu and caribou, for example, have been shown to lower predation risk by this strategy. Comparison of migrating and nonmigrating ungulates in the Serengeti, following a severe decline of buffalo (the largest nonmigrating herbivore there) and large predators to poaching, led to astonishing results: topi (*Damaliscus lunatus*), impala (*Aepyceros melampus*), Thomson's gazelle (*Gazella thomsonii*), and warthog (*Phacochoerus aethiopicus*) seem to be predator-controlled. The red hartebeest (*Alcelaphus buselaphus*), a close relative of topi (but one with a different feeding style) seem to be regulated by intra-specific competition, and giraffe (*Giraffa camelopardalis*) and waterbuck (*Kobus ellipsiprymnus*) declined

due to poaching. At least in Thomson's gazelle, but probably also others, the reason for the influence of predators on population performance seem to be more complex than simple mortality. Vigilance and flight increase, whenever predator pressure increases, and this of course affects all individuals, not only the unlucky ones being killed. These costs of anti-predator behavior (avoiding potentially dangerous feeding habitats, spending time alert or on the move, etc.) have to be carried by all members of the population, and they do not bring any benefit to the predators (contrary to the "direct costs" of killed animals). This example demonstrates again the complexity of the whole issue.

How can communities, groups of several or even many species, live together? The ecological term "guild" defines a group of species that use the same resources in a comparable way. Thus we would expect them to compete for these resources, and either ecological displacement or niche separation, at least along one or a few niche axes, should occur. In many guilds of species, a recurring phenomenon known as character displacement exists. This means that in areas where two or more competing species occur (sympatric occurrence), at least one trait should differ more pronouncedly than between populations of the same species in non-overlapping (allopatric) habitats. One example: the ermine, a small weasel, is smaller in Ireland than in Great Britain, where an even smaller species, the least weasel (*M. nivalis*) occurs sympatrically. Guilds of carnivores have been studied in many countries, and in many cases character displacement is evident. In areas of sympatry between two small cat species of South America, the margay (*Leopardus wiedii*) and the jaguarundi (*Herpailurus yaguarondi*), the margay is more aboreal. Degree of arboreality is also a frequent pattern in separate primate species, both in guilds of guenons and between sympatric lorisids such as the angwantibo (*Arctocebus*) and the potto (*Perodicticus*) in tropical Africa. In the case of the lorisids, one is a smaller, more slender-built species using thinner branches and the upper canopy, while the other is larger and more stoutly built, using the lower, thicker branches.

Female African lions (*Panthera leo*) are predators in their ecosystem. (Photo by David M. Maylen III. Reproduced by permission.)

New species of animals are still being discovered. Shown here is the recently described species of mouse lemur (*Microcebus griseorufus*). (Photo by Harald Schütz. Reproduced by permission.)

American mink (*Mustela vison*) or a recently immigrated one, such as the striped jackal (*Canis adustus*) in East Africa, was present. In those cases, character divergence and niche shifts or niche compression did become evident, and mostly led to displacement in one species (European mink [*Mustela lutreola*], silverback jackal [*Canis mesomelas*]).

A guild of granivorous (seed-eating) rodents was studied in the Sonora Desert of Arizona. One cricetid and four heteromyid species of different sizes did occur in this community, and all appeared to eat seeds of the same plant species (except for one large species eating a somewhat larger number of insects and one small species eating more seeds of one particular bush). A remarkable difference, however, was found in the spatial arrangement of their feeding places. The large species, a kangaroo rat, mostly exploited patches rich in seeds, such as near rocks or in depressions in the soil. The small species, a pocket mouse, also collected seeds from patches but only in about 6.6% of observed feeding bouts. For the rest, seeds were collected in a more systematic way, while searching in a "sauntering" manner. Both species used olfactory cues to search for food. However, as the kangaroo rats move bipedally in a rapid hop, and thus can easily move from one patch to the other, the smaller species walk quadrupedally. These differences in locomotion not only carry different energetic costs but also allow the animals to make use of olfactory gradients (gradual increases/decreases of concentration) differently: the faster an animal moves, the easier it can detect an olfactory gradient when a larger patch of seeds exudes some stronger smell. The slow-moving pocket mouse, on the other hand, can sniff out individual grains.

A four-striped grass mouse (*Rhabdomys pumilio*) sheltering in its burrow from the heat of the day. When temperatures above ground reach to 108°F (42°C), temperature inside the burrows will remain below 77°F (25°C). Such sheltering enables a variety of mammals to survive in desert and semi-arid regions throughout the African continent. (Photo by Rudi van Aarde. Reproduced by permission.)

Carnivore communities are special in that direct, aggressive competition can be observed. Nevertheless, there are many axes along which species can separate. One is body size, which directly relates to prey size. In African savannas, lions, leopards, hyenas, hunting dogs, cheetahs, and jackals all coexist, and direct competition for similar-sized prey is almost fully restricted to cheetahs versus hunting dogs. This relatively peaceful coexistence is due to many factors: social hunting allows some species to take prey of much larger size than their own; the leopard is more arboreal than the other predators; and some predator species migrate to follow their prey while others don't. In sympatric carnivore species of similar body size, gape size (the ability to open one's mouth more or less widely) often acts as a separating axis (as in the guild of cats in South America, mentioned above). Comparative studies of carnivore guilds in Israel (13 species, 4 families), the British Isles (5 species of mustelids), and East Africa (3 species of jackals) found that there is regular separation, apart from body size, in terms of degree of cursorial locomotion, in diameter or shape of canine teeth, and in skull length. Direct overlap of all these niche parameters mostly occurred when a newly introduced species such as

A black rhinoceros (*Diceros bicornis*) browses on low-growing shrubs. (Photo by Rudi van Aarde. Reproduced by permission.)

The most obvious and oft-cited guilds of larger mammals certainly are ungulate communities. Many studies have been made on groups of herbivore species both in temperate and tropical areas. Despite the fact that up to at least six species of native ungulates can coexist in temperate climes and more than 20 in some African savannas, it is surprising that most studies do not find obvious competition effects. Contrary to carnivores, where inter-species killing and direct competition over food are regular features, there is practically no evidence of direct aggressive competition. Even indirect, long-term effects on population density caused by the changing number of another species is mostly absent. Things only change as soon as newly introduced species come into the community. Thus it was found that feral muntjacs (*Muntiacus*) in Great Britain severely competed with roe deer (*Capreolus*). When cattle were introduced into an area where black-tailed deer (*Odocoileus*) were numerous, the deer retreated into other habitats, which cattle avoided. Competition between species of deer was documented in communities of deer in New Zealand, where all three species (red [*cervus elaphus*], fallow [*Dama dama*], and white-tailed deer [*Odocoileus virginianus*]) had been introduced. Thus it seems that guilds of ungulates having a long (co-) evolutionary history together can coexist, probably because there are enough dimensions along which to separate. The most famous example is the grazer-browser

division or, more accurately, the division into concentrate selectors, bulk-feeders, and intermediate feeders. Apart from selecting leaves, grass, or something in between, there are subdivisions in each set of species. Height preference is also a separating criterion for grazers: some species, such as zebra, feed on high and dry or lignified grass, while others feed only on lower, mostly fresh plants. Another separating axis is bite size, which is determined by jaw/snout size and tooth rows. Animals with a broader mouth are less selective in feeding. Body and gut size are also important criteria in deciding what to forage and how to digest. The smaller species have a higher energy demand and specialize on high quality leaves; larger species feed on lower quality grass such as stems or leaf sheaths, or older plants. The reason for this is that larger species, with larger fermentation chambers in their guts, can digest cell walls more effectively and need less energy per unit of body mass. Equids are better able to extract energy from large amounts of fiber-rich food, and ruminants do better with restricted food mass. Equids have a lower reproductive rate than ruminants but are better able to defend themselves against predators due to their sociality. Often in savannas there is a succession of different ungulate species foraging on the same spot, one after the other: zebras (*Equus*) start by eat-

The bamboo that the panda (*Ailuropoda melanoleuca*) eats provides it with very little nutrition, but it is available year round. (Photo by © Keren Su/Corbis. Reproduced by permission.)

The capped langur (*Trachypithecus pileatus*) is one of the rarest primates and lives in the trees of Bhutan. (Photo by Harald Schütz. Reproduced by permission.)

Humans have an effect on other mammals' ecology. Here an Antarctic fur seal (*Arctocephalus gazella*) has plastic packaging around its neck. (Photo by Paul Martin/Nature Portfolio. Reproduced by permission.)

ing the high, old, dry grass; wildebeest (*Connochaetes*) follow and eat the lower grass plants, but still select on the level of whole plants; Thomson's gazelle, with their narrow snouts, select only the freshest parts in the middle of grass plants; and kongoni feed on the long, lignified stalks that remain. Migrating versus non-migrating is another axis of niche separation. This is the famous Bell-Jarman principle first described for the Serengeti and other East African ecosystems, which allows up to eight species of grazers and about 20 species of ruminants in total to coexist.

Resources

Books

Dayan, T., and D. Simborloff. "Patterns of Size Separation in Carnivore Communities." In: *Carnivore Behavior, Ecology and Evolution*, vol. 2, edited by J. C. Gittleman, 243–266. Chicago: University of Chicago Press, 1996.

Halle, S., and N. C. Stenseth, eds. *Activity Patterns in Small Mammals*. Berlin: Springer, 2000.

Jedrzejewska, B., and W. Jedrzejewski. *Predation in Vertebrate Communities. The Bialowieza Primeval Forest as a Case Study*. Berlin: Springer, 1998.

Keith, L. B. "Dynamics of Snowshoe Hare Population." In *Current Mammalogy*, vol. 2, edited by H. H. Genoways, 119–196. New York/London: Plenuum, 1990.

Nevo, E. *Mosaic Evolution of Subterraneous Mammals*. Oxford: Oxford University Press, 1999.

Putman, R. J. *Competition and Resource Partitioning in Temperate Ungulate Assemblies*. London: Chapman & Hall, 1996.

Reichman, O. J. "Factors Influencing Foraging in Desert Rodents." In *Foraging Behavior*, edited by A. C. Kamil and T. D. Sargent, 195–214. New York: Garland, 1981.

Sinclair, A. R. E., and P. Arcese, eds. *Serengeti II*. Chicago: University of Chicago Press, 1995.

Periodicals

Duncan, P. "Competition and Coexistence Between Species in Ungulate Communities of African Savanna Ecosystems." *Advances in Ethology* 35 (2000): 3–6.

Oli, M. K. "Population Cycles of Small Rodents Are Caused by Specialist Predators: Or Are They? *TREE* 18 (2003): 105–107.

Udo Gansloßer, PhD

Nutritional adaptations

The dietary needs of mammals

Like the rest of the animal kingdom, mammals need food for energy and the maintenance of bodily processes such as growth and reproduction. The chemical compounds used to supply the energy and building materials are obtained by eating plants or organic material. Both plant- and animal-based sources of food are made up of highly complex compounds that need to be digested and broken down into simpler forms.

Four of the most common naturally occurring elements—oxygen, carbon, hydrogen, and nitrogen—make up 96% of the total body weight of an animal. The remaining 4% is made up of the seven next most abundant elements—calcium, phosphorus, potassium, sulfur, sodium, chlorine, and magnesium, in that order. Necessary for many physiological processes, any change in their concentrations may be deleterious or fatal.

An animal's major dietary components are fat, water, protein, and minerals. The main digestion products of these compounds are amino acids (from proteins), various simple sugars that are present in the food or derived from starch digestion, short-chain fatty acids (from cellulose fermentation), and long-chain fatty acids (from fat digestion). The oxidation of these digestion products yields virtually all the chemical energy needed by animal organisms.

Despite carbohydrates' essential role in animal metabolism, their total concentration is always less than 1%. The two major animal carbohydrates are glucose and glycogen.

Body lipids act as energy reserves, as structural elements in cell and organelle membranes, and as sterol hormones. Because lipids can be stored as relatively non-hydrated adipose tissue containing 2–15% free water, eight times more calories per unit of weight can be stored as fat than as hydrated carbohydrates. This is the reason fat storage is essential for active animals, while carbohydrates are a major energy reserve for plants. Hibernating mammals deposit fat and may double their body weight at the end of the summer prior to hibernation; the white adipose tissue reserves allow them to survive the winter.

In addition, there are 15 elements making up less than 0.01% of the body of a mammal. These elements occur in such small amounts that they became known as trace elements. Still, they have vital physiological and biochemical

roles. Iron, for instance, is a key constituent of hemoglobin in blood and several intercellular enzyme systems. While the amount of iron found in an adult human is only 0.14 oz (4 g)—70% in hemoglobin, 3.2% in myoglobin, 0.1% in cytochromes, 0.1% in catalase, and the remainder in storage compounds in the liver—growing animals need more iron, and adult females need to replace that which is lost in reproductive processes such as the growth of the fetus and menstruation. The dietary requirement for adult mammals is very small since iron (from the breakdown of hemoglobin) is stored in the liver and used again for hemoglobin synthesis. Other trace elements include copper, zinc, vanadium, chromium, manganese, molybdenum, silicon, tin, arsenic, selenium, fluorine, and iodine.

Animals can differ markedly in their vitamin requirements. Ascorbic acid (vitamin C), for example, can be synthesized by most mammals, but humans and a few other mammals such as non-human primates, bats, and guinea pigs, need to have it supplied in their diet.

Ruminants do not appear to need several vitamins in the B group since the microbial synthesis of vitamins in the ruminant stomach frees these animals from having to seek out additional dietary sources.

Adaptations in the digestive system

All carnivores, when fed a whole prey-based diet, consume proteins and fats from the muscle, vitamins from organs and gut contents, minerals from bones, and roughage from the hide, feathers, hooves, teeth, and gut contents. Felids are set apart from other, more omnivorous meat eaters because of their inability to effectively utilize carbohydrates as an energy source. They therefore depend on a higher concentration of fats and protein in their diet, as well as dietary sources of preformed vitamin A and D, arachadonic acid (an essential fatty acid), and taurine.

Herbivores, on the other hand, have adapted numerous methods of utilizing roughage-based diets. Plankton feeders such as the baleen whales have a filtering apparatus that consists of a series of horny plates attached to the upper jaw and then left hanging from both sides. As the whale makes its course through the ocean, water flows over and between

A cougar (*Puma concolor*) will take down prey larger than itself, usually breaking the animal's neck. (Photo by © Charles Krebs/Corbis. Reproduced by permission.)

the plates, and plankton is caught in the plates' hair-like edges.

Ruminants and some non-ruminant herbivores (e.g., sloths, hippos, colobines, large marsupials) utilize pre-gastric microbial fermentation to break down cell wall constituents, while the odd-toed hoofed animals, or perissodactyls, rely primarily on post-gastric fermentation.

Some small herbivores, like rodents and rabbits, have relatively higher nutrient requirements compared with larger herbivores. In order to meet these requirements, they must routinely practice coprophagy to obtain the protein, water, enzymes, vitamins, and minerals provided by the microbes. Coprophagy, which comes from the Greek *copros*, meaning "excrement," and *phagein*, meaning "to eat," is of great nutritional importance. If coprophagy is prevented in rabbits, their ability to digest food decreases, as does their ability to utilize protein and retain nitrogen. This process is reversible, however. When coprophagy is allowed again, the rabbits' ability to digest cellulose is restored.

The soft feces that a rabbit re-ingests originate in the cecum, or "blind gut," a large blind pouch forming the beginning of the large intestine. Upon ingestion, these feces are not masticated and mixed with other food in the stomach. In-

stead, they tend to lodge separately in the base of the stomach. A membrane coats the soft feces, and they continue to ferment in the stomach for many hours. One of the fermentation products is lactic acid.

For most herbivores, the gastrointestinal microbial population is an integral component of the feeding strategies, especially since most of them live on food that make cellulose digestion essential. Some of the most important domestic meat- and milk-producers (cattle, sheep, goats) have specialized tracts that are highly adapted to symbiotic cellulose digestion; they are known as ruminants.

The stomach of a ruminant consists of several compartments, or in more precise terms, the true digestive stomach, the *abomasum*, is preceded by several large compartments. The abomasum corresponds to the digestive stomach of other mammals. The first and largest compartment of this system is the rumen, which serves as the main fermentation center in which the food, after it has been mixed with saliva, undergoes heavy fermentation.

Both bacteria and protozoans reside in the rumen in large numbers. These microorganisms work to break down cellulose and make it available for further digestion. The fermentation products (mostly acetic, propionic, and butyric acids)

A camel's (*Camelus dromedarius*) hump is primarily fat, which is metabolized when food is scarce. (Photo by © Dave G. Houser/Corbis. Reproduced by permission.)

are absorbed and utilized, while gases (carbon dioxide and methane) formed in the fermentation process are released through belching. A cow fed 11 lb (5 kg) of hay a day can give off 1 qt (191 l) of methane each day.

What is rumination?

The act of rumination, or "chewing the cud," is the regurgitation and remastication of undigested fibrous material before it is swallowed again. As the food reenters the rumen, it undergoes further fermentation. The products of fermentation—in the form of broken-down food particles—then slowly pass to the other parts of the stomach, where the usual digestive juices of the abomasum perform their work.

Ruminants secrete copious amounts of saliva that serve to buffer fermentation products in the rumen. It also serves as a fermentation medium for the microorganisms. The total secretion of saliva per day has been estimated at 6–17 qt (6–16 l) in sheep and goats, and 105–200 qt (100–190 l) in cattle. Since sheep and goats have an average weight of 88 lb (40 kg) and cattle, 1,100 lb (500 kg), the daily production of saliva may reach about one-third of the body weight.

The obligate anaerobic organisms residing in the rumen include ciliates that occur in numbers of several hundred

thousand per fluid ounce (milliliter) of rumen contents. Laboratory extracts from pure cultures of rumen organisms have demonstrated cellulase activity, the enzyme that breaks down cellulose so that its byproducts become available to the host mammal.

Rumen microorganisms can also synthesize protein from inorganic nitrogen compounds such as ammonium salts. Dairy farmers have been supplementing the feed of milk cows with urea—normally, an excretory product eliminated in the urine—to increase protein synthesis, rather than through the use of more expensive high-protein feed.

In the rumen, urea is hydrolyzed to carbon dioxide and ammonia—the latter being used by the microorganisms for the resynthesis of protein. Since a camel fed a nearly protein-free diet of inferior hay and dates excretes virtually no urea in the urine, it can recycle much of the small quantity of protein nitrogen it has available this way. A similar reutilization of urea nitrogen in animals fed low-protein diets has been observed in sheep and, under certain conditions, rabbits.

Rumen microorganisms can also contribute to the quality of the protein that is synthesized. If inorganic sulfate is added to the diet of the ruminant, the microbial synthesis of pro-

The North American porcupine (*Erethizon dorsatum*) derives all of its nutrition from vegetation. (Photo by Corbis. Reproduced by permission.)

cow feces are smooth and well-ground up with few large, visible fragments.

Multi-compartment stomachs are not unique to the ruminants, or even the ungulates. Animals such as the sloth, the langur monkey, and even certain marsupials have rumen-like stomachs. The diminutive quokka, for instance, has a large stomach harboring microorganisms that participate in cellulose digestion. For an animal weighing in at 4.4–11 lb (2–5 kg), its stomach equals about 15% of its body weight, a number similar to that found in most ruminants.

The kangaroo and wallaby are ruminant-like large marsupials that utilize the same mechanism of microbial fermentation taking place anterior to the digestive stomach. At the beginning of the dry season, when the nitrogen content in the vegetation starts to decline, wallabies begin to recycle urea and continue to do so throughout the prolonged dry season. This way, they are relatively independent of the low quality of the available feed.

There are other species-specific anatomical morphologies that adapt to the animal's specific nutritional needs. The small intestine, for instance, is the primary site of enzymatic digestion and absorption. The mammalian small intestine is morphologically divided into a proximal duodenum looping around the pancreas, intermediate jejunum, and distal ileum. The length of all intestinal segments in mammals relative to body weight is longest in herbivores, intermediate in grain- and fruit-eaters, and shortest in carnivores and insectivores. Relative intestinal length, weight, and volume within each species vary with sex, age, seasonal food habits, as well as with changing nutritional requirements or food quality and the level of intake.

tein is improved and the sulfate is incorporated into the essential amino acids, cysteine and methionine.

Since the microbes in the ruminant stomach can synthesize all the essential amino acids, ruminants are nutritionally independent of these, and therefore the quality of the protein they receive in their feed is not of vital importance.

Some important vitamins are also synthesized by rumen microorganisms, including several of the vitamin B groups. The natural supply of B_{12} for ruminants, for example, is obtained entirely from microorganisms.

Rumen fermentation takes place in the anterior portion of the gastrointestinal tract so that the products of fermentation can pass through the long intestine for further digestion and absorption. This way, the mechanical breakdown of the food can also be carried much further, and coarse and undigested particles can be regurgitated and masticated repeatedly.

If a comparison is made of the fecal material of cattle (ruminants) and horses (non-ruminants), it will be found that horse feces contain coarse fragments of still-intact food, while

A jaguar's (*Panthera onca*) diet varies with its location. It has been known to eat cattle and horses, deer, peccaries, larger rodents such as capybara, paca, and agouti, and reptiles, monkeys, and fish. (Photo by Animals Animals ©Gerrard Lace. Reproduced by permission.)

The domestic cow's (*Bos taurus*) stomach has evolved into four chambers that allow the animal to derive the most nutrients from its vegetative diet. (Photo by © Richard T. Nowitz/Corbis. Reproduced by permission.)

The large intestine, to provide another example, varies in length depending on the species and its dietary regimen. Its length relative to the small intestine averages 6% in small carnivorous mammals, 33% in omnivores, and 78% in herbivores as fiber digestion, bulk, and a reduced rate of passage increase in importance.

The ceca and the large intestine work towards the fermentation of plant fiber and soluble plant matter, as well as the absorption of water and small water-soluble nutrients such as ammonia, amino acids, and volatile fatty acids. They also function to synthesize bacterial vitamins.

Seasonal changes in nutritional requirements

Changes in diet often follow the changes in seasons. When researchers at Sea World, Durban, traced the annual food consumption of their female dusky dolphin (*Lagenorhynchus obscurus*) over a 13-year period, they found that her annual food intake jumped from 4,784 lb (2,170 kg) when she was five years old, to nearly 6,393 lb (2,900 kg) the year after. This increase coincided with the installation of a cooling system that was used in the summer in the years thereafter, and after her sixth year, her food consumption fluctuated between 5,290 lb and 6,173 lb (2,400–2,800 kg) per year. In general, her food intake was above average during autumn and winter, and below average during spring and summer, and the average pool water temperature fluctuated seasonally.

A similar study of California sea lions (*Zalophus californianus californianus*) found that the voluntary decrease in food intake during summer was associated with increased aggres-

sive behavior in males, while the seasonal fluctuation in non-reproductive females was negligible. Seasonal fluctuations in male food intake was especially pronounced between the ages of four and eight, when sexual maturity was reached.

Territorial male California sea lions defended their territories in the breeding season, during which they do not feed, and remain in their territory for an average of 27 days. In captivity, they have been shown to lose as much as 198 lb (90 kg) during the breeding season—independent of food availability, suggesting the possibility of an endogenous rhythm. The simultaneous increase in aggression suggests testosterone involvement as well. The females, on the other hand, showed less profound fluctuations in monthly food intake than their male counterparts, possibly because females are non-territorial and do feed during the breeding season.

Seasonal variation in temperatures may also be important as male sea lions in particular ate less when air and water temperatures were high and a thick fat layer was less important for maintenance of constant body temperature.

Sheep have also been observed to alter their diets according to the shifting seasons. They are known to consume a more fibrous type of forage such as tussock grass during the winter season or seasons with a scarcity of resources. Two independent studies, in the semiarid rangelands of Argentina and Australia, respectively, reported that sheep preferred tussocks only in winter, and avoided them during the growing season. Although sheep behave generally as bulk grazers, they will also consume, when offered, a considerable amount of shrubs in the fall and winter seasons. This preference for evergreen shrubs corresponds with times of the year when grasses are less available or nutritious. Scottish sheep, in a comparable high-latitude oceanic climate, also displayed a similar feed-

A brown bear (*Ursus arctos*) feeds on salmon in the summer months. (Photo by © John Conrad/Corbis. Reproduced by permission.)

Barren-ground caribou (*Rangifer tarandus*) favor lichens, especially caribou moss, but they are not essential for survival. (Photo by Michael Gian-ncchini/Photo Researchers, Inc. Reproduced by permission.)

ing pattern, i.e., they consumed a high proportion of shrubs only in winter. Shrubs are also a type of forage that maintain a relatively high-protein content during the colder seasons.

Variation in the diet of a species over seasons may also be the result of habitation in different landscapes of the same region, thus linking prey use with availability. The swift fox (*Vulpes velox*), for example, occupies two distinct landscapes in western Kansas that are dominated by either cropland or rangeland. In spring and the fall, plants such as sunflower seeds, and birds were consumed more frequently in cropland than in rangeland. However, birds were more common in the swift fox diet in cropland during the fall.

Nutrition and the reproductive cycle

Energy requirements and food intake of pregnant females are about 17–32% higher than non-reproducing females, and yet only 10–20% of this additional energy is retained as new tissue by the developing uterus. The rest of the energy is lost as heat, slowing down the growth rate and thus lengthening the gestation period. A slower fetal growth rate may be advantageous in an environment with limited dietary protein or minerals, especially in the case of such animals as the fruit-eating or leaf-eating primate.

In females, the fetus represents 80% of the energy retained by the uterus. Most of the increase in mass in the mother occurs after 50–60% of the gestation period has elapsed. The water content of the developing fetus also decreases while the fat, protein, and mineral content increase during gestation. The mammalian newborn, or *neonate*, averages 12.5 ± 2.3% protein and 2.7 ± 0.8% ash at birth. Neonatal fat, water, and energy content vary between different species. For instance, neonatal seals, guinea pigs, and humans contain 4–8 times more fat than other mammals, whose content averages 2.1 ± 1.0%. The fat reserves of the guinea pig are broken down just a few days after its birth, while those of the seal are retained because of the cold environment and the short milk production period.

Because of the very low fat content of neonatal mammals, their high metabolism, and frequently, poor insulation, the chances of survival are only a few hours to days without care from the mother.

After giving birth, the production of milk by the mammalian female bridges the dietary gap between the passive, completely dependent fetus to the weaned and more or less nutritionally independent juvenile. Milk production, or lactation, enables the young mammal to continue its growth in an almost embryonic manner without having to remain

North American beavers (*Castor canadensis*) eat mostly cellulose, which is broken down by microorganisms in their cecum. (Photo by © Phil Schermeister/Corbis. Reproduced by permission.)

anatomically attached to its mother. The female, in this way, is freed from the locomotory, nutritional, and anatomical constraints of carrying the fetus.

According to Lopez and Robinson in 1994, nutrient requirements for pregnancy are moderate in comparison with the estimated nutrient requirements for maintenance. In the case of captive Atlantic bottlenose dolphin, food consumption in the females showed little increase during gestation, but was 58–97% higher during lactation than during similar periods in non-reproductive years.

For most mammals, milk production closely follows the nutrient requirements of the newborn animals. In the first few days, the requirements of the newborn may be substantially lower than the mother's potential to produce milk. As the needs of the growing animal increase, so does its requirements for milk and milk production.

Lactation itself can put an enormous nutritional burden on the female. In terms of energy expenditure, lactation is two to three times more costly than gestation, and the female's nutrient requirements may increase considerably. The total energy expenditure, including milk produced, of the lactating female is about 215% higher than her non-lactating counterpart.

The production of milk generally rises during early lactation and hits a maximal peak. This is the point where weaning takes place since the neonate has to increase its nutrient intake further by relying on nutritional sources other than milk.

The decline in milk production once the peak has been reached can last for as little as five days in mice to many months in large ungulates. While it seems to make evolutionary sense for larger species to have longer lactation periods, there are many exceptions to the evolutionary constraints. The hooded seal, for instance, has one of the shortest lactation periods despite a maternal weight averaging 395 lb (179 kg). The pup grows from 47.4 to 96.3 lb (21.5 to 43.7 kg), with 70% of the gain being fat, in just four days.

A basic rule of thumb, however, is that poorly nourished females and those nursing larger litters often reduce milk production faster than do well-nourished females.

Milk composition

According to Elsie Widdowson in 1984, of the 4,300 species of mammals, only the milks of 176 have been analyzed for protein, fat, and carbohydrate. Of these analyses, she said, only the figures for 48 of those species are considered to be reliable. The difficulty in the analyses lies with the fact that milk composition changes rather markedly during a lactation cycle.

The first milk, or "colostrum," contains a high concentration of maternal antibodies, or immunoglobins, active phagocytic cells, and bacteriocidal enzymes. While neonatal primates, guinea pigs, and rabbits acquire their circulating maternal immunoglobins in utero, other mammals such as ungulates, marsupials, and mink depend on the colostrum as their sole source of a passive immune system. Yet another group, intermediate to these two, acquire maternal immunoglobins, both in utero and from colostrum. Among these animals are rats, cats, and dogs. The differences in in-utero transfer of immunoglobins are determined by the number of cellular layers in the placenta that separates fetal and maternal circulation.

In order for the secretion of colostral immunoglobins to be effective, the neonatal gut needs to remain permeable to their absorption and minimize any upper-tract digestion of these proteins. The time the mammalian intestine remains permeable to the intact immunoglobins varies between species: 24–36 hours after birth in the case of ungulates; 16–20 days in mice and rats; eight days in mink; and 100–200 days in large marsupials. The marsupial's prolonged absorption capabilities relate to the time the young reside in the mother's pouch.

Other types of immunoglobins that occur in milk, after absorption of the first wave of intact molecules has ceased, protect the neonatal gut from infection.

The major constituents of milk are water, minerals, proteins (such as casein), fat, and carbohydrates. Protein concentration ranges from under 3.5 oz/qt (10 g/l) in some primates to more than 3.5 oz/qt (100 g/l) in hares, rabbits, and some carnivores. Fat concentrations vary from small amounts in the milk of rhinoceroses and horses to more than 17.6 oz/qt (500 g/l) in some seals and whales. The main carbohydrate of placental mammal milk is the disaccharide lactose, a polymer of glucose and galactose. Lactose content also ranges from trace amounts in the milk of some marine mammals and marsupials to more than 3.5 oz/qt (100 g/l) in some primates. Marsupial milk, nevertheless, can be very high in carbohydrates, but the sugars are primarily oligo- and poly-saccharides-rich in galactose.

Compromises have to be made between the physiological constraints to milk synthesis and selective pressures to maximize offspring survival. The variation in milk composition between species is one of them. Most aquatic mammals produce highly concentrated milks. The reduction in milk water content in aquatic mammals provides a high-energy, low-bulk diet that is useful in offsetting neonatal heat loss in cold environments. It also conserves water in the mothers of species (such as the northern elephant seal) that abstain entirely from eating or drinking during a relatively short, but intense, lactation. Similarly, seals that give birth on pack ice and have a short lactation period (e.g., hooded seals—four-day lactation) or those that leave the neonate for feeding trips lasting several days produce more concentrated, higher fat milks than do other seals.

For terrestrial mammals, the largest changes in milk composition over time occur in marsupials. Some marsupials can produce several kinds of milk simultaneously since they may have young of different ages. For the embryonic marsupial confined to the pouch, dilute, high-sugar milk provides nourishment similar to that occurring in the uterus of a placental mammal during its longer gestation. Once the young leave the pouch, the milk becomes more concentrated with more fat and protein and less sugar. Most terrestrial placentals produce milks that are intermediate in concentration to the nutrient-rich milk of aquatic mammals and the very dilute milks of primates and perissodactyls. The milks of domestic cattle, goats, and camels contain about one-half the protein and energy per unit volume that occurs in the milks of wild even-toed hoofed mammals (also known as artiodactyls). The dilute milks from domestic artiodactyls are more similar to that produced by humans than they are to wild artiodactyls. Because sugars, particularly lactose, and some minerals such as sodium and potassium are important regulators of the osmotic potential or water content of milk in the mammary gland, concentrated milks have either a low sugar (such as marine mammals) or mineral content, while dilute milks have a higher sugar (such as primates and perissodactyls) or mineral content.

Koalas (*Phascolarctos cinereus*) have developed bacteria in their stomachs to break down the toxins found in the eucalyptus leaves they eat. (Photo by © L. Clarke/Corbis. Reproduced by permission.)

The actual composition of the milk fat, protein, and sugar also differs between animals. For example, the fatty acids of most milk are dominated by palmitic and oleic acids. However, the main fatty acid of lagomorph and elephant milk is capric acid, which is synthesized in the mammary gland. Seal milk is composed of long-chain unsaturated fatty acids that are probably derived directly from the diet.

In carnivores, the amino acid, taurine, appears to be essential for the diets of most of their neonates. The taurine content of colostrum is usually higher than in mature milk. However, carnivores have a much higher concentration of taurine in the mature milk than do herbivores.

Resources

Books

Robbins, C. T. *Wildlife Feeding and Nutrition*, 2nd edition. San Diego: Harcourt Brace Jovanovich, 1993.

Schmidt-Nielsen, K. *Animal Physiology: Adaptation and Environment*, 4th edition. Cambridge: Cambridge University Press, 1990.

Vague, J. *Obesities*. London: John Libbey and Company, 1998.

Periodicals

Barry, R. E., Jr. "Length and Absorptive Surface Area Apportionment of Segments of the Hindgut for Eight Species of Small Mammals." *Journal of Mammalogy*, 58 (1977): 1978–1984.

Bishop, J. P., J. A. Froseth, H. N. Verettoni, and C. H. Noller. "Diet and Performance of Sheep on Rangeland in Semiarid Argentina." *Journal of Range Management*, 25 (1975): 52–55.

Blix, A. S., H. J. Grau, and J. Ronald. "Some Aspects of Temperature Regulation in Newborn Harp Seal Pups." *American Journal of Physiology*, 236 (1979): R188–R197.

Kastelein, R. A., N. Vaughan, S. Walton, and P. R. Wiepkema. "Food Intake and Body Measurements of the Atlantic Bottlenose Dolphins (*Tursiops truncatus*) in

Resources

Captivity." *Marine Environmental Research*, 53 (2002): 199–218.

Kastelein, R. A., C. A. van der Elst, H. K. Tennant, and P. R. Wiepkema. "Consumption and Growth of a Female Dusky Dolphin (*Lagenorhynchus obscurus*)." *Zoo Biology*, 19 (2000): 131–142.

Kastelein, R. A., N. M. Schooneman, N. Vaughan, and P. R. Wiepkema. "Food Consumption and Growth of California Sea Lions (*Zalophus californianus californianus*)." *Zoo Biology*, 19 (2000): 143–159.

Kurta, A., G. P. Bell, K. A. Nagy, and T. H. Kunx. "Energetics of Pregnancy and Lactation in Free-ranging Little Brown Bats (*Myotis lucifugus*)." *Physiological Zoology*, 62 (1989): 804–818.

Lopez, S., and J. J. Robinson. "Nutrition and Pregnancy in Sheep." *Investigacion Agraria Production y Sanidad Animales*, 9, no. 2 (1994): 189–219.

Mattingly, D. K., and P.A. McClure. "Nutrition and Pregnancy in Sheep." *Ecology*, 63 (1982): 183–195.

Posse, G., J. Anchorena, and M. B. Collantes. "Seasonal Diets of Sheep in the Steppe Region of Tierra del Fuego, Argentina." *Journal of Range Management*, 49 (1996): 24–30.

Robbins, C. T., and A. N. Moen. "Uterine Composition and Growth in Pregnant White-tailed Deer." *Journal of Wildlife Management*, 39 (1972): 684–691.

Stewart, R. E. A., and D. M. Lavigne. "Uterine Composition and Growth in Pregnant White-tailed Deer." *Journal of Mammalogy*, 61 (1980): 670–680.

Squires, V. R. "Uterine Composition and Growth in Pregnant White-tailed Deer." *Journal of Range Management*, 35 (1982): 116–119.

Widdowson, E. M. "Milk and the Newborn Animal." *Proceedings of the Nutrition Society*, 35 (1982): 116–119.

Worth, G. A. J., and D. M. Lavigne. "Changes in Energy Stores during Postnatal Development of the Harp Seal, *Phoca groenlandica*." *Journal of Mammalogy*, 64 (1983): 89–96.

Other

Dierenfeld, E. S. "Mammal Nutrition: Basic Knowledge and Black Holes." *Wildlife Conservation Society*. <http://www.zcog.org/>

Jasmin Chua, MS

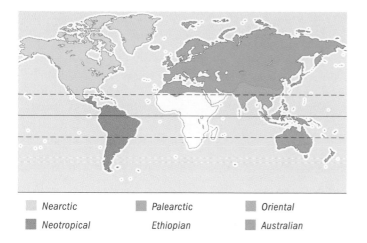

Biogeographical regions of the world. (Map by XNR Productions. Courtesy of Gale.)

Legend:
- Nearctic
- Neotropical
- Palearctic
- Ethiopian
- Oriental
- Australian

years ago until domesticated varieties were brought back by the Spanish conquistadors in the early part of the fifteenth century. Tapirs also evolved in North America and dispersed southwards into Central and South America and northwards into Asia. They subsequently became extinct in North America, leaving the current discontinuous distribution, with three species in Central and South America and one in Southeast Asia.

Offshore islands can be reached by swimming mammals, and a chain of islands allows species to reach the farthest islands on the stepping-stone principle, moving from one island to another. However, the farther away an island is from the nearest mainland, the fewer the number of species that succeed in reaching them. Bats can fly to them and small mammals may arrive by chance, carried there on logs or rafts of floating vegetation.

There are many other barriers to dispersal, such as large rivers, mountain ranges, deserts or other unsuitable habitat, the presence of predators, or competing species. Any factor or combination of factors that cause isolation may lead to further speciation (formation of new species). Variations in climatic conditions during the Pleistocene, especially the more than 20 major glaciations that occurred in the Northern Hemisphere, have also had a significant effect on the distribution of species.

Conditions during the Pleistocene also affected marine environments. For example, cold Arctic and Antarctic waters, rich in dissolved CO_2 and O_2, together with upswellings of mineral-rich deep ocean currents generated tremendous plantation productivity, providing a foundation for the evolution of baleen whale diversity.

Major regions

These differences are conventionally discussed in terms of the division of the world into six major regions as originally proposed in the nineteenth century and followed since with some modifications. These are the Nearctic, Neotropics, Palaearctic, Ethiopian or Afrotropical, Oriental, and Australian.

Nearctic

This region comprises North America up to northern Mexico and Greenland. Ten orders are present, including 37 families, and around 643 species. Two families are endemic, each containing a single species. These are the Antilocapridae (pronghorn) and the Aplodontidae (sewellel, or mountain beaver), endemic to western North America. There are a large number of endemic rodents. These include the woodrats (genus *Neotoma*), 17 species of ground squirrel (*Citellus*), three antelope squirrels, 16 chipmunks, 10 squirrels and flying squirrels, 12 pocket gophers, and 37 species of heteromyid rodents (pocket mice, kangaroo mice, and kangaroo rats). Characteristic larger endemic species include bison (*Bison bison*), mountain goat (*Oreamnos americanus*), bighorn sheep (*Ovis canadensis*), and thinhorn sheep (*O. dalli*).

The Nearctic shares many aspects of its mammal fauna with the Palaearctic and Neotropical regions. More than 80% of Nearctic families and 60% of the genera also occur in the Neotropical region. There are relatively few species of Neotropical origin in the Nearctic fauna. Only the southern opossum (*Didelphis marsupialis*), the nine-banded armadillo (*Dasypus novemcinctus*), and some bats have survived of those that entered the region from the south, in contrast to the much larger number of Nearctic species that entered the Neotropical region following formation of the Panamanian land bridge. Nearly half the families are shared with the Palaearctic region, a reflection of the length of time that the two regions have been connected across the Bering Strait either by a land bridge or a chain of islands. Twenty-one genera arrived from the Palaearctic at the end of the Pleistocene. Shared groups include several families of Carnivora (Felidae, Canidae, Mustelidae, Ursidae), deer (Cervidae), shrews (Soricidae), and moles (Talpidae).

A number of species in high latitudes have a circumpolar distribution in the Palaearctic and Nearctic regions, such as caribou (*Rangifer tarandus*), wolverine (*Gulo gulo*), gray wolf

A Galápagos sea lion (*Zalophus californianus wollebaecki*) basks in the sun near a brown pelican. (Photo by Harald Schütz. Reproduced by permission.)

Pronghorn antelope (*Antilocapra americana*) inhabit the Great Plains of North America. (Photo by Joe Van Wormer. Bruce Coleman, Inc. Reproduced by permission.)

Nineteen of the 50 families and about 80% of the almost 1,100 species that occur are also endemic. These include ten endemic families of rodents, including the guinea pigs, chinchillas, and agoutis; several families of bats; and two families of primates. Two genera of wild camelids (*Vicugna* and *Lama*) are endemic. The guanaco (*L. guanicoe*) is evidently the progenitor of two domesticated varieties, the llama and alpaca.

The Neotropical mammal fauna consists of three main strata. The ancestral fauna consisted of some very distinctive extinct orders, the order Xenarthra and early marsupials. These were augmented by intermittent "invaders" from the north, such as primates, rodents, and some carnivores. Then about 2 million years ago during the Pleistocene, formation of the Panamanian land bridge allowed the immigration of many new forms from North America. These included perissodactyls (tapirs and horses), artiodactyls (camelids, deer, and peccaries) and carnivores (felids, canids, and mustelids). This wave of mammalian invasions had a drastic effect on the existing fauna and resulted in many extinctions, including the large herbivore orders Notoungulata and Litopterna, and ground sloths. Ultimately, a unique array of mammals, perhaps as distinctive as those of Australia, disappeared alto-

(*Canis lupus*), Arctic fox (*Alopex lagopus*), and moose (*Alces alces*). Musk ox (*Ovibos moschatus*) also once ranged all around the tundra zone but is now extinct in Eurasia apart from some small introduced populations.

Many species in the boreal zones of both regions have a near counterpart in the other, for example pine marten (*Martes martes*) and stone marten (*M. foina*) in the Palaearctic and American marten (*M. americana*) and fisher (*M. pennanti*) in North America; European otter (*Lutra lutra*) and river otter (*L. canadensis*); American mink (*Mustela vison*) and European mink (*M. lutreola*).

Neotropical region

The Neotropics includes South America, Central America from southern Mexico southwards, and the West Indies. This region contains a very diverse fauna. Twelve orders are represented, two of them endemic. The Paucituberculata, shrew opossums, consists of three genera and five species, all found in the Andean region. The Microbiotheria has a single species, the monito del monte (*Dromiciops australis*), distributed in southern South America. The order Xenarthra (sloths, armadillos and anteaters) is mainly restricted to the Neotropical region, with one species occurring in the southern part of the Nearctic. The marsupial order Didelphimorphia has 63 species, all but one of which are restricted to the Neotropics.

A polar bear (*Ursus maritimus*) swimming underwater. (Photo by Mark Newman. Bruce Coleman, Inc. Reproduced by permission.)

Leopards (*Panthera pardus*) at sunset in the Masai Mara Game Reservation in Kenya, Africa. (Photo by Mark Newman. Bruce Coleman, Inc. Reproduced by permission.)

gether. This interchange has been mainly, but not exclusively, one way. Many more species moved south than in the opposite direction. Northern species entering South America have also proved more successful colonizers, penetrating to the southern tip of the continent.

The high degree of mammalian diversity is mainly a result of South America's isolation from other major land masses for long periods of geological history. Climate changes during the Pleistocene and alternating wet and dry periods caused the rainforest to contract and fragment, perhaps becoming in effect a series of forest islands separated by dry savanna or scrub that acted as barriers to dispersal and further leading to the development of new forms and evolution of new species.

There is considerable habitat diversity within the Neotropical region. In addition to the tracts of rainforest there are also extensive areas of dry scrub woodland, tropical savanna, temperate grasslands, desert, and high mountains. The Andes run almost the length of the continent of South America and reach an altitude of 22,830 ft (6,960 m) at their highest point. The rainforest does not cover a continuous extent but is divided into four main blocks by large rivers, mountains, and extensive areas of drier habitat types. The Atlantic rainforest of southeastern Brazil has been completely isolated

from the Amazon rainforest for a very long time and has its own endemic genera and species. During the Pleistocene, alternating wet and dry periods heavily influenced the extent of rainforest. At times this contracted to smaller patches isolated by expanses of arid habitats that acted as barriers to the dispersal of rainforest mammals. Many species subsequently evolved in these isolated forest refugia. Large rivers also act as dispersal barriers and related but separate species occur on opposite banks. For example, the fauna north and south of the Orinoco in northern South America show a number of differences, and the Amazon and Rio Negro also isolate some species.

Much of the characteristic fauna of the West Indies has disappeared; in fact, a disproportionate number of the mammals that have become extinct in recent times were endemic to the West Indies. These include an entire family containing eight species of shrews (*Nesophontes*), a species of raccoon (*Procyon gloveralleni*) formerly found on Barbados, 21 species of rodents, and the Caribbean monk seal (*Monachus tropicalis*). There is still one endemic family of large insectivores—the Solenodontidae—with two extant species, one each on the islands of Cuba and Hispaniola (Haiti and the Dominican Republic). These are thought to be island relics of a formerly much more widespread group.

An American pika (*Ochotona princeps*) living among the rocks in Colorado, USA. (Photo by John Shaw. Bruce Coleman, Inc. Reproduced by permission.)

Palaearctic region

The Palaearctic region covers Europe, North Africa, most of the Arabian Peninsula, and Asia north of the Himalayas, including Japan and Korea. The Palaearctic is the largest of the major faunal regions in terms of its geographical area and it contains a wide range of habitats. Despite this, it has the lowest rate of endemism of all the major faunal regions. Thirteen orders and 42 families are present, none of them endemic. About 30% of the 262 genera and 60% of the 843 species that occur are endemic. The relatively low level of endemism reflects the long periods of contact at various times with the Nearctic, Oriental, and Ethiopian regions. Around half the Palaearctic families also occur in the Nearctic region, a consequence of the intermittent existence of the Bering land bridge. About 60% of the families also occur in the Oriental region.

The Palaearctic lacks the great diversity of ungulates found in the Ethiopian region, though a few species occur, including an endemic genus of Central Asian gazelles, *Procapra*. Groups that evolved in southern Eurasia such as deer

and caprins are well-represented, with members of the latter group found in virtually all the regions' mountain ranges.

Equids and camelids both arrived from North America and still survive in the Palaearctic, though the distributions of all the surviving species are greatly reduced. Przewalski's horse (*Equus caballus*) was last seen in the wild in the 1960s, though reintroduction projects aimed at returning it to the wild in Mongolia began during the 1990s. Two species of wild ass still survive. Kulan, or onager (*E. hemionus*), is widespread in Mongolia, and is still found in fragments of its former range, in western India and in Turkmenistan. Kiang, or Tibetan wild ass (*E. kiang*), remains numerous on the Tibetan Plateau. Wild bactrian camels are now restricted to three small areas of the Gobi and Takla Makan deserts.

A high-altitude ungulate fauna has evolved on the Qinghai-Tibet Plateau, adapted to elevations above 13,000 ft (4,400 m) and prolonged periods of cold in winter. Component species include Tibetan antelope, or chiru (*Pantholops hodgsonii*), Tibetan gazelle (*Procapra picticaudata*), wild yak (*Bos grunniens*), Kiang (*Equus kiang*), Tibetan argali (*Ovis ammon hodgsoni*), and blue sheep (*Pseudois nayaur*) with white-lipped deer (*Cervus albirostris*) in the eastern third of the plateau. The diversity of these herbivores does not approach that of the East African savannas but they occurred in large numbers, with several nineteenth century travelers and sportsmen reporting vast herds of thousands of animals. These numbers have since been reduced, but in the eastern steppes of Mongolia, herds of several thousand Mongolian gazelles (*Procapra gutturosa*) can still be seen.

Of the smaller groups, there are 19 endemic species of pika (*Ochotona*) distributed through the mountains of Central Asia and the Himalayas. Two species of desmans, endemic insectivores, occur in the Pyrenees and Russia respectively. All but one of 20 species of dormice (Myoxidae) are endemic to the Palaearctic and there is a diverse assemblage of desert rodents in the region: 32 jerboas (Dipodidae), and 41 gerbils and jirds (Muridae).

The Hodgson's brown-toothed shrew (*Soriculus caudatus*) inhabits damp areas of the forests of Bhutan, China, northern India, northern Myanmar, Nepal, Sikkim, Taiwan, and Vietnam. (Photo by Harald Schütz. Reproduced by permission.)

Adult African elephant (*Loxodonta africana*) and calf drink water in the arid plains of Africa. (Photo by St. Meyer/OKAPIA/Photo Researchers, Inc. Reproduced by permission.)

In the eastern Palaearctic there are several island endemics. Japan has an endemic species of macaque (*Macaca fuscata*), serow (*Capricornis crispus*), and hare (*Lepus brachyurus*). The Ryukyu rabbit (*Pentalagus furnessi*) is endemic to two of the Ryukyu Islands.

Ethiopian region

This region includes Africa south of the Sahara, Madagascar, and the southwest corner of Arabia. Its mammal fauna exhibits the greatest diversity of all the major regions and 13 out of 26 orders are present. One of these is endemic, Tubulidentata, with a single species, the aardvark. There are 52 families (18 endemic), and over 1,000 species (more than 90% endemic). Endemic families include Giraffidae, Hippopotamidae, Chrysochloridae (golden moles), Tenrecidae (tenrecs), and Macroscelididae (elephant shrews).

Africa was once joined to the Oriental region so there are many elements in common. One order, Pholidota (pangolins or scaly anteaters) contains one genus (*Manis*) with four Ethiopian species and three in the Oriental region. There are also many families in common, but long periods of isolation

have led to the development of unique genera, for example elephants, rhinos, monkeys, apes, hyenas, and cattle.

The region is noted for the impressive array of large herbivores that occur in large numbers on the savannas of central, eastern, and southern Africa. There are also seven endemic genera of Old World monkeys, and two out of the four genera of great apes (chimpanzees and gorillas). There is also a great diversity of rodents and viverrid carnivores (23 out 25 genera are endemic). There are some noteworthy absences from the region, such as deer (Cervidae) and bears (Ursidae).

The unique nature of Madagascar's mammal fauna is well known and results from its long isolation from the mainland of Africa that now lies more than 250 mi (420 km) away. Madagascar finally broke away from Africa during the Middle Tertiary period and only a few groups of mammals appear to have been present at that time, namely lemurs, insectivores, small carnivores, and rodents. In the absence of competitors and later immigration, these ancestral groups were able to radiate into diverse arrays of new species. The lemurs developed into three families (Lemuri-

A coyote (*Canis latrans*) with a fish in Texas, USA. (Photo by John Snyder. Bruce Coleman, Inc. Reproduced by permission.)

dae, Indriidae, Daubentoniidae) containing nine genera and 38 species. The tenrecs provide an excellent example of adaptive radiation, the 27 extant species having a bewildering variety of forms and occupying a great diversity of niches. The early carnivores consisted only of mongooses and civets and both have evolved into distinctive forms. The four extant mongoose species form an endemic subfamily Galidiinae. The civets also show a remarkable development, with seven species belonging to seven separate genera in three subfamilies. There are ten endemic rodents and an endemic bat family (Myzopodidae) containing a single species. The other bats present were presumably able to fly to the island at a later stage. Almost equally striking is the absence of so many widespread African forms such as antelopes, zebras, larger carnivores, lagomorphs, and monkeys. At some point, the bushpig (*Potamochoerus porcus*) and a small species of hippopotamus (*Hippopotamus lemerlei*) reached the island; the bushpig survives but the hippo became extinct, probably in prehistoric times.

Oriental region

The Oriental region includes Asia south of the Himalayas, southern China, the Philippines, and Southeast Asia up to Wallace's Line, between the islands of Bali and Lombok. The region has two endemic orders, Scandentia (tree shrews), with 19 species, and Dermoptera (colugos). There are two species of colugos, often also called flying lemurs, a doubly confusing name as they are not lemurs and they glide, rather than fly. There are 50 families in the region, four endemic, and 260 genera, about 35% of them endemic. About two thirds of the more than one thousand species are also endemic. Endemic families include Kitti's hog-nosed bat (Craseonycteridae); tarsiers (Tarsiidae); gibbons (Hylobatidae); and tree shrews (Tupaiidae). The region has strong affinities with the Palaearctic and Ethiopian regions. There is a long land boundary with the Palaearctic along the Himalayas and through China, and almost 75% of the families are shared with the Palaearctic region. These include bears (Ursidae), deer (Cervidae), musk deer (Moschidae), and Felidae, includ-

ing the tiger (*Panthera tigris*). Wooded savannas formerly connected the Indian subcontinent with Africa, although these linking areas now consist largely of desert. Groups in common between the Oriental and Ethiopian regions include elephants (one species in each), rhinoceroses, big cats (lion, leopard, and cheetah), viverrids, and great apes. Orangutans (genus *Pongo*) occur on Borneo and Sumatra and the 14 species of gibbons are distributed from eastern India and southern China through Southeast Asia. There are seven endemic genera of monkeys containing 26 species. Two of these are restricted to Sri Lanka, and two to the Mentawai Islands off the coast of Sumatra. The Oriental region lacks the great diversity of antelopes and other herbivores present in Ethiopian region, though a few species are present. Three are endemic to India, blackbuck (*Antilope cervicapra*), nilgai (*Boselaphus tragocamelus*), and four-horned antelope (*Tetracerus quadricornis*). The herbivore niches are filled in part by several species of deer. Other artiodactyls endemic to the region include five species of wild pigs, and four species of wild cattle including the little-known kouprey (*Bos sauveli*). The best-known endemic is undoubtedly the giant panda (*Ailuropoda melanoleuca*), which is indigenous to western China.

Australian region

This region includes Australia, New Zealand, New Guinea, Sulawesi, and some islands in the southwest Pacific. The Australian fauna is the least diverse in terms of the number of orders and species present, but certainly the most distinctive of all the major regions. Excluding Cetacea, only eight orders of mammals are native to Australia, but five of these, Monotremata and four orders of marsupials, are endemic. The other three indigenous orders are Chiroptera, Rodentia, and Carnivora. The latter is represented only by seals, though one terrestrial species, the dingo (*Canis familiaris dingo*), was brought to Australia by early human inhabitants several thousand years ago. Further introductions have been made by European settlers. There are 17 endemic families, and 60% of the genera and nearly 90% of the species are also endemic.

This distinctive fauna is the result of long isolation after the region broke away from Gondwana around 55 million years ago. When this event took place, only the monotremes

Lion (*Panthera leo*) courtship in Kenya, where there is a long dry summer. (Photo by Harald Schütz Reproduced by permission.)

and marsupials were present. The Monotremata are an ancient evolutionary line composed of three species, the duck-billed platypus (Ornithorhynchidae) and two species of echidna or spiny anteater (Tachyglossidae). Marsupials appear to have reached Australia by a filter route from South America via Antarctica. When Australia and Antarctica separated, these original marsupials were isolated. During the 40 million years that followed they radiated to fill most of the niches occupied elsewhere by placental mammals. Larger species of kangaroos and wallabies occupy the large herbivore niche filled by ungulates in most of the rest of the world.

Rodents arrived in two separate migrations. It appears that rodents reached New Guinea from Southeast Asia and moved on into Australia by using islands as stepping stones. Upon reaching Australia, they radiated into many species. There are currently around 13 genera of rats and mice. Some bats are also thought to have reached the continent at a very early stage, with others entering the region later from the north. There are two endemic genera.

There are four orders of marsupials. Dasyuromorphia contains 17 diverse genera including the Tasmanian wolf (*Thylacinus cynocephalus*), Tasmanian devil, and the numbat or banded anteater. Order Peramelemorphia consists of small species, the bandicoots and bilbies. Order Notoryctemorphia has two species of burrowing marsupial "moles." The Diprodontia contains ten families and about 113 species including the familiar kangaroos, wallabies, and koala, as well as cuscuses and possums. The Australian region contains very few carnivores compared with other regions, only six species in Australia and five in New Guinea.

In more recent times, European settlers introduced several species that have succeeded in colonizing all or large parts of the continent. These include the house mouse (*Mus musculus*), rabbit (*Oryctolagus cuniculus*), red fox (*Vulpes vulpes*), and feral cat (*Felis catus*). Several domestic herbivores have also established feral populations. Rabbits have degraded vegetation over vast areas and introduced foxes are blamed for the destruction of much of the native fauna. House mice and feral cats are distributed across most of Australia, and the rabbit and fox over about 60%.

The island of Tasmania has many common forms and has acted as a refuge for others. The recently extinct thylacine or Tasmanian wolf (*Thylacinus cynocephalus*) was present here in historic times and is unconfirmed from the mainland. It is still thought by some to be possibly present. The Tasmanian devil (*Sarcophilus harrisii*) survives only on the island but was formerly found across Australia. Both may have suffered from competition with the dingo.

Australia and New Guinea have been isolated for most of the last 45 million years by the 100-mi (160-km) wide Torres Strait, and they have been only intermittently connected to each other. New Guinea has many marsupial species including endemic genera and one endemic monotreme species, the long-beaked echidna (*Zaglossus bruijni*). There are also a number of highly endemic mice and bats. The murid fauna has evolved in isolation from that of Australia. The fauna of the islands at the western end of the region, lying between

Red-necked wallabies (*Macropus rufogriseus*) are especially common in Queensland, northeastern New South Wales, and Tasmania but also inhabit the coastal forests of eastern and southeastern Australia. (Photo by E & P Bauer. Bruce Coleman, Inc. Reproduced by permission.)

the two continental shelves, is a mix of Australian and Oriental forms. The largest island, Sulawesi, contains a number of endemic species. These include the babirusa (*Babyrousa babyrussa*)—an aberrant type of wild pig, two species of anoa—the smallest of the wild cattle, and four species of macaques.

New Zealand's mammal fauna is a special case. The islands broke away from Gondwana and drifted across what is now the Pacific about 80 million years ago. Only 11 mammal species are indigenous—four bats and seven pinnipeds (seals and sea lions). However, many more have been introduced and its mammal fauna consists of 65 species. The first introductions, of rats and dogs, were made by Polynesian settlers who reached the islands around 1,000–1,200 years ago. European colonists arriving from 1769 onwards brought in many more, mostly for food or game. These include 23 marsupials and 14 ungulates (deer, chamois, and Himalayan tahr). Out of 54 known introductions, 20 came from Europe, especially Great Britain, 14 from Australia, six from Asia (three surviving), and 10 from North or South America (three are estab-

North American beavers (*Castor canadensis*) live in the wetlands of North America. (Photo by Windland Rice. Bruce Coleman, Inc. Reproduced by permission.)

lished locally). Introduced species far outnumber native mammals and there are now more large species than small mammals, the opposite of the usual situation.

The smaller Pacific Islands have generally impoverished mammal faunas reflecting the difficulties in colonizing them. The bats are the best represented order, which is unsurprising in view of their ability to fly to small oceanic islands. Sixteen species of endemic bats occur on the Solomon Islands and two more are extinct. There are five endemic species of bats on the Melanesian Islands. One endemic species on Guam (*Pteropus tokudae*) is extinct, but a new species of bat was discovered on Guadalcanal in 1990. Rodents are the next most frequent group and there are some endemic mice. Ancestors of some of these presumably reached the islands by chance, floating on logs or rafts of vegetation.

Antarctica

While the Arctic is covered within the Nearctic and Palaearctic, the fauna of Antarctica is usually omitted from descriptions of faunal regions. As noted above, no terrestrial mammals occur in the Antarctic but several species of cetaceans and seals use Antarctic waters. Crab-eater seals (*Lobodon carcinophagus*) occur around the coasts and pack ice of Antarctica and occasionally haul out on the shore. One has even been found on a glacier at an altitude of 3,600 ft (1,100 m). The Weddell seal (*Leptonycotes weddellii*), probably the most southern of the world's mammals, prefers land fast ice to pack ice and is usually found in sight of land. Ross seals (*Ommatophoca rossi*) also inhabit the Antarctic pack ice. Leopard seals (*Hydrurga leptonyx*) are distributed throughout Antarctic waters where they prey on smaller species. Orcas and a few other cetaceans are regularly seen in these southern waters. More cetaceans and pinnipeds, including Antarctic fur seals and sea lions, occur farther north in subantarctic waters and around islands such as South Georgia, Macquarie Island, and Kerguelen. Reindeer (*Rangifer tarandus*) have been introduced to South Georgia.

The marine realm

While seals and possibly sirenians may be included in descriptions of the conventional biogeographic regions, this is not usually the case with cetaceans. It seems worthwhile to consider all the above in a marine realm for the sake of completeness. Marine mammals are distributed from the Arctic Ocean to the Antarctic and occur in the deep ocean, coastal waters, estuaries, and in a few cases in rivers and lakes. Two orders are fully aquatic, Cetacea (whales and dolphins) and Sirenia (manatees and dugongs). The third order, the Pinnipedia, is sometimes classified as part of the Carnivora. It consists of three families, seals, sea lions and fur seals, and walruses. They are mainly aquatic but haul out onto land or ice to rest, mate, and give birth.

The 76 species of whales and dolphins collectively have a worldwide distribution. Several species have extensive individual ranges in all the major oceans. Freshwater species of dolphins live in the Amazon, Yangtze, Indus, and Ganges Rivers. The four living species of sirenians are found in coastal waters, estuaries, and rivers mainly in the tropical and subtropical zones. Steller's sea cow (*Hydrodamalis gigas*), the only cold-water adapted species, formerly lived in the Bering Sea but was hunted to extinction by 1768.

There are 33 extant species of pinnipeds, mainly distributed in temperate and polar waters but a few species are found in the tropics. The 14 species of fur seals and sea lions are mainly distributed in the Pacific and southern oceans. One species is endemic to the Galápagos Islands, and two more have restricted distributions on Juan Fernandez and nearby islands and islands off California. The walrus (*Odobenus rosmarus*), the only member of the family Odobenidae is an inhabitant of the Arctic. The 19 species of so-called true seals, Phocidae, are predominantly distributed in northern and southern waters. Six species occur in Arctic and subarctic waters and four in the Antarctic. The Hawaiian monk seal (*Monachus schauinslandi*) is endemic to those islands. Isolated species occur in the Caspian Sea (*Phoca caspica*) and in the freshwater of Lake Baikal in Siberia (*P. sibirica*).

In addition, the polar bear (*Ursus maritimus*) and the sea otter (*Enhydra lutris*) of western North America also spend much of their lives in a marine environment. Taken together, marine mammals amount to only about 2.5% of all mammal species, despite the fact that seas and oceans cover a greater proportion of the surface of the earth than land does. This reflects the relative homogeneity of the environment and a lack of natural barriers that allow species to evolve in isolation.

The influence of humans

Humans have had a profound negative impact on the distribution of the world's mammals, through hunting, habitat destruction, and introductions. The consequences of hunting may have begun with early humans, as the Columbian mammoth (*Mammuthus columbi*) went extinct after only 500 years of contact with Clovis humans. Many of the large mammals of the Mediterranean coasts of North Africa were exterminated by the Romans who exported huge numbers of them to Rome to be killed in public arenas. The invention of modern firearms drastically increased the destructive potential of hunting. An estimated 60 million bison (*Bison bison*) existed across North America at the beginning of the eighteenth century but 150 years later they had been hunted to the brink of extinction and their former distribution reduced to fragments. During the nineteenth and twentieth centuries Arabian oryx (*Oryx leucoryx*) and gazelles were severely depleted in Arabia and North Africa, and the tiger (*Panthera tigris*) disappeared from large swathes of its former distribution and from Bali, Java, and Central Asia altogether. Domestication of sheep, goats, and cattle and their subsequent spread around most of the world have led to them becoming the most widespread of all mammals. The total biomass of domestic animals exceeds that of wild species in many places. The domestic animals compete for grazing and through overgrazing, degrade the habitat so that it can no longer support the original wild populations whose ranges shrink accordingly. Some species such as rats and mice have been introduced accidentally through transport with ships' cargo; others have been introduced deliberately, for sport or amenity. In virtually all cases they have had an adverse effect on the indigenous faunas. The effects of introduced foxes and rabbits on the habitats and wildlife of Australia were mentioned above. In Great Britain, the North American gray squirrel (*Sciurus carolinensis*) was released in the nineteenth century; since then it has steadily expanded its range in the country at the expense of the native red squirrel (*S. vulgaris*).

This generally negative trend has been partially reversed by reintroduction programs that restore former distributions. In the European Alps, reintroductions of the Alpine ibex (*Capra ibex*) in the nineteenth century and of lynx (*Lynx lynx*) during the 1980s-1990s have proven very successful. In the Arabian Peninsula, the Arabian oryx (*Oryx leucoryx*), mountain gazelle (*Gazella gazella*), and Arabian sand gazelle (*G. subgutturosa marica*) have all been returned to the wild during the 1990s, albeit in limited areas.

Resources

Books

Eisenberg, J. F. *The Mammalian Radiations*. Chicago: University of Chicago Press, 1981.

Eisenberg, J. F. *Mammals of the Neotropics*. Volume 1. Chicago: University of Chicago Press, 1989.

Feldhammer, G. A., L. C. Drickhamer, S. H. Vessey, and J. F. Merritt. *Mammalogy: Adaptation, Diversity, and Ecology*. Boston: McGraw-Hill, 1999.

Flannery, T. *Mammals of New Guinea*. Ithaca: Cornell University Press, 1995.

King, C. M. *The Handbook of New Zealand Mammals*. Auckland: Oxford University Press, 1990.

Macdonald, D. W., and P. Barrett. *Mammals of Britain and Europe*. London: Collins, 1993.

Nowak, R. M. *Walker's Mammals of the World*. 6th edition. Baltimore: The Johns Hopkins University Press, 1999.

Simpson, George Gaylord. *In Splendid Isolation*. New Haven: Yale University Press. 1980.

Strahan, R. *Mammals of Australia*. Washington DC: Smithsonian Institution Press, 1995.

David P. Mallon, PhD

Behavior

When asking what is typically mammalian in behavior, we must first consider which adaptations and preconditions of a mammal normally shape its life and body. Mammals are warm-blooded, or endothermic, and their system of body temperature regulation through metabolism requires more energy than what is needed by ectotherms. Foraging is also an important aspect of behavior, and has to be considered as a decisive factor in shaping social systems. Also, mammals in general (and female mammals specifically) invest a lot more in terms of time, effort, energy, nutrition, and risk, into their offspring than most other vertebrates do. Again, this shapes social systems, in particular mating and rearing, but also puts severe demands on foraging strategies. Another characteristic is the highly developed brain, specifically in those areas that are necessary for behavioral plasticity and variability, such as the highly evolved forebrain and its hemispheres. This in turn allows the mammal to adapt to a diversity of ecological conditions, and also to form complex and individualized societies. In connection with the intensive and often long periods of infant care, not only by the parents but also other members of the group, this can lead, again, to highly variable and adaptable solutions to ecological as well as social problems and situations. In this chapter, we will cover two of those areas in which mammals are special: learning and behavioral plasticity, and social systems (which include mating and rearing as well as foraging and anti-predator systems). Each of these fields is currently the focus of scientific attention in many places, and by many different approaches. In order to fully understand any biological phenomenon, Tinbergen in 1963 proposed to answer four questions, and only after getting satisfactory answers to all four can we presume that we have "explained" this phenomenon. They are:

- Where did it come from in evolution?

- What selective advantage does an individual get from having this particular trait (the so-called ultimate reasons)?

- How does it work (physiology, so-called causal mechanisms)?

- How does it develop in an individual's life (so-called ontogeny)?

We shall use these four questions to structure our discussion of mammalian behavior. In order to answer these questions, a combination of different scientific approaches is necessary. Thus, we will draw data from long-term field studies as well as from laboratory and zoo research, from experimental trials as well as from purely observational approaches, and will also need support from other biological disciplines such as endocrinology and molecular genetics. Behavior in itself is at the interface of genetics and ecology, and its understanding is central also to questions of animal welfare, conservation biology, zoo management, and our relationship with pets and companion animals.

Behavioral plasticity

Learning in itself, of course, is by no means specific for mammals, or even higher animals. When asking the first Tinbergen question, we then have to look for those areas of behavioral plasticity that distinguish mammals from their reptilian ancestors. So-called higher forms of learning, which require certain degrees of neural complexity, are (among others) spatial memory and cognitive mapping. Predators that follow prey, primate bands that follow certain routes between sleeping and foraging sites, caribou that migrate over long distances, and other mammals on the move often display an astonishing ability to cut corners, find shortcuts over ridges, circumvent deep parts in rivers after nightly rainfall, and still arrive at their destination without delay. Caribou that are delayed by late snowfall in spring even use these shortcuts to save time in migration. In all of these cases, some sort of "map" must be represented in the animals' nervous systems, and each element of the map must not only have an "address," but also a possibility to relate it to other elements. Another form of behavioral plasticity is called "problem-solving by insight." In typical cases, an animal is confronted by a situation it cannot immediately solve, such as bananas hanging too high to reach, or food hidden in a box. Problem-solving by insight requires that the animal first familiarize itself with the situation and then start to act in a goal-directed way (such as using a tool, elongating one stick with another one, or opening the lid of the box with a lever). Tool use has been described for mammals from at least six orders. A tool here is defined as a movable object that is not a fixed part of the animal's body, is being carried shortly before or during usage, and is positioned in an adequate way for its subsequent use. Following this definition, mongoose use tools to crack eggs, sea

1. The silverback gorilla shows dominance through size and fur coloration; 2. Bighorn sheep posture and fight to establish dominance; 3. Wolves use subtle body language to show submission and dominance; 4. Kangaroos face each other, standing erect, crouching, and grooming themselves while challenging a competitor before fighting. (Illustration by Wendy Baker)

otters carry stones as anvils, elephants use twigs to swat flies, primates throw stones and branches not only to defend themselves but also to detach fruit from trees, chimpanzees angle for termites, etc. Remarkably, more forms of tool use have been described from captive than free-ranging animals, and only in some apes do we have sufficient evidence for observational learning of tool use from the field.

Even though some of these higher forms of learning and cognition can be found in some birds as well, they are not yet in any case described from reptiles, and we can thus safely assume that the ability for them evolved somewhere in mammalian phylogeny. Thus, question number one seems at least partly answered.

What about selective advantage and survival value? It is of course easy to state that animals that learn better will be better able to cope with environmental challenges and will thus be more apt to survive. Hard evidence from carefully designed studies, however, is scarce. In several vole species of the genus *Microtus*, there is a clear correlation between spatial learning ability and ranging behavior: only in species where males have larger home ranges than females do males fare better in spatial learning (maze-running) tests. In food choice trials with rodents as well as ferrets and other carnivores, decision time was significantly shorter between novel, or new, foods for animals reared with a more variable diet. When an animal is

quicker to reach a decision to eat something, it can eat more per given time, and the extra amount of nutrients certainly is an advantage. Feeding can also become more efficient when search-images have been developed, as demonstrated with hamsters and other rodents. Animals that learn about potentially dangerous predators, as ground squirrels do from hearing other colony members giving warning calls, are another example of learning with a direct survival value.

To address the third question, physiological correlates of learning are known for at least several learning phenomena: brain areas responsible for spatial learning are larger in males of those vole species whose spatial learning is better than females, but not in those without such a sex difference. The olfactory bulb in the brain of a young ferret during the critical period of olfactory food imprinting is larger than before or after this time. We also know that thyroxine, the hormone of the thyroid gland, is responsible for neurological changes during food imprinting in this species, and that oxytocin, a pituitary hormone, is necessary in the brain of monogamous animals to learn who their specific partner is during pair formation. Several areas in the limbic system of the brain, particularly the hippocampus, have been identified as being responsible for exploratory behavior and learning.

So, to address the fourth and last question, what data do we have about ontogenetic influences on behavioral plastic-

Individuals of many species occupy well-defined territories and attack trespassers. 1. A white-tailed buck rubs his horns on a tree, and paws and urinates in a patch of earth to mark his territory during breeding season; 2. Howler monkeys communicate territoriality over long distances with their voices; 3. Wolves use voice to alert others of their location and use scat and urine to mark their territory; 4. Lemurs rub scent from their anal glands as a territory marker; 5. Hyenas use scat, urine, fluid from anal glands, and pawing the earth to mark territory. (Illustration by Wendy Baker)

ity? When observing young mammals, play behavior is among the most obvious patterns performed regularly. There are many suggestions that during play, behavior is trained and general reactivity and adaptability is thus improved. Again, however, there are mostly plausibility arguments for this: "Because play occurs, and because it is costly in terms of time, energy, risk of injury, etc., it must have some positive effect. Otherwise, selection would have abolished it long since." Field studies of the same species under different conditions, with different amounts of juvenile play, often find less social cohesion in those individuals that played less. But this could also result from differences in other ecological conditions.

Nevertheless, we return to the question of learning and socialization in the discussion of social systems and social be-

havior. (There are, however, several studies on the influences of rearing condition and environmental factors on learning and problem-solving later in life.) From studies with laboratory rats and mice, we know, for example, that a well-structured environment, such as cages with climbing and hiding possibilities, is crucial to an animal's later ability to learn how to run through a maze, explore novel situations, climb over ropes, etc. The advantage of using laboratory rodents for these studies is that there are inbreeding strains that differ in learning ability. Thus we have "bright" and "dumb" mice, genetically speaking. However, rearing a "bright" mouse in a boring environment (standard lab cage) and a "dumb" mouse in an enriched, well-structured one leads to a near reversal of their genetic disposition—the "dumb" strain is now as good as, and

A polar bear (*Ursus maritimus*) rolls in the snow in Cape Churchill, Canada. Polar bears keep their coats clean by swimming and rolling in the snow. Clean coats keep the bears warmer. (Photo by Gary Schultz. Bruce Coleman, Inc. Reproduced by permission.)

Coyotes (*Canis latrans*) howl to defend their territory and inform others of their whereabouts. (Photo by Larry Allan. Bruce Coleman, Inc. Reproduced by permission.)

sometimes even better than, the "bright" one. Another approach to ontogenetic studies of learning and problem-solving was taken in studies with juvenile macaques and vervet monkeys. It was found that those monkeys who, as juveniles, were able to control their environment by deciding when to press a lever and get a food reward, later in life were more active in exploring and solving new situations that those that could press the same lever but received the same amount of food via random, computer-generated portions. Similar results are also described for domestic dog puppies raised in a chal-

lenging environment. Even a mild social stress, such as handling them a few times during early pup life, increased their activity levels, exploration, social initiative, and other environmentally directed activities considerably.

Social systems

Sexual reproduction in animals generally puts a heavier load on the female side. In mammals, however, this bias in cost of reproduction is far more extensive due to the period of gravidity (pregnancy) and the subsequent lactational period, both of which cannot be taken over by a male. Consider a female mouse suckling six young: shortly before weaning, each young has about half her weight. Thus, she has to nourish and support 400% of her body weight! There is an even higher evolutionary pressure on mammalian females in at least two aspects: females have to forage more intensively, and more effectively, in order to cover their energetic and nutri-

Bottlenosed dolphins (*Tursiops truncatus*) leap out of the water. (Photo by Hans Reinhard. Bruce Coleman, Inc. Reproduced by permission.)

The southern opossum (*Didelphis marsupialis*) covers itself with a foul saliva that discourages predators before it plays dead. (Photo by Joe McDonald. Bruce Coleman, Inc. Reproduced by permission.)

A jackal guards a lion kill while vultures wait. (Photo by © Peter Johnson/Corbis. Reproduced by permission.)

tional demands of reproduction. Secondly, as each young or litter forms a rather high proportion of her total lifetime reproduction, she is on heavy demand to select her potential mating partner. Male quality is thus very important, and female mate choice can be expected to be even more careful and elaborate than in other vertebrates.

Animal social systems are supposed to evolve in the context of providing each individual with a so-called "optimal compromise" regarding the demands of foraging, predator avoidance, reproduction, and sheltering. We have to accept the fact that a social group (or other social unit above the individual level) is not some sort of super-organism with its own demands and evolutionary history. Instead, each social unit is brought into existence simply and solely if it is catering to the demands of the individual members, and will remain stable only as long as all of its members do not have any option that, regarding this compromise, provides them with better conditions in total. This does not mean that the animals have to be aware of these choices and options. For natural selection to work, it is sufficient if they behave, based on at least some hereditary components of behavior, in the "correct" way, and their reward will likely be to have more, more viable, or otherwise advantageous young. This is the concept of Darwinian fitness—everyone has to put as many bearers of their own genetic heritage into the next generation, and the one with

most young reared successfully into the next generation's gene pool is the fittest. What we as humans use to colloquially call fitness (as in going to a fitness studio) is, in the terminology of behavioral ecology, called resource-holding power (RHP), the possibility to defend resources such as a territory, a mate,

Male lions fight for territory and supremacy. (Photo by Karl Ammann. Bruce Coleman, Inc. Reproduced by permission.)

A Japanese macaque (*Macaca fuscata*) grooming session. (Photo by Herbert Kehrer/OKAPIA/Photo Researchers, Inc. Reproduced by permission.)

mammals (the naked mole-rat and some other bathy-ergids), which means that reproductive suppression is irrevocable, leading to sterile worker castes and an overlap of several generations to be found. However, helpers can be found in many families (callitrichids, canids, marmots). These helpers normally are the young of previous years that remain within their parents' group and range, refrain from reproducing themselves, and help in rearing their parents' next offspring. The degree of helping often depends on the degree of relationship between helpers and the next litter (as demonstrated for the alpine marmot). Helping can be done by carrying them (callitrichids), feeding, guarding, and playing (canids), keeping the nest warm (marmots), or taking part in anti-predator vigilance or defense (dwarf mongoose).

Sociality in the framework of Tinbergen's questions

What do we know about phylogeny? It is not normally possible to find behavior in fossilized form, thus we have to take another, but also reliable approach, by comparing the phenomenon in question among as many living species as possible. When doing this with regard to social systems, the most basic one seems to be a sort of solitary or dispersed female system, foraging alone in undefended home ranges. This pat-

tern can be found in members of so many different taxa that we may assume it to be one that their common ancestors probably shared. Taking males into account as well, we can assume that a system of dispersed polygyny, one male overlapping the ranges of several females, probably was the basic male-female system. From the basic female system, evo-

A spotted hyena (*Crocuta crocuta*) feeds on a young elephant that a lion killed earlier. (Photo by Harald Schütz. Reproduced by permission.)

lutionary paths could have led either via territorial defense (then group defense and dispersed feeds) to social foraging in group territories, or without defense, via formation of ephemeral, and then later persistent groups.

What do we know about selective advantages of sociality? Behavioral ecology and sociobiology, those areas of behavioral biology that deal with this question, are among the most productive ones in about 20–30 years. Thus, only a few studies should be mentioned, to cover several aspects of this question. Anti-predator vigilance in the dwarf mongoose is, over a longer period of time, only guaranteed in groups of at least six adults; smaller groups sooner or later fell prey to raptors. Jackal pairs with one or more helpers had more success in rearing young—the energetic demand on parents for hunting and producing milk was significantly lower, and juvenile survival higher. Adult male sugar gliders that share dominance with an adult son have a higher proportion of time spent with young in the absence of the mother, which is helpful in defending as well as warning them. Pairs of klipspringer take turns in anti-predator vigilance, one feeding while the other watches out. Eastern gray kangaroos form larger groups in open areas and also during those times of the day when their main predator, the dingo, is likely to hunt. Lastly, survival of alpine marmots is higher when more young of the previous year hibernate together with their parents.

Physiological mechanisms that regulate mammalian social behavior are also currently subjects of intense studies. We already heard about the influence of oxytocin on development of social bonding, studies which have predominantly been conducted on the monogamous vole *Microtus ochrogaster*. Prolactin has been identified as the hormone of parental care,

and, excitingly enough, is not only maternal but also elevated in helpers, such as subordinate individuals in canid packs that help to rear the alpha pair's young. Testosterone in both sexes is connected with status/dominance position. Remarkably enough, testosterone levels often follow, not precede, an increase in status such as after winning a fight. Cortisole, one of the stress-related glucocorticoids, actively lowers status-related behavior and makes an individual more submissive, particularly in contest-related aggressive situations. Stressful reactions to potentially harmful or frightening situations are lower, or absent, if the situation is encountered in the presence of one's bonding mate.

Finally, some data related to the fourth Tinbergen question, ontogeny. The importance of complete socialization has been demonstrated in countless studies. Guinea pig males that had been reared in an all-female group were unable to integrate themselves peacefully into new colonies at sexual maturity due to a lack of two important behaviors: they did not behave submissively towards adult males, and they courted any female (even firmly bonded ones) that they might meet. However, young males reared in the presence of an adult male performed "correctly" immediately after introduction, and thus were integrated without any stress or aggression. Feral cats reared in the presence of other cats (or people) apart from their mothers and litter-mates, and coyote pups raised in presence of adult helpers at the den, became more gregarious than those without these influences. Monkeys reared in isolation were unable to perform socio-sexual behavior correctly, if they did not get at least regular play sessions with other juveniles. Female monkeys without experience in baby care (prior to giving birth themselves) were less competent in handling their own infants.

Resources

Books

Alcock, J. *Animal Behavior.* New York: Sinauer, 2001.

Broome, D., ed. *Coping with Challenge.* Berlin: Dahlem University Press, 2001.

Gansloßer, U. *Säugetierverhalten.* Fürth, Germany: Filander, 1998.

Geissmann, T. *Vergleichende Primatenkunde.* Berlin: Springer, 2002.

von Holst, D. "Social Stress in Wild Mammals in Their Natural Habitat." In *Coping with Challenge,* edited by D. Broome, 317–336. Berlin: Dahlem University Press, 2001.

Jarman, P. J., and H. Kruuk. "Phylogeny and Spatial Organisation in Mammals." In *Comparison of Marsupial and Placental Behavior,* edited by D. B. Croft and U. Gansloßer, 80–101. Fürth, Germany: Filander, 1996.

Pearce, J. D. *Animal Learning and Cognition.* New York: Lawrence Erlbaum, 1997.

Sachser, N. "What Is Important to Achieve Good Welfare in Animals?" In *Coping with Challenge,* edited by D. Broome, 31–48. Berlin: Dahlem University Press, 2001.

Shettleworth, S. J. *Cognition, Evolution, and Behavior.* Oxford: Oxford University Press, 1998.

Tomasello, M., and J. Calli. *Primate Cognition.* Chicago: University of Chicago Press, 1997.

Periodicals

von Holst, D. "The Concept of Stress and Its Relevance for Animal Behaviour." *Advances in the Study of Behaviour* 2 (1998).

Schradin, C., and G. Anzenberger. "Prolactine, the Hormone of Paternity." *News in Physiological Sciences* 14 (1999): 223–231.

Tinbergen, N. "On the Aims and Methods of Ethology." *Zeitschrift für Tierpsychologie* 20 (1963): 410–433.

Udo Gansloßer, PhD

Cognition and intelligence

In 1871 Charles Darwin's inclusion of mind and behavior in his theory of evolution gave scientific legitimacy to the investigation of animal thinking. Since that time, animal intelligence and cognition have been of interest to psychologists, anthropologists, ethologists, biologists, and cognitive scientists. The first published treatments of animal cognition were anecdotal observations that were richly interpreted to show the complexity of animal reasoning. Those anecdotal observations were criticized for their lack of objectivity, leading to the introduction of more objective techniques such as experimental studies and more careful interpretation of results. As the field of animal cognition has progressed, emphasis on scientific rigor and objectivity has characterized research. This essay provides an overview of animal cognition as well as suggesting directions that research in the early 2000s is taking. Most of the current research and theory in animal cognition focuses on the cognitive abilities of nonhuman primates. That emphasis is reflected here. The topics chosen for this review include those that are currently dynamic and are likely to show the most growth over the next several years. This overview provides a point of entry for those interested in exploring cognitive abilities in animals.

The question of animal intelligence

Intelligence has been a difficult characteristic to define and study in humans. Various definitions and theoretical perspectives abound, addressing such issues as whether intelligence is composed of one general factor or several factors and, if several, which ones. Regardless of theoretical stance, the measurement of intelligence has been difficult and fraught with controversy. As with the concept of intelligence in humans, intelligence in nonhuman animals has been difficult to define and investigate. A definition of intelligence applicable to nonhuman animals includes aspects of learning, memory, social cognition, conceptual ability, problem-solving ability, and cognitive flexibility. Measures of intelligence must include items or tasks appropriate to the organism being studied. The determination of intelligence in animals is complicated by the lack of verbal ability in nonhumans. It is difficult to pose a question to an animal who cannot respond with a verbal answer. Investigators in animal behavior and intelligence have developed several techniques designed to understand the animal mind through directly observable behaviors. Examples of these techniques are considered below.

Brain size and the phylogenetic scale

People often characterize animals of different species as more or less intelligent, and television programs delight in questions such as "What is the most intelligent species of animal?" Scientists have looked to brain size as a way to predict intelligence across species, but absolute brain size does not work, as body sizes vary so much across animal species. The encephalization quotient (EQ) was developed as a measure of relative brain size. The EQ is a calculation based on the size of a species' brain compared to the expected size based on body size. An EQ of 1.0 indicates that the brain size is the size expected for the species body size whereas an EQ over 1.0 indicates a brain that is larger than would be expected and an EQ less than 1.0 indicates a smaller brain than expected. Humans have the largest EQ (7.0), with monkeys and apes (1.5–3.0) and dolphins (up to 4.5) also high on the EQ scale of living species. Although the EQ (as well as other measures of brain complexity) is consistent with expectations of relative cognitive complexity, it indicates nothing about the types of cognitive abilities available to animals of different species.

Historically, the study of animal cognition began with attempts to arrange species along a phylogenetic scale in order of intelligence. This approach was guided by Aristotle's proposal that organisms can be arranged along a "ladder of life" with humans at the top and other animals at different rungs, or levels, down this ladder. This concept, the *scala naturae*, is no longer accepted in comparative studies of animal behavior or intelligence. Rather, evolution is perceived more as a tree-like structure with individual species or groups of species branching off as they evolve adaptations that distinguish them from ancestral forms. From this perspective, animals are studied in the context of their ecological niche, the specific environment in which the organism evolved, and comparisons are made across species that share evolutionary ancestors. Each species of animal has evolved sensory capacities and a behavioral repertoire that allow success in that species' particular ecological niche. What is "intelligent" behavior for one ani-

Barbary macaque (*Macaca sylvanus*) adults grooming, with baby. (Photo by Animals Animals ©J. & P. Wegner. Reproduced by permission.)

mal might not be so for an animal living in a different habitat. In this sense of intelligence, each species has its own type of intelligence. Investigators interested in animal intelligence have turned from global measures of intelligence and have focused research on various cognitive capacities such as basic processes like learning and memory, complex concepts like number, cognitive flexibility shown by tool use and construction, social cognition, and symbolic processing. The focus here will be on those cognitive processes.

Why study cognition in animals?

Scientific curiosity

A major reason for studying cognitive abilities in a particular species is to understand more about that species. Scientific curiosity drives many investigations of animal cognition. As we attempt to describe and understand natural phenomena, the question of the animal mind has increasingly been a subject of scientific investigation. New theoretical approaches and innovative methodologies have led to significant advances in the past 20 years in understanding how animals think. The study of animal cognition has been added to the study of animal behavior, ecology, and evolution as we continue to investigate the biology of our world.

To understand evolution

Comparing cognitive abilities across evolutionarily closely related species can provide hypotheses about how evolution occurred. Darwin's inclusion of minds in his principle of evolutionary continuity opened the area of nonhuman animal intelligence to scientific scrutiny. Because there are no fossils of behavior—only of physical structure and artifacts that imply behavioral capacities and inclinations—the relationship between structural and behavioral capacities in currently existing species is a major window into evolution. Studies of cognitive abilities across primate species provide understanding of human evolution. The search for understanding of our own species and its evolution guides much research on animal cognition. Evolutionary continuity also underlies the investigation of animal cognition in attempts to develop animal models for human phenomena. In some cases it is difficult to study a psychological process directly in humans. For example, although it is apparent that human memory relies on both verbal and nonverbal processes, it is difficult to study nonverbal memory in humans, who encode almost all information linguistically.

Conservation

Understanding the cognitive abilities of animals and how they use these capacities to solve daily problems of finding

A chimpanzee (*Pan troglodytes*) uses a stick to get termites in Sweetwaters Reserve, Kenya. (Photo by Mary Beth Angelo/Photo Researchers, Inc. Reproduced by permission.)

food, evading predators, avoiding other physical dangers, and reproducing can provide important information for conservation. As habitat destruction continues to threaten the existence of many animal species, the more information available about ecology and behavior, the more informed can be plans for delaying extinction. The design and location of protected reserves relies on understanding needs of those animals being protected. Such understanding is also vital to promote the welfare of those individuals who are housed in captivity, regardless of their endangered status. Providing captive animals with cognitive enrichment by challenging their cognitive skills contributes to their psychological well-being.

Finally, as we understand more about the cognitive complexity of the animals we are attempting to preserve, the importance of ensuring their survival is apparent. These three reasons for studying animal cognition (scientific curiosity, understanding evolution, and conservation) are exemplified by the Think Tank exhibit at the Smithsonian Institution National Zoological Park. This exhibit, which opened in 1995, is dedicated to the topic of animal thinking. As the first exhibit of its kind in any zoo or public forum, Think Tank combines basic research with cognitive enrichment for captive animals while educating the public about the cognitive complexity of animals. Daily live demonstrations of data collection with orangutans (*Pongo* spp.) show zoo visitors how orangutans solve complex problems such as acquiring language symbols and demonstrating numerical competence. The opportunity to observe an animal using complex cognitive abilities to solve a problem not only informs visitors of the capabilities of great apes, it also serves to illustrate the importance of preserving animals with such complex minds.

Basic processes: learning and memory

Learning is generally defined as a relatively permanent change in behavior as a function of experience. Most organisms show the capacity to learn. Early studies of intelligence in animals used learning tasks to attempt to characterize species differences in capacity.

In a series of studies investigating the learning skills of rhesus monkeys (*Macaca mulatta*), Harry Harlow showed in 1949 that, following extensive experience with large numbers of individual problems, monkeys were able to solve a novel problem in only one trial. The problems involved making a choice between two objects that differed from each other in several physical dimensions such as color, shape, material, size, and position. A reward such as a peanut was hidden under one of the objects. The monkeys gradually learned to choose consistently the rewarded object. At that point Harlow would introduce a new problem with novel objects. Over the course of many individual problems, the monkeys took fewer trials to reach a high level of performance.

Following a few hundred such problems, the monkeys were consistently correct on the second trial of each new

A goat uses a tree stump so she can reach high vegetation. (Photo by Antony B. Joyce/Photo Researchers, Inc. Reproduced by permission.)

An African leopard (*Panthera pardus*) stores its kill (an impala) in a tree. (Photo by Tom Brakefield. Bruce Coleman, Inc. Reproduced by permission.)

problem. That is, the monkeys no longer required a period during which they learned to choose the rewarded object through trial-and-error. Rather, their response on the first trial of a new problem (whether it was correct or incorrect) informed them which object to choose on subsequent trials. This understanding of the solution to the problem based on one experience with two novel objects was called "learning set," or "learning to learn" by Harlow. It is a good example of cognitive flexibility. Learning set is still used to study aspects of learning and cognitive flexibility in humans and nonhumans. Animals of a multitude of species are capable of learning set, including cats, rats, squirrels, minks, sea lions, and several species of monkeys. The investigation of learning set in rats demonstrates the importance of considering the species-typical sensory capacities of an animal when studying cognition. Rats, who have very poor vision but excellent olfactory ability, have some difficulty with visual learning set but easily achieve high levels of performance with olfactory discriminations.

Memory provides mental continuity across time by allowing information from one point in time to be used at a later point in time. That time span can be in seconds or minutes (short-term or working memory), hours, days, or longer (long-term or reference memory). It forms the basis for learning, since without memory the influence of past experiences would not exist. Memory involves three processes: encoding, storage, and retrieval. Encoding refers to the form or code in which items or events are stored; storage involves the way that memories are stored in the brain including where, how, and how long; and retrieval refers to the act of remembering, or accessing information that was previously stored in memory. Two ways of studying retrieval are through recall and recognition measures. In recall, an individual is asked to reproduce the items or information stored at a previous time. In humans recall usually involves verbal reports. Recognition measures simply ask "Have you seen (experienced) this item before?" and involve presenting various items that were and were not in the memory set being tested. The individual is required to respond in a way that implies "yes" or "no." Because it can be nonverbal, this form of retrieval is most commonly used with nonhuman animals.

The above distinctions and methods are relevant to laboratory research with animals, and they are addressed specifically below. However, memory clearly plays a role in an animal's daily life. Food-related behaviors such as foraging or storing food for later use require memory. To what extent memory is used in foraging, and how that process is used, has been addressed in field studies of nonhuman primates. The question is whether monkeys and apes use memory to guide

A lion stalking prey near a waterhole in Tanzania. (Photo by Robert L. Fleming. Bruce Coleman, Inc. Reproduced by permission.)

their travel when foraging. For animals like gorillas (*Gorilla gorilla*) or leaf monkeys (*Presbytis* spp. and *Trachypithecus* spp.) whose sources of food are easily found and available in large quantities at one site ("banqueters"), remembering the location of particular food sources may not be a crucial aspect of their foraging activity. For species like orangutans and capuchin monkeys (*Cebus* spp.) whose food is distributed in patches that change in availability as fruits mature ("foragers"), memory of where particular trees are located as well as when they fruit may be important. Researchers have investigated whether foraging paths in social groups can be better explained by opportunistic searching or by memory-guided paths to sites that are seasonally plentiful. They have found evidence that not only do monkeys remember where plentiful sources of fruit are located, they also remember when the trees will be fruiting.

Questions surrounding memory in nonhuman animals refer to capacity, duration, and organization. These questions are addressed in laboratory experiments, some of which are designed to provide similar challenges to those provided in the animal's natural environment. A good example of this is the radial maze, developed to study memory in rats. The maze consists of a center circular area to which are attached straight alleys or runways (arms) in a manner like spokes attached to the hub of a wheel. Goal boxes at the end of each runway provide incentives. This apparatus simulates the foraging challenge presented to rats in their natural ecology. The rat is released into the center area and observed to see how it will obtain all the food. The rat's task is to run down an arm, eat the food in that goal box, return to the center area, and choose another arm to enter, repeating this until all the food has been found. Returning to an arm that was previously visited is considered an error. The most efficient solution is for the rat to run down each arm only once. To do that, the rat has to rely on memory of where it has been. Rats show effective use of memory in the task, and researchers have demonstrated that this memory is based on the formation of a spatial map of the maze. Cues from the room containing the maze are used to remember the locations of the arms that have been explored.

Research with humans has shown that humans have a large capacity for remembering items and events over long periods of time. Monkeys and orangutans have shown that they remember photographic stimuli over a delay of at least a year. Even more impressive, a chimpanzee (*Pan troglodytes*) showed memory for symbols learned 20 years before. Memory for concepts has been shown by squirrel monkeys (*Saimiri* spp.) over a five-year period, by rhesus monkeys across seven years, and by a sea lion over a ten-year period. These findings are limited only by the test intervals imposed by researchers; the limitations of specific or conceptual memory in animals have

reference memory was discussed previously. Reference memory can also be divided into declarative (explicit, or conscious) and non-declarative (implicit, or unconscious) aspects. Declarative memory is further subdivided into episodic and semantic memory. Semantic memory refers to memory for information, in other words, generic knowledge. Episodic memory refers to memory for particular events or experiences and implies that the memory involves revisiting that event or experience. This aspect of episodic memory can be characterized as the distinction between knowing something versus recalling the specific event that provided the knowledge. Knowing that an incentive is located in a particular location without remembering the experience of seeing it hidden illustrates the distinction between semantic memory (knowledge) and episodic memory (memory for the event). Panzee, a language-trained chimpanzee, uses language symbols to indicate to an uninformed human caretaker that a particular item has been hidden at some previous time (as long as 16 hours before). Further, she will guide the human to the point where the item (that is outside of Panzee's enclosure and hence unavailable to her) is hidden. Panzee has shown the first attribute of episodic memory; the question is how to determine whether

After running a maze several times, a mouse takes less time to find the cheese. (Photo by © Don Mason/Corbis. Reproduced by permission.)

not yet been demonstrated. It is likely that animals have very long-duration memory capacity, especially for conceptual information.

Many experimental studies with nonhuman primates have demonstrated memory phenomena similar to those shown by humans. Historically, human memory has been studied in the laboratory by providing people with lists of items to remember. When humans are provided with an ordered list of items to be remembered, they will show what psychologists term the "serial position effect." The nature of the serial position effect is that items early in the list and late in the list are remembered better than those in the middle of the list. Monkeys show this same effect with visual stimuli. The tests of monkey memory involve recognition memory while the typical procedure with humans involves recall memory. However, Lana, a chimpanzee who had had language symbol training, was able to perform a recall memory task with a list of symbols by choosing the list items from her entire vocabulary of symbols. She, too, showed a serial position effect. The serial position effect is one of the most stable phenomena in memory research with humans. To find the same phenomenon in nonhuman primates suggests that memory processes are similar across species notwithstanding the difference in language ability.

A current area of investigation in nonhuman animals is the question of episodic memory. Memory can be categorized in many different ways. The distinction between working and

Chimpanzees (Pan troglodytes) are common subjects for memory and other research. Humans and chimpanzees have 98% of their genes in common. (Photo by ©Renee Lynn/Corbis. Reproduced by permission.)

this memory involves more than simple knowledge that a particular object is hidden outside the enclosure. That is, does Panzee's memory include "time traveling" back to the experience of seeing the object being hidden? At the present time, the answer is unclear. However, the limitations on demonstrating episodic memory in nonhuman animals are likely to be more those of procedure (how do we clearly demonstrate evidence for this capacity in nonverbal organisms?) than capacity (do animals have episodic memory?).

Number

How much do animals understand about number? Can they count? Although many animals will choose the larger of two amounts of a desirable substance, can they determine the larger of two quantities of objects? If so, what does that tell us about their understanding of number? The first animal reported to perform complex numerical tasks was Clever Hans, a horse that performed for audiences in Berlin during the early 1900s. Hans was able to answer complicated questions, including arithmetic problems involving addition, subtraction, fractions, and other arithmetic manipulations that were relatively sophisticated. When a question was posed, Hans used his right front foot to tap out the answer, and he was quite accurate. A scientific panel investigated Hans' numerical ability. They found that Hans was most accurate when his owner Mr. von Osten was present, but that he could also solve problems when Mr. von Osten was absent. This led them to the conclusion that Hans' ability was not based on fraud or tricks. However, Oskar Pfungst, a student of one of the panel members, went further in an investigation of Hans' numerical ability. He noted that Hans' accuracy depended on whether Mr. von Osten knew the answer to the question. When the answer was not known by the human questioner, Hans was inaccurate, usually tapping numbers higher than the correct answer. He also noted that there were three other individuals whose questioning of Hans was accompanied by high levels of accuracy in the horse.

With careful scrutiny of Mr. von Osten, Pfungst noted that after a question was posed to Hans, Mr. von Osten lowered his head slightly and bent forward, maintaining this position until the correct number was tapped, at which time he jerked his head upwards. The three other people for whom Hans performed well showed similar behaviors. Pfungst performed experiments in which Mr. von Osten (and the others) provided these behavioral cues at appropriate times (consistent with the correct answer) and at inappropriate times (consistent with a wrong answer). Hans' performance showed that he was responding to these cues. Pfungst suggested that Hans was sensitive to subtle cues from Mr. von Osten, beginning to tap when Mr. von Osten bent his head and continuing to tap until he detected a shift in Mr. von Osten's body posture. Hans' numerical ability was not based on an understanding of number, but rather on his reading of unintentionally provided subtle behavioral cues from his human questioner. The use of discriminative cues to control his tapping ruled out a high-level cognitive explanation for Hans' performance. Current studies of animal cognition include careful controls to

eliminate the possibility of a "Clever Hans" explanation for the results obtained.

Determining what animals understand about number has several levels. The simplest level of processing numerical information is in making judgments of relative numerousness. Rats can learn to respond differentially to stimuli that vary in number, responding with a particular response, for example, to the presentation of two light flashes or tone bursts and with a different response following the presentation of four lights or tones. Rats also discriminate number of rewards and can learn sequences of reward patterns.

Monkeys and apes make relative numerousness judgments shown by their choice of the larger of two quantities of food. Chimpanzees and orangutans can make such judgments even when the two choices are themselves divided into two groups of objects, suggesting some ability to combine quantities when making the judgment. Discrimination is very accurate for quantities that differ widely from one another, such as five versus two; the task becomes more difficult as the numbers get larger and approach one another, such as five and six. If overall mass is excluded as a possible solution by using items that vary in size (such as grapes or different candies), then a number-related cognitive ability is implicated. Most explanations of animals' success at such simple tasks include a noncounting mechanism known as subitizing, which refers to the ability to make quick perceptually based judgments of number. Subitizing is considered to be the likely way that most organisms, including human children and adults, make judgments about quantities of seven or fewer.

Understanding of ordinal relationships is a second level of numerical competence. Monkeys can order groups of items presented as abstract symbols on a video screen. They also learn to associate Arabic numerals with quantities up to nine (although this may not be the limit), and respond correctly to the ordinal positions of the Arabic numerals. Apes go further in their use of Arabic numerals as symbols by using them to demonstrate counting.

Counting implies symbolic representation of a collection of objects. The determination that an animal is counting includes three major criteria, developed initially in studying counting in human children. The one-to-one principle requires that each item is enumerated or "tagged" individually. In the stable-order principle each tag occurs in the same order (e.g., one-two-three rather than two-one-three). In the cardinal principle the final number in the count refers to the total number of items. Additional features include that any set of items can be counted, and that the order in which items are counted is arbitrary and irrelevant to determining the final count.

Chimpanzees have demonstrated all the criteria for counting. Several chimpanzees associate abstract symbols with specific quantities and act upon the symbols as they would the quantities represented by the symbols. Sheba, a chimpanzee who represents quantities with Arabic numerals, demonstrated tagging as she learned to assign Arabic numerals to collections of objects. She also spontaneously added quantities to report the sum of objects hidden across three locations.

An orangutan (*Pongo pygmaeus*) uses a finger to get at ants in a tree hole. (Photo by Tim Davis/Photo Researchers, Inc. Reproduced by permission.)

She was able to add quantities regardless of whether objects or numerals were presented. Ai, a chimpanzee who was trained using a touch screen to respond to symbols, has shown clear understanding of ordinal positions of numerals, arranging them in order even when there are numerals missing in the list presented to her.

An interesting demonstration of the power of cognitive abstraction with numbers was shown by chimpanzees Sheba, Sarah, Kermit, Darrell, and Bobby. The task was simple but not straightforward. A chimpanzee was shown two quantities of a desirable food such as candies and was permitted to choose one of the quantities by pointing to it. The quantity that was chosen was given to another chimpanzee (or taken away), and the quantity not chosen was provided to the chimpanzee who had made the choice. This procedure works against the natural predisposition to choose the larger of two quantities, and the chimps had a great deal of difficulty with the task, even though it was clear from their behavior that they understood that they were not getting the quantity chosen. Even with much experience with the task, these chimpanzees were unable to inhibit choosing the larger quantity. When the objects presented were non-desirable objects such as pebbles, they still could not inhibit the larger choice. However, when the quantities were represented by Arabic numerals, the chimps, who were able to use Arabic numerals to represent quantities, chose the smaller number and thus obtained the larger quantity. The animals were able to use the representation of the quantities to mediate their tendency to choose the larger of two piles of objects.

This tendency to take the larger of two quantities is consistent with the chimpanzee's ecology. Chimpanzees live in social groups and compete with one another for desired food objects. The ability to judge quantities rapidly and to grab the larger of two quantities would serve them in their natural environment. The finding that orangutans Azy and Indah were able to learn relatively rapidly to inhibit the choice of the larger quantity supports the suggestion that ecology plays a role in this behavior. Orangutans are solitary in the wild and would have little need for a food-getting strategy that requires quick choice of larger quantities. Animals of both species can clearly distinguish between different quantities of food, and both can learn to inhibit the choice of the larger quantity when it is to their advantage to do so. Orangutans are able to learn to inhibit relatively quickly, but chimpanzees must rely on an additional, cognitive step in the procedure to be able to inhibit their very strong tendency to choose the larger quantity. Species difference in cognitive abilities or in the way that cognition can mediate a response can tell much about how an animal is processing information and what kinds of cognitive judgments serve the animal in its natural environment.

Tool use and construction

Tools can be defined as detached objects that are used to achieve a goal. Achievement of the goal can involve manipulation of some aspect of the environment, including another organism. The making of tools involves modifying some object in the environment for use as a tool, but not all objects used as tools by animals are constructed.

Instances of tool use and tool construction have been observed in great apes and monkeys living in captivity. All species of great ape have been observed to make and use tools in captivity. Some monkeys have been observed to use tools and occasionally to construct tools in captivity. Observations of tool use and construction by nonhuman primates in the wild have been less widespread. This difference between demonstrations of tool use in captivity and the natural environment may exist because, for the most part, evolutionary adaptations of the animal may sufficiently address challenges in the natural environment. However, environmental contingencies in captivity may be unique in encouraging tool use. In studies of tool use, researchers set up problems that are best solved using tools. In captive environments, experience with such problems coupled with extensive experience with tool-like objects provide the environmental and cognitive scaffolding that facilitate the demonstration of tool use by animals who may not so readily demonstrate such capacities in their natural environment.

However, instances of tool use and construction have been observed in great apes in the wild, particularly in the chimpanzee. The most widely known of these is the chimpanzee's use of twigs to "fish" for termites in termite mounds at the Gombe Stream Reserve (now Gombe National Park) in Tanzania, East Africa. Following the report of that first observation by Jane Goodall in 1968, many instances of tool use and tool construction have been discovered in chimpanzees and other animals in their natural environment. For termite fishing, the choice of an appropriate branch, twig, or grass blade involves judgments of length, diameter, strength, and flexibility of the tool. Any leaves remaining on a branch are removed, and the tool is dipped into holes in the termite mound. Termites attack the intruding stick, attaching themselves to it, and the chimpanzee withdraws a termite-laden stick and proceeds to eat the termites.

Additional observations of tool use include the use of leaves that have been chewed to serve as a sponge to obtain water from tree hollows, the use of tree branches in agonistic displays, various uses of sticks, leaves, or branches to obtain otherwise inaccessible food, and the use of leaves to remove foreign substances from the body. Orangutans in the wild have been observed to use leaves as sponges and as containers for foods. They use tools to aid in opening large strong-husked fruits and to protect them from spines on the outside of fruits.

An interesting complicated instance of tool use is the use of rocks as hammer and anvil to crack open nuts reported for chimpanzees in West Africa. Coula nuts or palm-oil nuts with hard shells are placed on a hard surface and cracked open by hitting them with a hand-held rock. The supporting surface (or anvil) can be a tree root or a rock obtained from the forest floor by the chimpanzee. Rocks used as hammers have a size and shape that fit the chimpanzee's hand, and rocks chosen as anvils have a flat surface. The nut is placed on the flat surface of the anvil and pounded with the hand-held hammer rock. If an anvil has an uneven base that causes it to wobble when the nut is struck, the chimpanzee finds a smaller rock to use as a wedge under the smaller end of the anvil to balance it.

Infant chimpanzees learn tool use as they observe their mothers make and use tools. Observations of mother chimpanzees engaged in explicit teaching of tool use to their infants have been reported at only one site, and on only two occasions. On one occasion, a mother took a hammer rock from her daughter who was holding it in an orientation that was not conducive to nut cracking, and slowly rotated the orientation of the rock to a position that was useful. After the mother finished cracking nuts, her daughter used the rock in the same orientation as her mother had used it. On the second occasion, a mother re-positioned a nut that her son had placed on an anvil in a position that would not have permitted its being opened. The son then used a hammer rock to open the nut. Both of these observations are provocative, but in the absence of other such observations it is not clear that the mother's intent was to teach the infant. Indeed, most learning of tool use and tool construction by young chimpanzees appears to be from observation of the mother's behavior and manipulation of objects used as tools in the context in which they are used.

Culture in nonhuman primates?

In 1953 a young Japanese monkey named Imo did something remarkable. Her troop lived on Koshima Island in Japan, and was provisioned by Japanese researchers who studied them. Provisioning involves providing additional food to sustain the population in areas of limited resources or to encourage animals to remain in a particular locale for observation. Imo's group was provided with sweet potatoes placed on the sand at the edge of the water. Imo began to use the nearby water to wash sand from the sweet potatoes. This behavior spread through the group, with younger animals adopting it first and some older animals never adopting it. Four years later, Imo introduced another novel behavior. The monkeys were provided with grains of wheat scattered on the sand that were difficult to eat because they mixed with the sand. Imo placed handfuls of this mixture of wheat grains and sand into standing pools of water. The wheat grain floated to the top and could be scooped up and eaten, free of sand. Years and generations later the monkeys of Koshima Island continue to wash sweet potatoes and to place sandy wheat grains in water. The spread of these novel behaviors through individuals in the group appeared to be an instance of social transmission of a novel behavior. This interpretation began a discussion of culture in nonhuman primate groups.

Recent analyses of behaviors shown by communities of chimpanzees and of orangutans suggest the presence of cultural variations across groups within each species living in different geographical areas. Behaviors that varied included instances of tool use, social behaviors related to grooming or

A sea otter (*Enhydra lutris*) uses a rock to open a clam shell. (Photo by Jeff Foott. Bruce Coleman, Inc. Reproduced by permission.)

communicative signals, and food-related behaviors. For example, some chimpanzee communities crack nuts with hammer and anvil tools, but others do not, despite the availability in their environment of hard-shelled nuts and objects that could serve as tools for nutcracking. The absence of a behavior pattern in the presence of all necessary components (e.g., nuts and potential tools) rules out ecological factors to explain these differences. Similarly, genetic factors do not play a role in the variability of such behavior patterns. That is, a similar pattern of tool use may be shown by two groups who are genetically isolated from one another, or it may be present in one community but missing in another community of the same genetic background. Some form of social learning is thought to have promoted the spread and maintenance of these specific behavior patterns within certain communities.

Even in a species that is not typically group living, cultural differences are found across geographical areas. Orangutans observed in Sumatra use tools constructed from sticks to pry the seeds from *Neesia* fruits, a fruit with a very tough husk that also has spiny hairs protecting the seeds. Bornean orangutans do not use tools to acquire the seeds; rather, they tear a piece of the husk open to expose seeds. Only adult males can perform this latter method because of the strength required to force open the fruit. Otherwise, females and juveniles must wait until the husk opens and older, less desirable

seeds are naturally available. Geographic isolation leading to a genetic basis for these differences in technique is not a sufficient explanation. At a second Sumatran site orangutans do not use tools, ruling out the genetic explanation. *Neesia* is available to and eaten by orangutans in both Sumatra and Borneo, eliminating an ecological explanation. Social transmission through social learning is the most likely explanation for this phenomenon, through mother-offspring transmission and/or through social encounters among animals inhabiting the same area.

Social learning can occur at several levels, from simple to complex, all based on observation of one animal by another. Most simple is social facilitation in which one animal's interest in an object elicits interest in that object from another animal. By drawing another's interest to an object that the other then explores, the first animal has influenced the behavior of the second but has not explicitly transferred information. In observational learning, the observer learns something specific about stimuli and responses from watching another animal. Imitation is the most cognitively complex form of social learning and involves the observer copying the form and intent of a novel behavior demonstrated by another animal. Imitation is distinguished from a similar form of social learning termed emulation, in which an animal performs actions similar to another, with the same intent, but without mimicking the specific actions of the model. Great apes do show imitation but that capacity may be limited to great apes.

Whether the transmission of cultural variations within a group is based on imitation or some simpler form of social learning remains unresolved. However, ongoing investigations of nonhuman primate behavior in the laboratory and in the wild will provide better understanding about the origin and transmission of novel behaviors through social groups.

Self recognition and theory of mind

Mirrors provide a novel and rich source of information about social cognition in animals. The behavior of an animal toward its mirror image suggests much about the animal's understanding of the source of that image. Many animals such as cats and dogs, when first encountering their own mirror image, behave as though they have encountered a stranger of their own species. They may show aggressive behavior such as threats, or they may attempt to play and to reach around the mirror as though attempting to find the other animal. With time, the dog or cat will ignore the mirror image and no longer attempt to engage the reflection in social interaction. For the most part, monkeys behave in a similar way to their mirror image.

Great apes, however, appear to come to understand the nature of the mirror image. That is, with experience, they behave as though they understand that it is their own body that is reflected in the mirror. The phenomenon is referred to as mirror self-recognition (MSR) and has been of interest to researchers in primate cognition for decades. Chimpanzees were the first nonhuman species to show evidence of MSR. When provided with daily exposure to a mirror outside their enclosure, individual chimpanzees initially responded to the mirror image with social behaviors suggesting that the mirror image was perceived of as an unfamiliar chimpanzee. Threat behaviors such as head bobbing, charging the mirror, and vocalizing were common. After some time, however, the social behaviors waned and the chimpanzees began to direct behavior toward their own bodies while looking into the mirror. They groomed and investigated parts of their bodies, such as the face, that were invisible to them without the use of the mirror. Using the mirror to guide their hands, these animals groomed their eyes, picked their teeth, inspected their genital areas, and also made faces while watching in the mirror.

The behaviors directed to their own bodies, termed self-directed behaviors, appeared to indicate that the animals recognized themselves in the mirror. To test this interpretation, the chimpanzees were anesthetized and an odorless red mark was placed on one eyebrow and one ear, in a location where the chimpanzee could not see the mark without the use of the mirror. When the animals awoke from the anesthesia they were presented with the mirror, and all four animals touched the mark on their brow, using the mirror to guide their fingers to the mark. The importance of this response, directing their hand to the mark on their own body rather than to the mark on the mirror image, suggests that the animals indeed recognized themselves in the mirror.

Since the initial report in 1970 of this phenomenon in chimpanzees, individuals from the other great ape species have shown MSR by demonstrating self-directed behavior or by passing the mark test, but not all individuals of these species show the capacity. There are clear individual differences based partly on age. Like humans, chimpanzees develop the ability to show mirror self-recognition. Beginning at about 24–30 months of age young chimpanzees will touch a mark on their brow using the mirror to guide the touch and the ability is generally quite evident by the age of four. Human children studied under controlled conditions do not show the capacity for mirror self-recognition until about 15 months of age at the earliest, with most achieving this developmental milestone by 24 months.

There is some evidence that dolphins are capable of mirror self-recognition, and gibbons also may have this capacity. However, other animals have not clearly shown evidence of mirror self-recognition. It may be a species difference, or it may be the case that with additional research the apparent discontinuity will be resolved. The implications for evolutionary development of self in humans are apparent, although interpretations of this phenomenon are disputed. At the most extreme, a rich interpretation of self-recognition in animals suggests an understanding of the self as an entity, perhaps similar to humans' sense of self or self-concept. However, the ability to understand the nature of a mirror image and to direct behavior back to one's own body does not necessarily imply such a rich interpretation. The distribution of this capacity and its interpretation are open questions subject to ongoing empirical investigation and theoretical debate.

The ability to recognize oneself may be related to the ability to understand another individual's knowledge state, which represents a more complex cognitive ability. This phenomenon is called "Theory of Mind" and refers to an individual's ability to understand the perspective or the "state of mind" of another. It forms the basis for such complex social strategies as intentional deception. In order to deceive another by providing incorrect information, the actor must know something about the other's perspective and expectations and what information to provide (or withhold) as deception. Many instances of deception have been reported from observations of apes and monkeys in the field and in the laboratory, and the extent to which these deceptive incidents are based on the perspective-taking capacity implied in Theory of Mind is still an open question. It is clear that animals behave in ways that suggest that they are taking into account the knowledge state of others. Whether they are or not is still an active area of research in animal cognition.

The initial description of Theory of Mind was based on the ability of Sarah, a language-trained chimpanzee, to solve problems for a human who was in some state of distress. Sarah was experienced in many features of human life. For example, she had often observed her human caretakers using keys to unlock padlocks. She had often observed humans turning a faucet to provide water through a hose. She had observed humans plugging a cord for an electric heater into an electrical outlet. Sarah was provided with videotaped instances of humans in situations whose solutions were related to the above events as well as others. For example, she saw videotape in which one of her human caretakers was apparently locked in a cage and could not escape despite attempts to open the locked door. The videotape was stopped before the problem was solved, and Sarah was provided with photographs, one of which had the solution to the problem (in this case, a key for the padlock on the cage door). Sarah consistently chose the photograph with the appropriate solution to each problem. The interpretation of Sarah's behavior was that she was able to understand the state of the human in the videotape and she was able to choose the appropriate solution to the person's problem. Additional studies with Sarah suggested she could spontaneously show deception, withholding information or providing incorrect information about the location of a food item from a human who had previously failed to share food with her. In contrast, she provided information about the location of a food object to another human who always shared food with her.

Additional studies with Sarah and other chimpanzees provided results suggestive that chimpanzees can take the perspective of another and use it to solve problems. A number of studies have been unsuccessful at providing evidence of Theory of Mind; some that have been successful have been criticized on methodological grounds. However, such studies continue. The demonstration of this capacity in apes awaits a clear methodology that will adequately address the extent to which apes can project their understanding of states of mind to others.

Social cognition

Theory of Mind is a form of social cognition, or the ability to process cognitive information presented by social partners. In group-living animals species-specific social rules guide behavior. Understanding these rules and applying them appropriately is a complicated process for individuals in the group. Recognition of familiar versus unfamiliar conspecifics, of particular age and sex classes, and of particular individuals are necessary skills for each member of the group. Even animals that spend much of their time in solitude must recognize social features of other conspecifics, including individuals who may be living in the same area. Social cognition involves not only these forms of recognition, but also understanding of species-typical social communicative signals.

In many animal species, members of the group respond quickly to alarm calls from a group member. In some cases, such response may not require much processing of information and so may be based on simple associative mechanisms. In other cases, processing of alarm calls may provide some cognitive challenge. For example, vervet monkeys (*Chlorocebus aethiops*) in Africa have three types of alarm calls, elicited by three different predators. Each alarm call is followed by a particular behavior by members of the group. Following a "snake" alarm call the group members stand bipedally and visually search the grass around them, presumably to locate large pythons or poisonous snakes on the ground. Looking into the air and moving to the cover of bushes follow an "eagle" alarm call. A "leopard" alarm call sends group members to trees with branches that are too fragile to support the weight of a leop-

ard. These calls are made selectively and appropriately by adult members of the group, and receive selective and appropriate responses by group members. Infants learn the appropriate calls, sometimes making errors as they develop, for example, producing an eagle call to the sight of a harmless bird. It appears that production of the calls involves some learning, and it may be that appropriate response to the calls also is learned through observation of group members. These alarm calls are deemed cognitive rather than reflexive because production of each type of call is voluntary and the calls are referential. That is, each type of predator call refers to only one type of predator. The calls elicit different responses specific to each type of call, and production and response to each class of call show a developmental course.

Visual social signals provide information in social interactions. The simplest signals are threat or appeasement gestures. Visual social signals that guide another animal's attention to an object or event require more complex cognitive skills. For example, monkeys and apes will join another animal to investigate jointly an object of interest. Referential pointing and referential gazing are social signals that call attention to an object or event removed from the actor. Chimpanzees and orangutans can interpret pointing in humans. They also point and vocalize to draw a human's attention to a distant object or event. However, these apes do not typically use pointing to communicate with one another and their use and interpretation of pointing varies with the amount of human contact they have had.

Although dogs do not seem to respond appropriately to pointing (the usual response by the dog is to sniff the finger of the individual pointing), they do respond to gaze direction in humans and have been shown to use gaze as a cue for the location of hidden food. Similarly, chimpanzees and monkeys are able to use gaze as a cue, and monkeys will follow the direction of gaze of a conspecific, even one presented on videotape. Chimpanzees and orangutans use humans' gaze direction to locate hidden food, and as in pointing, those animals who have had extensive contact with humans are more likely to use and are more adept with gaze cues than are those with less human contact.

Language

One of the major distinctions between human and nonhuman animals is language. At the beginning of the animal language projects, the object was to show that an individual from a nonhuman species could acquire and use language. Apes have been the primary participants in these projects. One of the first attempts to teach language to an ape was that of Keith and Cathy Hayes, who reared Viki the chimpanzee in their home from 1947 to 1954. With much effort, Viki learned to say four words, "mama, papa, cup, and up." She had great difficulty producing these words, and they were barely intelligible. We now know that great apes lack vocal structures necessary to produce speech. However, they appear to have the cognitive ability to acquire aspects of language using symbols in communicative interactions with humans.

Mitchell, R. W., N. S. Thompson, and H. L. Miles, eds. *Anthropomorphism, Anecdotes and Animals.* Albany, NY: State University of New York Press, 1997.

Parker, S. T., and M. L. McKinney. *Origins of Intelligence.* Baltimore: The Johns Hopkins University Press, 1999.

Parker, S. T., R. W. Mitchell, and H. L. Miles, eds. *The Mentalities of Gorillas and Orangutans in Comparative Perspective.* Cambridge, UK: Cambridge University Press, 1999.

Pearce, J. M. *Animal Learning and Cognition.* 2nd ed. Hove, UK: Psychology Press, 1997.

Premack, D., and A. J. Premack. *Original Intelligence: The Architecture of the Human Mind.* New York: McGraw-Hill/Contemporary Books, 2002.

Roberts, W. A. *Principles of Animal Cognition.* New York: McGraw-Hill, 1998.

Shettleworth, S. J. *Cognition, Evolution, and Behavior.* New York: Oxford University Press, 1998.

Tomasello, M., and J. Call. *Primate Cognition.* New York: Oxford University Press, 1997.

Wynne, C. D. L. *Animal Cognition.* Basingstoke, UK: Palgrave, 2001.

Karyl B. Swartz, PhD

Migration

Migrations are mass movements from an unfavorable to a favorable locality and in plains-dwelling, large herbivores such as reindeer, bison, zebras, wildebeest, or elk they can be quite spectacular. In large sea mammals migrations are no less important, but to us are merely less visible and it has taken much effort by science to document at least a part of their extent, leaving much that is still shrouded in mystery. Mass-migrations occur when individuals flood to distant birthing or breeding grounds, to seasonal food sources, or to escape winter storms. Close associations of large numbers of individuals during migration is merely a security measure, capturing the great advantages of the "selfish herd" against predation. Long distance movements are also undertaken by individuals or by small groups, particularly in the oceans.

Migrations take their origin in minor seasonal movements between habitats, such as the movements by mountain-

dwelling deer, elk, or moose between winter ranges at low elevations and summer ranges at high elevations. By moving to lower elevations in fall the deer avoid the high snow levels that develop on their summer ranges at high elevation in late fall and winter. High snow levels hamper movements, at times severely. Also, at high elevations plants, and therefore food density per unit, are low. When high snow levels hamper mobility as well as increase the cost of locomotion, feeding in areas with low food density is uneconomical and individuals move to where there is more food, preferably such that can be acquired with little cost. Movement to lower el-

Aerial view of migrating wildebeest (*Connochaetes*). (Photo by © Yann Arthus-Betrand/Corbis. Reproduced by permission.)

A pod of spotted dolphins (*Stenella attenuata*) migrating. (Photo by © William Boyce/Corbis. Reproduced by permission.)

Migrating killer whales (*Orcinus orca*) in San Juan Islands. (Photo by © Stuart Westmorland/Corbis. Reproduced by permission.)

evations is the answer because forage density is higher and the cost of locomotion between mouthfuls of vegetation is much less. There is a high cost to scraping away snow from the forage it covers, and snow removal is easier and needs to be done less often at low elevations, at localities often outside the mountains. A fourth factor is the shelter provided by trees and forests in winter, which are more likely to grow in valleys than on high slopes and mountain tops. Nevertheless, there is no rule without its exceptions! In the high mountains of western Canada and Alaska, once the snow hardens in late winter and its crust can support the weight of herbivores, mountain caribou and black-tailed deer may ascend the mountains and feed in the sub-alpine coniferous forests on the now accessible tree lichens. Mountain sheep and mountain goats may also ascend in order to feed on high alpine ridges where the powerful winds have blown away the snow, exposing the short, scattered, but highly nutritious grasses, sedges, and dwarf shrubs. Here the packed snow allows them to roam far and wide, be it in search of forage or to escape the occasional visit by wolves or wolverines. Also, some bull elk and bull moose may opt to stay, usually widely dispersed, at high elevation in very deep, loose snow along creeks because wolves are very unlikely to reach or harm them under such snow conditions. When hard "sun crusts" develop on the snow in late winter, however, the tables may be turned temporarily. Wolf packs can then travel freely on

Caribou (*Rangifer tarandus*) migrating in the snow. (Photo by © Kennan Ward/Corbis. Reproduced by permission.)

Wildebeest (*Connochaetes taurinus*) massed at river's edge during migration, in Kenya, Africa. (Photo by Tom Brakefield. Bruce Coleman, Inc. Reproduced by permission.)

the crusted snow surfaces during the night. During the day "sun crusts" may soften, trapping all travelers temporarily in deep snow.

Elevation draws herbivores into seasonal movement and change of venue by the fact that vegetation sprouts in spring

A group of migrating polar bears (*Ursus maritimus*) in Churchhill, Manitoba, Canada. (Photo by © Larry Williams/Corbis. Reproduced by permission.)

at low elevation and that this sprouting creeps upward as spring progresses. Herbivores follow this line of sprouting vegetation because sprouting vegetation it is the most nutritious and most easily digested forage there is. Consequently, there are movements by all sort of large herbivores from the valley to the mountain tops in spring and summer. Less dramatic, but equally important are the movements of herbivores in response to flooding in river valleys and the withdrawal of the flood because the areas just flooded have been fertilized with silt, and rich plant growth follows the withdrawal of water.

Migrations can become more complex if special nutritional needs must be met, or if social factors intervene into the movement pattern. American mountain sheep illustrate this well, because interspersed into the long seasonal movements between summer and winter ranges are extended visits to mineral licks in spring, to separate socializing areas of males in fall and again in spring where they meet and interact socially, and to special birthing and child-rearing areas for female sheep. These localities of increased social interaction have a powerful pull on the males, just as the lambing and lamb-rearing areas have as strong pull on females. Consequently, males and females move and live separately for all of the year except for about a month during the mating season. Mating occurs on the core wintering areas of females, which the adult males leave right after mating, possibly so as not to compete for the limited food supply available to the mothers of their prospective offspring. Thus the males move to separate wintering areas of their own. The distances between sea-

Zebras tend to be the first species to migrate to a new area due to their toleration of tall grass. (Photo by Mitch Reardon/Photo Researchers, Inc. Reproduced by permission.)

sonal ranges for mountain sheep may be tens of miles (kilometers) apart.

Megaherbivores such as elephants, rhinos, and giraffes may move over huge annual home ranges between small dry season ranges close to water and large wet season ranges. The size of these ranges may be hundreds of square miles (kilometers). Areas within are used opportunistically depending on forage and water availability. The ability to move rapidly over long distances, coupled with excellent memory of locations, allows elephants to live in vast desert areas. While the wet season areas are large, travel may be restricted at that time by the abundance of food. Movement may be more extensive in the dry seasons due to decreased forage density.

Food availability also dictates mass movements of African plains animals such as zebras, wildebeest, and gazelles in the Serengeti plains. The great productivity of the plains during the wet season attracts them, and they leave the woodlands they use during the dry season. Zebras, wildebeest, and gazelles form a grazing succession in which the zebras first remove the coarse grasses. The re-growth of grasses then creates good foraging conditions for wildebeest, which in turn graze down the grass exposing dicotyledonous plants that attract gazelles. The variations in timing of the rainy seasons

thus affect the mass-movements of these herbivores to and from the woodlands.

Migrations are most spectacular when masses of animals unite in closely coordinated movements. As indicated earlier, the proximity of individuals to one another and the coordination that keeps them together arise as an anti-predator adaptation in species adapted to open landscapes—be they coverless terrestrial terrain or open oceans. In cover there is little possibility of group cohesion, but great opportunity for individual hiding and thus escape from predators. In the open, uniting into large formations has powerful advantages for prey. Joining with others of one's kind is a purely selfish act aimed at self-preservation. Consequently, this grouping has been labeled "the selfish herd." It is very common. To begin with, close herding empties the landscape of the species, forcing resident predators on alternate prey—except where the species is bunched. Here the chances of any individual being taken by a predator depend on the number of others of its kind it is with. In a larger herd the chances of being attacked are smaller than in a small herd. This is called the dilution effect. Thirdly, the individuals in the center of the herd are shielded from predator attacks by individuals on the periphery. This is the position effect. Fourthly, when a herd flees there is for each individual a greater chance in a large herd than in a small herd that someone is a slower

An aerial view of migrating caribou (*Rangifer tarandus*). (Photo by Ted Kerasote/Photo Researchers, Inc. Reproduced by permission.)

runner. Selfish herds are thus dangerous places for slow runners, be they ill or merely heavily pregnant. However, for a healthy individual of average abilities, joining with others and staying close together is an excellent way to minimize the risk of being caught by predators. As a consequence, uniting into huge herds is a possibility in large expanses of open landscape.

However, the very gathering into a huge mass raises severe problems in food acquisition. A large mass of herbivores will quickly deplete the available forage and must move on. That is, huge herds must move just to stay fed and it matters little where they go as long as they stay with food. This leads, of course, to unpredictable movements, which incidentally reduces predation risk still further as it does not permit predators to anticipate the direction or time of movements. This random movement, excepting predictable occupation of grasslands freed of snow in winter by warm katapatic winds along mountain fronts, or of grasslands recovering from large wildfires, was suggested to be the model of bison migration in North America up to the late nineteenth century. That is, food availability governed the mass-movements of the bison herds over the huge expanse of prairie in the center of the continent.

The mass migrations of reindeer in Eurasia and barren ground or Labrador caribou in North America have as their primary focus the calving ranges in spring. Pregnant females lead this migration, which is closely focused in both space and time. This is the most predictable migration and it leads to barren, snowed-over areas with a minimum of predators. These large, dense herds also reduce predation by "swamping" existing predators with a great surplus of calves on the calving grounds. Reindeer bulls lag behind in the migration. From the calving grounds the herds move predictably to summer ranges with high productivity and some protection from insects, which in large numbers can debilitate caribou. Males and females mix on the summer ranges. From the summer feeding grounds the herds drift towards the wintering areas below the timberline. They mate during a short rutting season while traveling to the winter ranges. Caribou can travel with remarkable speed and shift location within any seasonal range. Nevertheless, their movements are predictable as they follow the same landscape features including favorite locations above waterfalls and rapids on large rivers. These and large lakes are no serious barrier to caribou, which are excellent swimmers. Caribou may travel over 3,100 mi (5,000 km) in their annual migrations.

A feature common to migratory terrestrial or marine species is that they become very fat on their summer feeding grounds and then depart with a fat-load to subsidize their maintenance and reproductive activities elsewhere. Fattening —which is metabolically a very expensive process since for

Migrating blue wildebeest (*Connochaetes taurinus*) and Burchell's zebras (*Equus burchellii*). (Photo by © Fritz Polking/Frank Lane Pictures/Corbis. Reproduced by permission.)

every unit of fat stored, as much as another unit is lost as metabolic heat—evolved in response to both seasonality of climates and high diversity of habitats within any one season. Extremes in seasonality as well as in terrestrial and marine habitats developed during the major glaciations of the Ice Ages during the Pleistocene.

Some extreme examples of fattening and long distance migrations were evolved by baleen whales. For instance, in the short Antarctic summer such whales congregate about the Antarctic pack ice encircling the Antarctic continent, particularly in regions where the cold waters rich in dissolved oxygen and carbon dioxide are fertilized by mineral-rich, warm deep-current upwellings to produce explosive plankton growth. Whales feed gluttonously on this superabundant food, mainly on krill, which consist of small crustaceans, and lay down massive fat deposits. Some whales reach about 40% body fat at the end of summer. As the Antarctic summer turns to winter the pack ice extends outward from the continent, cutting off whales from their feeding grounds. The whales depart for warm waters thousands of miles (kilometers) away where the pregnant females give birth to their young and maintain lactation while usually fasting until the following summer. These whales thus live by a "boom and bust" economy and their huge size and reduced metabolism per unit mass allow them to exist without

feeding through the greater part of the year. Similar patterns of seasonal feeding and dispersal are followed by whales in the Northern and Southern Hemispheres except that the plankton food supply in the Antarctic appears to be richer on average and whales reach there larger sizes. Clearly, being able to fast during the reproductive periods allows whales to choose areas of increased security for their newborn and growing offspring. The phenomenon of gigantic migratory whales is predicted on the existence of polar mineral-enriched waters near the freezing point that, unlike tropical mineral-poor waters, are superlative producers of phyto and of zooplankton, as well as on the existence of fast-freezing polar ice shelves that after summer exclude the gluttonous behemoths.

Terrestrial migration has not been for small-bodied creatures, the myth of northern lemming migrations not withstanding. Small-bodied mammals tend to deal with seasonality by growing fat in summer and hibernating during the winter, as do some northern carnivores such as bears and badgers. Hibernating is a withdrawal in time from severe climatic and foraging conditions; migration is a withdrawal in space. Thus these are both adaptations that deal with the severity of seasonal winter climates.

The presence of large, very fat herbivores must have been an irresistible attraction to human hunters for untold millen-

Walruses (*Odobenus rosmarus*) migrate by following the ice packs. (Photo by Tui De Roy/Bruce Coleman, Inc. Reproduced by permission.)

nia. To exploit such migratory species, however, requires either following them continuously—an unlikely prospect—or developing a way to regularly intercept their migrations, killing in excess of need, and finding means to store the surplus for the future. Intercepting migratory species cannot be done with out understanding chronologic time. Migrants tend to reappear in the same localities much at the same calendar time, particularly if migration is closely tied to reproduction. Reproduction is timed in some of the migrants by the annual

light regime which, via stimuli through the optic nerve to the hypothalamus, regulates the hormonal orchestration of reproduction. The variation in annual reproduction tends to be narrowly confined about specific calendar dates, and so are the migratory movements associated with reproduction. The same individuals reappear in the same locality within a few days of the same calendar date each year.

There is some evidence to suggest that Upper Paleolithic people in Eurasia discovered chronologic time and used it to predict and exploit migratory reindeer. They marked what appear to be lunar calendars on tines cut from reindeer antlers. That would have enabled them to anticipate reindeer migrations, move to localities where reindeer were vulnerable, and prepare for the kill. Reindeer bones comprise over 90% of the bones from Upper Paleolithic archeological sites. There is evidence for large flint-blade processing sites. Long, very sharp flint blades were struck skillfully from large flint cores. These long, thin flint flakes are ideal knives for the many hands needed to carve up reindeer meat for drying. Wood ash, ideal for smoking and drying meat, predominates in Upper Paleolithic hearths. The pattern of bone fragments indicates that de-boning of reindeer carcasses was done away from processing sites, so that meat was brought in for processing with just the bones required to hold the meat together or those used for marrow fat or tools. The superlative physical development of these large-brained cave-painters indicates that they enjoyed abundant food of the highest quality. Caribou meat has balanced amino acids for humans and caribou fat has an optimum essential fatty acid balance for brain growth. Caribou are valued highly as food by indigenous people and Arctic travelers and their fur is unsurpassed as a material for clothing and shelters in Arctic climates. Eventually salmon replaced some of the reindeer in the Upper Paleolithic, but salmon runs are also chronologically timed and can be predicted by calendar dates. To exploit a reindeer or salmon run it is crucial to be at the right place in the right time. And the same applies to migrations of bowhead whales along the coast, an important traditional food source for northern coastal people. Without an idea of chronologic time and how to keep track of it, no large-bodied mammalian migrants could be hunted systematically. With the discovery of chronologic time, however, these very rich, predictable, high-quality food sources became available for exploitation, leading to the spread and dominance of modern humans from the late Pleistocene onward.

Resources

Books

Geist, V. *Deer of the World.* Mechanicsburg, PA: Stackpole Books, 1998.

Geist, V. *Mountain Sheep. A Study in Behavior and Evolution.* Chicago: University of Chicago Press, 1971.

Irwin, L. L. "Migration." In *North American Elk: Ecology and Management,* edited by Dale E. Toweill and Jack Ward Thomas. Washington, DC: Smithsonian Institution Press, 2002.

Maddock, Linda. "The 'Migration' and Grazing Succession." In *Serengeti: Dynamics of an Ecosystem,* edited by A. R. E.

Sinclair and M. Norton-Griffiths. Chicago: University of Chicago Press, 1979.

Owen-Smith, R. N. *Megaherbivores.* Cambridge: Cambridge University Press, 1988.

Slijper, E. J. *Whales.* Ithaca, NY: Cornell University Press, 1979.

Periodicals

Fancy, S. G., L. E. Pank, K. R. Whitten, and W. L. Regelin. "Seasonal Movements of Caribou in Arctic Alaska as Determined by Satellite." *Canadian Journal of Zoology* 67 (1989): 644–650.

Valerius Geist, PhD

. . . . .

Mammals and humans: Domestication and commensals

What is domestication?

Domestication is a process by which certain species of wild animals have been brought into close relationship with humans and thereby significantly changed the animals' ways of life. The process of domestication has been long and complicated, with unforeseeable consequences for both concerned sides. The consequences resulted in significant economic, social, cultural, and political changes.

The process of domestication is similar to that of evolution, except that the natural choice was made with artificial selection. Humans violently separated the ancestors of domestic animals from their wild relatives and step by step, generation after generation, also changed their appearence and features. This selection probably proceeded at first only accidentally. Only later, when humans noticed certain connections, did they start to use purposeful selection for different economic, cultural, or aesthetic reasons.

Which species were domesticated?

One of the main questions is why only a few species were domesticated from such a huge number of wild animal species. For example, only 14 species were domesticated from the large group of terrestrial mammalian herbivores and omnivores. The horse and the donkey were domesticated, but four zebra species and the Asiatic ass were not. There are many existing testimonies that people almost 20,000 years ago were keeping bears in captivity. The ancient Egyptians (in the Old Kingdom 2500 B.C.) were keeping tamed addax antelope, hartebeest, oryx, gazelle, and cheetahs (for hunting). The ancient Romans kept and bred dormice (for meat). None of these animals, however, was domesticated.

The answer is as follows. Wild animals must match several important conditions to be domesticated. If one of the conditions is not met, domestication will not occur. A candidate for domestication must not be a narrow food specialist (e.g., the anteater or panda) because nourishment must be easy to supply. It should also have a strong herd or pack instinct, which secures authority recognition and therefore simplifies comunication with a human. A social carnivore like a wolf is much easier to tame than a solitary hunter like a leopard. Likewise, sheep and goats, which have a social system

based on a single dominant leader, are much easier to tame than deer and most antelopes, which live in herds without dominance hierarchies. And the candidate animal should be "in the right place at the right moment."

A distinctive barrier to successful domestication is food competition (one of the reasons why the bear was not domesticated was that humans were not able to feed both themselves and the bear). Other obstacles are nasty disposition, reluctance to reproduce in captivity (cheetah), and the tendency to panic in enclosures or when faced with predators (antelopes, deer), or a long reproductive cycle and slow growth rate (which obviously has prevented real elephant domestication). The reason why zebras were not domesticated, even though the colonists tried it in South Africa from the seventeenth century onwards, was due to their biting habits and dangerous behavior (they kick a rival until it is killed).

When and where?

The beginnings and the progress of domestication of most "classical" domestic animals are not known in detail because they depend on archaeological discoveries of human settlements, for example, bone fragments from waste holes, cave paintings, or statuettes of animals. In the latest periods, it is possible to obtain domestication information using genetic comparative analysis. According to archaeologists, the beginnings of domestication were at a period when human gatherers and hunters became sedentary farmers, a period in which the domestication of wheat, barley, and peas also occurred. This change occured at the turning point of the Stone Age (Paleolithic and Neolithic) and it was so radical that it is referred to as the "Neolithic revolution." It is a temporal border that occurred more than 12,000 years ago. In archaeological findings from western Asia, which are 11,000 and 9,000 years old, it is possible to clearly follow changes in the way of life according to the structure of food, which changed very distinctively during that interval. While the remains of wild animal species including cattle, pigs, gazelle, deer, foxes, rodents, fish, and birds predominate in the older findings, there is a distinct predominance of sheep and goats in the more recent findings. It is very difficult to determine if the bone remains come from already domesticated animals or wild ones because at the beginnings of domestication the skeletal

The Yorkshire is a common breed of domesticated pig. It was created by crossing the large white pig with the smaller Chinese pig. (Photo by Lynn M. Stone. Bruce Coleman, Inc. Reproduced by permission.)

changes were very small. If there are predominatly male bones in the findings (females were left for the production of offspring) or if the bones are markedly smaller than those of wild animals, archaeologists assume that they belonged to domestic animals.

The domestication of the majority of the traditional domestic animals usually occurred in areas where the human populations had reached a certain level of cultural development and where there was a suitable wild ancestor. These areas are designated as a "center of domestication." The oldest (10,000 to 6000 B.C.) domestic centers were located in western Asia and in the Middle East (the area of the "Fertile Crescent") and were related to the beginnings of sedentary settlements and the first successful experiments with breeding grain. In that area, goats and sheep were domesticated for the first time, followed by cattle and pigs. Nevertheless, a very narrow relationship was created a few thousand years earlier between tamed wolves and humans, so that the first domestic animal was a dog. This period became a sort of "start" and "instruction" period for the next domestication processes. The next significant domestic centers occurred in the Indian

continent (zebu), in China (goose, duck, pig, silkworm moth), in Central Asia (horse, camel), and in Southeast Asia (domestic fowl, pig, buffalo). In these areas the common domestic animals of today were domesticated. A small percentage of domestic animals were bred on the American continents, in Middle America, the turkey and musky duck, and in the western part of South America, the llama and guinea pig. From these centers domestic animals spread to other areas. Some species expanded all over the world (dog, cat, cattle, horse, sheep, goat, domestic fowl) while others remained only in the original area of domestication (yak, Bali cattle, llama).

Why and how?

We have to appreciate that the first breeders of domestic animals did not have any instructions and they were not able to imagine where domestication would lead. It is assumed that the initial reasons motivating domestication were frequently different from the animals' subsequent use. However, the main reason was very simply to access a supply of food. Exceptions are cats and dogs, which became partners to people, and later assumed many other roles, such as dogs becoming guardians. Even though of various origins, the domestication scenarios of most animals were analogous and evolved in three main steps. First came capturing and holding a wild animal in captivity (mostly young animals, when their mothers were killed in the hunt), followed by gradual taming and herding. The third phase was breeding, where humans started to generate animals according to their needs or beliefs that they were improving certain desirable qualities (intensive livestock husbandry). Sometimes it was a spontaneous process when the animal connected to a human on a voluntary basis (dog, cat, pigeon), or when the human connected to the animal (rein-

Black miniature donkey and foal. People have used donkeys for riding, driving, and pulling for centuries. (Photo by Animals Animals ©Renee Stockdale. Reproduced by permission.)

deer). Other species were barely tamed (wild ox, horse, ass). Domestication was a lengthy process; it is generally believed that the shorter the developmental cycle of an animal, the quicker the change of generations and, therefore, the shorter the domestication process.

What is a domestic animal?

A domestic animal can be defined as one that has been bred for a long time in captivity for economic profit in a human community that maintains total control over its territorial organization, food supply, and breeding, which is the most important issue. Because domestic animals did not develop in a process of natural evolution, they are not considered distinct animal species, and in the zoological terminology they are catergorized as forms.

Humans breed and use many other animals that have not passed through the domestication process. A typical example is the elephant. The working elephant has been used for some 5,000 years and is still an important part of the work force in Southeast Asia, even though they were never domesticated. Every individual is caught in the wild, violently tamed, and educated to perform specific work. It almost never reproduces in captivity. The Indian elephant (*Elephas maximus*) is used mostly for work, but the ancient martial elephants, for example those in Hanibal's army, were African elephants (*Loxodonta africana*). Other nondomesticated animals humans have employed include birds of prey used for hunting (falconry), cormorants used for fishing, macaques used to pick cocoa nuts on the beaches, and dolphins and pinnipeds used by some militaries.

Laboratory animals used for investigative and experimental purposes are also a special group. Even though they have a shorter history of coexistence with humans than the more common domestic animals, these other species are also considered domestic. Almost all domestic animals have been used as laboratory animals, but not all laboratory animals are domesticated. In the last decades, the spectrum of animals used for laboratory purposes has expanded to include wild animals.

A special group of domesticated animals is pets. People have been breeding them in their surroundings for several centuries or millennia, for example, the peacock or the dancing mouse.

Species that humans breed and change also include semidomesticated animals, such as the fallow deer kept in enclosures. They reproduce with no problems and have several color forms. Animals used for fur also belong to this group (mink, fox, coypu, chinchilla), as do ostriches, which are bred in farms.

Another group of animals called "commensal," live with humans. Commensalism is defined as the reciprocal coexistence of two or more organisms. One of them benefits from this relationship and the other is neither harmed nor benefits ("no harm parasitism"). This relationship is very free, close to symbiosis. Humans provide many opportunities for commensalism. For example the house mouse (*Mus*

Sometimes mammals are used for entertainment, such is the case with horse racing. (Photo by © Kevin R. Morris/Corbis. Reproduced by permission.)

musculus) exploits human hospitality, and obtains food profit from human coexistence as well as a safe hiding place. Rats, rooks, seagulls, and many other animal species benefit from rich food allocation in human wastedumps. Another example of commensalism is the pariah dogs in Asia and Africa. These live by scavenging around human towns, settlements, and roads. They are tolerated because they contribute to tidiness.

Domestic changes

When a given species of animal is bred in isolation from its wild habitat and at the same time protected against unfavorable conditions, specific traits start to appear that disadvantage the animal in a natural environment and would keep it out of the reproductive process in the wild—either because markedly different individuals are easier victims for predators or because no partner will accept them. These different traits are not kept in the wild populations or are very rare. Conversely, these individuals were of interest to humans because of their different appearance or their submissive nature. After some time, the changes in the nature, behavior, and in the reproduction cycle become distinctive in domestic animals. They also become stratified in their genetic make-up.

Coloring

One of the first signs of domestication is variability of color. The individuals all have white spots or all white or all black. It is interesting that white coloring is usually connected with lower performance (there are few white racing horses and even fewer winners) or with different defects (white noble cats have a high incidence of deafness).

Sheep have been domesticated for both their meat and their wool. (Photo by Andris Apse. Bruce Coleman, Inc. Reproduced by permission.)

Changes in hair and feathers

The quality of hair also markedly changes with domestication—the difference between guard hair and undergrowth vanishes. Long, wavy, or curly hair appears (sheep, donkey, horse, rabbit, cat, pig, goat, dog) or conversely the individuals lose hair (dog, cat). In horses, the hair of the tail and mane became visibly longer—wild horses have a short standing mane, and domestic horses have a mane that falls down to the neck.

Changes in the size and proportions of the body

Dwarf and giant forms have developed as a result of domestication. In the early phases of domestication the overall size of the body decreased (cattle, goat, sheep). This change is attributed to premeditated selection (easier control and housing of smaller individuals) combined with inadequate food. Extra long ears also appeared (dog, sheep, pig, cattle, cat, rabbit), curled tails (dog, pig), or very short tails (dog, cat, sheep). Horns also became variable in size and curvature, totally disappearing in some species (cattle, sheep, goat).

Changes in the skeleton and internal organs

Striking changes occurred in the skull; for example, there was a shortening of the snout and jaws and at the same time a reduction of the number of teeth (dog, cat, cattle, pig). The shortened snout and accentuated rounded eyes induced the juvenile appearance of the "eternal cub." It also resulted in lower brainpower and smaller brain volume. The skeleton became less resistant than that of wild animals as a consequence of the "comfortable life" with its lack of movement. For the same reasons, the size of some of the physical organs decreased, for example, the heart (35% lower by weight in the domestic rabbit when compared to its wild ancestor). Of course there are exceptions. The English thoroughbred, trained for several centuries for racing, has a heart about one fifth heavier than that of other horses of the same size. The fat storage mechanism has also been modified by domestication. In wild animals, fat is stored in the surroundings of the

internal organs and under the skin, while in domestic animals it is stored among muscle filaments and around the tail, especially by pigs or sheep.

Physiological changes

There have been changes in the reproduction cycle, prolonged lactation (cattle, sheep, goats), and more numerous litters or eggs. Domestic animals reproduce themselves practically throughout the whole year, and sexual adolescence starts earlier that it does in wild animals.

Behavioral changes

Domesticated animals lost shyness and many of them lost totally the ability to survive in the wilderness (sheep). Some instincts have changed or totally vanished and the rhythm of given activities has completely changed. For example, some dusk animals and night animals have become strictly day animals (pig) while some changed from monogamous to polygamous (goose). Many domestic offspring are not taught behavior because they are quickly weaned from their parents,

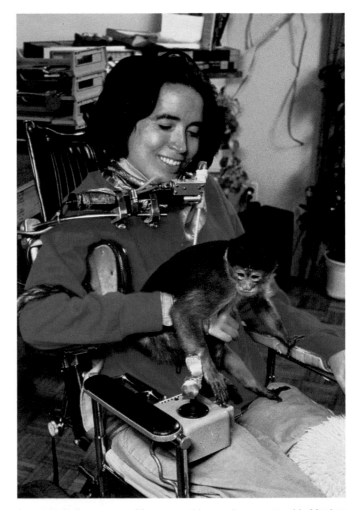

A quadriplegic woman with a capuchin monkey as an aid. Monkey helpers perform simple, every day tasks such as getting food and drink, retrieving dropped or out of reach items, turning lights on and off, and other chores. (Photo by Rita Nannini/Photo Researchers, Inc. Reproduced by permission.)

if they come in contact with them at all. They are denied the learning that parents teach their offspring in the wild.

What is the breed?

The basic category of domestic animals is the species, as is the case with wild animals. The species of domestic animals are differentiated into breeds, while that of wild animals are differentiated into subspecies. A breed is defined as a group of animals that has been selected by humans to possess a uniform appearance that is inheritable and distinguishes it from other groups of animals within the same species. Domestication and selective breeding have changed some species of domestic animals (camel, reindeer) very little. But in others, including the oldest and the most important species of domestic animals, tens and hundreds of breeds have been developed; compare for example the pocket-sized Chihuahua with the huge mastiff or Saint Bernard dog. Some specialized breeds are not able to carry out an independent existence. Others become wild without any problem and are able to set up feral populations under suitable conditions. People used artificial selection from the early days of breeding even though they did not know the rules of heredity and the selection of features was varied. For example, European cattle were bred to increase milk and meat production and working capacity, while African pastoralists preferred to increase the size of horns.

Threatened breeds

Enormous numbers of different breeds have developed during many millenia, most very well acclimatized to local conditions. It is estimated that 5,000–6,000 breeds exist today. Four thousand belong to the so-called "big nine" (cattle, horse, donkey, pig, sheep, goat, buffalo, domestic fowl, duck). However, the trend has moved to renewed selection efforts during the latest decades, for example, the present specialized daily milk production capacity of a Holstein Friesian averages 10.6 gal (40 l) of milk compared to the African N'Dama, produces 1.1 gal (4 l). These highly productive but very vulnerable breeds are now found all over the world, and in many places they have totally replaced the original breeds or have been crossbred with them. In this way, the original breeds disappear, and with them go extremly important genetic variations. In 1993, the Food and Agriculture Organization (FAO), started a project called Global Strategy for the Management of Farm Animal Genetic Resources, which is responsible for the preservation of livestock of no economic value. The categories "extinct" or "critically threatened breed" have been introduced, as has been done for wild animals.

Disadvantages of domestication

The number of domestic animals greatly exceed the number of wild or related species. In some cases, their wild ancestors have been completely exterminated. The breeding of domestic animals has provided people with many indisputable advantages, but it has its downside. The grazing of large livestock herds diminishes food and water resources of local wild

Domesticated cats have helped people keep homes free of rodents for many thousands of years. (Photo by Ernest A. James. Bruce Coleman, Inc. Reproduced by permission.)

animals in Africa and leads to the total devastation of the landscape. Bison almost became extinct because pasture lands were required for domestic cattle in North America. Large areas of rainforest in South America are being converted to pasture for cattle today, presenting conservation difficulties. Herds of sheep and goats completely devastated large areas of Mediterranean and central Asia. The infestation by domestic rabbits nearly devastated the breeding of sheep, another domesticated animal, in more than half of the Australian continent. Enormous ecological damage was committed by wild populations of goats and pigs that were abandoned by sailors in many Mediterranean islands. Together with feral dogs and cats they liquidated enormous amounts of local fauna and flora. They are responsible for more than a quarter of extinct species and subspecies of vertebrates. Feral goats present a similar problem on a number of the Galápagos Islands, threatening the wildlife there.

Domestic horse (*Equus caballus* f. *caballus*)

The horse was the last of the five most common livestock animals to be domesticated. After a short period when it was used only as a source of meat, it became established as a perfect means of transport until the recent past.

The history of the wild horse in Europe and Asia from the end of the Pleistocene until its domestication in perhaps 4000 to 3000 B.C. is poorly understood. According to prevailing opinion, wild horses during the domestication belonged to two species. These were the Przewalski horse (*E. przewalskii*) from semidesertic central Asia and the tarpan horse (*E. caballus*, syn. *E. ferus*) with two subspecies, the forest tarpan (*E. c. silvestris*) and the steppe tarpan (*E. c. ferus*), which lived in an area ranging from western Europe to the Ukrainian steppes. The two species are the only ancestors of all recent breeds of domestic horse. The last tarpan was exterminated in 1879 in the Ukraine and the Przewalski horse was no longer found in the wild as of 1960, surviving only in zoos. However, its breeding is so successful now that the process of its reintroduction has started in Mongolia.

Draft horses have made it possible for farmers to plow much larger fields. (Photo by Animals Animals ©Charles Palek. Reproduced by permission.)

It is assumed that the area of steppes between the Dnieper and the Volga Rivers is the oldest domesticating center for horses. The region was populated by a culture called the Sredni Stog in the years 4000 to 3400 B.C. The horses were bred doubtless not only for meat but also for riding. It is also assumed that, in the same period, horses were domesticated by other people who lived in the steppe corridor of Eurasia from southeast Europe to Mongolia. From the second millenium B.C., the combative tribes of nomadic Skyt, Kimmer, Hun, and Mongol started out from Asian steppes in regular intervals on the backs of their hardy and tough ponies to the west, east, and south and, until the fifteenth century, propagated not only terror and dread but also the fame and genes of their horses. In ancient civilizations, horses were first considered "luxury goods." However, they spread very quickly and at the turning point of the second and first millenium B.C. they we're a common phenomenon. During the first millenium B.C., horses were also commonly used for farming, transportation, and sport.

After almost 6,000 years of human service, the appearance and many features of the horse have changed, though less than that of other domestic animals. The first primitive horse breeds developed naturally with the influence of different climate, food, and prevailing working usage. Today there are more than 200 breeds of horses, in a range of different sizes. The smallest horse is the Falabella, at 28 in. (71 cm) high and

with a weight of 44 lb (20 kg), and the largest is the Percheron, which is 6 ft (1.8 m) and 2,600 lb (1,180 kg). All the breeds have unique performance, working abilities, and stamina. In the sixteenth century, horses went back to the American continent, where they had lived more than 10,000 years ago. Most of them ran wild and constituted large feral populations of mustangs (North America) and cimmarons (South America). Correspondingly, the same situation occurred later in Australia, where feral brumbies live today.

Domestic donkey (*Equus africanus* f. *asinus*)

It is said that the donkey is the horse of the "poor people" and undeservingly it remains in the shadow of its more famous relatives. It is not actually headstrong, dumb, and lazy. It has only a more evolved instinct of self-preservation, which allows it to preserve itself from human service. The donkey does not as a rule bond emotionally to humans as horses do. If it has good treatment and a warm stable, it is a priceless helper, especially in stony terrain. It does not mind hot weather or miserable food, it hates only dirty water and rainy weather.

The domestic donkey is the descendant of two subspecies of the African ass. The Nubian wild ass (*E. a. africanus*) was domesticated in 5000 B.C. in the Nile Valley and in the area

Almost 14,000-year-old paintings from the La Pileta cave in Spain show sheep and goats in a corral. It takes a long time for wild sheep, which were kept in simple corrals, to become truly domestic animals. The oldest findings of domestic sheep come from the north Iran mountains (Zawi Chemi Shanidar) and date from 9000 B.C. However, there were certainly more areas of domestication in western Asia at that time. In 4000 B.C., domestic sheep were bred throughout the civilized world. At first they gave only meat, milk, and leather, and only later did wool sheep appear, though only with short and rough wool (in Mesopotamia around 3000 B.C.). In the first millenium, sheep spread all over Europe, Africa (except in primeval forest areas), and Asia (to Sulawesi). At that time sheep with white, longer wool were common, with four horns or without horns (known from ancient Egyptian frescos). Sheep from antique Greece and Rome resembled modern breeds. The number of sheep breeds today ranges between 550 to 630. They are categorized according to wool type, tail length, fat deposits, or utility. There are no existing feral sheep populations except that of the European mouflon and the Soay sheep.

Goat (*Capra aegagrus* f. *hircus*)

Goats live throughout the world but they flourish in humid tropical forest areas. Their number is still increasing, especially in the desert and the semi-desert areas where it is not possible to keep other domestic animals. Goats give meat, milk, leather, hair, wool, horn, and dung and have very low food requirements.

The domestic goat progenitor is the bezoar goat (*C. aegagrus*), which lived in western and central Asia. According to radiocarbon dating, the sheep was domesticated first but goats were more numerous than sheep in the early domestic period. The oldest domestic center of both species was Sierra Zagros at the border of Iraq and Iran. Determination of whether remains belong to wild or domestic goats is possible according to horn shape. Archaeologic findings show that the originally scimitar horns of wild goats changed during several hundred years (8000 to 7000 B.C.) to the twisted horns of domestic goats. We do not know why people preferred these animals with twisted horns or if the horn shape related to the behavior of goats, or their productivity.

The domestic goat quickly spread to all inhabited areas of Europe, Asia (to Sulawesi), and Africa. It preserved its animation, shrewdness, and obstinacy from its progenitors, which passes for happy malevolence. There are many feral domestic goat populations, for example, in New Zealand, in Australia, and unfortunately in the Galápagos and other islands where they cause severe ecological problems. There are now some 200 to 350 goat breeds.

Pig (*Sus scrofa* f. *domestica*)

The domestic pig was likely domesticated after cattle. It is omnivorous (as humans are), and is exceedingly intelligent. The number of domestic pigs is estimated at nearly 913 million worldwide. The only places they are not bred are in unsuitable climate areas (tropics, polar areas) and in Israel and Islamic countries, where eating pork is forbidden by religion. The pig is bred only for its meat and fat, although leather is a secondary product.

The progenitor of the domestic animal is the wild boar (*Sus scrofa*), which lived in a large area from western Europe and north Africa to Southeast Asia. A range of subspecies evolved, of which two were domesticated, the European wild boar (*S. s. scrofa*) and the Asian banded boar (*S. s. vittatus*). The next two species domesticated in Southeast Asia were the Sulawesi wild boar (*S. celebensis*) and the Philippine warty pig (*S. phiilippensis*).

Wild boar domestication dates from 7000 to 6000 B.C. One of the important preconditions of this process was sedentary civilization because, unlike sheep, goats, and cattle, pigs are not able to live a nomadic life. Pig domestication occurred independently in two or maybe more places, partly because of how relatively easy pigs are to tame. The first domestic centers (6500 B.C.) were in western Asia, India, and some islands. From there, domesticated pigs were moved into China, Egypt, and farther into Africa. The second center of domestication from 5000 B.C. lies in northern Europe by the Baltic Sea. The third important area was the Mediterranean.

Domestic pigs came to the New world with European settlers. Sailors also left them on islands. Other feral populations developed in South and Central America, Australia, and New Zealand. As feral goats do, the pigs destroy specialized island fauna and flora.

Camels and llamas

The camel was and is an excellent transport vehicle in desert areas where horses and donkeys cannot survive. In deserts, camels can survive ten times longer than humans and four times longer than donkeys. They provide meat, milk, blood, leather, and hair, and excrements are used as fuel. The wild Bactrian camel (*Camelus ferus*) is the ancestor of the Bactrian or two-humped camel (*C. ferus* f. *bactrianus*). It comes from east and central Asia. The remaining populations live in the periphery of the Gobi Desert. It was domesticated by nomadic tribes, probably in the second or first millenium B.C. in Iran or in the Gobi Desert.

There is no known evidence about the wild progenitor of the dromdary or one-humped camel (*Camelus dromedarius*). It was proposed that the progenitor could be the extinct *Camelus thomasi*, which lived in the interface of the Tertiary and Quaternary periods in north Africa and in adjoining areas of Asia. The one-humped camel was probably domesticated in 3000 B.C. in the Arabian Peninsula or in the steppe areas of western Asia. The oldest testimonies of domestic camels come from Egypt and the Sinai, 5,000 years ago. Almost 19 million domestic camels are bred worldwide, of which almost 90% are one-humped camels. They live in desert areas from the western Sahara to India and as feral populations in Australia, where they were introduced one hundred years ago. The two-humped camel is bred in Mongolia, China, Afghanistan, Iran, and Turkey.

Two species of camelids live in South America: the guanaco (*Lama guanicoe*) from semidesert mountain areas, and the vicuña (*Vicugna vicugna*) from high mountain areas. Only the guanaco was domesticated, perhaps in 3000 B.C. Its descendants are the llama (*Lama guanicoe* f. *glama*) and the alpaca (*L. guanicoe* f. *pacos*). The llama has been used as a transport animal, and for meat and wool, and the alpaca primarily for wool.

Reindeer (*Rangifer tarandus*)

Not much is known about the domestication of the reindeer. It is assumed that it was domesticated at only one place in the Sayan mountains from 3000 to 1000 B.C. Chinese sources from the sixth century describe the reindeer as a domestic animal. The reindeer may have become used to humans when it came to human settlements to lick salt from human urine. It served in place of cattle and horses in harsh northern conditions. Domestic reindeer live in northern Eurasia and Canada. The Canadian caribou (*R. tarandus caribou*) was never domesticated. Domestic reindeer do not differ from wild reindeer, except that they may vary in size and coloring. Those who keep reindeer must live the nomadic way of life because the reindeer still migrate along ancient paths. The domestic reindeer serves as a riding and draft animal, and provides milk, meat, leather, hair, and antlers.

Dog (*Canis lupus* f. *familiaris*)

The dog was the first domestic animal; however, the beginning of its coexistence with humans is still unclear. The wolf (*Canis lupus*) is the progenitor of all domestic dog breeds and feral populations. Fifteen thousand years ago, the wolf lived in all of Eurasia, in north Africa, and in North and Central America. It evolved into many subspecies, which differ in size and color. The small Indian wolf (*C. l. pallipes*) and the larger Eurasian wolf (*C. l. lupus*) are the most likely dog ancestors.

Humans were interacting with wolves in the end of Pleistocene. They were hunting the same type of prey and in the areas where they coexisted, both species got used to each other over thousands of years. Food lured the wolves to human settlements and humans counted on the watchfulness of wolves. Taming an adult wolf is almost impossible, but small cubs are easier to tame. The problem of feeding the cubs was solved very simply: women nursed them. If the cubs were too aggressive, the humans killed them for food. Only submissive individuals were allowed to reach adulthood and reproduce. The ease of this domestication process was confirmed by recent experiments with foxes. Individuals were selected according to their level of aggressiveness. The behavior changed during a mere twenty generations and the white spots and curled tail present in domestic dogs appeared in that time as well.

The skeletal remains of the first domestic dogs come from different places around the world, from Israel, Turkey, Iran, Japan, England, Denmark, Germany, and North America. The oldest findings verifying the existence of the domestic dog is an 11,000-year-old grave, found in northern Israel in the Ein Mallaha settlement. Feral populations of dogs developed in places where wild populations of wolves never lived. The dingo (*C. l. dingo*) spread in Australia at least 4,000 years ago. The forest dingo (*C. l. halstromi*) spread over New Guinea and Timor. Feral dog populations exist in many other places all over the world. Around 420 domestic dog breeds are registered but that number may still change.

Domestic cat (*Felis silvestris* f. *catus*)

The cat is one of the most recent domestic animals. In spite of its coexistence with humans, it still has an independent nature and the perfect hunting instincts of a solitary hunter. It is highly individualistic and should not have undergone successful domestication at all.

The progenitor of the domestic cat is the African subspecies of wild cat (*Felis silvestris lybica*). Domestication of the cat occurred in Egypt from 4000 to 2000 B.C. Preserved cat mummies provide dometication evidence. The oldest are of tamed cats from the 4000 B.C. period while mummies from the end of the Middle Kingdom period are of domesticated cats. The domesticated cats have shortened skulls and often irregular denture. The cat likely started its coexistence with humans voluntarily and to the benefit of both sides. Wild cats were drawn together into the Nile Valley because of the number of rodents that accompany human settlements. They quickly became common domesctic animals and they achieved the status of sacred animal of the goddess Bastet. The domestic cat has spread from Egypt throughout the Mediterranean and reached southern Europe in 500 B.C. The cat came to the east with merchants to Turkey, Persia, and along the silk road to China and Southeast Asia. It penetrated Central Europe at the beginning of the Middle Ages. Benedictine monks were the first true cat breeders (they also bred the first rabbits and pigeons in Europe). Feral cat populations exist on many islands and threaten the populations of local insular animals.

Rabbit (*Oryctolagus cuniculus* f. *domesticus*)

The rabbit is the "youngest" domestic animal. The wild rabbit (*Oryctolagus cuniculus*) is the only progenitor of the domestic rabbit. The monks in Benedictine monasteries started its domestication at the beginning of the Middle Ages in the south of France. Rabbit meat was eaten during Lent and so the monks kept the rabbits in closed tiled monastery yards. Rabbit breeding spread from France to England, Belgium, and Holland in the seventeenth and eighteenth centuries and the biggest boom in rabbit breeding began in the nineteenth century. Rabbits were already the source of cheap and readily available meat for a large number of people in Europe. Several hundred breeds exist today because of the rabbit's popularity as a pet, as well as a source of food and other material.

Guinea pig (*Cavia aperea* f. *porcellus*)

The guinea pig is one of the few animals domesticated in the New World. It was a sacrificial animal and a pet for the local inhabitants. The wild guinea pig (*C. aperea*) may be the progenitor but some zoologists think the montane guinea pig (*C. tschudii*), which lives only in the mountainous areas of Peru, southern Bolivia, northwestern Argentina, and northern China, is the progenitor. The period of its domestication has not been determined exactly, but the remains of domestic guinea pig were discovered in deposits dating back to 3000 B.C. in the Peruvian Andes. The guinea pig became the most popular and the most often bred rodent. It is very easily tamed, almost never bites, and it is able to communicate very well.

Resources

Books

Budianski, S. *The Covenant of the Wild: Why Animals Chose Domestication*. New York: William Morrow, 1992.

Clutton-Brock, J. *A Natural History of Domesticated Mammals*. Cambridge: Cambridge University Press, 1999.

Harris, D., ed. *The Origin and Spread of Agriculture and Pastoralism in Eurasia*. London: UCL Press, 1996.

Hemmer, H. *Domestication: The Decline of Environmental Appreciation*. Cambridge: Cambridge University Press, 1990.

Serpell, J., ed. *The Domestic Dog: Its Evolution, Behaviour and Interaction with People*. Cambridge: Cambridge University Press, 1995.

Periodicals

Bradley, D. G. "Genetic Hoofprints." *Natural History* (February 2003): 36–41.

Diamond, J. "Evolution, Consequences and Future of Plant and Animal Domestication." *Nature* 418 (2002): 700–707.

Grikson, C. "An African Origin of African Cattle? Some Archaeological Evidence." *The African Archaeological Review* 9 (1992): 119–144.

Alena Cervená, PhD

Mammals and humans: Mammalian invasives and pests

Introduction

When all the crop losses and control and containment costs are added up, non-native invasive species (including weeds and insects) cost the United States alone an estimated $137 billion annually. The more intangible effects of invasive species on natural ecosystems are also serious. Invasive species are sometimes termed "biological pollutants," because predation and competition by invasive species can reduce populations of native species and cause extinctions. Indeed, half of the known cases of bird extinctions on islands are linked to introduced mammalian predators, such as cats.

The top invasive mammal pests worldwide are rats, mice, cats, dogs, cattle, burros, horses, goats, hedgehogs, foxes, gray squirrels, coypus, pigs, possums, rabbits, deer, weasels, mink, and the mongoose. Globally, rodents like rats and mice consume an estimated 5–15% of grains like rice, wheat, and corn in the fields before harvest. East African countries have lost as much as 80% of the crop during severe twentieth century rodent outbreaks. After the harvest, the combined actions of insects, rodents, fungi, and other organisms may destroy another 5–15% of stored grains, though some areas experience 20% losses. Stored grain losses are estimated at $5 billion per year in India alone, and may run as high as $1 billion per year in the United States. According to one estimate, rodents destroy enough food to feed 200 million people. These losses may be just the tip of the iceberg, as there is little economic documentation on invasive mammals.

Although they are considered invasive pests when feral in the wild, many of the top invasive mammals in the world lead a double life, as they are also desirable as pets and valuable as agricultural livestock. For example, cats, dogs, and rabbits are favored domestic pets and companions, but can be both a nuisance and a menace to ecosystems when turned loose in the wild. Similarly, horses and burros are used as pets and livestock, but in the wild they can damage ecosystems and deprive other species of food and water.

An understanding of what constitutes a pest helps clear up this seemingly contradictory duality of an animal being both beneficial and a pest. The term "pest" is not an absolute term, rather it is subjective. When an organism is in the wrong place at the wrong time and is unwanted it is deemed a pest.

When humans want the same organism around, it is no longer labeled a pest. With the definition of a pest so much in the eye of the beholder, reasonable people can, and often do, disagree about whether a particular organism is or is not an invasive pest.

For example, wild horses and wild burros, which were introduced on the North American continent as a consequence of the European colonization, are viewed positively, often nostalgically, by many people as a historical living legacy of America's frontier days when the wild West had miners with burros and cowboys on horseback. But to ranchers grazing public lands, wild horses and wild burros are often viewed as little more than hoofed locusts, stealing valuable forage from livestock.

Indeed, at one time wild horses were herded into dead-end canyons and shot, though now they are captured and adopted. Even environmentalists can alternately wax positive or negative, depending upon whether the wild horses or wild burros are befouling a sensitive area and threatening the food sources and drinking water of native species such as mountain sheep.

Thus, viewing an invading species in a negative light and designating it as pestiferous or alternatively viewing an invader in a positive light is the product of underlying assumptions, ideologies, and value judgments. Sometimes these underlying ideologies, assumptions, and value judgments are implicit, below the surface, and hard to discern. Other times, as with the case of the American mink (*Mustela vison*) in the United Kingdom, the rhetoric is as heated and open as the most contested political and ideological issues debated in the British Parliament.

In the United Kingdom, a European island nation, the American mink was imported for small-scale fur farming in 1929, and sometime thereafter began escaping into the wilds of England and Scotland. Being a polyphagous predator, American mink have gobbled up ground-nesting seabirds in the firths and lochs of northwest Scotland, as well as an endangered native mammal whose riparian habitat is threatened, the northern water vole, *Arvicola terrestris*. In some countries water voles could easily be depicted as just another water rat, and the case for their preservation dismissed as another example of radical environmental lunacy. But in the UK, water

The Maori may have hunted moas to extinction in New Zealand. (Illustration by Wendy Baker)

voles are a beloved icon of English literature, populating novels like Evelyn Waugh's *Scoop* and starring as Ratty, Toad's companion in *The Wind in the Willows*.

While the water vole has deep roots in English culture and supporters in high places direct Heritage Lottery Funds its way, getting rid of the invading American mink to aid the endangered vole is a cause that runs into the ideological concerns of animal rights activists who want to free the minks. When Great Britain passed the Mink (Keeping) Regulations of 1975, which mandated measures to stop the further escape of American minks from fur farms into the wild, little thought was given to activists in the animal rights movement invading fur farms and setting minks free. Not only have law-breaking animal rights activists invaded fur farms and illegally set loose more American mink into the British wilds, where they threaten endangered water voles, but many also question the premise that the American minks are an ecological threat.

Not all invasive mammal stories are quite as dramatic or filled with such emotional passion, even in the UK, where during the past century introduced North American gray squirrels, *Sciurus carolinensis*, have quietly displaced the UK's only native squirrel, the red squirrel, *S. vulgaris*. The red squirrel, whose British lineage dates back to the last Ice Age,

is now rare in southern England, with only remnant populations remaining in places like the Isle of Wight and Poole Harbor. The picture is equally bleak in central England, with remnant red squirrel populations in East Anglia, Staffordshire, Derbyshire, and Merseyside. Even in their current strongholds of northern England, Wales, and Scotland, red squirrels are losing territory to gray squirrels.

Local red squirrel extinctions have always been relatively common. But before the intentional introduction of gray squirrels began in the 1870s (an era when species were still freely moved from continent to continent with little concern for ecological consequences), red squirrels almost always recolonized areas several years after local extinctions. The outcome may have been different if there was just one release of gray squirrels. However, there were repeated releases of gray squirrels from the 1870s until the practice became illegal in 1938. No doubt these well-intentioned human releases of gray squirrels to recolonize habitats put the red squirrels at a great disadvantage and contributed to their relatively rapid loss of home range.

Nevertheless, the best explanation put forth today for the continuing displacement of native red squirrels by introduced gray squirrels is ecological competition. In other words, gray

This Norway rat (*Rattus norvegicus*) shows how easily rats are able to board ships and migrate from one area to another. (Photo by Tom McHugh/Photo Researchers, Inc. Reproduced by permission.)

squirrels are believed to be outcompeting and thereby replacing red squirrels in their former ecological niches such as parks, gardens, broadleaved woodlands, and conifer forests. Biodiversity and timber values (gray squirrels damage the bark) may ultimately be affected by this shift in squirrel species, which is still in progress.

The importation of the coypu, or nutria (*Myocastor coypus*), from South America to the North American continent is, like the gray squirrel and American mink, another case of seemingly good human intentions gone awry. Business people originally imported the herbivorous coypus from southern Argentina to the United States for fur farming in 1899. Coypu farms rapidly spread from California to Oregon, Washington, Michigan, Ohio, Louisiana, and other states. But the coypu fur craze and the demand for coypu meat eventually collapsed. Depending on the locale, and local stories vary, the coypus either escaped on their own or were intentionally let loose by failed fur farmers in the late 1930s.

Coypus are now well-established in the United States, including key coastal states like Louisiana, Texas, and Maryland. Without the predators, diseases, and other natural factors controlling their populations in South America, coypus are running amok in North America. Along Maryland's Chesapeake Bay, coypus gobble up patches of marsh plants and accelerate the conversion of rich wetland habitats into eroded ponds and bays. When not chewing up wetland vegetation in Louisiana and Texas, coypus are attacking rice and sugarcane fields.

Attempts to turn coypus into gourmet fare have yet to rekindle an interest in harvesting them for either their fur or meat. In any event, trapping is not a feasible solution for the North American continent. Hopefully, humans will devise a solution that mitigates North American coypu wetland damage, which, like most mammalian invasive species problems, was created as a consequence of human activities.

Humans, the most successful invasive species

The magnitude of the invasive species problem in agricultural and natural ecosystems prompted U. S. President Bill Clinton to organize the heads of eight federal agencies into the National Invasive Species Council in 1999. Actually, humans rank among the most successful invasive mammal species. Humans are believed to have spread from Africa to Europe and Asia over 100,000 years ago, and reached the island continent of Australia between 40,000 and 60,000 years ago. But humans are apparently relative newcomers to the Americas, having arrived between 15,000 and 20,000 years ago. The human invasion did not reach many Pacific Ocean islands until 1,000 to 2,000 years ago. A small human presence on the continent of Antarctica is a twentieth-century phenomenon.

Between 20 and 40 bird species have become extinct in North America over the past 11,000 years, a period when the presence of humans is well documented and not controversial. Many of these bird species likely disappeared because they had narrow ecological niches dependent upon now extinct large mammals like mammoths, mastodons, horses, tapirs, camels, and ground sloths. Human hunting likely played a role in the extinction of large mammals like the mammoth. But the magnitude of the prehistoric human role in extinctions involves conjecture and is still being debated.

Several species of long-legged, flightless moas and an eagle, *Harpagornis moorei*, are among the birds possibly hunted to extinction by New Zealand's first human inhabitants, the Maori. The loss of 62 endemic bird species in the Hawaiian Islands is associated with the arrival of the first human inhabitants from Polynesia. In North America, more recent European immigrants hunted the passenger pigeon to death at the end of the nineteenth century.

Rhinoceros species were hunted to the brink of extinction for their horns in Africa by the end of the twentieth century. Thanks to a new ecological consciousness sweeping the planet, small rhino populations still exist in protected reserves at the start of the twenty-first century. Humans have also extinguished species via habitat loss. This is among the problems addressed in the United States by legislation like the Endangered Species Act of 1973.

Invasive mammal species seem to be most serious on isolated islands where the native organisms have not evolved defenses against the mammals common to the major continents. In Great Britain, an island close to mainland Europe, only 22% of the mammal species are considered exotic. But on New Zealand's islands, which were only settled by humans within the last 2,000 years, 92% of the mammal species are recent introductions. Indeed, New Zealand and some other islands were free from mammalian predator pressure for so long that flightless bird species evolved.

In North America and the islands of the Pacific, other mammal species accompanied the human invasions. The first Asians entering the Americas brought along dogs. When the Polynesians set sail for new Pacific islands, they brought along their pigs, plants, and stowaways like lizards and rats. European colonialism from the fifteenth to twentieth centuries was a major driving force behind biological invasions of North America, Australia, New Zealand, and other areas. Like the ancient Polynesian voyagers, European colonists brought along their plants and animals to help settle these "new worlds." Besides livestock like sheep, goats, cattle, pigs, and horses, there were stowaway species like rats. Later, predators like the mongoose were deliberately introduced to help control the rats.

Feral cats in Australia

A good example of the pest/not-pest duality is the domestic cat on the island continent of Australia. Between 4 million and 18 million feral cats (*Felis catus*) live wild in Australia. Until recently most of these cats were believed to be descendants of European cats brought to the continent in the late eighteenth century, with a few earlier arrivals via trading ships and shipwrecks. However, Australia's aboriginal people regard cats as native. Genetic analysis indicates that Australian feral cats may have more in common with Asian than European cats, supporting the aboriginal view for an earlier arrival of cats on the continent.

But the debate of more practical consequence is whether feral cats threaten native species such as tammar wallabies (*Macropus eugenii*). If viewed as an invasive pest, then feral cats need to be hunted down, poisoned, given birth control, or otherwise controlled. If viewed as beneficial predators helping control other pests such as rabbits, rats, and mice, then feral cats should at least be tolerated.

In the late nineteenth century feral cats were viewed as beneficial. Cats were deliberately acclimatized and released into the Australian wild to hunt pestiferous (nuisance) European rabbits (*Oryctolagus cuniculus*). Indeed, Australia's Rabbit Nuisance Bill of 1883 supported releasing feral cats to help control rabbits that were damaging agricultural grazing lands.

But later in the twentieth century, feral cats were no longer welcomed as rabbit killers. Feral cats began to be viewed as invasive pests, threatening to native birds and mammals. Anti-cat forces pointed to the case of Marion Island, where five domestic cats released in 1949 had become a colony of over 2,000 by 1975. The Marion Island feral cat colony was destroying nearly half a million burrowing petrels per year.

On the Australian continent and on Australian offshore islands, feral cats were blamed for the regional extinction of several native bird and mammal species. The vanishing species were ground dwellers living in open habitats (favorable to cat hunting) and were the right size to be cat prey. The anti-cat forces also suspected that toxoplasmosis, a disease vectored by cats, may have played a role in mammal and carnivorous marsupial population declines many years earlier.

The hedgehog (*Erinaceus europaeus*) often invades hen houses to take eggs. (Photo by Hans Reinhard/OKAPIA/Photo Researchers, Inc. Reproduced by permission.)

Feral cats were suspects in Western Australia, where a red fox (*Vulpes vulpes*) removal program did not stop the population decline of native fauna. Feral cats were suspected of filling the niche vacated by the red fox, and adding native species to their predominately rabbit and rodent diet. To determine whether feral cat control is a necessary policy, researchers like Robyn Molsher set up studies in New South Wales to evaluate the ecological relationships among feral cats, red foxes, and other fauna.

Since feral cats are difficult to follow in the wild, their scats (fecal droppings) were analyzed for dietary clues. Rabbits were the primary feral cat prey in New South Wales; rabbit remains were present in 82% of the scats and constituted over 68% of scat volume. The majority of the other prey was carrion, primarily sheep and eastern gray kangaroo (*Macropus giganteus*). Even after rabbit populations plummeted following application of a biological control agent known as Rabbit Calicivirus Disease, rabbits remained the dominant prey of feral cats. Feral cats consumed more of the house mouse (*Mus musculus*) following the rabbit population decline.

Foxes and feral cats tracked via radio collars shared similar habitats and prey. Foxes displayed aggression and killed some of the feral cats competing for the same food resources. When foxes were present, feral cats ate mostly rabbits and left the carrion for the foxes. In fox removal experiments, feral cats ate more carrion and hunted more at night in the same prey-rich grassland habitats favored by foxes.

Molsher concluded that integrated rabbit control programs needed to also consider fox and cat control to prevent native fauna from becoming prey in the absence of rabbits. However, rabbits are so well established across such a vast

A striped skunk (*Mephitis mephitis*) invades home garbage. (Photo by Joe McDonald. Bruce Coleman, Inc. Reproduced by permission.)

area that the goal of rabbit eradication in Australia has been abandoned in favor of long-term population suppression. This seems to vindicate a continuing beneficial role for feral cats as rabbit and rodent predators in Australia.

Integrated rabbit control

Prolific reproductive potential has helped the European rabbit become a very successful invasive species. Originally from North Africa, the European rabbit spread north through Italy to the British Isles and then around the world, causing ecological havoc in some countries. On the Hawaiian island of Laysan, the rabbit is credited with wiping out 22 of 26 native plant species at the beginning of the twentieth century.

Since its mid-nineteenth century introduction into Australia, the European rabbit has been a major plague. Vegetation is grazed from vast stretches of land that become more desert-like and less suitable for livestock grazing despite the killing of millions of rabbits every year. Over a century after starting rabbit mitigation programs, Australia still spends an estimated $373 million per year on rabbit control.

Eradication is deemed feasible only on small islands or in small localized areas where rabbit populations are newly established. The few small islands off the coast of Western Australia where rabbits have been eradicated are the exception, not the rule. More typical is 46 mi² (120 km²) Macquarie Island and the main Australian continent, where rabbits are so well-established that eradication has been replaced with the more realistic goal of population suppression.

Population suppression is accomplished using a suite of varied biological, mechanical, and chemical control techniques. This integrated pest management approach includes predators, microbial control agents, warren ripping, and electrified and wire-net fences. Wild rabbits are also hunted and

"harvested" as a commercial product. Barrel or soft catch traps are still used against small isolated rabbit populations. (Humane considerations have largely precluded continued use of the traditional steel-jawed leg-hold trap.) But rabbit populations are so high and the species is so prolific that shooting and trapping have no significant impact on populations.

A variety of poisons and fumigants are still used against feral rabbits, though safety, environmental, and humane concerns have been raised. The most widely used vertebrate control pesticide is 1080 (sodium monofluoroacetate), which is formulated into paste, pellet, food cube, grain, and carcass baits to poison animals such as rabbits, feral pigs, wallabies, wombats, dingos (wild dogs), possums, rats, mice, and foxes. Food chain risks are inherent in poison baiting, particularly when individuals lack baiting expertise. Another drawback is that sheep, cattle, horses, goats, cats, dogs, some native wildlife, and humans are also very susceptible to 1080, and there is no known antidote to the poison.

Destroying warrens by ripping or plowing is a less controversial alternative to poisons, though the two techniques are sometimes combined. Sometimes rabbit kill is maximized by using dogs to drive rabbits into their warrens before burrow destruction commences. But rocky areas, riversides, and steep sandbanks with rabbit warrens are impossible to rip up and destroy, short of explosives.

In some locales rabbits prefer surface refugia rather than warrens. This means habitat management may be needed. However, the same shrub, blackberry, and log debris habitats favored by rabbits are also home to desirable species of birds, reptiles, amphibians, and other small mammals. So, it is not always desirable to modify the habitat to fight rabbits.

Common house rats (*Rattus rattus*) drinking milk in a temple in India. Though often considered pests, some religions consider rats to be holy. (Photo by M. Ranjit/Photo Researchers, Inc. Reproduced by permission.)

The eastern mole (*Scalopus aquaticus*) eats helpful invertebrates such as earthworms, which turn and aerate the soil. (Photo by L. L. Rue, III. Bruce Coleman Inc. Reproduced by permission.)

Biological control using predators, parasites, or microbes can be part of integrated rabbit control programs and help overcome the limitations of poisoning and habitat modification. One of the more famous instances of biological control was the introduction of the myxoma virus to fight rabbits.

Viruses for rabbit control

The myxoma virus was imported into Australia in 1936, and extensively studied before being released into the environment in 1950. The virulence of the myxoma virus is rated on a scale of one to five, with one being most virulent. The original myxoma virus strain released into the environment was rated one, and provided a spectacular 99% rabbit mortality when first released into the environment in the early 1950s. Without rabbits grazing on the landscape, the amount of forage available for sheep production soared and farmers were very happy with the financial windfall.

But the spectacular success did not last. There is an ecological interaction over time involving the pathogenicity or virulence of a microbe and the genetic suseptibility or immunity of the host population. A microbe that is 100% successful in killing its host would go extinct along with its host. So, ecological theory favors the evolution of a less virulent microbe that does not kill the entire host population. Indeed, the evolution of reduced myxoma virus virulence and increased rabbit immunity became a classic epidemiological case.

Over time the virus attenuated, and the virulence of field strains of myxoma virus declined from one to three. At the same time, the rabbit population developed greater immunity to myxoma virus. Releasing the highly virulent myxoma virus strain rated one into the environment no longer produces the 99% rabbit kill seen in the 1950s. At the end of the twentieth century, myxoma virus was producing a more sustainable rabbit kill of between 40% and 90%. This reduced rabbit kill is still very important to integrated control programs. Indeed, in the absence of myxoma virus rabbit populations can still soar to very high levels.

In theory, integrated pest management can incorporate multiple techniques, each providing a percentage of pest population suppression. In practice, studies in both central and south Australia demonstrate the advantage of combining multiple techniques into an integrated pest management program for rabbits. When myxoma virus is used alone, as is typically the case, rabbit populations rebound to high levels in subsequent years. However, when rabbit warrens are ripped or plowed after myxoma virus has already reduced the rabbit population, the area can remain almost devoid of rabbits for many years.

Bolstered by the half century of continuing success with myxoma virus, it has only been natural to look for additional rabbit pathogens to introduce into the environment. In 1984 Chinese scientists identified an acute infectious rabbit disease called Rabbit Calicivirus Disease (RCD). RCD was subsequently identified in Europe, Asia, Africa, and Mexico. Australia began studying RCD on wild and laboratory rabbits and non-target species in 1991. Several months after a 1995 field trial on South Australia's Wardang Island, RCD was detected on the mainland. In 1996, RCD was officially recognized as

Brown rats converge at a garbage dump. Rats can spread diseases that affect livestock and people. In addition, they eat and/or contaminate feed and their gnawing destroys buildings. (Photo by Jane Burton. Bruce Coleman, Inc. Reproduced by permission.)

The Mexican ground squirrel (*Spermophilus mexicanus*) destroys crops while foraging in the spring. (Photo by Anthony Mercieca/Photo Researchers, Inc. Reproduced by permission.)

a biological control agent under the Commonwealth Biological Control Act.

In some parts of South Australia RCD has killed over 90% of the rabbits, and longer term rabbit populations are down 17%. RCD has been spreading at the rate of 250 mi (400 km) per month. An insect vector traveling in the wind is believed responsible for this rapid spread. However, humans and implements in contact with carrier rabbits may also spread the disease. More time is still needed to see how well the combination of myxoma virus and RCD work against rabbit populations.

Less lethal fertility control agents, a humane solution favored by animal welfare groups, are also under development for rabbits, foxes, and mice. One idea is to engineer a virus like the myxoma virus with an antigen causing animals to produce antibodies that reduce fertility. An estimated 60% to 80% of female rabbits need to be stopped from breeding in order to reduce rabbit populations. Immunocontraception will likely be tested on wild rabbits sometime before 2010.

Even if not wildly successful by themselves, new techniques like immunocontraception will likely play a role in integrated pest management systems targeting rabbits and other pest mammals. It is clear from over a century of rabbit control efforts that no single pest control technique by itself will work everywhere. Even myxoma virus, which looked so promising with its initial 99% rabbit control, is no longer viewed as a stand-alone solution. The integrated pest management paradigm of combining multiple control techniques and ecological principles is the wave of the future for combating invasive pests.

Invaders in paradise

The 6,393 mi² (16,558 km²) Hawaiian Islands chain, a collection of 132 islands, reefs, and small shoals, has only one native land mammal species, the Hawaiian hoary bat (*Lasiurus cinereus semotus*). Thus, the Hawaiian Islands are a good "laboratory" for the studying the ecological effects of introduced mammals.

Even such usually ubiquitous land organisms as ants and mosquitoes were absent from the Hawaiian Islands' native fauna. For much of its geological history the world's largest ocean, the Pacific Ocean, acted like a giant moat, keeping the Hawaiian Islands relatively isolated and deterring invading species. The winds, ocean currents, and migrating birds brought species to the Hawaiian Islands, but very few became established. On average over the last 70 million years, only one invading species per 35,000 years successfully established in the Hawaiian Islands.

However, a diverse topography, a warm tropical climate and an absence of predator species during most of the past 70 million years made the Hawaiian Islands a good evolutionary locale for new species formation via adaptive radiation. A spectacular example of new bird species formation by adaptive radiation in the Hawaiian Islands is the 54 species of Hawaiian honeycreepers (Drepanididae). Compared to the Galápagos Islands and its 14 Galápagos finch species that inspired Charles Darwin's theory of evolution, the Hawaiian Islands have more habitat diversity and a longer evolutionary history.

A mammalian invasion began transforming the Hawaiian Islands approximately A.D. 400, when the first Polynesian sailing canoes arrived. The Polynesian voyagers brought animals and

Raccoons (*Procyon lotor*) have adapted to urban living and thus become pests to humans. (Photo by Steve Maslowski/Photo Researchers, Inc. Reproduced by permission.)

plants with them, and changed the landscape with their agriculture. New species in the Hawaiian Islands increased to three to four per century after the Polynesians arrived, a huge increase from the pre-human one species per 35,000 year average.

Even before the Europeans arrived in the late eighteenth century, the Hawaiian Islands had a few hundred thousand people of Polynesian descent. The fauna introduced by humans included the Polynesian (or Pacific) rat (*Rattus exulans*), dogs, pigs, fowl, and reptiles. The Polynesian rat is a native of Southeast Asia that spread across the Pacific Ocean islands to the Hawaiian Islands with the Polynesians, but never reached the mainland of the United States.

Polynesian rats attract a lot of attention as pests of plantation agricultural crops like sugarcane and pineapple, though a broad range of crops are attacked. Polynesian rats are omnivorous, and studies show adverse impacts on coastal tree and lizard species in New Zealand and on seabirds on several Pacific islands. There are little data available on Polynesian rat ecological effects on now extinct Hawaiian Island birds.

In the Hawaiian Islands, about half the land bird species predating human arrival have vanished. Direct human impacts from hunting and gathering and indirect human impacts are

strongly implicated in the decline or extinction of native species, particularly flightless birds and ground-nesting winged species. The magnitude of ancient human impacts on specific species is still the subject of vigorous debate and inquiry. However, there is little doubt that the rate of worldwide ecological change and species extinctions directly and indirectly attributable to human beings began increasing in recent centuries.

The European ships of the eighteenth and nineteenth centuries brought exotic mammals from around the world to the Hawaiian Islands, including new rat species, European pig genotypes, cattle, goats, sheep, the house mouse, and the mongoose. Goats and cattle trampled and grazed native plants and generally degraded habitats. Hawaii's native bird species suffered additionally when early nineteenth century whalers introduced the first mosquitoes and avian malaria. New diseases like smallpox and syphilis were transferred from the European arrivals to the Polynesian population. Clearly, as the top mammal species, human impacts increased as human populations increased and spread.

Globalization and the expansion of ship and airplane commerce in recent centuries accelerated the rate of new species

Invasive animals damage local ecosystems. 1. Wooded area; 2. Wild pigs root and wallow in the wooded area; 3. The pigs uproot plants and leave the area open to erosion; 4. New growth in the disturbed area includes invasive plants such as briars, burs, and other species carried as seeds in the pigs' manure. (Illustration by Wendy Baker)

introductions. When all the newly introduced plant and animal species were added up, the rate of new species introductions into the Hawaiian Islands was estimated to have accelerated to several dozen species per year in the twentieth century. Ecological upsets and pest problems from introduced mammals became more noticeable in the Hawaiian Islands during the nineteenth and twentieth centuries.

Black rats, also known as roof rats (*Rattus rattus*), and Norway rats (*Rattus norvegicus*) disembarked on Pacific islands as stowaways aboard European sailing ships and spread rapidly in the nineteenth century. Norway rats and black rats are omnivorous, attacking agricultural crops and feasting on the young and eggs of seabirds like petrels, shearwaters, gulls, terns, and tropicbirds.

Birds on isolated islands like New Zealand and Hawaii evolved in pre-human times when there was no need for defenses against mammalian predators like rats. Bird depredations by rats are less common nearer the equator. One untested hypothesis is that birds nearer the equator developed better predator defenses useful against rats, because of land crabs preying on eggs and chicks.

Black rats are suspected in the demise of many native Hawaiian birds during the nineteenth century. But the bio-

logical documentation from that period is not considered conclusive by modern standards. Very recent technological advances like night vision videos provide more conclusive evidence. For example, night vision videos revealed beyond doubt that black rats were a major predator of New Zealand's endangered Rarotonga flycatcher (*Pomarea dimidiata*). Consequently, rat control became part of the program to save that endangered forest bird species.

Mongooses for rat control

The Indian mongoose, *Herpestes auropunctatus*, was deliberately introduced into Pacific and Caribbean islands in the late nineteenth century for biological control of rats in plantation crops like sugarcane. This was a rare attempt at biological control by introducing a mammal to prey on another introduced mammal. One of the more colorful stories from the Hawaiian Islands is that a man known as "Mongoose" Forbes sold mongooses to sugar plantations as rat catchers in the 1870s. More scientific accounts suggest that the Indian mongoose was introduced to the Hawaiian Islands via the West Indies in the early 1880s. Whatever the story, the deliberate introduction of the mongoose for rat control was a badly flawed idea.

Modern twentieth century biological control programs screen potential introductions to make sure that desirable flora and fauna are not destroyed. This was not part of the scientific protocol in the nineteenth century when the mongoose was introduced for rat control. Even though mongooses eat rodents in sugarcane fields, they do not provide adequate rat control. One problem is that the mongoose is a diurnal hunter, whereas the rats they were introduced to control are nocturnal. Also, mongooses do not stay put in agricultural plantations. So there have been serious consequences for native species in native ecosystems.

From the islands of Fiji to the Caribbean, mongooses are predators of a wide range of native wildlife and a potential reservoir for diseases like rabies and leptospirosis. The ground-nesting quail dove (*Geotrygon mystacea*) was nearly eliminated from the Virgin Islands by mongoose predation. Hawaii's endangered state bird, the nene or Hawaiian goose (*Nesochen sandvicensis*) is attacked by the mongoose, as are the endangered Hawaiian dark-rumped petrel (*Pterodroma phaeopygia*), Newell's shearwater (*Puffinus newelli*), and the Hawaiian crow (*Corvus hawaiiensis*). Turtles, native lizards, snakes, and poultry are also mongoose prey. Live traps, hunting, and poison baits are among the mongoose control methods available.

The bad experience with mongooses did not extinguish Hawaii's interest in biological control of rats. In the late 1950s, Hawaii introduced barn owls (*Tyto alba*) for rodent biocontrol. Even dogs have been used to hunt rodents in sugarcane fields. Trapping tends to be too labor intensive for large outdoor areas with rats. Early in the twentieth century almost 150,000 rats were trapped annually in Hawaii's sugarcane plantations with no noticeable effect on rat populations or crop damage. Shooting is also of questionable value for rat control. So, poison baits have been the fallback for keeping rat populations under control.

The opossum (*Didelphis virginiana*) has adapted to life near humans, and has been known to raid hen houses and attics of homes. (Photo by Steve Maslowski/Photo Researchers, Inc. Reproduced by permission.)

Feral pigs

Feral European pigs, a late eighteenth century introduction, may be the largest mammalian threat to native forest ecosystems in the Hawaiian Islands. Feral pigs dig up young trees to eat the roots and spread weed seeds. Native plants and crops like sugarcane are attacked. From New Zealand to the Galápagos Islands, feral pigs also have a reputation for digging into burrows to consume seabirds like petrels. Feral pigs also feast on the eggs and young of surface-nesting seabirds like boobies, shags, and albatrosses.

The original pot-bellied pigs introduced from Polynesia into the Hawaiian Islands in the fourth century are smaller, more docile animals and a different genotype than the larger, much more aggressive European pigs. Polynesian pigs are also less inclined to roam and go feral, and are less of a threat to the native ecosystems of the Hawaiian Islands.

Most popular accounts date the arrival of the European pig genotype in the Hawaiian Islands to the arrival of British ships under the command of Captain James Cook in 1778. Apparently the British were disappointed by the small size and tougher texture of the meat of the Polynesian pig, and so introduced the larger, more succulent European pig to the Hawaiian Islands.

In the nineteenth century, the introduction of new species was viewed as a positive and encouraged. Indeed, mid-nineteenth century advertisements exhorted sea captains to import and release their favorite songbirds in the Hawaiian Islands. Even in the mid-1800s, settlers in the Hawaiian Islands noticed forest habitat destruction and native bird losses, and felt that the introduction of their favorite European species would help compensate. Never mind that previous introductions like European pigs and goats were among the culprits destroying the habitat. It was an age when feral European species were welcomed by hunters, albeit opposed by agricultural interests suffering crop damage.

By the early twentieth century European pigs had completely displaced Polynesian pigs in the wild, and an eradication program was started because of feral pig damage to the Hawaiian Islands' native rainforests. By the mid-twentieth century, 170,000 feral pigs were killed. But feral pig popula-

A house mouse (*Mus musculus*) raids a sack of seeds. (Photo by Stephen Dalton/Photo Researchers, Inc. Reproduced by permission.)

tions were still trampling, uprooting, eating, and otherwise destroying the Hawaiian Islands' native ecosystems.

Though damage by goats, feral pigs, and other grazing mammals seemed obvious to observers, there was little quantitative baseline data against which to measure the ecological impacts. Near Thurston Lava Tube in Hawaii Volcanoes National Park, the National Park Service set up an experiment in which some forest plots were fenced to keep out feral pigs. Compared to unfenced plots where pigs roamed freely, fenced plots had fewer exotic plant species and more native plant species. Fenced plots also had less pig damage, as measured by fewer exposed plant roots and less exposed soil. Not surprisingly, fencing is used as an integrated control measure to keep pigs out of sensitive areas.

The National Park Service has also been following the spread of feral pigs from lower to higher elevations on the Hawaiian island of Maui. Feral pigs were removed from Maui's Kipahulu Valley in the 1980s. Removing the pigs allowed the forest understory to recover and slowed the invasion of exotic plant species, a positive ecological outcome. Nonetheless, there has been opposition to using snares and hunting to remove the pigs. Indeed, one person's invasive mammal pest may have redeeming positive qualities for another person. Hunting groups want the pigs to remain as game, and some indigenous groups oppose eliminating the pigs for cultural reasons.

Pigs threaten foxes

The case for removing feral pigs (*Sus scrofa*), has less opposition in the Channel Islands National Park and other islands off the coast of southern California. This is a case where one invasive exotic mammal species, the feral pig, has changed the ecological relationships among several native

predators. On four islands, feral pigs introduced into the ecosystem are indirectly leading to the extinction of four subspecies of island fox (*Urocyon littoralis*), a tiny animal smaller in size than a house cat. The smallest member of the family Canidae in North America, island foxes show little fear of humans and were probably once kept as pets by Native Americans.

There are no island foxes left in the wild on San Miguel and Santa Rosa Islands, where captive breeding programs are aiming to save the island fox subspecies from extinction. Fewer than 200 island foxes are left on Santa Catalina Island, in part because of canine distemper virus vectored by domestic dogs. The island fox subspecies on Santa Cruz Island has seen its population decline from 1,300 to fewer than 100. The U. S. Fish and Wildlife Service proposed the four rare island fox subspecies for protection under the Endangered Species Act, and eliminating feral pigs from the Channel Islands is part of the plan to save the island foxes.

Feral pigs and other introduced grazing mammals like rabbits, sheep, goats, cattle, deer, elk, and horses contribute to degradation of the island habitat. The presence of feral pigs also introduces an indirect ecosystem food chain effect contributing to the demise of the island fox. Feral pigs serve as prey, allowing golden eagle (*Aquila chrysaetos*) populations to flourish. Historically, golden eagles are a mainland species that has neither bred nor overwintered on the islands. The islands have historically been nesting grounds for the bald eagle (*Haliaeetus leucocephalus*), which preys mostly on marine mammals and fish.

Basically, the introduction of feral pigs allowed the golden eagle, a mainland species, to become established on the islands. Unlike bald eagles, golden eagles prey on island foxes. One study found that with 90% golden eagle predation, island fox numbers declined to zero. Released from competition with island foxes, island spotted skunk (*Spilogale gracilis amphiala*) populations also increase as an indirect consequence of feral pigs supporting golden eagle populations. In other words, feral pigs feed the golden eagles which eat the foxes which frees up space for the skunks.

Golden eagle removal and relocation is part of the plan to save the island fox and restore the ecosystem. However, satellite telemetry studies indicate that golden eagles relocated to the mainland will try to return to the islands as long as feral pigs are available as a food source. Reintroduction of the bald eagle is also being considered to help restore the ecosystem. The bald eagle is very territorial, and may deter the golden eagle from nesting. However, removing feral pigs as a food source is the more important factor in preventing reestablishment of golden eagle populations, restoring the ecosystem, and saving the island fox.

Conclusion

There are many more stories of invasive mammals to tell, and more details untold about feral pigs, foxes, rabbits, cats, squirrels, rats, mice, horses, burros, and other invasive mammals than can easily fit on the printed page. But this overview

Researchers use low flying planes to do aerial surveys and to count animals. (Photo by Rudi van Aarde. Reproduced by permission.)

Sociological techniques offer their own challenges, as the barriers to an effective survey tend to be cultural rather than technological or biological. For example, who is interviewed, who does the interview, and even what dialect is used for the interview all affect the survey results. There are strong cultural pressures on the interview process that are difficult to comprehend or anticipate without extensive experience in the target community. Interview methods are effective for relatively large species that are regularly encountered or eaten by local people. Paying a bounty for local hunters to produce specific animals is tempting, but not recommended without consideration of the long-term impact of creating a market for wildlife.

Creating a species list is an open-ended activity; the longer researchers look, the more they find. A species accumulation curve, which graphs the number of new species detected for every additional unit of time or effort, can be used to gauge when a species list is complete. It should be noted that the method used to measure the number of species may only provide the number of species that can be detected.

Beyond the need to do a thorough survey, the most difficult component of a species list is comparability. If lists are to be compared between areas or to lists collected previously, there must be a way to measure the amount of effort used in the survey. Two biologists will not conduct surveys exactly the same way, and differences in lists are always subject to individual protocols.

Population index

Population indexes are usually species-specific and are based on the number of animals, or their sign, detected for each unit of effort. The effort can be measured in many ways, including number of miles/kilometers walked, number of traps set, number of trees examined, or number of hours observed. The index itself is usually tied to specific traits of the species: for example, claw marks on trees made by Asiatic

black bears (*Ursus thibetanus*), calls by howler monkeys (*Alouatta* spp.), night nests of gorillas (*Gorilla* spp.), and latrines of rhinos are indexes that are not broadly applicable to other species. Indexes based on sign are conducted along transects of known length, with all sign within a set distance of transect recorded, or a predetermined area is searched for sign. When interpreting signs, one must have observed how such signs are actually created by the animal, as well as which activities do not leave any signs at all (i.e., a deer feeding on dead leaves shed by trees or small bits and pieces of conifer branches or lichen or small fruit leaves no trace at all). When the signs are not permanent, such as fecal pellets or tracks, and the area is to be resurveyed at a later date, the boundaries of the area can be marked and all sign cleared from the area at the end of each survey. When traps or cameras are used during a survey, the measure of effort is usually expressed as trap-nights (or camera-nights). This is the number of traps (or cameras) set each night multiplied by the number of nights. Using this criterion, a small number of traps used for a long period would be roughly equivalent to the effort of a large number of traps used for a short period.

It is the rare index for which the number of signs can be directly converted into the number of animals. This is because there are assumptions that must be used for each conversion. For instance, to convert the number of pellet groups detected to the number of deer, there must be estimates of how many pellet groups each deer produces daily and how rapidly the pellet groups degrade. Both these measures have a variance that is so large any resulting density estimate is meaningless. For most indexes, it is also difficult to determine if increased signs indicate increased numbers or increased activity. For example, would the detection of six pellet groups mean six deer used the area once, or that one deer used the area six times?

As it is tailored to a specific species, a good index can be quantified and it is likely that two biologists can compare results. The power of an index is to give relative comparisons between sites or periods for a minimal amount of work. Some effort must be made to verify that the index used reflects changes in density over the range measured. There is also the danger that an increased number of sign does not reflect more animals but rather shifts in habits such as diet or habitat. One must question whether seasonal increases in deer pellet groups in an old field represent an increase in the number of deer or a shift in habitat use by the same number of deer. If all habitats are being monitored simultaneously, it is possible to differentiate between shifts in habitat use and shifts in abundance.

No index can work under all circumstances, so a pilot study that measures both density and the index is preferable to making assumptions of correlation. Few indexes have a linear relationship with density over its entire range, as usually an index flattens out as density increases beyond a certain range. For example, an increase in the number of subadult or nonreproductive individuals may not be reflected in an index based on the number of morning calls by adult howler monkeys. It is important for the research or monitoring to demonstrate that the index is responsive to changes in density over the range. For many species, verification of an index has already been accomplished and a review of relevant literature

A biologist weighs a bear cub to record its weight. (Photo by © Raymond Gehman/Corbis. Reproduced by permission.)

is recommended before undertaking an index survey. In short, good preparation by observing animals is required before collecting data and making inferences from them.

Density estimate

Density estimates have two components, and both cause difficulties to biologists: the amount of area surveyed and the number of animals. As opposed to an index, the density estimate relies on a measure of area surveyed and not on effort. If the survey area is a true island, then the measurement is straightforward. If the survey area has an arbitrary boundary between "inside" and "outside," assumptions have to be made as to how the animals move with respect to the boundary. If the area surveyed is a mosaic of favorable and unfavorable habitat for a species, sampling protocol must take this into account. Placing survey lines to estimate the number of animals within the favorable area while using both favorable and unfavorable habitat to estimate the survey area will overestimate the number of animals. Unfavorable habitat may not contain a large number of animals during the survey, but may be used at other times of the year or when environmental conditions change.

It is difficult to estimate density from a line of traps, as it is difficult to estimate the distance that animals are drawn into the traps. Removal trapping suffers from this handicap, as removal of animals creates a vacuum that will eventually be filled by immigrating animals. Most studies that rely solely on trapping to estimate density construct trapping grids or webs to estimate the size of the survey area. Without prior knowledge of an animal's home range size and movements, it is difficult to accurately calculate the area sampled using most capture techniques.

The second component of the density estimate is the number of animals. To obtain a density estimate, a species usu-

ally needs to be observable or caught on a regular basis. It also helps if natural or added markings allow individuals to be identified. The two main techniques to estimate absolute density without changing the density are either mark/recapture or distance sampling. Mark/recapture uses an initial capture period that results in a known number of marked animals. A second capture period then looks at the ratio of marked to unmarked animals and estimates how many animals the population contains. Mark/recapture is based on the two assumptions that all animals can be captured and capture does not influence the probability of future captures and that animals can be marked and the marks will not fall off or influence the probability of the animal to be recaptured. There is also an assumption that no mortality, natality, or immigration occurs between capture sessions, but it is possible to relax this assumption and still estimate density.

For species that are readily captured, such as terrestrial rodents, it is a matter of trapping the animals in live-traps on a regular basis and estimating density. Commercial traps are available and techniques are well developed for small mammals. With all trapping, it is important to minimize trauma

A veterinarian examines a sedated chimpanzee (*Pan troglodytes*). (Photo by © Martin Harvey/Gallo Images/Corbis. Reproduced by permission.)

Scientists study rodents as surrogates for assessing vegetation condition in conservation areas. (Photo by Rudi van Aarde. Reproduced by permission.)

to the animal not just because it is humane but also so that the initial capture does not influence the probability of recapturing the animal. When the intent is to recapture the animal, some period of pre-baiting (with the trap baited but locked open) will increase the probability of capture. By providing adequate food and bedding and checking traps regularly, most small mammals can be trapped repeatedly and their density estimated. If the first priority is to ensure that all species have been detected, then using a variety of traps, both live and kill, would result in a better survey.

There is no terrestrial mammal that cannot be captured and released in a humane manner. However, for many animals such as large predators, the first trapping is traumatic enough to preclude the animal being recaptured by the same method. This does not necessarily preclude density estimates, as there is no assumption in the capture/recapture model that the capture method is the same for each period. The animals can be captured with snares, affixed with a mark, and then recaptured by cameras at bait stations. The first capture can be as a neonatal in the nest, and the recapture in a trap. For instance, an animal's DNA can be obtained in a blood sample attained at the initial capture in a snare, and the DNA later recaptured in a hair sample snagged on barbed wire strung around a bait station. If the animal has unique markings, as do many large cats, both the initial capture and subsequent recaptures can be with trip-cameras.

Removal sampling allows the observer to estimate a population's density post-hoc. The number of animals removed for each unit of effort, or a change in the ratio of two classes of animals (i.e., antlered and antlerless deer), is the basis for most surveys of harvested large mammals. For both calculations, the advantage is that the researcher is not necessarily the one doing the removal. Large areas and large populations can be estimated using volunteer hunters that agree to follow simple rules such as harvesting only antlered animals. The disadvantage is that there are limited species that attract a large number of volunteer hunters and the removal is limited to populations that are robust enough to sustain a harvest. It is possible to use the catch-per-unit-effort technique without

harvesting animals, but the observer must be able to identify animals that have been previously sighted. There are several statistical programs that are commonly used to analyze this type of data and details of the programs and their assumptions are available at <http://www.cnr.colostate.edu/~gwhite/software.html>.

For animals that are readily observable, or at least easier to observe than catch, strip transects or distance sampling techniques are used. In a strip sample, one counts all animals along a strip of known length and width. An assumption of this method is that all animals within the strip are counted. For most species, if the strip is wide enough to encounter a significant number of animals it is also wide enough to miss animals at the boundaries of the strip. There are ways to estimate the number of animals missed, such as double counting, but they do not solve the problem. Aerial surveys often involve strip samples where all animals within a strip along a side of the plane are counted by observers. These surveys are diurnal and usually occur in grassland or open habitats. Thermal imaging allows the heat signature of animals to be used to count animals at night. Use of thermal imaging for

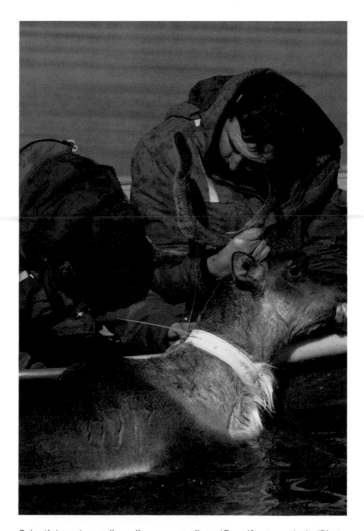

Scientists put a radio collar on a caribou (*Rangifer tarandus*). (Photo by © Natalie Fobes/Corbis. Reproduced by permission.)

density estimates is problematic because abiotic objects can radiate heat and be misidentified as animals. Also, canopy closure in forest settings is often dense enough to block the sensor's view of animals.

Distance sampling takes into account the assumption that animals can be missed along the transect, and the farther the animal is from the transect the more likely it is to be missed. If the researchers know how far they surveyed, the number of animals observed, and their perpendicular distance to the transect line, they can estimate the area that they surveyed and the number of animals that they did not see within that area. The number of animals seen plus the number of animals missed per unit area is a density estimate. The assumptions are that all animals along the line are observed; animals are randomly distributed relative to the line; and no animals are counted twice along the same line. This technique was originally developed to survey marine mammals, but has quickly been adapted for many large terrestrial mammals. There are many nuances to distance sampling, and it is difficult to cover them all in this space, but additional information is available <http://www.ruwpa.st-and.ac.uk/distance>.

Natural history information

Before a species can be managed, it must be understood. A species has requirements for food, shelter, and habitat that directly impact management decisions. Understanding social structure, interspecific competition, predator pressures, movements, disease transmission, mating, and behavior help humans realize the impact of their actions on mammals. Most, though not all, of these factors cannot be determined from laboratory studies of captive individuals. Effective means for studying natural populations is a key ingredient to effective management of wild mammals.

Direct observation

The most straightforward means to derive natural history information is direct observation. Its value should not be underestimated, as direct observation is often the most effective way to place the trait in context of the animal's physical and social environment. The observer might be able to learn more about a species from a few hours of direct observation, than from a year of examining trip-camera photos or radio telemetry locations.

When designing a direct observation study, all terms must be understood and quantified, especially when more then one observer is used. The term "feeding" is readily understood at a basic level, but there are often many types and gradations of feeding in a natural setting. As with village interviews used to create a species list, the attitude and background of the observer sometimes influences how a behavior is recorded. An observer has to guard against anthropomorphic biases and also against interpreting events through preconceived theories. For example, which animal is considered dominant during an interaction should not be a qualitative measure, but based on quantifiable criteria.

As with indexes, direct observations often have a unit of effort. How many attempted matings, bark strippings, or so-

A research scientist funded by Conservation International to conduct research in Madagascar on the elusive fossa (*Cryptoprocta ferox*). Fossa are trapped to gather information on their movement patterns and population sizes. (Photo by Rudi van Aarde. Reproduced by permission.)

cial grooming events observed is always a function of how much time was spent observing the animal. This is complicated when animals are part of a social group. Time spent monitoring the entire group is not the same as time spent observing one individual in the group. Usually too many activities are being conducted simultaneously to watch an entire group and researchers identify focal individuals that are monitored for a set period of time. This set period of individual observation might be punctuated by a sample of the behavior of each group member, referred to as a scan sample.

Direct observations are usually inexpensive; a good pair of binoculars or spotting scope and a watch are the main expenses. However, time is usually the limiting factor with direct observations. Many hours can be spent to obtain a few minutes of direct observation. With more cryptic or more diffuse animals, there is a point at which the observations are not worth the time spent to obtain them. Video cameras and recording equipment can be used to continuously monitor a location, with the tapes reviewed by the researcher at a later date; but

Conservation scientist from South African national parks place a collar fitted with a GPS unit onto an elephant to study movement patterns. (Photo by Rudi van Aarde. Reproduced by permission.)

this would only be effective at feeding or watering sites that attract animals. In addition to time considerations, there is also the issue of how the observer's presence impacts the animal's behavior and movements. If the behavior being recorded is the result of the animal's movements away from the observer, or toward a concentrated food site, then the value of the observation is reduced. A period of habituation of an individual or group to the observer is standard practice. But this needs to be considered with some care. For instance, too much familiarity can trigger attacks, or habituated animals may be vulnerable after the study is terminated.

Indirect observations

Many behaviors can be indirectly inferred from sign indexes and density estimates. Habitat selection is often measured through indexes that compare an animal's use between habitats or seasons. When indexes are used as a surrogate for direct observations, the closer the index is to the target behavior the less chance of error. For example, a browse index based on the number of buds clipped per tree is directly re-

lated to ungulate feeding, while pellet or track counts are more indirect indexes.

Radio telemetry, a means of indirect observation, is the most important advance in the last 50 years for the study of wild mammals. Before radio telemetry, researchers were limited by their ability to follow animals or detect their presence. Indirect indexes of activity and movement, such as sign counts, were the best means to measure habitat selection. Behavioral observations were often limited to sites near feeding or watering holes where animals could be observed from established blinds; when an animal disappeared it was often impossible to know if it had died, dispersed, or merely stopped coming to the observation site. Radio telemetry allows a researcher to locate a specific animal when it needs to be recaptured, observed, or its movements monitored. Radio telemetry allows a researcher to remotely track a specific animal's movements and survival with a minimal disturbance to its behavior. These abilities have opened observations into animal ecology that were unavailable to earlier workers.

Two components of radio telemetry are the receiver and transmitter. The receiver can pick up a range of frequencies and can be set to detect each unique transmitter. Traditionally, the animal carries the transmitter and the receiver is either a hand-held device or an orbiting satellite. The hand-held receiver can be moved on foot, by car, boat, or plane. To locate the animal, it can be approached directly, triangulated from a number of bearings from known locations, or by using the principles of the Doppler effect in the case of the satellite. More recent global positioning system (GPS) collars have the animal fitted with the receiver and the transmitters are aboard orbiting satellites. The receiver calculates its position based on the known position of the satellites and the time it takes a signal to travel between the satellite and receiver. The optimal arrangement is a package that contains a combination of units, so the animal's position can be determined through multiple means.

Traditional tradeoffs in radio telemetry are between power output and battery weight. Attempts to increase the range or duration of the signal are matched by increases in unit weight. Combining types of radio telemetry units into one collar also increases package weight. A general rule is that a package weight's should be less than 5% of the animal's body weight, but there are enough exceptions to this rule to warrant a pilot study before attaching packages to wild animals. Advances in battery technology and computer software have made weight considerations less important, especially for larger mammals. When the battery power cannot be sustained for the length of the project, microchips can regulate when the unit is active. Weight limitations are still serious concerns for species weighing less than 2.2 lb (1 kg).

A limitation of radio telemetry is that the observer must capture the animal to attach the telemetry unit. Most animals can be captured, but the time and labor involved can consume a large part of a project's budget. Once a telemetry unit is attached, the logistics of recovery are simpler. The telemetry unit can lead to the animal for application of anesthesia, or a remotely triggered tranquilizer dart can be inserted in the collar. When the study has terminated, the unit can be released,

A game warden collects fluids from a dead elephant bull to determine its cause of death. (Photo by Rudi van Aarde. Reproduced by permission.)

(i.e., arboreal, volant, terrestrial, subterranean, aquatic), and its behavior (i.e., cryptic, nocturnal, social, vocal). General guidelines can be given to the limitations imposed by each type of mammal.

Large terrestrial mammals

Visible, large mammals can have density estimates derived from distance sampling techniques. When large mammals are not easily visible due to dense habitat, nocturnal activity, or low density, they all leave sign that is observable and can be used as an index. Bears leave claw marks in trees, gorillas create nightly nests, elephants deposit conspicuous dung, and ungulates leave tracks in most soils. Removal techniques are suitable for large mammals that are harvested. Large animals are also most easily captured with trip-cameras. Indeed, camera surveys of large predators are the norm in most forest habitats. It is difficult to convert these camera indexes into density estimates, with the exception of animals with unique markings that are surveyed using trip-cameras. Most large mammals have large home ranges, which make it difficult to estimate the area being sampled by any technique. Direct observations are most frequently used with large mammals, and telemetry units can be large enough to contain most features needed. A limitation might be that the capture of the animal for attachment of telemetry unit will take special skills and equipment.

Volant mammals

All volant mammals are relatively small and most are nocturnal. The most readily surveyed are the communal species that can be found in caves for all or part of their annual cycle. Bats can be counted in the roost, or individuals captured with nets, or counted visually when the bats enter or exit the cave. For bats with solitary roosts or to sample foraging sites of communal species, nets that span natural passageways or watering holes are the norm. However, it is difficult to erect mist-nets and harp-nets to match the space being used by the

either by providing a weak link or a remote-release magnetic mechanism in the collar.

Once locations have been collected for an animal, it is possible to determine its home range, habitat use, and multiple other natural history parameters. An additional benefit of radio-collared animals is the ability to construct life history tables. It is difficult to determine if wild animals have died or migrated when they stop being detected by other means. However, which fate occurred is very important for most modeling of animal populations. With radio-collared animals, it is possible to differentiate between mortality and migration, and to determine the timing and cause of mortality.

Limitations

Of the two considerations when designing a field study, the level of information needed and the limitations imposed by the animal itself, it is the latter that usually dictates what is possible. When deciding the proper field technique, consideration has to be given to the size of the animal, its niche

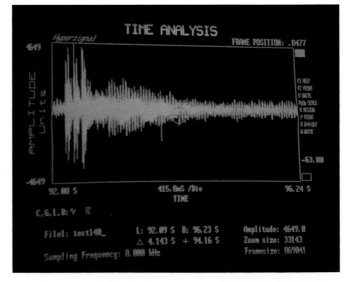

A sonogram of a northern minke whale (*Balaenoptera acutorostrata*). (Photo by Flip Nicklin/Minden Pictures. Reproduced by permission.)

Researchers measure the head of a Tasmanian devil (*Sarcophilus laniarius*) during a study in Australia. (Photo by © Penny Tweedie/Corbis. Reproduced by permission.)

bats, particularly bats that forage above the canopy or within complex foliage.

Direct observation is usually not possible, especially during foraging. An exception is small light tags that can be placed on bats to observe short-distance foraging. It is possible to observe maternal behavior in communal nesting species either directly or with video equipment. Radio telemetry units are just reaching the size where movements can be monitored for longer than a day, but weight is still a serious concern.

A technique with potential is using the bats' ultrasonic call to identify species and possibly individuals. Devices can record and catalog the calls to a computer where a number of call parameters can be measured. At this time, there are quantifiable means to differentiate some, but not all bat species based on call parameters. Usually the calls can be used to predict which suite of species generated the call, but not the exact species. There is individual variability in the calls and it may be possible in the future to capture calls, as one would pictures of animals with unique markings. At this time, not enough call parameters can be measured to recommend use.

Aquatic mammals

All are relatively large and most are cryptic, in that a good portion of their time is spent below the surface of the water. Most marine mammals are too large for extensive trapping and are best surveyed with distance sampling, though trapping can be conducted on some dolphin, manatee, or otter species that inhabit coastal waters. Direct observations are limited to the animal's time above the surface of the water or using scuba equipment. Radio telemetry can provide important movement data, but difficulties include the loss of signals when the animal is below the surface of the water. Sonar can be used to track movements and sounds of larger whales.

Freshwater mammals are generally smaller and more readily trapped. These mammals also usually spend some portion of their time on land and leave obvious sign such as gnawing on trees or construction of homes, nests, and dams. Radio telemetry and direct observations are more effective with this group because of their time out of water. Small semi-aquatic insectivores are cryptic, difficult to capture, and leave no observable signs. With this subgroup, neither direct observations or radio telemetry have proven effective. They can best be sampled with pitfalls or some form of kill trap.

Small terrestrial mammals

Most of these species are too cryptic to be observed reliably, and live or kill trapping are the most common survey means. Traps can be obtained from several commercial sources and can be scaled to the size of the animal. Food baits are used to lure animals into traps. If done properly, animals can be captured repeatedly and all segments of the population can be sampled. Direct observation has not been effective, with the exception of those species at the larger end of the range, such as ground squirrels. Radio telemetry units are small enough for most species, but the range of the small units is below the dispersal distance of most rodents and limits their utility in some studies. For species less than 0.4 oz (12 g), most commercial traps are ineffective and radio transmitters are short-lasting due to weight considerations. Pitfall arrays are often used to record a species presence; pitfalls are buckets that are either deep enough that the animal cannot jump out or are partially filled with liquid to kill the animal. Animals are not lured into the buckets with bait, but rather barriers funnel all passing mammals into the buckets.

Subterranean mammals

As with terrestrial small mammals, subterranean mammals are cryptic, but unlike terrestrial small mammals they are usually difficult to capture alive. Commercial killing traps can be obtained and modified to capture most species. Evidence of digging activity can be used to indicate a species' presence, but these signs of activity are not often correlated with density and are a poor index. Some subterranean mammals have a portion of the year, or day, that they spend aboveground and this is the period during which they are most easily captured and surveyed. Direct observations and radio telemetry are ineffective because the soil blocks both sight and radio signal transmission.

Medium-sized mammals:

Most can be captured in traps, but only a few can be captured repeatedly, either due to large home range or behavioral wariness. Traps are either large holding traps or snares and leg-holds. Wariness prevents some species from readily entering box traps. With proper training, snares with springs and leg-holds with padding can humanely capture animals for release. When animals are not captured, most are large enough to be readily detected through tracks, sign, or cameras. For predators, animals can be attracted to specific sites with food baits or scent lures, which, derived from glands or urine, are commercially available and can be used to bring animals over a site prepared for tracks or monitored with a trip-camera. These lures were originally developed for snare or leg-hold trapping and are still effectively used for that purpose. Radio telemetry is used extensively with this group; the unit is attached during the first capture.

Arboreal mammals

It is difficult to capture arboreal mammals because of the logistical complications in working in the canopy. As with subterranean mammals, if arboreal species have a period of the day or year that they use the ground or come close to the ground, they can most readily be sampled at that time. Sometimes species can be anesthetized with drugs delivered in a dart, but still there is the problem of the animal falling from the canopy when the drug takes affect. Diurnal species can be surveyed with distance sampling, or mark/recapture techniques when there are unique markings. Nocturnal species are more difficult to survey visually because the darkness increases their cryptic ability and there is difficulty in estimating distance; indexes can be derived from calls and visual detections along transects. For nocturnal species, use of a spotlight to detect "eye-shine" makes transect surveys feasible. Many species do have unique calls that can be used for index surveys. Radio telemetry can be used if the animal can be captured. For animals that forage high in the canopy, the utility of radio telemetry is diminished because it is difficult to estimate the third dimension in an animal's space use.

Clearly, one technique cannot be used to study all mammals. A study that attempts to record the density and behavior of all mammal species within an area has a daunting task that will take time and money to accomplish. A population study for a single species or suite of species must be tailored to match the habits and attributes of the animal. An advantage to mammal studies is that there is a rich literature of mammal field research conducted over the past 100 years that can provide guidelines. With increasing human densities, the affairs of mammals and humans can no longer be separated, and understanding wild mammals is the first step to ensuring their survival.

Resources

Books

Bookhout, T. A., ed. *Research and Management Techniques for Wildlife and Habitats.* 5th ed. Bethesda, MD: The Wildlife Society, 1996.

Boitani, L., and Fuller, T. K., eds. *Research Techniques in Animal Ecology: Controversies and Consequences.* New York: Columbia University Press, 2000.

Buckland, S. T., D. R. Anderson, K. P. Burnham, J. L. Laake, D. L. Borchers, and L. Thomas. *Introduction to Distance Sampling—Estimating Abundance of Biological Populations.* Oxford: Oxford University Press, 2001.

Elzinga, C., D. Salzer, J. Willoughby, and J. Gibbs. *Monitoring Plant and Animal Populations.* Oxford: Blackwell Publishing, 2001.

Feinsinger, P. *Designing Field Studies for Biodiversity Conservation.* Washington, DC: Island Press, 2001.

Laake, J. L., S. T. Buckland, D. R. Anderson, and K. P. Burnham. *DISTANCE User's Guide V2.1.* Fort Collins, CO: Colorado Cooperative Fish & Wildlife Research Unit, Colorado State University, 1994.

Southwood, T. R. E., and P. A. Henderson. *Ecological Methods.* 3rd ed. Oxford: Blackwell Publishing, 2000.

Sutherland, W. J., ed. *Ecological Census Techniques: A Handbook.* Cambridge: Cambridge University Press, 1996.

White, G. C., and R. A. Garrott. *Analysis of Radio-tracking Data.* New York: Academic Press, 1990.

Wilson, D. E., F. R. Cole, J. D. Nichols, R. Rudran, and M. S. Foster, eds. *Measuring and Monitoring Biological Diversity: Standard Methods for Mammals.* Washington, DC: Smithsonian Institution Press, 1996.

Periodicals

"Guidelines for the Capture, Handling, and Care of Mammals as Approved by the American Society of Mammalogists." *Journal of Mammalogy* 79 (1998): 1416–1431.

William J. McShea, PhD

Mammals and humans: Mammals in zoos

A brief history

Evidence from cave wall drawings suggests that wild mammals were kept in the company of humans as early as the Stone Age. In fact, this was probably the beginning of animal domestication. By 2500 B.C. Egyptian kings were keeping antelopes for amusement and to impress foreign visitors. Only ruling classes could afford to keep and care for exotic mammals. Queen Hatshepsut of Egypt could be credited with organizing the first mixed collection of exotic mammals. In approximately 1500 B.C., she sent collectors to Somaliland. They imported greyhounds, monkeys, leopards, cattle, and the first giraffe into captivity. In 1000 B.C., Emperor Wu Wang of China assembled the first zoological park. This was a sophisticated 15,000 acre (6,070 ha) Ling-Yu or "Garden of Intelligence." It included tigers, rhinos, and rare giant pandas. Between 1000 and 400 B.C., many small zoos were created in north Africa, India, and China and domestication of mammals had become an art. These collections were mostly symbols of power and wealth, definitely not for public enjoyment. By 400 B.C. the ancient Greeks established a zoological garden in most if not all Greek cities. They used their collections for serious scientific study. Only students and scholars were allowed to visit these collections. It was at this time that Aristotle wrote *The History of Animals*. In this encyclopedia, he described hundreds of species of exotic vertebrates from these collections.

The Romans were perhaps the first to open exotic mammal collections to the public at large. From about 100 B.C. to A.D. 600 the Romans used these collections for scholarly study but they also used many of these animals for entertainment. Much of the entertainment consisted of cruel and bloody spectacles in public arenas, with large carnivores attacking or being attacked by men as well as other mammals. Often these were slow, agonizing and gruesome battles to the death. With the fall of the Roman Empire zoos went into a decline. Most exotic mammal collections consisted of small private menageries and traveling exhibitions.

By A.D. 1400 global exploration sparked public interest in zoos once again, with strange creatures from the New World. During this time, Hernando Cortes visited the zoo of Montezuma, chief of the Aztecs. It was a huge collection and employed over 300 zookeepers. Over the next few hundred years many European zoos opened and the collections grew larger. Most of these collections consisted of individual specimens of as many species of mammals as possible. The animals were from many different countries and represented many continents. They were often called postage stamp collections. They were open to the public and attracted many visitors. Although they have been renovated and have updated their policies, some of these zoos are still open today. The oldest is the Schonbrunn Zoo in Vienna, Austria. It was built in 1752 by Emperor Franz Josef for his wife. It opened to the public in 1765. Others that still exist are the Madrid Zoo in Spain, which opened in 1775 and the Jardin des Plantes collection in Paris which opened in 1793. For comparison, the oldest zoos in the United States are the Philadelphia Zoo, which opened in 1874; the Central Park Zoo in New York, which opened in 1862; the Lincoln Park Zoo in Chicago, which opened in 1868; and the Cincinnati Zoo, which opened in 1875.

Why do we keep mammals in zoos?

Zoos have been popular for centuries and this trend continues today. In an urban setting, a zoological park offers a rare opportunity for visitors to view and learn about exotic mammals. This is the only type of opportunity many people have in their lifetime to see living specimens in person. Many of these mammals would be difficult and costly to see in the wild.

The goals of many zoos are the same. Usually they involve entertainment and education for the community they serve. Conservation and scientific research have also become major concerns for modern zoological parks. In an annual survey, the American Zoo and Aquarium Association reported that almost 135 million people visited 194 member facilities in 2001. By comparison, this attendance is greater than all professional sporting events combined in the United States for 2001.

Entertainment

Entertainment is a major goal for every public zoo. These attendance numbers indicate that indeed zoos are fun to visit. There are many theories as to why people enjoy zoos. It could be that current human cultures feel isolated from nature and observing exotic wildlife in zoos bridges a primal connection and meets our psychological needs. Mammals are certainly interesting to watch and their exotic shapes and colors appeal to our artistic nature. The whimsical behavior of primates or the grace and power of elephants, appeal to visitors of all age groups.

Education

Perhaps the greatest service a zoo offers to the community is education. Most zoos have programs that teach basic zoology, animal behavior, geography, and natural history. The classroom setting sometimes involves safaris, campouts, scavenger hunts, and participation in animal care. Usually students have the opportunity to observe, interact with, and often touch exotic mammals about which they are learning. Another unique educational and emotional experience is outreach programs that target audiences unable to physically visit the zoo. Examples include the transport of education animals and staff to audiences in nursing homes or hospitals. Increasingly important are programs teaching responsible use of habitat and conservation of natural resources. These programs often target younger audiences, in an effort to sustain the long-term quality of life for all life forms.

Conservation

Zoological parks are becoming important conservation organizations. They contain a large resource base in both staff training and fund raising. Thousands of staff members employed by zoos participate in numerous in situ and ex situ conservation projects worldwide. Some of these projects have resulted in saving wild populations of mammals from certain extinction. Good examples of this are the golden lion tamarin (*Leontopithecus rosalia*) and Arabian oryx (*Oryx leucoryx*) programs. Both wild populations lacked enough animals for long term survival. A consortium of zoos working together moved and released captive born animals back into their historical ranges in hopes of re-establishing wild populations. Both of these re-introductions have succeeded and these now wild populations continue to reproduce and their numbers are increasing.

Scientific study

Zoos are often called "living laboratories"; this term reinforces the fact that animal observations and experiments in captivity can be very valuable. In fact, most of our knowledge of mammals has come from captive studies. For instance, black-footed ferrets (*Mustela nigripes*), extinct in the wild, are being re-introduced through artificial insemination. The oryx have also recently been poached to near extinction again and their numbers are just starting to increase in the wild. Also, it is often more practical, easier to control external factors, and more scientifically useful to study mammals in captivity. In an effort to encourage scientific study in zoological parks and provide funding for these programs, the American Zoo and Aquarium Association and Disney's Animal Kingdom have established the Conservation Endowment Fund. It pays for many conservation and research studies in captivity and in the wild each year. The Zoological Society of London in the United Kingdom has supported research since the late nineteenth century.

Rehabilitation pools are built to temporarily house ill or injured marine animals, such as this killer whale, for treatment. The intent is to return the animals to the wild once they have recovered. (Photo by © Steve Starr/Corbis. Reproduced by permission.)

Exhibit design concepts

Early zoos consisted of numerous animals exhibited in small cages made of concrete or wood with metal bars for security. These were usually large charismatic mammals exhibited mainly for shock value. While a few zoos remain with this type of exhibit design, the majority of modern zoo exhibits are one of two types. A third philosophy is included because it could represent the future of mammal exhibits in zoological gardens.

The Hagenbeck concept

Carl Hagenbeck was an animal entrepreneur. He supplied animals to zoos and was also an animal trainer. He pioneered many display and exhibit techniques. Hagenbeck initially gained his reputation by exhibiting people and animals in traveling exhibits. On October 6, 1878 over 62,000 people visited the Berlin Zoo to see his traveling exhibit of Nubians from the Sudan, Laplanders, Eskimos, Kalmucks, Tierra del Fuego natives, and Buddhist priests. There were also elephants, camels, giraffes, and rhinos, but it was the people who were most popular. These human zoos made Hagenbeck a fortune. In 1900 he bought a potato farm on which he wanted to build a wild animal park. Hagenbeck is credited as being the inventor of the cage without bars. He put his animals in moated enclosures. The enclosures were planted with trees and shrubs and decorated with artificial rockwork which was very pleasing to the visitor and gave the illusion that the animals were free-ranging. Their captivity was well hidden. Hagenbeck was a master at the placement of moats and hedges to create an exhibit illusion that placed predator and prey together. He was not that concerned with scientific study or educating the audience, his priorities were aesthetics and beauty. These exhibits showcased large mammals with a geographic theme. Hagenbeck's concepts are still used in exhibit design today.

A zoo educator teaches children about endangered cats. (Photo by © James L. Amos/Corbis. Reproduced by permission.)

Immersion concept

This philosophy was an attempt to involve or place the visitor in a naturalistic backdrop for a specific theme. These types of exhibits are often characterized by paths that wind through areas built to resemble natural habitat. Small mammals and birds are sometimes free-ranging in the entire exhibit, including public areas. Species of similar geographic origin and habitat are mixed in the exhibit, interacting with one another if they peacefully cohabitate. Larger mammals are housed within the same space, but for safety and function are enclosed within the exhibits with visually hidden barriers. Therefore, some of Hagenbeck's techniques are used in immersion exhibits. The primary purpose of this exhibit is to give the visitor an appreciation of an animal's natural habitat and educate them on the natural history and need for conservation of that habitat. One of the first good examples in the 1970s was the gorilla exhibit at Woodland Park Zoo in Seattle; more recent examples of this technique can be found in the Congo exhibit at the Bronx Zoo and the Amazonia exhibit at the National Zoo.

Biopark concept

This technique is beginning to attract interest. The basic idea is to create exhibits that explain, elucidate, and exemplify the interconnectedness of life. More specifically, it is the idea of putting humans and our biology in the context of the rest of life. Robinson (1996) describes this technique as emphasizing "the complex specializations in a host of dependencies, interdependencies and interactions with invertebrates, plants, protozoa, bacteria, viruses and so on to which mammals have evolved." He further states "it is time to end the isolationism of simply exhibiting mammals against a naturalistic backdrop."

Management of captive mammal populations

Why manage zoo populations?

Gone are the days of extensive collecting expeditions to replace mammals that have died during transport or exhibition. Zoos must now manage their captive populations to the fullest. It is no longer ecologically responsible or ethical to capture wild mammals for the sole purpose of entertainment. In fact in 1992, 82% of the worldwide mammal population living in zoological parks was born in captivity, according to data collected in 2002 by the International Species Information System (ISIS).

When compared to wild mammal populations, most captive populations are very small. Small populations tend to be

In some cases, orphaned or abandoned animals are taken to and raised at zoos. (Photo by © Carl & Ann Purcell/Corbis. Reproduced by permission.)

adapted to captive environments, or they are managed to preserve genetic diversity.

Record keeping

Animal record keeping is the foundation of captive animal management. Zoo professionals depend on detailed direction from animal records. Mammals with missing or unknown ancestry or other life history information are of very limited use in long term management strategies. The International Species Information System (ISIS), first developed in 1973, collects animal data from over 560 institutions in 72 countries on 6 continents and stores them in a computerized database. These data are kept by a computerized program called ARKS (Animal Record Keeping System). Much of the data are entered into a computer software program called SPARKS (Small Population Analysis and Record Keeping System) by a studbook keeper, in order to produce a studbook. A studbook is an inventory of the life history and ancestry of an animal and SPARKS can perform mathematical analyses of studbook data. From SPARKS one can get age class and sex class graphs as well as survivorship and mortality data for a population. It can also generate many useful reproductive

Modern zoos try to recreate the animals' natural environment, as with this cheetah (*Acinonyx jubatus*) cub. (Photo by © Lynda Richardon/Corbis. Reproduced by permission.)

biologically unstable and their long-term survival is unlikely if events occur at random. Genetically only one-half or 50% of a mammal's alleles are passed to its offspring. Alleles are mutable genes that are responsible for inheritable traits. Therefore, genetic diversity is reduced by 50% with each generation. In other words, the relatedness or mean-kinship of these individuals increases by 50% with each generation. This increase in relatedness also increases the chance that alleles that produce a fatal birth defect or reduced survivability will be expressed. These are often called lethal genes. Demographic factors also influence the long term survival of small populations. If an excessively disproportionate number of females or males occurs in a population, only a limited number of animals will be represented in the population. This will result in a reduction of genetic diversity and increased mean kinship. This may eventually cause a small population to crash without the addition of new founders (genetically unrepresented animals in the population). Therefore, a population with a sex ratio that is nearly equal has a better long-term chance for survival. Age distribution in a small population is also very important. If there is an excessively disproportionate number of older post-reproductive or younger pre-reproductive mammals in a population, reproduction will intensify in a single cycle and decline in a single cycle. Without the addition of new founders, this will also reduce genetic diversity and cause a small population to crash.

Frankham (1986) and Foose (1986) very eloquently document two extreme reasons for managing captive zoo populations. Individuals are either intentionally selected to be well

This lar gibbon (*Hylobates lar*) takes advantage of the climbing system installed in its zoo habitat to allow more natural behavior. (Photo by © Robert Holmes/Corbis. Reproduced by permission.)

analyses. In order to formulate a population management plan, SPARKS data are exported to a software program called PM2000. This is a powerful computer program that can generate, among other things, target population size, generation time, growth rate, and current percent of genetic diversity. It also allows the user to build preferred breeding pair calculations, so recommendations can be made by the population manager.

Techniques for small population management

Incorporating profound knowledge and documentation of genetics and principles of population biology, Ballou and Foose (1996) describe specific guidelines for efficient demographic and genetic management of small populations. These guidelines are used extensively by zoological gardens in an attempt to achieve long-term survival of small populations. Their basic techniques are summarized here.

FOUNDER REPRESENTATION

The long-term survival of a small population depends upon obtaining a sufficient number of founders to maximize allelic diversity and heterozygosity. The goal here is to obtain enough unrepresented animals to build a population that will represent a cross-section of the genotype and phenotype of the source population. Unfortunately, one cannot predict the quality of the sample because it cannot immediately be measured. Founder numbers are considered adequate for effectively sampling allelic diversity based on the most likely allele

distributions. Genetic variation over the range of the source population should also be considered, with between 25 and 50 founders considered sufficient in most cases.

REACH CARRYING CAPACITY

Carrying capacity in zoos is the entire number of spaces available for a particular species among all program participants. In order to maximize genetic efficiency the population size should be increased as rapidly as possible in order to meet the carrying capacity. Genetic diversity is lost when growth rates are slow, because the chances of all animals in that population successfully reproducing and being represented in the first generation decreases with time.

STABILIZE THE POPULATION

The population should be stabilized once it is near the carrying capacity. The current population size and growth rate are used to determine the proximity to carrying capacity and the population is stabilized by regulating birth control. Birth control of mammals in zoos is achieved both biologically and by the physical separation of animals.

EXTEND GENERATION LENGTH

Mean generation length is defined as the average age at which the females in a population produce offspring. Since genetic diversity is lost with each successive generation, extending the generation length reduces the degree of diversity lost in a small population over a given number of years. In other words, if the females in a population do not breed until later in life the loss of genetic diversity is delayed. Risk is incurred with this strategy since the animal's reproductive potential may be lost with time, due to age related fertility problems, health problems, and accidental injury or death.

ADJUST FOUNDER LINEAGES

In order to maximize the genetic diversity and survivability of a captive mammal population the representation of founder lineages should be as proportional as possible to the distribution of founder alleles surviving in that living population. Initially in a captive population some animals will reproduce well and be highly represented genetically while others will not. Therefore, the population will not be evenly represented genetically. In order to compensate for this, preferential breeding pairs should be formed. Descendants of underrepresented founders should be preferentially bred and the reproduction of overrepresented animals should be limited.

REDUCE THE REPRESENTATION OF EXTREME TRAITS

Reproduction should be limited in animals that produce traits that are not typical or that would offer a selective disadvantage to survival in the natural environment. Albinism is an example of such a trait.

SUBDIVIDE THE POPULATION AND REGULATE GENE FLOW

Division of a large captive population into geographically isolated subunits offers some advantages. The exchange of animals as well as gametes between these subunits should be regulated. This method offers increased protection to the population from communicable diseases and natural disasters such as fire, earthquake, tornado, and hurricanes. Also, isolated subunits of the population will be exposed to a wider range of selective pressures. This offers the advantage of a slowed reduction of genetic diversity.

INTRODUCE NEW FOUNDERS

The addition of new founders should, in theory, increase genetic diversity. If a program can be devised to exchange wild caught animals for captive born ones, this can sustain a population. However, great care must be taken to eliminate the possibility of disease transfer between the populations.

INCORPORATE REPRODUCTIVE TECHNOLOGY

Emerging advances in reproductive technology can be used to increase genetic diversity and sustain a captive population. Genetic material such as semen, ova, embryos, and tissues can be stored and used at a latter date. This can increase the chances of a genetically underrepresented animal being utilized in a population. It can also increase the practical aspect of introducing new founders from wild populations. It is much easier and safer to move an animal's gametes rather than to physically transfer the entire animal.

Types of populations to manage

Frankham (1986) documents four types of captive populations of interest to zoos. They are summarized in the following groups:

COMMON DISPLAY SPECIES

The goal for these species is to selectively breed for traits adapted to captivity in an effort to establish a tractable, easily managed population. This could include animals that are docile and do not stress easily.

ENDANGERED SPECIES IN CAPTIVITY FOR LONG TERM CONSERVATION

The goal for these species is long-term maintenance of a viable population and the preservation of genetic diversity. Species of this designation require extensive genetic planning and captive management.

The North American gorilla population is one example. In 2002 there were 361 animals in that population, of the 171 potential founders, 111 are represented by living descendants. The population is doing very well, in fact 99% of genetic diversity has been retained.

ENDANGERED SPECIES BEING PROPAGATED FOR RELEASE INTO THE WILD

Reproduction for these species should be maximized for rapid population growth. Also, the captive environment should be as similar as possible to the natural environment in which the species is designated for release. External disruption should be held to an absolute minimum.

The golden lion tamarin population is a classic example. This captive population has been overseen by an international committee since 1981. Nearly all of these animals are owned by the Brazilian government. In 1990, the government of

Zoos that promote breeding programs for rare or endangered animals, such as the giant panda (*Ailuropoda melanoleuca*), help to defend against the extinction of a species. (Photo by AP Photo/Xinhua, Wu Wei. Reproduced by permission.)

Brazil gave the committee jurisdiction over the management of both the wild and captive populations. Therefore, this committee establishes policy for and manages the wild and captive populations of golden lion tamarins by making recommendations to the Brazilian government. The committee has no implementation authority but the government tends to accept its recommendations. As a result, the captive population supplements the survival of the wild population through reintroduction and management policy.

RARE SPECIES NOT CAPABLE OF SELF-SUSTAINING REPRODUCTION IN CAPTIVITY

Intensive efforts should be made with these species to develop successful husbandry and propagation techniques in order to achieve self-sustaining populations. When this is accomplished, genetic diversity should be maximized in the captive population.

An Asian elephant (*Elephas maximus*) mud bathing in a zoo. Zoos provide opportunities for people to see animals that they would not typically have a chance to see. Photo by Animals Animals ©Doug Wechsler. Reproduced by permission.)

and dental cleanings. Appropriate vaccinations against life-threatening diseases are also given. A fecal analysis to check for the presence of parasites is usually performed every three months on each mammal in a zoo collection. In the event of a serious illness, many zoos have access to and perform the same advanced procedures on captive mammals as a human would being treated by a surgeon or other specialist. Veterinarians depend on the input of zookeepers for evaluating when an animal is sick and what would be the least stressful mode of treatment.

Training

Training is another technique that zookeepers often use to improve the quality of care of mammals in captivity. Although the term "training" probably conjures up images of tigers jumping through hoops of fire or elephants in ornate dress standing on their heads, training for the purpose of management in modern zoological parks has a very different purpose. When an animal caregiver needs to get very close to, touch, or otherwise interact with a mammal, that specimen must be conditioned to allow their presence. Target training is the method by which this conditioning behavior is learned. It can be a simple or complex program depending on the species and the training goal.

The training of mammals in zoos usually involves two techniques. One is the theory of shaping. This involves the immediate reinforcement of a desired behavior with some sort of reward to the animal. When a mammal receives positive reinforcement for a particular behavior, it repeats that behavior. Shaping several small behaviors initially and then forming them into a series, is the method by which a training program is built. A second method often used is target training. In the simplest form a target is used to focus the animal's attention. The target is often a colorful ball on a stick or a colorful card. When an animal touches the target or performs the behavior that is required by the trainer, a bridge occurs. The bridge is a form of non-verbal communication and feedback between trainer and animal. The bridge is usually a whistle, clicker, or other device that makes a sound that is audible to the animal being trained. The bridge is often accompanied by a reward. Therefore, when the animal hears the bridge it knows it has performed the correct behavior and may get a reward. Training in zoos is very helpful for captive management. An animal can be trained to position in a specified area or hold a stationary position. This can allow a zookeeper or veterinarian to perform a physical exam, and collect blood, as well as other medical procedures without the danger of anesthesia.

Ethical considerations

Keeping wild mammals in captivity has been a controversial issue since the industrial revolution created more time for such interests. As late as the 1970s, most zoos kept species that were easy to acquire at the time, in substandard conditions. Apes, monkeys, tigers, and bears were kept in concrete and steel cages reminiscent of prison cells. This outraged many visitors. Animal welfare is a movement based on the desire to reduce animal suffering and to minimize negative impacts that might result from human interactions. The problem with this philosophy is that to some degree it depends on humans being able to assess if they themselves are having a negative impact and are causing an animal to suffer. Suffering can be very subtle, occurring without humans having the capability for observation. The biology of a species must be understood very well to make such a decision. Animal rights is based on the philosophy that each individual animal has rights that morally should not be violated. The problem with this philosophy is that it focuses on the individual exclusively, and also requires a human judgment. In other words, humans must decide on acceptable morals based on human values not the values of each species being regulated. These are very heavily debated subjects that will influence the management of animals in captivity.

Future challenges

Today, many animals and their habitats are threatened with extinction. Most zoos use naturalistic habitats and put animal welfare and conservation at the top of their duties. Ironically, captivity is the only hope for the survival in the wild of an always increasing number of species. Due to this responsibility accepted by zoos today, the famous author, naturalist, and zoo builder Gerald Durrell described them as "stationary arks." If the stationary ark is to remain in place, zoos must continue to produce results in preserving the welfare of captive and wild animal populations. Much of this work cannot be done within the perimeter gates of a zoo. Zoos must house and store healthy animal populations, and they must also give visitors instruction and direction on the mission so it will expand and move beyond the gates. Staff must also be active in worldwide field missions to teach conservation messages. A zoo must also be able to compete and have their voice heard in the global political arena. Ironically, if zoos are successful enough, there will be no need for them.

Resources

Books

Ballou, J. D., and T. J. Foose, "Demographic and Genetic Management of Captive Populations." In *Wild Mammals in Captivity, Principles and Techniques*, edited by D. Kleiman et al. Chicago: University of Chicago Press, 1996.

Coe, John C., et al. *Keepers of the Kingdom, the New American Zoo*. New York: Thomasson-Grant and Lickle, 1990.

Durrell, Gerald. *The Stationary Ark*. London: Collins, 1976.

Hagenbeck, C. *Beasts and Men*. London: Longmans, Green, and Company, 1909.

Hancocks, D. *A Different Nature: The Paradoxical World of Zoos and Their Uncertain Future*. Berkeley: University of California Press, 2001.

Kleiman, Devra G., Mary E. Allen, Katerina V. Thompson, and Susan Lumpkin, eds. *Wild Mammals in Captivity, Principles and Techniques*. Chicago: University of Chicago Press, 1996.

Norton, Bryan G., et al. *Ethics on the Ark*. Washington, DC: Smithsonian Institution Press, 1995.

Shepherdson, David J., Jill D. Mellen, and Michael Hutchins. *Second Nature, Environmental Enrichment for Captive Animals*. Washington, DC: Smithsonian Institution Press, 1998.

Tudge, Colin. *Last Animals at the Zoo: How Mass Extinction Might Be Stopped*. London: Hutchinson Radius, 1991.

Periodicals

Foose, Thomas J., et al. "Propagation Plans." *Zoo Biology* 5 (1986): 139–146

Frankham, R., H. Hemmer, O. A. Ryder, E. G. Cothran, M. E. Soule, N. D. Murray, and M. Snyder. "Selection in Captive Populations." *Zoo Biology* 5 (1986): 127–138

Hutchins, M. "Zoo and Aquarium Animal Management and Conservation: Current Trends and Future Challenges." *International Zoo Yearbook* 38 (2003): 14–28.

Hutchins, M. and W. G. Conway. "Beyond Noah's Ark: The Evolving Role of Modern Zoological Parks and Aquariums in Field Conservation." *International Zoo Yearbook* 34 (1995): 84–97.

Hutchins, M., and B. Smith. "Characteristics of a World Class Zoo or Aquarium in the 21st Century." *International Zoo Yearbook* 38 (2003): 130–141.

Organizations

American Zoo and Aquarium Association. 8403 Colesville Road, Suite 710, Silver Spring, MD 20910 USA. Phone: (301) 562-0777. Fax: (301) 562-0888. Web site: <http://www.aza.org/>.

Center for Ecosystem Survival. 699 Mississippi Street, Suite 106, San Francisco, 94107 USA. Phone: (415) 648-3392. Fax: (415) 648-3392. E-mail: info@savenature.org. Web site: <http://www.savenature.org/>.

World Association of Zoos and Aquariums. PO Box 23, Liebefeld-Bern CH-3097 Switzerland. Website:<http://www.waza.org>.

Ken B. Naugher, BSc

Conservation

We are living in an era of unprecedented loss in biodiversity. The most optimistic projections forecast the loss of several thousand species over the next few decades; less sanguine conservationists fear that almost a million species may vanish before the end of the present century. The earth has suffered mass extinctions before, but the present episode is qualitatively different for two reasons: first, it is extremely rapid; second, it is caused by one mammalian species, a large, primate of African origin. Ironically this creature has named itself *Homo sapiens*, the Wise or Knowing Man.

The human activities that threaten wildlife and ecosystems worldwide include deforestation, pollution, over-exploitation of native species, introduction of non-native species, acceleration of climatic changes, and spread of infectious diseases. The essential problem is that more and more people are using more and more resources, leaving less and less for other animal species.

Threats to biodiversity

For more than 99% of its species history, *H. sapiens* existed in small groups of hunter-gathers, a highly intelligent primate that learned to exploit virtually every terrestrial environment that existed on Earth. About 8,000–12,000 years ago, however, people largely ceased living within the constraints of given ecosystems and became ecosystem-creators. The rise of agriculture drastically altered the earth's carrying capacity for *H. sapiens*, and human populations could increase. For several thousand years after our species became essentially an agricultural granivore, populations of *H. sapiens* were held in check by occasional famine and by the infectious diseases that co-evolved with densely packed agricultural humanity. However, beginning in the eighteenth century, scientific and technological advances led to increases in agricultural productivity and (temporary?) conquest of many infectious diseases. During the nineteenth and twentieth centuries, human populations grew exponentially, expanding from approximately 1 billion in 1850 to about 6.3 billion in July 2003. Furthermore, at the same time that human populations were increasing so dramatically, each person was, on average, using a greater portion of the world's natural resources. As a result, Wise or Knowing Man has had a devastating impact on most natural systems throughout the world.

Humans dominate the global ecosystem in four primary ways:

- *Direct transformation of about half the earth's ice-free land for human use.* Houses, cities, roads, strip-mines, shopping malls, and highways involve obvious land transformations. Even more surface area is occupied by agricultural systems hostile to almost all living organisms except the monocultural domesticates being produced for food or fiber.

- *Alteration of the nutrient cycles within the world ecosystem.* Globally, the release of nitrogen—through the consumption of fossil fuels and use nitrogen-based fertilizers—is the most critical, though introduction or extraction of other nutrients may be locally important.

- *Disruption of the atmospheric carbon cycle, particularly through the consumption of fossil fuels.* This is the primary cause of anthropogenic climate change.

- *Introduction of pollutants into the world ecosystem.* To this point pollution has had far less impact than the three factors listed above. However, some scientists believe that the accelerated release of pesticides, industrial wastes, and other bioactive chemicals may have increasingly severe ecological consequences. Furthermore, in a world of economic inequality and political instability, a massive infusion of nuclear pollutants is a possibility that should not be discounted.

Mammals at risk

Although the least speciose of tetrapod classes, mammals are also of particular interest to many people. Even if we leave our own species aside, mammals dominate many terrestrial ecosystems as well as some aquatic habitats. Mammals include the largest animal the world has ever known, blue whales (*Balaenoptera musculus*). Mammals reside atop many food chains, comprise vital links in most terrestrial-vertebrate food webs, and consume primary or secondary production everywhere they occur. Furthermore, perhaps more than any other class of organisms, mammals are threatened by changes occurring at the hand of Wise or Knowing Man. World Conservation

Red wolves (*Canis rufus*) have been reintroduced into the wild by the Nature Conservancy in North Carolina, USA. (Photo by Stephen J. Krasemann/ Photo Researchers, Inc. Reproduced by permission.)

Union (IUCN) data from 2002 suggest that almost one in four mammalian species may be in danger of extinction within the foreseeable future.

When considering threats to biodiversity, we should remember that different species face different threats, and many, if not most species, are menaced by multiple factors. For example, Miss Waldron's red colobus (*Procolobus badius waldroni*), an African monkey declared extinct in September 2000, suffered intensely from both deforestation and hunting for the bushmeat trade. The black-footed ferret (*Mustela nigripes*) approached extinction because of habitat conversion, destruction of its prey, and infectious diseases contracted from domestic animals. Sea otters (*Enhydra lutris*) were exploited for the fur trade and killed by fisherman who considered them competitors for shellfish. And presently, their recovery has been slowed by depredations by killer whales (*Orcinus orca*). Despite the admitted complexity of extinction processes, we review four principle threats to mammalian species: habitat destruction, direct exploitation, introduction of exotics, and infectious disease.

Habitat fragmentation and destruction

Despite the potential importance of altered nutrient-cycles and pollution, the primary immediate threat to biodiversity is from habitat loss that results from expanding human populations and their economic activities. Habitat loss includes outright destruction as well as habitat disturbance due to fragmentation and localized pollution. In many areas of the world including Europe, China, south and Southeast Asia, Madagascar, Oceania, and much of the United States, most original habitat has already been destroyed. Today the highest annual deforestation rates are in developing and tropical countries, where a large proportion of the world's biodiversity is found. The countries having the highest rates of deforestation in the early twenty-first century are Costa Rica (3.0% annual loss of remaining forest cover), Thailand (2.6%), Vietnam (1.4%), Ghana (1.3%), Laos P.D.R. (1.2%), and Colombia (1.2%). Threats to savannas, grasslands, and freshwater aquatic systems are less well documented, but they are as important as deforestation—and, like deforestation, are concentrated within economically poor countries that are rich in biodiversity. This is a matter considered later in the present essay.

One typical result of environmental alteration is the isolation of habitat fragments, or patches. Fragmentation can reduce or prohibit the dispersal of individuals, and local extinction of some species becomes more likely. Later this essay considers some consequences of habitat fragmentation.

A small koala clings to the back of a German shepherd dog in a koala park near Brisbane, Australia. (Photo by © Kit Kittle/Corbis. Reproduced by permission.)

Direct exploitation

Many of the world's natural resources have been over-exploited. Some resources such as fossil fuels are non-renewable, and the best we can do is to slow our depletion of these important commodities. Other resources—such as water, timber, and wildlife—are renewable, and, if used wisely, they may last indefinitely. Unfortunately, people have often been careless about the conservation of renewable resources. The gigantic Steller's sea cow (*Hydrodamalis gigas*) was first encountered by Europeans in 1741; 27 years later the last individual was killed. The near-extinction of American buffalo (*Bison bison*) is an example of over-exploitation known to most U. S. school children. The last wild European cow (*Bos taurus*) was killed in Poland around 1630, and cattle-breeders still decry the loss of important genetic information. Similarly, today in Southeast Asia, at least three species of forest "cow" are in peril of extinction, and these are renewable resources that could still be preserved. During the early and mid-twentieth century, most species of baleen whales were hunted to the edge of economic extinction. Belated protection has allowed species survival, though recoveries have been slow. Great apes, such as gorillas, chimpanzees, and bonobos, are being hunted to extinction for commercial bushmeat in the equatorial forests of west and central Africa. In 2003, it is projected that some 2,000 bushmeat hunters supported by the timber industry infra-structure will illegally shoot and butcher over 3,000 gorillas and 4,000 chimpanzees.

Introduction of exotics

Humans have accidentally and intentionally introduced species into new areas. In a sense, agricultural production itself is a replacement of natural species by domesticated species, under human control. Other examples also abound. Domestic cats (*Felis catus*) brought by Europeans to Australia, out-competed—or out ate—many small, native marsupials. A small, Australian marsupial (*Trichosurus vulpecula*, the brush-tail possum) was introduced into New Zealand in 1837. The varmint prospered beyond all expectations, threatening New Zealand's fragile domestic wildlife.

Other familiar examples of unfortunate mammalian introductions include rabbits and foxes into Australia; goats and cats into the Galápagos and Hawaii; rats, goats, and cats into Cuba and Hispaniola (pity the poor *Solenodon*); mongoose into Jamaica; hogs into too many parts of the southeastern United States; and horses in North America.

Infectious disease

Infectious disease can spread across people, wildlife populations, and domesticated animals. Transmission of infectious agents from domesticated species (*e.g.*, dogs, cattle, water buffalo) to sympatric wildlife can result in a range of potentially fatal infectious diseases. Canine distemper virus is believed to have caused several fatal epidemics among African wild dogs (*Lycaon pictus*), silverback jackals (*Canis mesomelas*), and bat-eared foxes (*Otocyon megalotis*) in the Serengeti. Populations of African lions (*Panthera leo*) and cheetah (*Acinonyx jubatus*) have probably been affected by diseases transmitted from domestic cats. Problems of disease transmission from wildlife to domesticated animals and human beings may also be severe, but they are beyond the scope of this essay.

The nature of conservation biology

Conservation biology is a multidisciplinary science whose overall mission is to conserve biological diversity. The discipline can be subdivided into three primary areas: documenting the world's biodiversity; understanding the nature, causes and consequences of the loss of genetic diversity, populations and species; and developing solutions for the preservation, restoration, and maintenance of biodiversity.

Three ecological postulates that underlie conservation biology

Modern conservation biologists must often transcend the traditional boundaries of academic disciplines, for these scientists increasingly need to know about politics, economics, philosophy, anthropology, and sociology in order maintain or restore the health of ecosystems. However, because conservation biology is fundamentally concerned with the dynamics of wildlife populations, a solid understanding of biology and ecology is paramount for workers within the discipline. Three ecological principles are fundamental for understanding the relationship between population dynamics and conservation.

MANY ORGANISMS ARE THE PRODUCTS OF COEVOLUTION

If most species in an ecosystem were generalists, then in the absence of one generalist species, another generalist species would broaden its niche slightly, and the system would continue to function without important changes. However, if species tend to be specialized, then they are not interchangeable parts in the system; when one is lost, the local ecological community may be affected. For the conservation biologist, interdependent specialization is particularly important, and interdependent specialization often arises through coevolution.

Coevolution involves a series of reciprocal adaptive steps during which two or more interacting species respond to one another evolutionarily. A study of mammalian grazing ecology offers many classic examples of coevolution. Ruminant artiodactyls have evolved fermentation chambers that shelter legions of microscopic flora and fauna. These microbial symbionts extract the energy and nutrients they need from the vegetation consumed by the host-ruminant. In return, the gut-flora ferment cellulose, providing energy and repackaging nitrogen for their hosts. Grazing mammals, in turn, structure the vegetative communities of their grassland habitats. Higher-order coevolution has been demonstrated among

At the Animal Welfare and Conservation Organization in Okonjima, Namibia, lions are encouraged to play to help stay fit. (Photo by Nigel J. Dennis/Photo Researchers, Inc. Repoduced by permission.)

species of grazing ungulates, particularly in Africa. Thomson's gazelles, or "tommies" (*Gazella thomsonii*), for example, are so small that they cannot effectively exploit the tall grass that grows rapidly after the first rains of the wet season. So, just as other grazers depart the depleted grasslands surrounding a recovering waterhole, tommies move in to exploit the flush of tender, new grass. Eventually tommies disperse to exploit grasslands grazed low by other ungulates (particularly zebra, *Equus burchellii*, and wildebeest, *Connochaetes taurinus*). The seasonal ecology, anatomy, and gut flora of *G. thomsonii* evolved in response to the seasonal ecology of the larger ungulates; without these other animals, populations of the little tommies would be much smaller indeed.

Some species, called keystone species, are especially important for the interdependent functioning of an ecosystem. Keystone species may comprise only a small proportion of the total biomass of a given community and yet have fundamental impacts on the community's organization and survival. The loss of such species may have dramatic and far-reaching consequences in the broader ecological community. Primates and bats are believed to play key roles in maintaining ecosystems through dispersing seeds (some primates), pollinating plants (bats and some primates) and serving as prey items. The loss of these species from ecosystems would be predicted to have deep impacts on ecosystem health. For example, throughout many areas in Trinidad, large mammalian species such as deer,

An African wild cat (*Felis libyca*) in the Kalahari Gemsbok Park, South Africa. Game preserves have been established to provide natural habitats for wild animals. (Photo by Animals Animals ©Ingrid Van Den Berg. Reproduced by permission.)

paca, agouti, and peccaries have been extirpated—and yet the ecosystems still remain functional. On the other hand, within these ecosystems, Trinidad's bat and primate populations may be fulfilling ecological roles for which few other occupants remain. Thus the monkeys and bats may now be keystone species, whose presence is vital to the now-fragile existence of the Trinidad ecosystems. Similarly, in pre-European South Carolina, cougars (*Felis concolor*) and a large, social canid (*Canis* sp.) structured the forest herbivores. Now, in the absence of these top predators, whitetail deer (*Odocoileus virginianus*) obliterate populations of several species of forest herb.

IN ECOLOGICAL SYSTEMS SOME CRITICAL VARIABLES HAVE THRESHOLD LEVELS

Changes in one of these variables may make very little difference in ecosystem operation—until a threshold is crossed, and then dramatic systems-alterations will occur. The mathematical study of nonlinear "threshold relationships" is the province of bifurcation theory, which has been used to model catastrophic phenomena ranging from domestic violence to human heart failure. Many conservation biologists emphasize a particular corollary of this general threshold postulate: some ecological processes may suddenly fail when the landscape patch in which they operate is reduced below a threshold size.

Biologist-activist Paul R. Ehrlich has written several books on ecology and conservation, a recent one is *A World of Wounds: Ecologists and the Human Dilemma*, Ecology Institute, Oldendorf-Luhe, 1997. He illustrated potential dangers of ecological non-linearities by the following metaphor. Pretend that the world ecological system is an aircraft and that species within the system are rivets holding the aircraft together. If one or two rivets are lost, the aircraft continues to fly as if nothing had happened. More rivets are lost and the airplane still flies. But eventually the loss of "just one more rivet" may bring the flight to a sudden, disastrous end.

GENETIC AND DEMOGRAPHIC SYSTEMS HAVE THRESHOLDS

Like ecological systems, genetic and demographic systems can be nonlinear and have thresholds below which nonadaptive, random processes begin to displace adaptive, "statistically deterministic" processes.

One example of this is the loss of alleles in small populations because of genetic "drift." Another is the extinction of a small population through random binomial processes. This point can be illustrated by an extreme demographic example. Consider a hypothetical species that does not breed during the dry season and suffers high dry-season mortality. More specifically, assume that each female entering the dry season has a 50% probability of surviving until the end of the dry season. Now consider the probable fates of two different populations:

- 10,000 females enter the dry season. The chances are about 95% that the population at the end of the dry season will include 4900 to 5100 females. In other words, the chances of population extinction are almost exactly 0%. (These statements can be demonstrated by an approximation of the binomial theorem.)

- Two females enter the dry season. The chances are about 25% that the population at the end of the dry season will include 0 females. In other words, the chances of extinction for this small population are about 1 in 4.

Conservation biology and three value statements

The three ecological principles listed above form part of the biological foundation for the discipline of conservation biology. In addition, many conservation biologists accept three value statements—which by their very nature are not subject to scientific confirmation or disproof. In other words, conservation biology is inherently a value-laden discipline, and the following assumptions of worth define the ethical positions of many conservation biologists.

DIVERSITY OF ORGANISMS IS ASSUMED TO BE GOOD

Whenever possible conservation biologists defend diversity on utilitarian grounds—and make statements like, "Some little tropical plant may contain a cure for cancer." Furthermore, evidence exists that biological diversity within an ecosystem contributes to the ecosystem's persistence, stability, and productivity. Nevertheless, even without utilitarian support, many conservation biologists would assume that diversity is good in itself (*an sich*, as the German philosophers used to write) and therefore needs no means-toward-an-end justification. This assumption of intrinsic value is beautifully expressed by Archie Carr in his book *Ulendo: Travels of a Naturalist in and out of Africa*.

As a corollary to this value principle, most conservation biologists believe untimely extinctions (in general, defined as extinctions that result from human activities as opposed to extinctions that result from natural processes) should be prevented. Most conservation biologists also believe that local biodiversity is a universal good. Thus if desperately poor Madagascar cannot afford to protect the 50 endangered and vulnerable species living within her borders, then perhaps wealthy nations (or individuals) are morally obligated to assist with this conservation enterprise. This idea is returned to later in the present essay.

ECOLOGICAL COMPLEXITY IS ASSUMED TO BE GOOD

Clearly this is related to the first value principle above, but it is not exactly the same thing. Consider, for example, a botanical garden with its specimen trees and its greenhouses. Such an installation might contain more different species than a tropical rainforest (and thus would satisfy the principle that "diversity of organisms is good"), but it would not manifest the complex web of inter-organism relationships that characterize a tropical rainforest. The conservation biologist would likely prefer the rainforest to the botanical garden. Or consider this value principle in the form of a question. Some authorities believe that fewer than 1,000 species of large mammals can be preserved from extinction only in captivity. Will a typical conservation biologist be completely satisfied if these mammals survive only in zoos?

EVOLUTION IS ASSUMED TO BE GOOD

The diversity of organisms and the ecological complexities of their interrelations are products of evolution. Most conservation biologists affirm not only the value of the product but also the value of the process that made it. Let us see how this value principle might affect the political agenda of a conservation biologist. What if the wildlife-refuge systems of the world were sufficiently extensive to preserve every living species: would the conservation biologist be satisfied if refuges were not large and diverse enough to allow continued speciation (evolution)?

What areas are the most important to preserve?

Questions of "conservation triage" are difficult, but they must be faced in a world of limited support for conservation agendas. Faced with this problem, British ecologist Norman Myers devised the concept of biological "hotspots," which he defined as regions particularly rich in endemic species and immediately threatened by habitat destruction. Myers is the author of 17 books on the environment, among them *Gaia: An Atlas of Planet Management*, 1993. Myers listed 25 particularly important hotspots, which total only 1.4% of the earth's land surface, but contain 44% of all plant species and 35% of all terrestrial vertebrate species. The Indo-Burma hotspot covers

An ecotourist sits watching a highly endangered baby mountain gorilla (*Gorilla beringei beringei*) as it rolls on its back over flattened shrubbery. (Photo by © Staffan Widstrand/Corbis. Reproduced by permission.)

approximately 795,000 mi² (2,060,000 km²) in south and Southeast Asia, and is home to such threatened species as tigers (*Panthera tigris*), red-shanked douc langurs (*Pygathrix nemaeus*), Sumatran and Javan rhinoceros (*Dicerorhinus sumatrensis* and *Rhinoceros sondaicus*), and Eld's deer (*Cervus eldi*). However, only 61,780 mi² (160,000 km²), or 7.8% of the total area, is protected. Myers's favored strategy would be for conservation organizations to focus their efforts for fundraising and biodiversity conservation upon these areas, and such organizations as Conservation International and the MacArthur Foundation now largely subscribe to the hotspot approach.

However popular it may become, hotspot triage is not without its problems and detractors. For example, some readers may be surprised to learn that rainforests in the Amazon and Congo Basins do not make the magic Top-25. These areas, of course, are rich in endemic species—but they maintain over 75% of their forest cover and are in no immediate danger of complete destruction. Hotspot advocates would argue, "We should spend scarce dollars on species-rich real estate that's about to be destroyed." Opponents might reply, "I'd rather spend scarce dollars on species-rich Amazonia while I can still afford a really big chunk of it."

Regardless of her or his affection for (or disaffection with) hotspot triage, any mammalogist concerned with long-term conservation should become familiar with the types of habitats most important to mammalian diversity and most severely threatened with destruction. Such habits include tropical rainforests, tropical deciduous forests, grasslands, mangroves, and aquatic habitats.

During the 1980s and 1990s, tropical rainforests captured an increasing share of public attention. Rainforests cover less than 2% of the earth's surface, yet they are home to over 40% of all macroscopic life forms on our planet—as many as 30 million species of plants and animals. Rainforests are quite simply the richest, oldest, most productive, and most complex ecosystems on Earth. Furthermore, many of them, particularly in Asia and Oceania, are increasingly threatened by destruction.

An entangled northern fur seal (*Callorhinus ursinus*) is held down by an Aleut teen and seal biologist to remove fishnet that would kill the animal. (Photo by YVA Momatiuk & John Eastcott/Photo Reserachers, Inc. Reproduced by permission.)

Because they are more amenable to sedentary agriculture, some tropical deciduous forests are even more severely threatened than rainforests. Tropical savannas, with their magnificent relics of the Pleistocene mammalian megafauna, are easily converted into pasturelands—and unspoiled tropical savannas scarcely exist at all outside of formal National Parks. Mangroves, which shelter a number of important mammalian species, such as proboscis monkeys (*Nasalis larvatus*), are threatened by pollution, conversion for intensive aquaculture, and destruction for firewood. Lakes, rivers, estuaries, seacoasts, and other aquatic habitats are also under increasing threats of multiple dimensions.

A problem faced by most conservation biologists working in the international arena is that countries with the highest degree of biodiversity (and this is particularly true for threatened and endangered mammals) are usually the countries least able to afford the conservation of their natural resources. For example, in some areas the median per capita income is less than $1US per day. People living under these conditions often consider conservation to be an unaffordable luxury.

Of course conservation biologists have long recognized that the futures of tropical peoples and of tropical wildlife are inextricably mixed. And for more than a decade almost all conservation action plans have emphasized the fact that local people should have an economic stake in the protection of their wild resources. Zimbabwe's "Campfire" program provides a classic example of the local benefit philosophy in action. Village councils were given authority to manage wildlife resources. Then, for example, when Europeans or Americans came to Zimbabwe to kill elephants, the villages could profit from the substantial expenditures of the wealthy hunters. Unfortunately, Campfire (and related "eco tourism" plans) is selling a high dollar luxury activity—which is at the mercy of international market forces and local interference. International economic downturns or national instability (both of which have beset Campfire) can undermine value-added conservation programs.

Under some circumstances sustained-yield harvest programs that return valued wild products directly to local users can be successful. Nevertheless, among people who are desperately poor, the odds against such programs are high. Many impoverished people naturally think of wild mammals as meat—not as an abstract food-resource to be harvested on a long-term, sustained-yield basis but as meat, now, for children who will otherwise be far too hungry before nightfall. The sale of bushmeat is rapidly becoming a substantial source of income, as well.

Because many conservation biologists believe that local biodiversity is a universal good, some argue that wealthier nations (and individuals) have a moral duty to assist poorer nations in conserving humanity's general biodiversity heritage. Even if one subscribes to this idea, it is difficult to determine (particularly on personal, financial levels) the degree of sacrifice that is morally obligatory. Furthermore, in recognizing that severe poverty threatens conservation, there are two more fundamental facts:

- First, severe poverty is in part a function of inequality. In 2003 the United States contains about 5% of the world's people—but is responsible for 30% of the world's resource-consumption. It is difficult for Americans to preach conservation to the rest of the world until the United States begins to clean up its own house.

- Second, severe poverty is in part a function of population-size. If the economic "pie" is finite in size, then even if the pie were equitably shared, "more people" would mean "smaller pieces per capita." At some point, population control becomes a prerequisite to effective conservation policy.

A conservationist with chimpanzees (*Pan troglodytes*). (Photo by © Yann Arthus-Bertrand/Corbis. Reproduced by permission.)

Conservation biology, habitat fragmentation, and island biogeography

The theory of island biogeography was formulated to explain how rates of colonization and extinction affects species diversity observed on actual islands. Currently, protected areas (such as national parks, to which many threatened mammalian species are increasingly restricted) are beginning to resemble habitat-islands in vast seas of agricultural or even urban development. Therefore, island biogeography is increasingly considered an intellectual tool with which conservation biologists should be familiar.

The basic theory of island biogeography grew out of two empirical observations: (1) larger islands often have more species than small islands, and (2) an island's distance from the nearest continent is inversely related to the island's species diversity. These observations were eventually brought together into the equilibrium theory of island species diversity. Conservation biologists use insight from this theory in the management of fragmented landscapes. In particular they often ask how small a refuge "island" can be, before threshold effects arise and species-extinctions dominate community dynamics.

Basically, if a habitat-patch is too small to include home ranges for a viable population of a mammalian species, then the long-term survival of that species is improbable. Information about extinction rates of small mammals in habitat fragments is difficult to evaluate, in part because biologists lack comparable data from undisturbed habitats to serve as controls. However, two studies on forest fragments provide disturbing evidence that mammalian diversity can decline quickly:

- Short term, Thailand. In Surat Thani Province approximately 100 islands were created in 1986 when the Saeng River was dammed to create a hydroelectric reservoir. Rapid changes occurred in the small mammal assemblages on these new islands. Within five years, two of the 12 species (a murid rodent, *Leopoldamys sabanus* and an insectivore, *Hylomys suillus*) were lost. Further extinctions are likely.

- Long term, Panama. Early in the twentieth century, several forest hilltops were isolated during the damming of the Chagras River during the construction of the Panama Canal. After 80 years of isolation, only one out of 16 rodents species remained on islands smaller than 42.3 acres (17.1 ha). The rate of mammalian species-loss from these small island-fragments was approximately one species per 3–11 years.

The fate of large mammal communities in small habitat-fragments is even grimmer. Most big mammals must have a great deal of space. For instance, the home range of a Southeast Asian rhinoceros (*Rhinoceros sondaicus annamiticus*) in Cat Tien National Park, Vietnam has been estimated at 1,480–2,470 acres (600–1,000 ha). Some solitary carnivores require areas an order of magnitude larger. A tiger, for instance, might roam across more than 24,700 acres (10,000 ha), and a single wolverine (*Gulo gulo*) would probably need twice that much room.

These are area-requirements for individual mammals, while of course, viable populations are comprised of many individuals. These populations need even larger patches of habitat. For example, many species of African grazing artiodactyls can exhibit their natural social behavior only in large groups. Large groups require enormous areas, sometimes with widely separated dry-season and wet-season ranges. The

Wardens fix water pipes in Khaudum Game Reserve, in Namibia, as elephants in the backround wait for water. (Photo by Rudi van Aarde. Reproduced by permission.)

ritorial behavior. Such information is useful for predicting how a target species will respond to habitat fragmentation, how edge-effects will impact a given species, or whether the species will use habitat corridors.

Genetic concerns about small populations

The discussion of thresholds bemoaned the fact that small populations were at risk of extinction by demographic stochastic processes. Efforts to maximize intra-specific genetic diversity arc a high priority for conservationists. In general, genetic diversity is correlated with population size. Thus larger populations should manifest a greater variety of phenotypes—and should therefore be better able to respond to variations in environment. However, with habitat loss and fragmentation, populations of many mammalian species are declining and are being fractured into small, disconnected units. And as populations shrink, genetic variation may be lost.

Loss of genetic variability from a population is primarily a function of three interrelated phenomena: "bottlenecking," random genetic drift, and inbreeding depression.

Bottlenecks are events that greatly reduce a population's size. Such reductions can have many causes, including habitat alteration or loss, introduction of competitors or predators, and the spread of epidemic disease. The individuals that survive a bottleneck are the founders for all future generations, and only the genetic variability that is represented in these founder-individuals (plus subsequent mutations) can be preserved within the species. The cheetah (*Acinonyx jubatus*) is the classic example of a bottleneck species. A population crash some 10,000 years ago radically reduced genetic variability. Even today all cheetahs remain so similar, genetically, that skin transplants from one animal to another are not rejected. More significant, cheetahs may lack the genetic variability to respond, evolutionarily, to new diseases confronting the species. But a large population size can usually overcome problems of low genetic variation, as is the case with the current populations of elephant seals, for example.

In evaluating a bottleneck, conservation biologists are especially concerned about the effective population size, or N_e, which is (roughly) the number of breeding animals in a population. N_e is generally smaller than the number of individuals in the population—and if breeding animals are closely related, it can be much smaller indeed. For example, two sets of identical twins do not count as four complete animals, for genetic purposes. The rate of a population's heterozygosity-loss, per generation, is largely a function of effective population size. That is to say, the amount of genetic variability preserved from one generation to the next is approximately proportional to $(1 - 0.5\ N_e)$. Obviously, when N_e is large, the majority of genetic variability will be maintained, and when N_e is small, heterozygosity can be lost very rapidly.

Clearly, even the tightest bottlenecks need not be fatal to a population's survival. Every mammalian species began from a minimal founder-size, and if a bottlenecked population is allowed to increase greatly in numbers, any decay in genetic diversity can be balanced by new variability added through mutation. The most serious problems arise when the bottleneck-

annual migration of east African wildebeest (*Connochaetes taurinus*) covers hundreds of miles (kilometers) and crosses national borders. Of course wildebeest can be kept alive in modest pastures, and tigers can be maintained in zoo-cages. But these conditions are not fully satisfactory and clearly only the largest national parks allow viable populations of most mammals to exist in natural social conditions.

Conserving such large tracts of habitat is often difficult. One approach is to connect habitat fragments by means of corridors, or protected habitat-strips that allow animals to move between patches. In Africa it has been observed that some mammals (as well as reptiles and birds) use corridors as inter-patch bridges. However, some conservation biologists question whether this phenomenon is at all general.

It should be clear that as conservationists contemplate the establishment, enlargement, or maintenance of a refuge, they should be aware of the particular needs of those target organisms that the refuge is designed to shelter. Behavioral ecologists, for example, often gather data on a species' activity patterns, foraging behavior, group size, home range, and ter-

squeeze is maintained over multiple generations, because in this case (1) loss of variability by genetic drift vastly exceeds replacement by mutation, and (2) this process is accelerated by inbreeding. An example is the lowered allozyme and DNA variability observed in the brown hares (*Lepus europaeus*) of New Zealand and Britain and attributed to bottlenecks.

Random genetic drift, sometimes called the Sewall Wright effect, designates changes in a population's allele frequencies due to chance fluctuations. Random genetic drift becomes important only when populations are small. Cross-generational transfer of alleles is then subject to sampling error, and a given allele can be lost (decline to 0% frequency) or fixed (increase to 100% frequency). In other words, when small populations of a species are isolated, out of pure chance the few individuals who carry certain relatively rare genes may fail to transmit them. The genes can therefore disappear and their loss may lead to the emergence of new species, although natural selection has played no part in the process. And the smaller a bottleneck, the more rapidly genetic drift can operate. The longer a bottleneck persists, the greater the potential cumulative effects of genetic drift.

Inbreeding depression can result from matings between close relatives and is more likely to occur in a small population confined to a small habitat-patch. The deleterious effects of inbreeding have been repeatedly documented in zoo populations, before the implementation of genetic management programs. For example, inbred calves of Dorcas gazelle (*Gazella dorcas*) suffered from high juvenile mortality and delayed sexual maturity of females. Among wild populations, inbreeding depression in Florida panthers (*Felis concolor coryi*) may have lowered reproductive rates and reduced the species' capacity to respond to disease.

Przewalski's horse (*Equus caballus przewalskii*) is the only surviving variety of wild horse. This animal is considered extinct in nature (wild individuals were last observed in 1969) and survives only because of captive breeding. The founder-stock for the captive herd was limited in number. Therefore, genetic drift and inbreeding depression have led to a loss of genetic diversity in *E. c. przewalskii* and are reflected by high juvenile mortality and a reduced lifespan. International management programs aim to retain 95% of the current average individual heterozygosity for at least 200 years. And a program is underway to reintroduce these animals into Mongolia.

Wildlife

Conservation medicine focuses on the changing health-relationships between people, other animals, and shared ecosystems. Of course every variety of mammal has been affected by diseases and parasites throughout its species-history. And across evolutionary time, mammalian species have established accommodations with pathogenic organisms: immune and behavioral defenses evolve in the host; responses (often including transmission "improvement" and reduced virulence) co-evolve in the pathogen. However, the present age—with its fragmented populations of wild mammals and anthropogenic mixing of previously separate species—has destabilized these

The Florida panther (*Felis concolor coryi*) is an endangered species. International efforts are being made to encourage the survival of all identified endangered species. (Photo by Animals Animals ©John Pontier. Reproduced by permission.)

dynamic equilibria. And the results are of concern to conservation biologists.

In recent years several emerging infectious diseases (EIDs) have been identified in wildlife populations. Among immune-naive populations, these EIDs may inflict direct mortality beyond a species' evolved capacity for demographic response. In addition, EIDs may affect reproduction, susceptibility to predation, and the competitive fitness of infected hosts. When these detrimental effects extend to the population level, they may in turn affect community structure by altering the relative abundance of species.

Habitat fragmentation often places wildlife species in closer proximity to domesticated animals because habitat patches may be adjacent to farms, villages, and even urban areas. Such situations can facilitate cross-species contagion of disease. People have transmitted tuberculosis, measles, and influenza to gorillas (*Gorilla gorilla*), orangutans (*Pongo pygmaeus*), ferrets (*Mustela putorius furo*), and other mammals. The slaughter of primates for bushmeat in Africa was prob-

On French Island, Australia, koalas (*Phascolarctos cinereus*) are captured for conservation measures while they rest in eucalyptus trees. (Photo by Richard T. Nowitz/Photo Researchers, Inc. Reproduced by permission.)

to become an important area of research for conservation biologists.

Ex situ conservation issues: demand, consumption, and captive breeding

Much of this essay has focused on *in situ* conservation problems because the most important conservation battles will likely be fought on the home grounds of the target species. Nevertheless, the importance of *ex situ* conservation issues should not be underrated. *Ex situ* issues are of two very different types.

Issues of demand and consumption

The impact of economic factors on conservation must be recognized as well. As mentioned earlier, human poverty undercuts conservation programs near many of the world's biodiversity hotspots, but economic factors can affect conservation even from a distance. Research on west Africa's bushmeat trade shows that if markets for meat are exclusively local, the impact of hunting is relatively limited. However, if bushmeat becomes a commodity in a nation's general capitalist economy (if, for example, a market for bushmeat develops in a large city), then demand for forest animals becomes practically unlimited, and vulnerable species may be hunted to extinction. Similarly (here a non-mammalian example), the perilous condition of Southeast Asia's hitherto magnificent chelonian fauna is primarily a function of China's emergence as the regional economic superpower—and of China's insatiable demand for turtle products. Sometimes economic influences can be somewhat less direct. An analysis of the Japanese whaling industry in the 1950s and 1960s indicated that commercial species could be harvested at reasonable profits indefinitely, on a sustained-yield basis. However, the rate of whale-replenishment (r in the population growth equations) was slower than the rate of Japan's economic growth. Therefore, it made good business-sense for commercial whalers to "liquidate their investments" in whales (i.e., to hunt them out) and reinvest their yen in sectors of the Japanese economy yielding higher rates of "interest."

It is hoped that many conservation-and-economics dilemmas may eventually yield to analysis by economically sophisticated conservation biologists—or even by "conservation economists." That is, sustainable development programs combining people, profits, and wildlife may yet save the day. However, in the long run Wise or Knowing Man must develop a new conservation ethic—of sharing, sacrifice, appreciation, and awe—if an appreciable portion of mammalian biodiversity is to be preserved. In other words, why read Phillips and Abercrombie instead of re-reading Aldo Leopold (*A Sand County Almanac* first published in 1949) and Archie Carr?

Captive breeding and reintroduction

In recent years ex situ conservation efforts have become increasingly important. Zoos, botanical gardens, wildlife parks, and conservation trusts now work in collaboration to maintain captive assurance colonies of threatened plants and animals. Studbooks on target species are kept by participating institutions, and breedings are scheduled in consultation with conservation geneticists.

ably the first exposure of people to the precursor of human immunodeficiency virus (HIV). Lions in the Serengeti have been affected by diseases of cats and dogs from Tanzanian villages. In 1984 a disease caused by calicivirus was detected among rabbits in China. The origin of the disease is not definitively known, but the pathogen soon spread westward into Europe, where it affected perhaps 90% of the rabbit populations. Rabbit calicivirus is now used as the primary means for controlling feral rabbits in Australia. (Earlier control relied mainly in the introduction of myxomatosis virus among Australian rabbits. In the 1950s, myxomatosis had a kill-efficiency of about 99%. Over time, however, Australian rabbit populations evolved immunity.) Efforts to re-establish wild populations of black-footed ferrets (*Mustela nigripes*) in western North America have been hampered by the spread of cat and dog diseases among the ferrets, and because enormous populations of ferret prey (prairie dogs) had been destroyed by sylvatic plague.

Because human health may be increasingly affected by diseases transferred from wildlife, conservation medicine is likely

The proximate mission of ex situ colonies is to maximize genetic diversity within a captive population of affordable size. Long-term, however, the fundamental conservation goals of captive breeding are release-in-habitat with the skills necessary for survival—and eventual reestablishment of viable wild populations within target species' historical ranges. Captive propagation definitely works, and more threatened species are now bred in zoos, etc., than ever before. On the other hand, reintroductions (as well as related programs such as translocation of wild animals) have enjoyed only mixed success.

The case of the giant panda (*Ailuropoda melanoleuca*) may be instructive. Today only about 1,000 of these magnificent animals survive in the mountain forests of central China. Years of environmental degradation, disease, and depredation have taken their toll on *A. melanoleuca*. Furthermore, within the animals' fragmented habitat, post-flowering bamboo die-offs have added malnutrition to the pandas' tale of woe. Presently, about 140 giant pandas are maintained in Chinese zoos and in other breeding facilities around the world. However, despite the infusion of massive amounts of money, captive breeding programs have met with only limited success. Of the 226 giant pandas born in captivity between 1963 and 1998, only 52% survived for as long as a month, and others died before they reached reproductive maturity. At present the captive population is scarcely self-sustaining, and it may not produce appreciable numbers for release within the foreseeable future.

By contrast, captive propagation and reintroduction have been more successful with the golden lion tamarin (*Leontopithecus rosalia*), a small primate endemic to the Atlantic coastal forests of eastern Brazil. Because of over-exploitation and (particularly) habitat destruction, this tamarin had become endangered by the late to mid twentieth century. Beginning in 1984, scientists from Brazil and the United States began reintroducing zoo-born golden lion tamarins back into their habitat in the wild—primarily onto private lands that could be protected against the ravages of timber harvest. The combined efforts of governments, nongovernmental organizations (NGOs), local communities, zoological parks, and conservation scientists have more than doubled the size of the wild golden lion tamarin population.

A review of 116 reintroductions (89 involved mammals; all were carried out between 1980 and 2000) concluded that 26% succeeded and 27% failed. (The remaining 47% were classified as uncertain.) Many factors influence reintroduction success. These include habitat quality, the number of individuals released (there is no magic number, but 100 has become the rule of thumb), and the density of predators. Before any reintroduction is attempted, however, conservationists should identify and eliminate the cause of the target species' initial decline. If this can be done, then a reintroduction has a reasonable chance of success. Otherwise, a reintroduction may provide opportunities for feel-good press releases, but it is unlikely to result in establishment of a viable wild population.

Translocation involves moving wild animals from one place to another and therefore is, in a sense, an ex situ activity. Sometimes translocations are attempted without adequate preparation. For example, in French Guiana, howler monkeys (*Alouatta seniculus*) were translocated because their habitat was to be flooded by the construction of a hydroelectric dam. These individuals were moved to an area where howler numbers had been reduced by hunters—and where hunting still occurred. In other words, the cause of the original population decline had not been addressed, and the success of the translocation is still in doubt.

The translocation of Asian rhinoceroses (*Rhinoceros unicornis*) in Nepal is a happier story. Nepali government biologists, assisted by World Wildlife Fund for Nature (WWF), U.S. Fish and Wildlife Service (USFWS), and the King Mahendra Trust, captured rhinos in Royal Chitwan National Park and transported them 220 mi (350 km) cross-country to Royal Bardia National Park. The receiving park was within the Asian rhino's historic range. Habitat was excellent (in quality and quantity), and although *R. unicornis* had been extirpated from Bardia by poachers, the park was well protected by the time the reintroduction project began. Rhino translocations have continued for a decade, and now conservation biologists believe that Nepal has successfully established a second viable population of *R. unicornis*.

Translocations of wild animals are often attempted for reasons unrelated to conservation. While conservationists may not generally support the movement of wildlife to solve human-animal conflicts, they can sometimes learn valuable lessons from such activities. For example, nuisance brushtail possums (*Trichosuros vulpecula*) moved from the city of Melbourne into the Australian outback suffered heavy depredation. Similarly, small, endangered mammals, raised in a zoo, might require some sort of predator-avoidance training before they were released into areas with high predator densities. White-tailed deer (*Odocoileus virginianus*), captured for translocation when the Florida Everglades were in flood, generally did not survive the ordeal. From this experience some biologists learned a great deal about the importance of minimizing capture-trauma when dealing with mammals already under severe stress. Translocations of nuisance raccoons (*Procyon* spp.) and black bears (*Ursus americanus*), though perhaps unwise, reinforce the lesson that some animals know how to get home—and will walk a very long way to get there.

Evaluating the success of conservation projects

This essay is not intended to offer instructions on how to conduct a conservation program. Decisions about supporting conservation are important and they should be made conscientiously on the following basis: "Evaluate with a critical mind, and then support with an open heart."

IT IS MOST IMPORTANT TO EVALUATE THE CANDIDATE'S OR ORGANIZATION'S EFFICIENCY AND INTEGRITY

In this day and age, almost every political candidate will claim to be a great supporter of conservation. Every "Save the Whatever" organization employs experts who design mail-out appeals that are aesthetically elegant and read as sincerely as the Sermon on the Mount. Most readers of this essay are sophisticated enough to see beyond political hype. Also, almost

Monotremata
(Monotremes)

Class Mammalia

Order Monotremata

Number of genera, species
3 genera; 3 species

Photo: A short-beaked echidna (*Tachyglossus aculeatus*), pair in Queensland, Australia. (Photo by J & D Bartlett. Bruce Coleman, Inc. Reproduced by permission.)

Introduction

The three living species of monotreme have always been regarded as zoological enigmas, so much so that when the first specimen of a duck-billed platypus (a single skin) was shipped back to Europe from Australia at the close of the eighteenth century, it was widely believed to be a hoax, and an unconvincing one at that. It would not have been the first time a ship arrived from the East Indies with a cleverly faked "marvel." Previous hoaxes had included supposed mermaids (half fish, half monkey) and wildly exotic birds of paradise, so the scientific community at the time can be forgiven for being skeptical. However, the duck-billed platypus is one of those cases where fact is at least as strange as fiction. It is an unlikely looking amalgam of parts taken from other animals, including a stout, cylindrical body covered in fur, huge webbed feet, a flat paddle-shaped tail, and a unique rubbery bill. The English naturalist George Shaw examined the skin, and was initially dubious as to the validity of the specimen. After examining it carefully and failing to find any evidence of forgery, he formally described it in 1799. He named it *Platypus anantinus*, which literally translated means "flat-footed duck-like." The name was changed to *Ornithorhynchus anatinus* when it later transpired that the genus name, *Platypus*, had already been used to describe a species of flat-footed beetle. The platypus was not the first known monotreme. The short-beaked echidna (*Tachyglossus aculeatus*) had been described seven years earlier, also by Shaw, but had caused much less of a stir, presumably because its superficial similarity to the European hedgehog made it easier to accept. It was not until 1802 that an entire preserved specimen of the platypus

arrived in Europe, providing unequivocal proof that it was a genuine animal. But this was just the first in a long series of zoological conundrums presented by this highly distinctive group. Closer examination of platypuses and echidnas has showed that these are far from regular mammals. The suggestion that they reproduced by laying eggs instead of giving birth to live young left most Victorian zoologists incredulous, and it was almost 100 years before this could be proved. These and many other disconcerting new discoveries have earned the monotremes an enigmatic reputation among mammals, and they are still far from being fully understood.

Evolution and systematics

The first reaction of many taxonomists on examining the monotremes was to classify them as an unusual group of fur-bearing reptiles. It took almost 200 years for monotremes to be declared unequivocally as mammals—indeed, some people remain unconvinced. The characteristics that qualify them as mammalian include a single bone in the lower jaw, three small bones in the middle ear, a high metabolic rate and warm blood, a body covering or hair, and the ability of females to produce milk to feed their young. This last feature is the most important. Milk is secreted by mammary glands; hence, the class name, Mammalia.

In order to emphasize the differences between monotremes and other mammals, the order is often placed in its own subclass, the Prototheria. The name, meaning "first beasts," is unfortunate, since it implies that monotremes are in some way

The platypus (*Ornithorhynchus anatinus*) is one of only three egg-laying mammals. (Photo by J. & D. Bartlett. Bruce Coleman, Inc. Reproduced by permission.)

ancestral to other mammals. This is a common misconception. Many people still insist on describing monotremes as primitive, or inferior. This is demonstrably not the case. However, the Monotremata are certainly an ancient group—they split from the main branch of the mammalian phylogenetic tree sometime during the Cretaceous period, probably about 125–130 million years ago. This was before the divergence of marsupials and placental mammals, but at least 80 million years after the split between reptiles and mammals. Monotremes are no less advanced than any other living mammal group. Far from being ancestors of other mammals, they are cousins that have simply evolved in a different direction. They have retained many characteristics attributed to mammal ancestors, but they have also developed sophisticated adaptations lacking in other mammals. For example, they possess a remarkable "sixth sense" that enables them to sense the minute electric fields generated by other animals.

Seemingly, the monotremes have never been a particularly large or predominant group. The fossils identified to date include just eight extinct species, most of which have been found in Australia. However, these are quite diverse, including two species representing extinct families, the Kollokodontidae and Steropodontidae. There are also three extinct species of echidna and three long-dead species of platypus. The earliest known monotreme is also the earliest known mammal in Australia, a fossil known as *Stenopodon galmani*. It was discovered in rocks about 110 million years old during the excavations of an opal mine at Lightning Ridge in New South Wales. Only a fragment of jaw and a few teeth have been recovered, but this is enough to show that monotremes were as much a feature of Cretaceous Australian landscape as dinosaurs. Another important fossil find includes a 62-million-year-old platypus tooth from Patagonia. This fossil, known as *Monotrematum sudamericanum*, the South American monotreme, is significant

because it suggests that the ancestors of modern monotremes may once have been widespread on the prehistoric southern landmass of Gondwana.

The most informative monotreme fossil discovered to date comes from the extraordinarily rich fossil beds of Riversleigh in Queensland, Australia. Estimated at 13 million years old, the specimen is an almost perfectly preserved skull that closely resembles that of a modern platypus, except that the jaw is full of developed teeth. The modern platypus only has baby teeth (milk teeth), which are, in the adult, replaced by flat horny pads that are used like millstones to crush and grind food before swallowing. The long-extinct relatives of the duck-billed platypus probably had a broad insectivorous diet much like that of modern-day hedgehogs and shrews. The Riversleigh platypus was given the generic name *Obdurodon*, meaning "enduring tooth." It is considered one of the mammalian fossil finds of the century, and is one reason Riversleigh has been designated a World Heritage Site.

Physical characteristics

Monotremes have retained a number of skeletal characteristics possessed by reptilian ancestors, most importantly the structure of the shoulder girdle and some features of the skull. The skull has a fairly large, rounded braincase and an elongated muzzle. Adults of the living monotremes have no teeth. Vestigial teeth are present in the jaws of juvenile platypuses, but they never erupt from the gums. The fact that they are present at all is an example of what evolutionary and developmental biologists call ontogeny-recapitulating phylogeny, which means the embryonic development of a young platypus follows a similar pattern to evolutionary development of the species. The same phenomenon is seen in frogs develop-

A short-beaked echidna (*Tachyglossus aculeatus*) forages for food. (Photo by Ann & Steve Toon Wildlife Photography. Reproduced by permission.)

A duck-billed platypus (*Ornithorhynchus anatinus*) swims underwater. (Photo by Dave Watts/Naturepl.com. Reproduced by permission.)

ing from fish-like tadpoles, or land-dwelling crabs starting life as aquatic shrimp-like larvae. This theory is supported by the discovery of several fossil monotremes with fully developed dentition.

Living monotremes lack sensory whiskers. They have small, beady eyes and no external ears. Internally, however, the ears are much like those of conventional mammals, with three tiny ear bones—the incus, malleus, and stapes. These three bones, which help transmit vibrations from the eardrum to the inner ear, evolved in mammal ancestors from part of the jaw after the split from other reptiles.

The lack of development in certain sense organs such as ears and whiskers is more than amply compensated by the presence of another sense, which is unique to this order. In all living monotremes, the snout is covered in soft, rubbery skin, and pitted with tiny pores. These are lined with thousands of highly sensitive receptors that detect and transmit sensory information directly to the animal's brain. The shape of the snout varies considerably and is clearly adaptive. The snouts of the two echidna species are narrow and cylindrical, ideal for probing among leaf litter or into anthills. The bill of the platypus is flat and shovel-shaped for sweeping through the top layer of sediment on lake and river beds. It resembles that of a duck in shape alone; in living specimens,

it is soft and moist, more like a dog's nose than a bird's hard beak.

All monotremes are hairy. The platypus has a particularly well-developed pelt of fine, dense hairs. The coat is an adaptation to the animal's semi-aquatic lifestyle, and serves to keep it warm by trapping a layer of air close to the skin. In the echidnas, as in placental hedgehogs, porcupines, and some insectivores, the body hairs are interspersed with spines. In fact, the spines are themselves enlarged hairs. They are made of the protein keratin and grow from follicles in the skin.

All montremes have short, powerful legs. Those of the platypus are adapted for swimming. Each of their large feet has five long toes, connected by a leathery webbing. The legs and feet of echidnas are adapted for digging and breaking open anthills and rotten logs in search of food. Both echidna species possess very well-developed claws. Male monotremes also have characteristic horny spurs on their ankles. In adult male platypuses, these are large and sharp with longitudinal grooves connected to ducts from glands in the thigh that secrete a highly potent venom. The spurs of male echidnas are smaller and less well developed.

Unlike most of the world's mammals, the digestive, excretory, and reproductive tracts of monotremes, in both males

A short-beaked echidna (*Tachyglossus aculeatus*) in a defensive ball. (Photo by C. B. & D. W. Frith. Bruce Coleman, Inc. Reproduced by permission.)

and females, all exit the body via a single opening, called the cloaca. Females have mammary glands but no teats as such; the mammary ducts open in pores on the female's furry abdomen. Male platypuses and echidnas do have a penis—it is forked like that of some marsupials, but is used only for delivering sperm and not for urination. In male and female monotremes, urine from the bladder passes via the cloaca.

Monotremes are warm-blooded, but they maintain their body temperature at a slightly lower level than placental mammals—usually somewhere between 86°F and 91.4°F (30–33°C). The blood is pumped by a four-chambered heart, which differs from that of other mammals in having an incomplete separation between the right atrium and ventricle.

Distribution

Modern monotremes, and all but one of the fossil species so far described, are confined to the continent of Australia and the island of New Guinea. The short-beaked echidna is the most common and widespread of the three living species. It occurs throughout Australia and Tasmania and in central and southern New Guinea. The platypus is more restricted—it occurs only in eastern Australia in the states of Queensland, New South Wales, Victoria, and Tasmania. There is an introduced population on Kangaroo Island off the coast of South Australia. The long-beaked echidna (*Zaglossus* spp.) is endemic to New Guinea and is increasingly restricted to remote areas that remain inaccessible to humans.

Habitat

For a small group of specialized animals, monotremes occupy a surprisingly wide range of habitats. The duck-billed platypus is semi-aquatic and is dependent on permanent rivers or freshwater pools. Thus, it is restricted to parts of Australia

with a relatively high rainfall. Both males and females construct simple burrows on the banks of rivers or pools and catch most of their food underwater. Polluted waterways and those that have undergone severe bankside development or canalization are generally not suitable. However, platypuses are increasingly common in suburban settings, due to legal protection and environmental restoration projects.

Short-beaked echidnas are among the most ubiquitous Australian mammals. They have no specialist habitat requirements other than an adequate supply of ants for food, and live everywhere from tropical rainforest to suburban gardens and city parks. There is enough moisture in their diet to sustain them even in the arid central desert of Australia, although there they are much more sparsely distributed. The long-beaked echidna occupies a more restricted range of habitats, mainly montane forests and damp alpine meadows in the higher parts of New Guinea. It is much less tolerant of dry conditions than its short-beaked cousin.

Behavior

As general rule, monotremes are nocturnal or crepuscular, and the best time to watch them is around dusk and dawn. They are active by night in order to avoid the heat of the day. However, daily activity is dictated to some extent by climate and the degree of disturbance. The two species whose range extends into temperate parts of Australia, the duck-billed platypus and the short-beaked echidna, are often active by day, especially in winter when the nights can be quite cold.

Cold weather and food shortages can induce the short-beaked echidna to enter short periods of torpor—a deep sleep, during which metabolic processes slow down and energy is conserved. The species is also the only monotreme capable of full hibernation. In most parts of the species' range, this is never necessary, but in the Snowy Mountains of New South Wales, winters can be sufficiently long and harsh so that echidnas spend up to four months asleep. They may wake periodically to investigate their surroundings, and even move to another den before going back to sleep. During hibernation, the echidna's body temperature may drop as low as 39.2°F (4°C), much lower than during shallow summer torpor. A hibernating animal uses very little energy, but even so, a long winter can take a serious toll on an individual's reserves of fat, and it may emerge in spring some 10–20% lighter than when it first began hibernating.

Detailed studies of the duck-billed platypus and the long-beaked echidna are hampered by the fact that these are rather shy and elusive animals. The same cannot be said of the short-beaked echidna, which is bold by comparison and especially tolerant of humans. In fact, Australian echidnas have few enemies and little to fear from any predator. They are too big to be threatened by cats or even foxes, and their sharp spines are usually sufficient to deter large dogs such as the dingo or birds of prey. If a large animal approaches too close for comfort, the echidna curls its body into a tight ball and raises its spines for protection. Should evasive maneuvers be necessary, the echidna can burrow extremely quickly, literally sinking into the ground until all that remains are a few protruding spines. Despite their lack of spines, adult platypuses are large enough to have few natural predators.

The platypus spends much of its time hidden away in waterside burrows, and emerges to feed only in the quiet hours after dusk and just before dawn. It tends not to travel far from home and usually slips straight into the water. When moving on land, the platypus uses a brisk but relatively inefficient waddle, but it is a superb swimmer and spends much of its time beneath the surface. Its perfectly streamlined body is propelled swiftly and silently through the water with the large webbed front feet. The back feet act as rudders and brakes, and the animal is able to twist and turn with a speed and agility comparable to that of a bird in flight. The platypus returns to the surface every minute or so to take a breath, but it does so silently and without a splash. At the start of each dive, it rolls forward in the water, its sleek form barely breaking the surface. Echidnas can swim, too—their spines help make them surprisingly buoyant and they can make rapid progress in water using a steady doggy paddle.

The monotremes are for the most part solitary animals, except for mothers with young. Single animals occupy a home range that may overlap with those of several others, but they are not territorial and show very little interest in each other except during the breeding season.

Less is known of the New Guinean long-beaked echidna than its two Australian cousins. It too is nocturnal and generally lives alone.

Feeding ecology and diet

Monotremes are highly specialized feeders on invertebrate prey, but the diets and foraging behaviors of the living species are all very different. The short-beaked echidna specializes in feeding on ants, an abundant food resource exploited by relatively few other Australian animals, which is another reason for the species' great success. The echidna's long narrow snout or "beak" is thought to be equipped with an additional sense that enables the animal to detect electrical activity, but probably the most important sense when it comes to feeding is smell. The echidna shoves its snout into ant nests and rotten logs. If ants are detected, the animal uses its claws to rip open the nest and begins lapping up the insects with its long, sticky tongue. Mouthfuls of ants are mashed between the tongue and the hard palette of the echidna's mouth before being swallowed.

Like the short-beaked echidna, the duck-billed platypus faces very little competition for food. The platypus hunts underwater in the dark, but the gloomy conditions are no handicap. The platypus has reasonable eyesight and hearing, but it closes both its eyes and ears when underwater and relies wholly on the information transmitted to its brain by nerves serving the snout. Not only is the snout sensitive to touch, it also contains about 850,000 tiny receptors, able to detect the minute electrical fields generated by the bodies of other living animals, even very small ones such as those of insect larvae. It is difficult to imagine how this extra sense works—it is similar to the lateral line sense of fish, but more finely tuned. Larger prey animals are snapped up and crushed against the

A duck-billed platypus (*Ornithorhynchus anatinus*) in the wild. (Photo by Dave Watts/Naturepl.com. Reproduced by permission.)

hard palate, while smaller ones are strained out of the water or sediment. The food is then pushed into large cheek pouches and stored while the platypus searches for more. The platypus returns to the surface periodically to process and eat its catch. Food is crushed and ground between horny plates that line each jaw, and swallowed. It may take several minutes to finish such a meal, during which time the platypus drifts at the surface with its feet spread wide.

The third species, the long-beaked echidna of New Guinea, is though to feed mainly on earthworms, which it unearths in humid forests. Its extra long nose is used to probe deep into the humus layer or into the topsoil, and once detected, prey is quickly excavated with the front feet, collected with the aid of a long, mobile tongue that is armed with hooklike spines in a central groove. The worms are lightly mashed in the mouth, then swallowed.

Reproductive biology

One of the most remarkable montreme features, and the one that initially seemed to be the biggest obstacle to their inclusion in the class Mammalia, is the fact that females lay eggs instead of giving birth to live young. The eggs are subsequently brooded and hatched outside the mother's body, as in reptiles and birds.

Monotremes typically breed slowly. It takes a mother platypus about six months to raise a small litter of one or two young to independence, seven months in the case of the short-beaked echidna, which typically has only one baby at a time. Long-beaked echidnas have larger litters of four to six young, but still only breed once a year. By investing a large amount of parental care in a few young, the young have a high rate

The world's first platypus (*Ornithorhynchus anatinus*) twin puggles born in captivity are shown together for the first time during a full health check at Taronga Zoo's veterinary clinic in Sydney, Australia, March 28, 2003. (Photo by AFP PHOTO/Torsten BLACKWOOD. Reproduced by permission.)

of survival. They are also surprisingly long lived. Wild platypuses usually survive into their early teens, whereas long-beaked and short-beaked echidnas may live well into the 20s, while captives have lived 30 and 50 years, respectively.

By early spring, courtship and rivalry among platypuses is well underway and males become very aggressive. They will fight for dominance and the right to mate with the females living within their range. They have no teeth and their claws are blunt, but the sharp spurs on the ankles are deadly. Normally, they are kept folded away to avoid snagging, but during battle they are raised. Fights occur in the water, where the animals are most agile, and combatants swim in tight circles, each attempting to spike the other and inject a debilitating dose of venom. The venom is toxic enough to kill a dog and cause agonizing pain and prolonged paralysis in humans. The male duck-billed platypus is the world's only venomous mammal. The spurs in male echidnas are small and sharp, but lack the deadly venom. Having seen off his rivals, the victorious male woos the female with a courtship involving a slow circular dance, during which he holds her tail in his bill. Both courtship and mating take place in the water.

Rivalry among male echidnas is equally intense, though not quite as violent. At the start of the breeding season, male echidnas begin following females around. After two or three weeks, some females have attracted a following of six or seven suitors, that follow her every move in a line known somewhat

whimsically as a "love train." As the female comes into breeding condition, the males begin circling her, creating a circular trench from which each male attempts to evict his rivals. The last male left in attendance claims the right to mate.

Beyond courtship and mating, male monotremes have nothing more to do with the rearing of their young. The females, on the other hand, are diligent parents. After mating, females are busy preparing their nests, which are built in deep burrows. The echidnas excavate burrows for nesting or take advantage of natural dens such as rock crevices or hollow logs. Platypus burrows are simple oval tunnels with a sleeping chamber at the end. Breeding females also build more extensive nesting burrows. These may extend as far as 65 ft (20 m) into the bank, with branching tunnels that twist and turn, some leading to living chambers, others to dead ends. Unlike most other mammals, which do their best to keep nesting areas snug and dry, the atmosphere inside a platypus nest is very humid. The nest is made of damp leaves and other vegetation collected from the water or the banks and carried to the burrow clasped under the body by the tail. The breeding tunnel is blocked every few feet with loose earth, which the female shifts and replaces every time she comes and goes.

As in marsupials, most development takes place outside the mother's body and pregnancy itself is very short—just two weeks in both the platypus and the short-beaked echidna. The eggshells are rubbery, not brittle, and surround each embryo while it develops in the uterus. Each egg contains a very large yolk to sustain the embryo until it has developed sufficiently to hatch out and sustain itself on milk. The eggs are small and almost spherical. Those of the short-beaked echidna are laid directly into a temporary fold of skin, like a marsupial pouch, and can be carried with the mother. The platypus has no pouch and, once she has laid her eggs, she stays with them in the nest, her body curled around them, never leaving them for more than a minute or two, for fear they become chilled. The eggs are flexible and slightly sticky, so once laid they tend not to roll around.

The young of both species hatch after an incubation period lasting about 10 days. They cut their way out of the egg using a special milk tooth to pierce and tear the leathery shell. For the echidnas, this single tooth is the only one they will ever possess. In the platypus, baby teeth do develop, but they never become functional.

Newly hatched monotremes are barely 1 in (2.5 cm) long. The body is pink, naked, and almost transparent. Their skin is so delicate that they would shrivel and die in minutes if exposed to the sun. But in the humid environment of the mother's pouch or the nest, young echidnas and platypuses are safe from desiccation as long as they can find milk. As for all mammal babies, the first urgent task for a newly hatched monotreme is to reach the mammary ducts on the mother's abdomen, when they start leaking milk about 10 days after birth. The development of lactation in mammals is one evolutionary mystery on which the monotremes have been able to shed some light. Mammary glands are thought to have evolved from sweat glands. In the ancestors of mammals, the young of animals that laid eggs like monotremes must have benefited from the secretion of a sweat-like substance from cutaneous glands on their mother's brood pouch. To begin with, they may have simply absorbed extra salts or moisture, but once this small nutritional advantage was established, natural selection favored lineages in which the glands became more and more active. Lactation in mammals was obviously well established by the time the monotremes diverged from the placental and marsupial lineages, but the former apparently never developed specialized structures for the delivery of milk to the offspring, namely teats. In marsupials and placental mammals, the release of milk is triggered by giving birth and is sustained by the stimulus of young sucking on a teat. The situation in monotremes is different, since milk is not needed until 10 days after giving birth and there are no teats for the young to latch on to. Instead, the milk seeps into the mother's fur. Young platypuses lap up the milk as it accumulates in the fur, while baby short-beaked echidnas suck vigorously at the mammary pores.

A young echidna may ride in its mother's makeshift pouch for up to three months, but not surprisingly it is evicted as soon as its spines begin to grow. Then it will be left in the nest while its mother goes out to feed. Likewise, as young platypuses become able to maintain their own body heat, their mother can leave for longer periods. The young are well protected, even when left home alone—each time she leaves, the

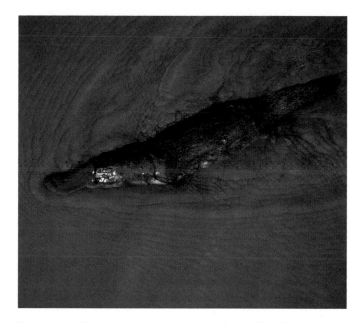

The platypus (*Ornithorhynchus anatinus*) surfaces after filling its cheek pouches with food. (Photo by Tom & Pam Gardner/FLPA–Images of Nature. Reproduced by permission.)

mother carefully replaces a plug of earth as a deterrent to predators and to prevent her offspring from getting out.

Young platypuses are weaned at three or four months. Young short-beaked echidnas first venture outside the pouch at about the same age, but are not capable of feeding themselves for a further three months.

Conservation

The long-beaked echidna is listed as Endangered by the IUCN. Never as widespread or abundant as its two Australian relatives, *Zaglossus* is now threatened by habitat loss and severe over-hunting for meat. The logging industry in New Guinea is not only responsible for the destruction of huge areas of forest, it also leaves the remaining habitat much more accessible to hunters. The latest estimates put the total population at about a quarter of a million.

In contrast, the short-beaked echidna is apparently thriving. It is one of few native Australian mammals for which the arrival of European settlers and introduced wildlife has not resulted in a serious decline. It is not hunted for meat, and its spines are ample protection from most animal predators. Its diet of ants means it can survive in a wide variety of habitats and is not adversely affected by many forms of development. Both echidnas and their prey can even survive bush fires, by burrowing underground and waiting for the flames to pass above.

The duck-billed platypus is something of a conservation success story. It was hunted extensively for its fur, which is thick and silky like that of an otter. The platypus also suffered indirectly from the actions of humans, as rivers were polluted by industrial and mining effluent and waterways were modified around human settlement. Concrete banks are not good for burrowing, and human-made structures such as weirs, drain guards, and dams are all potential platypus death traps. Many thousands have been drowned in fishing nets. However, the story has a happy ending. Unlike many other threatened mammals, the decline of the duck-billed platypus did not go unnoticed and, since the 1960s, it has been a well-protected species. Many waterways have been restored specifically to meet platypus needs and it is an increasingly common species, even in some towns.

Significance to humans

Echidnas and platypuses are charismatic animals, which do no harm to human interests. They are also of considerable novelty value. "Platypus-spotting" is one of the many wildlife encounters offered by the highly lucrative Australian eco-tourism industry. Sightings are rarely guaranteed, but for many people, a glimpse of one of these enigmatic creatures slipping silently into view and disappearing once more will remain a treasured highlight of a visit down under.

Resources

Books

Macdonald, D. *The New Encyclopedia of Mammals.* Oxford: Oxford University Press, 2001.

Nowak, R. "Order Monotremata." In *Walker's Mammals of the World*, Vol I, 6th edition. Baltimore and London: Johns Hopkins University Press, 1999.

Quirk, S., and M. Archer. *Prehistoric Animals of Australia.* Sydney: Australian Museum, 1983.

Strahan, R. *The Mammals of Australia.* Carlton, Australia: Reed New Holland, 1995.

Periodicals

Flannery, T. F., M. Archer, T. H. Rich, and R. Jones. "A New Family of Monotremes from the Cretaceous of Australia." *Nature* 377 (1995): 418–420.

Amy-Jane Beer, PhD

Echidnas
(Tachyglossidae)

Class Mammalia
Order Monotremata
Family Tachyglossidae

Thumbnail description
Small to medium, terrestrial, invertebrate-feeding, egg-laying mammal, characterized by a head with large brain and narrow beak-like snout covered with leathery skin, minute mouth opening under the tip of the beak, no teeth, worm-like tongue, small external eyes, large ear slits, stocky, rounded body covered with spines and fur, powerful front digging limbs, and backwards rotated hind limbs

Size
16–40 in (40–100 cm); 5.5–35.2 lb (2.5–16 kg)

Number of genera, species
2 genera; 4 species

Habitat
Forests, grasslands, heath, shrublands, and woodlands

Conservation status
Endangered: 3 species; Lower Risk/Near Threatened: 1 species

Distribution
Australia and New Guinea

Evolution and systematics

Fossil records for echidnas are scarce. The first tachyglosssid fossil, a long-beaked echidna (*Zaglossus robusta* = *Megalibgwilia*), found in a gold mine at Gulgong, New South Wales in 1895 was about 15 million years old. *Tachyglossus* remains have been found in Pleistocene sediment (about 100,000 years old) at Mammoth Cave, Western Australia and in the Naracoorte Caves of South Australia. Fossil records show that long-beaked echidnas became extinct on mainland Australia in the late Pleistocene.

The divergence of present day echidnas and their relationship to other fossil monotremes from the Cretaceous (120 million years ago), are unknown.

The family Tachyglossidae is divided into two genera. *Tachyglossus*, the short-beaked echidna, consists of one species, *T. aculeatus*, and five subspecies. *Zaglossus*, the long-beaked echidna, has three species, with four subspecies of *Z. bartoni*. *Zaglossus* taxonomy is based on world museum collections.

Physical characteristics

Together with the platypus, echidnas are the world's only monotremes, or egg-laying mammals. They are sometimes referred to as "spiny anteaters" because they look like the hedgehog and the porcupine in that they are covered by sharp spines.

Echidnas have a domed shaped back with short stubby tail, no obvious neck, and a flat belly. Back and sides are covered with spines (modified hairs) of varying sizes and lengths. Fine to course hair covers the legs and belly and surrounds the spines. The panniculus carnosus, a muscle located under the skin and around the body allows echidnas to assume contortionist shapes from very round to nearly flat. This muscle also permits movement of individual spines and helps form the pouch in reproductively active females.

Short and long-beaked echidnas are easily distinguished by differences in size, body mass, and length of the beak. Appearance of ears and eyes are also different. Adult *Zaglossus* range 24–40 in (60–100 cm) in length, weigh

A short-beaked echidna (*Tachyglossus aculeatus*) forages. (Photo by Animals Animals ©Roger W. Archibald. Reproduced by permission.)

13.2–35.2 lb (6–16 kg), and have a 4.2 in (10.5 cm) beak, often displaying a downward curve. Adult *Tachyglossus* are 12–20 in (30–50 cm) long, weigh 5.5–15.4 lb (2.5–7 kg), and have a 2.1 in (5.5 cm) straight beak. Whereas *Zaglossus* often have small distinct pinna, *Tachyglossus* generally have no external ear. Eyes of *Tachyglossus* are nearly obscured by hairs, but surrounded by bare wrinkled skin in *Zaglossus*. Contrary to lore, echidnas see well and can learn using visual cues.

The thick, woolly hair and short spines of *Zaglossus* are most like the Tasmanian subspecies of *Tachyglossus*, *T. a. setosus*. Pelage density, color, and spine length differ between subspecies. The arid dwelling *T. a. acanthion* tend to have longer, thinner spines and less hair than the other mainland subspecies *T. a. aculeatus*. Pelage of Kangaroo Island, *T. a. multiaculeatus*, and Tasmania, *T. a. setosus*, subspecies varies from light straw-colored to very dark whereas mainland subspecies are uniformly dark. Albinism has been reported in most subspecies.

Front and back limbs have five toes, with one to three grooming claws on each hind foot. Articulation of the rotated hind limbs in the pelvic girdle gives echidnas extreme dexterity to scratch between spines on any part of the body. Arrangement of the muscles in relation to the short, stout

limbs gives echidnas enormous strength for digging and climbing. The limbs of *Zaglossus* are twice as long as those of *Tachyglossus*.

Echidna body temperature is low compared to other mammals, 87.8–91.4°F (31–33°C) and individuals use torpor (lowering of body temperature and metabolic rate) at any time of the year. There is only one opening, the cloaca, for excretion of urinary, fecal, and reproductive products out of the body. It is not possible to tell the gender of an echidna by external features. All genitalia are located internally. Both males and females may retain a spur on the inside of the hind foot and both can contract abdominal muscles to form a pseudo-pouch. Echidnas can live in excess of 50 years.

Distribution

Short-beaked echidnas are found throughout Australia and in some parts of New Guinea. Long-beaked echidnas occur only in New Guinea and Salawati. They once inhabited parts of Australia, but died out about 20,000 years ago.

A short-beaked echidna (*Tachyglossus aculeatus*) feeds on termites. (Photo by Animals Animals ©K. Atkinson, OSF. Reproduced by permission.)

An echidna burrowing for protection. (Photo by Bill Bachman/Photo Researchers, Inc. Reproduced by permission.)

Habitat

Short-beaked echidnas are found in all types of native and exotic Australian habitats from sea level to alpine and from arid through tropical. Little is known about the distribution of *T. a. lawesi* in New Guinea; most records are from the south and southwest of the country. Long-beaked echidnas have been found from sea level to 12,500 ft (4,150 m), primarily in areas of higher rainfall.

Behavior

Echidnas are solitary living, extremely mobile and have home ranges up to 494 acres (200 ha). Home ranges of several animals overlap and are not defended as territories. Individuals do not interact, forage communally, or use the same shelter sites. Short-beaked echidnas are active both day and night, depending on time of year and locality. They avoid the heat of the day because they do not sweat or pant. Long-beaked echidnas are thought to be totally nocturnal, but little is known about their natural history.

Echidnas rarely vocalize. Apart from audible snuffing sounds, there are a few reports of soft "cooing" or "purring."

Feeding ecology and diet

Whereas short-beaked echidnas feed on all types of invertebrate species found in the soil or rotting wood; long-beaked echidnas feed primarily on earthworms. The tongue of the long-beaked echidna is grooved and has three rows of backward directed keratinous spines at the tip, that help extract worms from the ground. The tongue of the short-beaked echidna is lubricated with a sticky secretion, extends up to 7 in (18 cm) beyond the tip of the beak, and has an agile tip for drawing insects into the mouth. Echidnas have no teeth, but grind their food between a set of tiny keratinized spines located on the base of the tongue and the roof of the mouth. Echidnas find their prey using acute senses of smell and hearing. They also sense vibrations with their beak.

Reproductive biology

Short-beaked echidnas are sexually mature at five to seven years of age. Courtship and breeding occur during the Australian winter through spring, June to September. At the beginning of the courtship period, male echidnas abandon their solitary life style in search of a female. A group of males following a single female is called an echidna train. Courtship lasts between one and four weeks, with up to 10 males accompanying, prodding, and following a female until she is receptive. Males then compete, by shoving each other head on head, to dig a trench beside the female. When only one male remains, he completes the mating trench that prevents him from rolling over as he lifts the female and places his tail under hers, cloaca on cloaca. Copulation, which lasts between 30 and 120 minutes, is the only time that the penis is outside the body. A female mates only once during the reproductive season. After mating, males and females return to a solitary life style.

During the 22-day gestation the mammary glands of the female begin to swell and form a longitudinal pouch on the belly. In a sitting position, the female extends her cloaca and lays a single egg directly into the pouch. The 0.6 in (15 mm)

Long-beaked echidnas (*Zaglossus bruijni*) are believed to be solitary animals. (Photo by Tom McHugh/Photo Researchers, Inc. Reproduced by permission.)

The short-beaked echidna (*Tachyglossus aculeatus*) curls up into a protective ball when threatened. (Photo by Laura Riley. Bruce Coleman, Inc. Reproduced by permission.)

The long-beaked echidna (*Zaglossus bruijni*) has tooth-like projections on its tongue. (Photo by Pavel German. Reproduced by permission.)

egg, about the diameter of 10-cent coin, has a soft, leathery shell. After 10.5 days of incubation, the young echidna, called a "puggle" hatches. Weighing only 0.0105 oz (300 mg), it takes 10 puggles to weigh as much as a 10-cent coin. There are no teats or nipples for the puggle to attach to. It clings to the hairs on the mother's belly with minuscule but well developed front limbs. Milk is suckled from the milk patches, areolae with specialized hairs, located anteriorly and on either side of the pouch. The puggle remains in the pouch for about 50 days and increases its body mass 85,000%. When too large to carry, the female leaves the young in a secure nursery burrow and returns for two hours every five days to suckle it. At weaning, seven months of age, the young weigh 1.7–3.3 lb (800–1,500g), depending on the size of the mother. There is no mother/offspring relationship after weaning. Young leave the natal area at about one year of age and travel up to 25 mi (40 km) to establish a home range. Most sexually mature females produce only one young every three to five years.

There have been no observations on the reproductive ecology of long-beaked echidnas. Their urogenital systems are identical to *Tachyglossus* and their reproductive biology is believed to be similar. No one has ever seen a pouch or burrow young *Zaglossus* in the wild and there are no specimens of young in world museum collections.

Conservation status

Although *T. aculeatus* have a wide distribution, populations are not large. Throughout Australia numbers have declined due to loss of habitat, predation by feral foxes, cats, dogs, and pigs, as well as roads, electric fences, and herbicide/pesticide use. IUCN lists this species as Lower Risk/Near Threatened. *Zaglossus* spp. are listed as Endangered throughout their range. Populations have disappeared primarily due to hunting which increased with human densities and a breakdown in traditional taboos.

Significance to humans

Some aboriginal groups hunted and ate short-beaked echidnas while other groups revered them as a totem. Early Europeans used echidna fat as a harness dressing or lubricant. Today short-beaked echidnas are an Australian icon used by many grass roots organizations to represent their down-to-earth, get-on-with-it work ethics. In 2000 the echidna was an Olympic mascot. Long-beaked echidnas are hunted for food in some areas of New Guinea. As a member of the oldest surviving group of mammals echidnas are symbolic of species sustainability.

1. Long-beaked echidna (*Zaglossus* sp.); 2. Short-beaked echidna (*Tachyglossus aculeatus*). (Illustration by Barbara Duperron)

The short-beaked echidna (*Tachyglossus aculeatus*) uses its flat claws for digging. (Photo by C. B. & D. W. Frith. Bruce Coleman, Inc. Reproduced by permission.)

Common name / Scientific name/ Other common names	Physical characteristics	Habitat and behavior	Distribution	Food	Conservation status
Short-beaked echidna *Tachyglossus aculeatus*	Stocky, rounded body covered with spines and fur. 12–20 in (30–50 cm) long; weight 5.5–15.4 lb (2.5–7 kg); has a 2.1 in (5.5 cm) straight beak. Generally have no external ear and eyes are nearly obscured by hairs.	Found in all types of Australian habitats—sea level to alpine, and arid to tropical. Solitary; active both day and night.	Throughout Australia and in some parts of New Guinea.	All types of invertebrate species found in soil or rotting wood.	Lower Risk/ Near Threatened
Long-beaked echidna *Zaglossus* spp.	Stocky, rounded body covered with spines and fur. 24–40 in (60–100 cm) long; weight 13.2–35.2 lb (6–16 kg); has a 4.2 in (10.5 cm) beak, often displaying a downward curve. Often have small distinct external ears, and eyes surrounded by bare wrinkled skin.	Found from sea level to 12,500 ft (4,150 m), primarily in areas of higher rainfall. Solitary; thought to be totally nocturnal.	New Guinea and Salawati.	Primarily earthworms.	Endangered

Resources

Books

Augee, M. L., ed. *Monotreme Biology*. Mosman: Royal Zoological Society of New South Wales, 1978.

Augee, M. L., ed. *Platypus and echidnas*. Mosman: Royal Zoological Society of New South Wales, 1992.

Augee, M. L., and B. Gooden. *Echidnas of Australia and New Guinea*. Kensington: New South Wales University Press, 1993.

Baille, J., and B. Groombridge, eds. *IUCN Red List of Threatened Animals*. Gland, Switzerland: International Union of Conservation for Nature, 1996.

Duck-billed platypus
(Ornithorhynchidae)

Class Mammalia

Order Monotremata

Family Ornithorhynchidae

Thumbnail description
Amphibious predator in freshwater habitats, characterized by a broad tail, flat head and body, short limbs adapted to digging and swimming, and conspicuous duck-like bill

Size
16–24 in (0.4–0.6 m); 1.5–6.6 lb (0.7–3 kg)

Number of genera, species
1 genus; 1 species

Habitat
Rivers, lakes, and streams

Conservation status
Not threatened

Distribution
Eastern Australia, including Tasmania

Evolution and systematics

The family Ornithorhynchidae includes just one modern species, the duck-billed platypus. No subspecies or races are known to occur. Several extinct ornithorhynchid species have been described, mainly from fossils found in Australia. At least one type of ancient platypus is also known to have lived in the Patagonian region of South America some 61–63 million years ago (mya), when South America was still physically joined to Australia as part of the giant southern supercontinent, Gondwana. The nearest living relatives are the echidnas (family Tachyglossidae). Based on genetic evidence, it is believed that the platypus and echidna lines have been evolving separately since the late Cretaceous or early Tertiary periods, 63–78 mya.

The taxonomy of this species is *Ornithorynchus anatinus* (Shaw, 1799), New Holland (Sydney), New South Wales, Australia.

Physical characteristics

The platypus has a flattened, streamlined head and body, well suited to its aquatic lifestyle. The animal's color pattern ensures that the platypus blends in with its watery environ-

ment when viewed from either above or below. The fur is dark brown above (apart from a small light-colored spot just in front of each eye) while the chest and belly are silvery cream, sometimes marked with a tawny or reddish streak running along the animal's midline. Interestingly, the platypus relies almost exclusively on its front limbs to propel itself through the water. The end of each front foot is equipped with a broad expanse of webbing, forming a highly effective paddle when the animal swims and dives. In contrast, the back feet are only moderately webbed and mainly used for grooming the fur.

Its most striking feature is undoubtedly its bill. This structure is superficially duck-like—so much so that George Shaw, the first professional zoologist to examine a platypus (a dried skin arrived in England in 1799), felt compelled to probe at the line where the bill joins the rest of the head to see if the specimen had been forged by a clever taxidermist. While the animal's bill may look like a duck's beak, it is actually more like a human thumb in terms of its physical attributes and the way it is used. Like a thumb, the platypus's bill is fleshy and covered by soft, sensitive skin, and is used by the animal to provide essential information about the surrounding environment as well as grab and hold objects.

Duck-billed platypus (*Ornithorhynchus anatinus*). (Illustration by Joseph E. Trumpey)

The platypus is also remarkable in being one of the few mammals known to be poisonous. From a gland (the crural gland) located in the upper thigh, adult males secrete venom, which runs through a duct to a hollow, pointed spur (measuring 0.5–0.8 in [1.2–2 cm] in length) located on the ankle of each hind leg. Platypus venom is produced most abundantly just before and during the annual breeding season, suggesting that it has mainly evolved to help adult males compete for mates. While platypus venom is not considered to be life-threatening to humans, it can cause excruciating pain for a number of days after a person is spurred.

The platypus is a relatively small animal; males are typically 15–20% longer and weigh 60–90% more than females at any given locality. The largest animals measure about 24 in (0.6 m) in total length and weigh 6.6 lb (3 kg). To help reduce its rate of heat loss in the water, a healthy platypus maintains its body temperature at around 89–90°F (32°C), which is about 9°F (5°C) cooler than that of humans. Additionally, there are two layers of fur: a dense, wooly undercoat covered by longer, coarser, waterproof guard hairs. The undercoat traps a layer of air next to the body when a platypus is in the water, helping to keep the animal warm even in freezing winter conditions.

Distribution

The platypus inhabits waterways along the eastern and southeastern coast of mainland Australia (to about as far north as Cooktown, Queensland), and on Tasmania and King Island. An introduced population is also found on Kangaroo Island in South Australia, where 19 individuals were released in Flinders Chase National Park in the years be-

tween 1928 and 1946. The absence of platypus populations in central and western Australia reflects the rarity of permanent lakes or rivers in these areas, while predation by crocodiles may plausibly limit its distribution at the northern end of its range.

Habitat

The platypus occupies a wide range of freshwater habitats, including ponds, lakes, rivers, and streams at all elevations. The animals are not adapted to feed on dry land, and so are most commonly found in permanent water bodies. The species will also use humanmade reservoirs as long as the water is not too deep, mainly feeding at a depth of less than 16.5 ft (5 m). The platypus is known to occur in both urban and agricultural areas. The animals are also occasionally seen in river estuaries, though there is no evidence that they occupy saltwater habitats on a permanent basis.

Behavior

Direct observational studies of platypus behavior are exceedingly difficult to undertake: the animals are active mainly at night, and spend most of their time either feeding underwater or resting in underground burrows. Accordingly, much of what is known about the species' movements, habits, and activity patterns has been gained through radio-tracking studies. By fitting animals with special miniature radio-transmitter tags, their location and behavior can be monitored in a consistent manner both during the day and at night.

The platypus is essentially solitary in its habits, though three or four animals may occasionally be found foraging within a few dozen yards (meters) of each other at a spot where food is abundant. Animals residing along a stream or river typically have a home range comprising 0.5–6 mi (1–10 km) of channel. Home-range size varies with an individual's sex

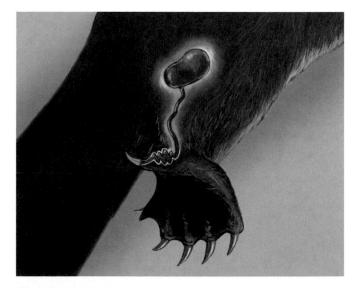

The platypus is one of only a few mammals known to be poisonous. Males inject venom with their spurs. (Illustration by Katie Nealis)

A single platypus may have several underground burrows, including one where it incubates its eggs. (Illustration by Katie Nealis)

A duck-billed platypus (*Ornithorhynchus anatinus*) swimming underwater in Tasmania. (Photo by Erwin & Peggy Bauer. Bruce Coleman, Inc. Reproduced by permission.)

(males have bigger home ranges than females) and habitat productivity. Home-range size shrinks as waterways support more of the small organisms eaten by the platypus, presumably because the animals do not have to travel as far to find enough food.

When a platypus is not feeding, it spends up to 17 hours a day resting in a dry, snug burrow located in a bank at the edge of the water. The animals' front toes are tipped with stout claws, and observations made in captivity have shown that a platypus is capable of digging a new burrow at the rate of around 1.5 ft (0.5 m) per hour. An adult without dependent offspring normally occupies several different burrows (up to about a dozen) within a period of a few weeks. By having numerous burrows scattered along the length of its home range, a platypus is always reasonably close to a safe refuge while feeding.

The platypus rarely vocalizes but, when feeling threatened, the animal can produce a high-pitched growl.

Feeding ecology and diet

The platypus is a predator, mainly feeding on bottom-dwelling aquatic insects such as caddis-fly and mayfly larvae. The platypus is also partial to worms, snails, freshwater shrimps and crayfish, and pea-shell mussels. The size of its prey is limited by the fact that platypus teeth are lost quite early in development and replaced by flat, molar-like grinding pads at the back of the mouth. Unlike most mammalian teeth, these pads grow constantly to compensate for surface wear.

A platypus may find food by digging under banks or snapping up morsels floating on the water surface, as well as searching along the bottom sediments. Small prey is stored temporarily in cheek pouches while an animal is submerged. A foraging platypus typically remains underwater for 10–60

seconds before returning to the surface to breathe and chew its meal with a side-to-side motion of the jaws.

Its eyes and ears are located within shallow, muscular grooves on the sides of the head that automatically pinch shut when an animal dives. The platypus mainly relies on its bill to find food underwater. The surface of this remarkable organ is densely packed with tens of thousands of specialized sensory receptors, sensitive to either touch and vibration (push rods) or electrical currents (mucous sensory glands). It has been shown experimentally that the platypus is capable of registering the tiny amount of electricity generated in the water by the tail flick of a shrimp at a distance of around 2 in (5

The duck-billed platypus (*Ornithorhynchus anatinus*) fills its lungs with air, enabling it to stay underwater for several minutes. (Photo by Dave Watts/Naturepl.com. Reproduced by permission.)

A duck-billed platypus (*Ornithorhynchus anatinus*) walking in the grass in Australia. Usually platypuses are active after dusk and before dawn. (Photo by J & D Bartlett. Bruce Coleman, Inc. Reproduced by permission.)

cm). In turn, this information presumably is used to detect as well as track down the location of prey.

Reproductive biology

Together with the echidnas, the platypus is distinguished as a monotreme, or egg-laying mammal. Males and females have a single physical opening (the cloaca) that is used both for reproduction and excretion. To help maintain a streamlined shape, the male's penis and testes are carried inside the body; mating occurs in the water. In the female platypus, the right ovary is small and nonfunctional. At the time of ovulation, the platypus egg is about 0.16 in (4 mm) in diameter. After being fertilized, the first of three shell layers is formed in the fallopian tube before the egg moves into the uterus. There, the egg is supplied with additional nutrients and two more shell layers are secreted, so the egg is about 0.6 in (15 mm) in diameter when it is laid. Though the time required for gestation has never been determined precisely, it is believed that it takes around three weeks.

Platypus eggs are produced in late winter and spring (August–November), with some evidence that breeding occurs later in southern populations as compared to those found in Queensland. The eggs are laid in a burrow typically measuring 10–20 ft (3–6 m) in length, though sometimes much longer. Throughout incubation and juvenile development, the female keeps the burrow's entry tunnel blocked by several plugs of soil. Besides discouraging access by predators such as snakes and Australian water rats (*Hydromys chrysogaster*), the plugs reduce the likelihood that juveniles drown in the event of a flood. A few days before laying her clutch of one to three eggs, a female drags a large quantity of wet leaves and other vegetation into the rounded burrow chamber to make a nest. It is believed that the female incubates the leathery-shelled eggs for about 10 days, clasping them between her curled-up tail and belly as she lies on her back or side. When they hatch, the young are less than 0.5 in (9 mm) long. Their emergence from the egg is assisted by the presence of a prominent bump (caruncle) at the tip of the snout, an inward-curving egg tooth, and forelimbs armed with tiny claws. When they hatch, the young are less than 0.8 in (20 mm) long.

Juveniles develop in the nursery burrow for about four months before entering the water for the first time. Through-

out this period, they are nourished solely on milk. The female does not have nipples. Instead, milk is secreted directly onto the mother's fur from two circular patches of skin located about halfway along her abdomen. An orphaned platypus will drink milk from a human hand by sucking up the liquid while sweeping its short bill rhythmically back and forth against the palm. In the wild, such sweeping movements may help to stimulate the flow of milk.

Both males and females are physically mature at the age of two years, though some females may delay having offspring until they are four years old or more. Courtship involves two individuals swimming alongside or circling each other, sometimes accompanied by nuzzling or rubbing. One animal may use its bill to grasp the tip of the other's tail and be towed or swim behind. Little is known about the platypus breeding system, apart from the fact that the animals do not appear to form long-term pair bonds. Instead, it is believed that males tend to move about widely during the breeding season, trying to mate with as many females as possible. By the same token, adult females appear to rear the young without any help from their mates.

Conservation status

The platypus is a difficult animal to census or survey: burrow entrances are generally well hidden and the animals rarely leave evidence of their activities in the forms of tracks, scats, or food scraps. Live-trapping nets are time consuming to set and must be monitored closely through the night. Accordingly, knowledge of how the species is faring is sketchy in many parts of its range. In broad terms, it is known that the platypus remains fairly common along some waterways, but

The bill of a duck-billed platypus (*Ornithorhynchus anatinus*) is fleshy and covered by soft, sensitive skin. (Photo by © Peter Marsack/Lochman Transparencies. Reproduced by permission.)

A duck-billed platypus (*Ornithorhynchus anatinus*) exits the water. (Photo by Mitsuaki Iwago/Minden Pictures. Reproduced by permission.)

The duck-billed platypus (*Ornithorhynchus anatinus*) lives in a freshwater habitat, but also forages on land. (Photo by Dr. Lloyd Glenn Ingles © California Academy of Sciences. Reproduced by permission.)

has declined or vanished from others. The species is fully protected by law throughout its range.

In most areas, the key factor limiting platypus numbers is likely to be the quality of habitat. Waterways supporting large platypus populations generally have plenty of trees and smaller plants growing on the banks; a varied array of pools, shallow riffles, rocks, and woody materials in the channel; and reliably flowing fresh water throughout the year. All of these attributes favor the small aquatic invertebrates that in turn feed the platypus. Conversely, factors implicated in the decline of platypus populations include erosion, overgrazing by rabbits and livestock, altered water flow regimes, overclearing of native vegetation, and the systematic removal of logs and large branches from the channel.

Besides habitat degradation, the platypus is vulnerable to drowning in nets and traps set illegally for fish and crayfish. Many individuals also die each year after becoming entangled in garbage such as abandoned loops of nylon fishing line.

Significance to humans

The platypus was hunted in the nineteenth and early twentieth centuries for its soft fur, which (due to the thickness of the skin) was mainly used to make slippers, blankets, and rugs. Today, the animals have no direct economic value apart from their role in attracting tourists to Australia. People living in Australia generally regard the species with great interest and affection. Hence, the platypus also has an important role to play in encouraging landholders and the general community to protect freshwater environments.

Resources

Books

Augee, Michael L., ed. *Platypus and Echidnas*. Mosman, New South Wales: The Royal Zoological Society of New South Wales, 1992.

Grant, Tom. *The Platypus: A Unique Mammal*. Sydney: University of New South Wales Press, 1995.

Moyal, Ann. *Platypus: The Extraordinary Story of How a Curious Creature Baffled the World*. Crows Nest, New South Wales: Allen & Unwin, 2001.

Periodicals

Evans, B. K., D. R. Jones, J. Baldwin, and G. R. J. Gabbott. "Diving Ability of the Platypus." *Australian Journal of Zoology* 42 (1994): 17–27.

Fenner, P. J., J. A. Williamson, and D. Myers. "Platypus Envenomation—A Painful Learning Experience." *The Medical Journal of Australia* 157 (1992): 829–832.

Resources

Musser, A. M. "Evolution, Biogeography and Palaeocology of the Ornithorhynchidae." *Australian Mammalogy* 20 (1998): 147–162.

Proske, U., J. E. Gregory, and A. Iggo. "Sensory Receptors in Monotremes." *Royal Siety of London Philosophical Transactions* 353 (1998): 1187–1198.

Serena, M. "Use of Time and Space by Platypus (*Ornithorhynchus anatinus*: Monotremata) along a Victorian Stream." *Journal of Zoology, London* 232 (1994): 117–131.

——— "Duck-billed Platypus: Australia's Urban Oddity." *National Geographic* 197, no. 4 (April 2000): 118–129.

Serena, M., and G. Williams. "Rubber and Plastic Rubbish: A Summary of the Hazard Posed to Platypus *Ornithorhynchus anatinus* in Suburban Habitats." *The Victorian Naturalist* 115 (1998): 47–49.

Serena, M., M. Worley, M. Swinnerton, and G. A. Williams. "Effect of Food Availability and Habitat on the Distribution of Platypus (*Ornithorhynchus anatinus*) Foraging Activity." *Australian Journal of Zoology* 49 (2001): 263–277.

Temple-Smith, P., and T. Grant. "Uncertain Breeding: A Short History of Reproduction in Monotremes." *Reproduction, Fertility and Development* 13 (2001): 487–497.

Organizations

Australian Platypus Conservancy. P.O. Box 84, Whittlesea, Victoria 3757 Australia. Phone: 613 9716 1626. Fax: 613 9716 1664. E-mail: platypus@vicnet.net.au Web site: <http://www.platypus.asn.au>.

Other

The Complete Platypus. 2000 [2003]. <http://www.platypus .org.uk>.

Melody Serena, PhD

Didelphimorphia
New World opossums
(*Didelphidae*)

Class Mammalia
Order Didelphimorphia
Family Didelphidae
Number of families 1

Thumbnail description
Small- to medium-sized terrestrial mammal with long, naked tail, opposable thumbs both in the hands and feet, long, pointed snout, naked ears that range from small to large, and medium to large eyes; color varies from nearly pure white to blackish; some species are unicolored, whereas others have distinct light and dark blotches and bands

Size
3–22 in (8–55 cm); 0.9 oz–11 lb (25–5,000 g)

Number of genera, species
15 genera; 61 species

Habitat
Dry and moist tropical forests, temperate forest, woodland, grasslands, scrub, and mangroves

Conservation status
Critically Endangered: 3 species; Endangered: 3 species; Vulnerable: 15 species; Near Threatened: 18 species; Data Deficient: 2 species

Distribution
North America from southern Canada and New England to southern Mexico, Central America, and South America to southern Argentina and Chile

Evolution and systematics

This group of mammals was traditionally included in the marsupials until 1993 when it was placed in its own order, Didelphimorphia, together with two other New World opossum-like families, the Microbiotheriidae (one species) and the Caenolestidae (five species). Here Didelphidae is treated in its own order, Didelphimorphia, and Microbiotheriidae and Caenoletidae are each covered in an order as well. The fossil record suggests that didelphids are relatively primitive, unspecialized mammals that emerged some 75–100 million years ago (mya) in North America. Their evolutionary and biogeographical history is complex; they colonized Europe, Asia, and Africa some 60 mya, but disappeared from those continents some 20 million years later, presumably due to competition and predation with the appearance of placental mammals. Twenty mya they were restricted to South America, where they underwent an impressive radiation in the absence of other ecologically similar or predatory placental mammals. It was not until the Isthmus of Panama rose and North and South America joined, some three million years ago at the beginning of the

Pleistocene, that the didelphids again entered North America. Throughout this time they retained a remarkably stable morphology. The connection between North and South America also allowed the entrance of other placental mammals into South America. For the first time in 20 million years, marsupials again faced the faster, larger-brained placental competitors and predators. Groups like the Borhyaenids (wolf- or hyena-like marsupials) and Thylacosmilids (marsupial saber-toothed cats) disappeared and gave way to true canids and felids. Some factors that may have contributed to their disappearance are the smaller encephalization quotient, lower metabolic rate and overall speed, and lower cursorial abilities of the marsupials compared to their placental counterparts. Didelphids have relatively low evolutionary rates and a strong stabilizing selection that prevents greater morphological diversification in the group.

The family Didelphidae is arranged into two subfamilies, Caluromyinae and Didelphinae. As of 2002, 15 genera are recognized: woolly opossums (*Caluromys*), black-shouldered

The bare-tailed woolly opossum (*Caluromys philander*) is a marsupial. (Photo by Jany Sauvanet/Photo Researchers, Inc. Reproduced by permission.)

opossums (*Caluromysiops*), water opossums (*Chironectes*), common opossums (*Didelphis*), bushy-tailed opossums (*Glironia*), gracile opossums (*Gracilinanus*), Patagonian opossums (*Lestodelphys*), lutrine opossums (*Lutreolina*), mouse opossums (*Marmosa*), slender mouse opossums (*Marmosops*), brown four-eyed opossums (*Metachirus*), woolly mouse opossums (*Micoureus*), short-tailed opossums (*Monodelphis*), gray four-eyed opossums (*Philander*), and fat-tailed mouse opossums (*Thylamys*). In 1993, Gardner recognized a total of 61 species, but there have been at least three additional species described since then.

Physical characteristics

New World opossums are small- to medium-sized mammals. The tail can be almost completely covered with hair or almost naked, depending on the species, and frequently is bicolored with the distant half whitish and the base dark. When the tail is naked, it is covered with scales. Tail length is as long as or longer than the head and body in all genera except *Monodelphis*, in which it is about 50% of the head and body length. In most genera, the tail is at least partially prehensile. Ears are always medium to large, rounded and naked, except in the genera *Lutreolina* and *Lestodelphys*, in which ears are short. The snout is long and pointed, and eyes are large, sometimes bulging, and black or brown. The mouth is large and can open to a remarkably wide gape. Tooth and skull morphology is notably consistent, indicating the relatively low level of specialization in this group. The dental formula is I5/4 C1/1 PM3/3 M4/4 × 2 = 50. The thumbs are opposable in the hind feet, giving extremities their characteristic grasping ability, although the water opossum has webbed feet to aid in swimming action and its thumbs are only slightly opposable. Arms and legs vary with the genus. Some genera,

like *Metachirus*, have relatively long legs, whereas others like *Monodelphis* and *Lestodelphys* have relatively short legs. Except for a few species such as *Monodelphis dimidiata*, there is no sexual dimorphism.

Coloration varies widely. Some species are uniformly blackish, while others are almost completely whitish; other species are rusty reddish, gray, brown, tan, or yellowish brown. Underparts are nearly always paler than the dorsum. The venter of the water opossum is silvery white. Two genera have distinct dark blotches above the eyes; they are called four-eyed opossums. Some genera have characteristic dark and pale patterns, sometimes broad, dark saddle-like bands across the back, sometimes a longitudinal stripe along the dorsal spine and continuing along the top of the snout and to the tip of the nose. The hair can be short or long depending on the species, but it is always dense. In females of some genera, there is a distinct ventral pouch where young are kept in the developmental stages. The pouch opens circularly and can be almost completely closed. Mammae number 12 to 18 and are arranged in a circle with one in the center. One species, the water opossum, has a pouch that seals hermetically with an oily substance so that females can dive under the water surface without drowning the young that are attached to the nipples. Likewise, males of this species have a pouch, which protects the scrotum and testicles from contact with the water.

Distribution

Individuals belonging to this family inhabit only the New World, from Patagonian Argentina and Chile north to

A Virginia opossum (*Didelphis virginiana*) peers out from tree trunk nest. (Photo by Animals Animals ©Carson Baldwin, Jr. Reproduced by permission.)

A Virginia opossum (*Didelphis virginiana*) feeds on wild muscadine grapes. (Photo by Animals Animals ©Fred Whitehead. Reproduced by permission.)

southern Canada and the northeastern United States. They are much more diverse in the tropical and subtropical regions between Mexico and Argentina, while only a single species, the Virginia opossum (*Didelphis virginiana*), extends into the United States and Canada. Because of their long period of evolutionary isolation in South America, many genera and species can only be found in that subcontinent. The genera *Thylamys*, *Lestodelphys*, one species of *Didelphis* (*D. aurita*), three of the six species of *Gracilinanus* (*G. aceramarcae*, *G. agilis*, and *G. microtarsus*), two species of *Marmosops* (*M. dorothea* and *M. incanus*), one of Micoureus (*M. constantiae*), and 10 species of *Monodelphis* (*M. americana*, *M. dimidiata*, *M. domestica*, *M. iheringi*, *M. kunsi*, *M. osgoodi*, *M. rubida*, *M. scallops*, *M. sorex*, and *M. unistriata*) are restricted to South America south of the Amazon river. Several species of *Marmosa*, *Marmosops*, and *Thylamys* are known only from the type localities, and several genera such as *Caluromysiops* and *Glironia* and species such as *Gracilinanus emiliae*, *Marmosa rubra*, and *Monodelphis americana* are restricted to the Amazon basin.

The Virginia opossum is one of the most widespread species in the family, and the only one with a distribution extending well beyond that of any other species. Virtually all other species coexist with one or more additional species of didelphids, but in all of the United States and southern Canada, the Virginia opossum is the only species present. Most species inhabit tropical habitats, but a few species, remarkably in the genera *Thylamys* and *Lestodelphys*, are adapted to temperate ecosystems, and inhabit only the southern latitudes of Chile and Argentina.

Habitat

The family Didelphidae can be found in a wide variety of habitats, from moist and dry tropical forests to cloud forests, mangrove swamps, grasslands, scrub, and even into temperate forests. One species, the lutrine opossum or thick-tailed opossum (*Lutreolina crassicaudata*) is considered to be strongly adapted to life in the South American grasslands or pampas,

and readily enters lakes and streams where it swims remarkably well. Another species, the water opossum (*Chironectes minimus*), has as a primary habitat of streams and lakes in moist forests, making its dens in the banks.

Many species are able tree climbers and also move around on the ground; some individuals have been found high in tropical moist forest canopy trees. Other species are predominantly terrestrial, such as the members of the genera *Monodelphis* and *Metachirus*, and appear clumsy if placed in trees, even low to the ground.

Because of their long period of evolutionary isolation in South America, opossums were able to invade virtually every habitat available at every latitude and from sea level to almost 13,100 ft (4,000 m) above sea level. Most species, however, have a relatively restricted habitat and geographic range. Only species such as *Didelphis virginiana* and *D. marsupialis* can be found in several varying habitats, from temperate forests to tropical moist and dry forests to mangrove swamps.

Some species are able to exist in human-modified habitats and a few may even benefit from these, by invading banana, coffee, and citrus plantations, corn fields, and other agroecosystems. Other species seem particularly adept at using forest edges and secondary vegetation. Finally, many other species depend on undisturbed habitats and do not tolerate disturbance or deforestation.

Behavior

Nearly all species in the family are nocturnal, although occasionally diurnal sightings of mouse opossums and water opossums have been reported, and some species of *Monodelphis* are reportedly primarily diurnal. Many scansorial species take to the trees when threatened, whereas terrestrial species run with a characteristic gait. No species is particularly fast during escape behavior. One species, the Virginia opossum, feigns death when threatened by a predator, lying on its side, gaping its mouth spasmodically, and emitting a strong musky smell. Other defense behaviors found in the family include gaping and snapping at intruders while hissing loudly and secreting musk from

A southern opossum (*Didelphis marsupialis*) wards off a dog. (Photo by Joe McDonald. Bruce Coleman, Inc. Reproduced by permission.)

A red mouse opossum (*Marmosa rubra*) shows aggressive behavior. (Photo by Art Wolfe/Photo Researchers, Inc. Reproduced by permission.)

the anal region. The water opossum dives under the surface and then propels itself with strong strokes from the hind legs.

All opossums are primarily solitary, avoiding contact with each other. After dispersal, juveniles do not keep contact. Males and females come in contact only during the female estrus for a short period of time. Individuals generally remain in a home range, but this is almost never defended. Instead, when two animals coincide in space and time, they avoid each other. At least in the bare-tailed woolly opossum, *Caluromys philander*, social dominance is clearly established on the basis of age and body mass. Older, heavier males dominate younger, lighter ones, and agonistic behavior is exacerbated by the presence of females. Interspecific aggression may occur only as a result of the generalized opportunistic behavior to procure food; one *Didelphis* reportedly attacked, killed, and partially consumed a *Philander* opossum after an encounter.

Opossums are silent animals the vast majority of the time; sounds are produced only when they are threatened and these are only hisses and explosive gasps. Foraging behavior is exploratory and continuous. Opossums use primarily their sense of smell to locate food. Stalking is seldom used to capture animal prey, but sight and hearing are continually used in the

search for food. Young opossums, particularly mouse opossums, emit a loud chirping cry when detached from the female's nipple. This induces the female to approach and grasp the young, and push it under the venter, where it reattaches itself to the nipple.

Predators of opossums include a variety of snakes, foxes, owls, ocelot (*Leopardis pardalis*), puma (*Puma cencolor*), and jaguar (*Panthera onca*). Indigenous human groups in the Neotropical region often include the larger species of opossums in their diet. *Didelphis albiventris* has been shown to have an antibothropic biochemical factor in its blood and milk that neutralizes the venom of poisonous snakes.

Longevity records in the wild rarely reach two years, with many species barely surpassing one year. In captivity, longevity is extended with reports of three to five years for species such as *Marmosa robinsoni*, *Caluromys philander*, and *Chironectes minimus*. The record is seven years for a captive *Caluromysiops irrupta*.

A murine mouse opossum (*Marmosa murina*) eats fruit after dark on a branch 15 ft (4.6 m) above ground in the lowland Amazon rainforest of northeast Peru. (Photo by Gregory G. Dimijian/Photo Researchers, Inc. Reproduced by permission.)

Feeding ecology and diet

New World opossums are generalist omnivores. Food items include insects, insect larvae, worms and other invertebrates, bird's eggs and nestlings, fruit, carrion, reptiles, amphibians, birds, and small mammals. Almost all species are omnivorous with varying degrees of frugivorous or carnivorous tendencies. Species in the genera *Caluromys* and *Caluromysiops* are considered primarily frugivorous, although they do include animal matter in their diet. Smaller species such as those in the genera *Marmosa*, *Marmosops*, *Gracilinanus*, etc., tend to be primarily insectivorous with some eggs, fruit, and meat also included. Mouse opossums kept in captivity will not hesitate to attack large moths or grasshoppers, grasping them with their hands and quickly administering killing bites all over the body of the insect. They normally discard the wings and legs, and other chitinous, undigestible pieces. Caterpillars are rapidly rolled and rubbed against substrates to remove stinging hairs. Fruit can be plucked off tree branches directly or can be eaten after the fruit has fallen to the ground. Sweet, juicy fruits are preferred, such as zapotes, blackberries, guavas, chirimoyas, etc., although other drier fruits such as wild figs and wild cacao, as well as introduced, human-grown varieties like bananas, figs, citrus fruits, apples, and cherries, are also consumed. Common and four-eyed opossums take large amounts of fruit from secondary forest species such as *Cecropia*. Common opossums (*Didelphis*) seem to be better seed dispersers and to invest more in seeking fruits of *Cecropia* than gray four-eyed opossums (*Philander*).

The water opossum (*Chironectes minimus*) is particularly interesting in that it seems to be the only New World opossum species that is completely carnivorous. This species feeds mainly on fish, crustaceans, mollusks, and frogs. Its long fingers and bulbous fingertips are held outstretched while the animal swims. Under water, their hands are used to feel under rocks and logs for potential prey. They coexist and overlap locally with the common Neotropical otter (*Lontra longicaudis*). In the Old World, in Africa and Asia, common otters coexist with clawless or small-clawed otters (genera *Aonyx* and *Amblonyx*), which have similar habits and hand morphology to the water opossum. This allows them to coexist by partitioning food resources. The water opossum and the otter may be in the same situation, with the otter feeding primarily on fish and less on crustaceans and mollusks, whereas the opossum probably takes more invertebrates than fish.

In Mediterranean habitats of South America, *Thylamys elegans* has been shown to regulate the frequency of torpor periods, and therefore energy expenditure, by the availability of food. When food is plentiful, animals never enter torpor, and frequency of torpor periods increases with decreasing food availability.

Reproductive biology

Information on reproduction is available for only a few species. Some species show one defined reproductive event during the year that coincides with the greatest seasonal abundance of food. Other species, particularly those inhabiting

A southern opossum (*Didelphis marsupialis*) with its young. (Photo by Laura Riley. Bruce Coleman, Inc. Reproduced by permission.)

more tropical climates, may have two reproductive periods every year or even have an indistinct pattern in which reproductive females can be found in every month of the year.

New World opossums are polygamous. Like other marsupials, have a very short gestation period, followed by a long developmental period. After a gestation between 12 and 15 days, the embryo-like newborn, naked, with eyes and ears closed but strong arms and well-developed claws, crawls along a path between the cloaca and the marsupium (or mammary region in those species with no marsupium) that has been licked by the female. Once in the mammary region, each individual attaches to a nipple, where it will remain for four to eight weeks, depending on the species. Offspring can then detach from the nipples for the first time. When they are too large to fit in the female's pouch, they crawl on her back or are simply dragged behind her, while still attached to the mammae. Young remain dependent on their mothers until they are two to four months old, after which they are weaned and proceed to disperse.

Litter size varies greatly in different genera. Five to 12 and up to 16 offspring are born in *Monodelphis*, two to five in *Caluromys*, four to 12 in *Marmosa*, two to five in *Chironectes*, six to 15 in *Didelphis*, and one to nine in *Philander*. Likely, many more young are born than those found by scientists attached to the mammae, but they die before they can attach themselves to a nipple.

Sexual maturity is attained at three to nine months in different genera. Many species construct nests inside rotten trees both standing and fallen, while others have nests on the ground or, in the case of the water opossum, in tunnels excavated in stream banks. Some species of mouse opossum utilize hummingbird nests as their own resting places. New World opossums use primarily plant matter to construct the spherical nests. These materials are transported in the partially rolled-up tail while the animal moves to the nest.

Many species are semelparous with males dying shortly after mating and females after weaning their first and only lit-

A Virginia opossum (*Didelphis virginiana*) uses its tail as an aid in climbing down. (Photo by Animals Animals ©Ted Levin. Reproduced by permission.)

ter produced. Females of some species, however, can breed twice in the same year with only a few months apart between births, and individuals of a few species can have two years of reproductive activity.

Conservation status

Of the species recognized by Gardner in 1993, three (*Gracilinanus aceramarcae*, *Marmosa andersoni*, and *Marmosops handleyi*) are considered Critically Endangered by the IUCN, and three other species (*Marmosa xerophila*, *Marmosops cracens*, and *Monodelphis kunsi*) are considered Endangered. All three Crit-

ically Endangered species have extremely restricted distributions, limited to one or two localities in Bolivia or Colombia. Although none of these species face direct threats from humans, all are facing rampant habitat destruction that has strong negative conservation implications. The three Endangered species face the same threats of habitat destruction and restricted distributions in Colombia, Venezuela, Bolivia, and Brazil, albeit slightly larger than the Critically Endangered species. Fifteen additional species are considered Vulnerable, 18 more are under the category of Lower Risk/Least Concern, and two are Data Deficient. There are 20 species that have not been ranked by the IUCN. Undoubtedly other species are facing conservation threats, especially those with restricted distributions and found in habitats affected by high rates of deforestation, but much more information is necessary to correctly assess their status.

In Mexico, two species, the water opossum and the woolly opossum, are included in the list of species at risk as endangered. These two species are considered sensitive to habitat disruption and their populations have been severely decreased as a result of deforestation and water pollution by discharge of fertilizers and pesticides.

Significance to humans

Species such as the common opossums and even some four-eyed and mouse opossums frequently benefit from human-induced habitat changes. Some humans find opossum species attractive as pets, and their tanned pelts used to have some value in the fur market, especially at the end of the nineteenth century. In the tropics, mouse opossums and short-tailed opossums are valued for controlling of cockroaches, scorpions, and other unwanted animals, especially in rural settlements. Virginia, common, and four-eyed opossums are sometimes used as food by indigenous and other human populations. Colonies of some species, notably *Monodelphis*, are kept for developmental and biomedical research.

Opossums are sometimes considered pests because of their raids on commercially valuable fruits in orchards and agricultural fields, as well as on poultry farms. The southern opossum, *Didelphis marsupialis*, has been identified as one of the key hosts of the parasitic protozoan *Trypanosoma cruzi*, which causes Chagas' disease. Chagas' disease is transmitted to humans when an infected kissing, or assassin, bug (Hemiptera: Reduviidae; genus *Triatoma*) bites a human to feed on the blood and then defecates on the skin. The person then scratches the bite and transports the protozoans through an open wound into the body. Sixteen to 18 million people are infected, and 50,000 of these die annually. Other species of mammals have also been identified as hosts.

1. Bare-tailed woolly opossum (*Caluromys philander*); 2. Bushy-tailed opossum (*Glironia venusta*); 3. Water opossum (*Chironectes minimus*); 4. Patagonian opossum (*Lestodelphys halli*); 5. Virginia opossum (*Didelphis virginiana*); 6. Thick-tailed opossum (*Lutreolina crassicaudata*). (Illustration by Jonathan Higgins)

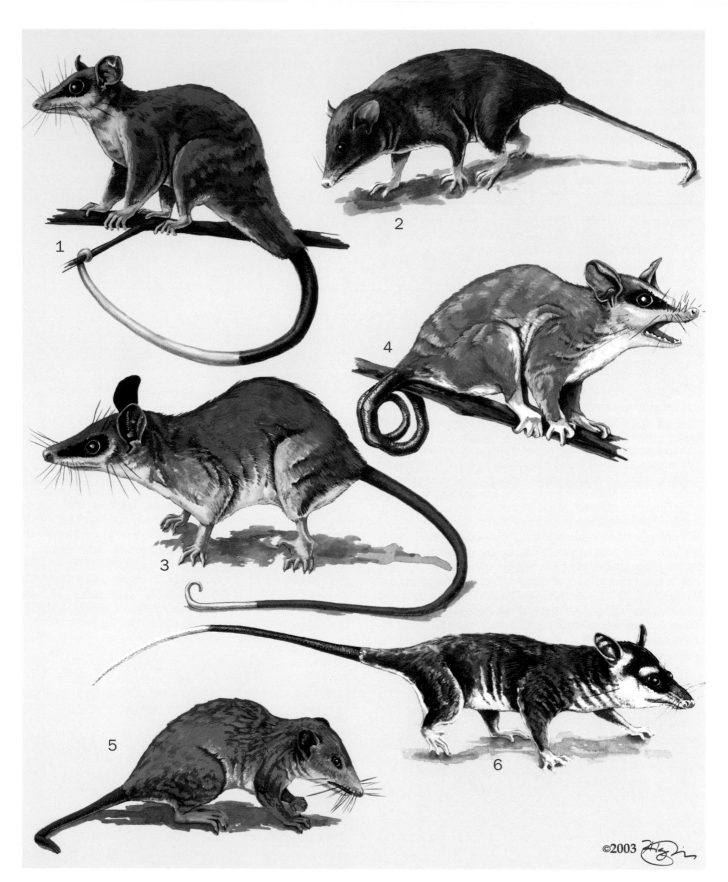

1. Alston's woolly mouse opossum (*Micoureus alstoni*); 2. Pygmy short-tailed opossum (*Monodelphis kunsi*); 3. Gray slender mouse opossum (*Marmosops incanus*); 4. Mexican mouse opossum (*Marmosa mexicana*); 5. Red-legged short-tailed opossum (*Monodelphis brevicaudata*); 6. Gray four-eyed opossum (*Philander opossum*). (Illustration by Jonathan Higgins)

Bushy-tailed opossum
Glironia venusta

SUBFAMILY
Didelphinae

TAXONOMY
Glironia venusta Thomas, 1912, Huánuco, Peru.

OTHER COMMON NAMES
German: Buschschwanzbeutelratten; Spanish: Comadreja de cola peluda, zarigüeya de cola peluda.

PHYSICAL CHARACTERISTICS
Length 6.3–8.3 in (16–21 cm). The hair on the back is dense and soft, uniformly cinnamon brown. The venter is grayish to brownish white. The tail is long and completely furred with only part of the ventral surface naked, which gives this species its common name. There are two large blackish patches surrounding the eyes, separated by a brown stripe along the top of the snout.

DISTRIBUTION
Eastern Amazonia in western Brazil, Ecuador, Peru, and Bolivia.

HABITAT
It has been found only in intact lowland tropical moist forest, up to an altitude of 2,600 ft (800 m).

BEHAVIOR
Not much is known about this species. Considered arboreal on the basis of the specimens found in trees and the morphology of the hand with well-developed grasping abilities.

FEEDING ECOLOGY AND DIET
Primarily a insectivorous species, but likely it also feeds on fruit, eggs, and small vertebrates.

REPRODUCTIVE BIOLOGY
Polygamous, but nothing else is known.

CONSERVATION STATUS
Considered Vulnerable by the IUCN. Habitat destruction is the primary threat. There is no information on effects of habitat loss.

SIGNIFICANCE TO HUMANS
None known. ◆

Patagonian opossum
Lestodelphys halli

SUBFAMILY
Didelphinae

TAXONOMY
Notodelphys halli (Thomas, 1921), Santa Cruz, Argentina.

OTHER COMMON NAMES
French: Opossum de Patagonie; German: Patagonien-Beutelratten; Spanish: Comadrejita patagónica.

PHYSICAL CHARACTERISTICS
Length 5–6 in (13–15 cm). The dorsal hair is dense and soft, dark grayish brown with paler sides. Males have an orange patch on the throat. The face is paler than the rest of the body. There are dark patches on shoulders and hips, and the underparts, hands, and feet are white. The tail is clearly shorter than the head and body, and seasonally it appears thick from fat reserves. Tail furry only at the base and covered with fine hairs the rest of its length. Canine teeth are relatively long.

DISTRIBUTION
Occurs in a relatively small region of southern Argentina in the provinces of Río Negro, Neuquén, Santa Cruz, La Pampa, Mendoza, and Chubut.

HABITAT
It has been reported from the South American steppe grasslands (pampas), and also from shrublands; often associated with streams and other water bodies.

BEHAVIOR
Seems to be a primarily terrestrial species. It is solitary and active at night.

FEEDING ECOLOGY AND DIET
A specimen was captured in a trap baited with a dead bird. This species is considered a carnivore but more likely it is insectivorous. Its diet may also include fruit, eggs, and small vertebrates.

REPRODUCTIVE BIOLOGY
Polygamous, but nothing else is known.

CONSERVATION STATUS
The distribution is restricted to a small region of southern Argentina. Classified as Vulnerable. Some portions of its habitat have been modified.

SIGNIFICANCE TO HUMANS
None known. ◆

☐ *Glironia venusta*
☐ *Marmosops incanus*

Thick-tailed opossum
Lutreolina crassicaudata

SUBFAMILY
Didelphinae

TAXONOMY
Didelphis crassicaudata (Desmarest, 1804), Asunción, Paraguay.

OTHER COMMON NAMES
English: Little water opossum; French: Opossum á queue grasse; German: Dickschwanzbeutelratte; Spanish: Comadreja colorada, coligrueso.

PHYSICAL CHARACTERISTICS
Length 10–16 in (25–40 cm); weight 7–19 oz (200–540 g). The dense, soft, and relatively short hair is uniformly light cinnamon to dark brown and paler below. The legs are relatively short and the body is elongated with a long neck; the ears are short; the tail is long and almost completely furred, except for the ventral surface. There is a well-developed pouch.

DISTRIBUTION
As understood in 2002, the distribution is disjunct, with one population occurring in eastern Colombia, southern Venezuela, and Guyana, and another in eastern Bolivia, northeastern Argentina, southern Brazil, Paraguay, and Uruguay, from an altitude of 1,970–6,560 ft (600–2,000 m).

HABITAT
Found in lowland and mid-elevation tropical moist forests, grasslands, and shrublands, as well as in forest edges. Always associated with streams and rivers.

BEHAVIOR
Roosts in hollows in trees, dens of other animals, and nests constructed among the vegetation. An excellent swimmer and also a good climber, this is a nocturnal species. It is apparently the only species of didelphid that can be accommodated in captivity in small groups, with a weak social structure that permits coexistence of two to three animals.

FEEDING ECOLOGY AND DIET
Primarily carnivorous, feeding on small vertebrates on land and in the water, as well as crustaceans, insects, and other small animals. An antibothropic biochemical factor has been isolated from its blood, indicating some level of immunity to the venom of snakes.

REPRODUCTIVE BIOLOGY
Polygamous. Gestation lasts about two weeks. Females give birth to the young in a very undeveloped state. These crawl into the pouch where they attach themselves to a nipple. Births occur twice during the year.

CONSERVATION STATUS
Not listed by the IUCN. It seems to be adaptable to a certain degree of disturbance and can be locally common in some areas.

SIGNIFICANCE TO HUMANS
Sometimes considered a nuisance because of its occasional raids on henhouses. ◆

Mexican mouse opossum
Marmosa mexicana

SUBFAMILY
Didelphinae

TAXONOMY
Marmosa murina mexicana Merriam, 1897, Oaxaca, Mexico.

OTHER COMMON NAMES
Spanish: Ratón tlacuache, tacuazín.

PHYSICAL CHARACTERISTICS
Length 4.7–6.7 in (12–17 cm); weight 0.9–3.2 oz (26–92 g). The back is uniformly reddish brown to grayish brown. The tail, which is prehensile, is about as long as the head and body and 90% of its length is naked. The feet and undersides are paler, sometimes white. There are two large black patches surrounding the large, black eyes. The ears are large, rounded, and naked.

DISTRIBUTION
From eastern and southern Mexico south through central America to western Panama, from sea level up to about 5,900 ft (1,800 m).

HABITAT
The main habitat is tropical moist forest, but it can also be found in dry tropical forest, secondary forests and disturbed vegetation, and mangrove forests.

BEHAVIOR
A nocturnal species that is primarily arboreal, although it can also be found on the ground. The mouse opossum readily eats in captivity, quickly attacking and consuming any large insects, eggs, or small vertebrates. It has been found resting inside abandoned hummingbird nests.

FEEDING ECOLOGY AND DIET
Primarily insectivorous but also eats some fruit, bird eggs, and nestlings, and other prey similar in size.

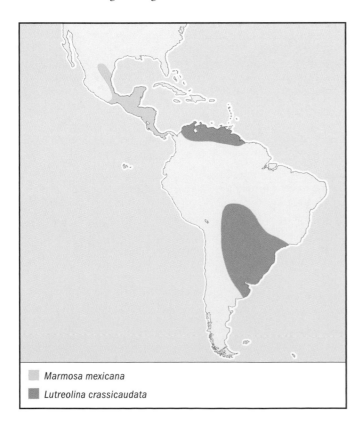

Marmosa mexicana

Lutreolina crassicaudata

REPRODUCTIVE BIOLOGY
Polygamous. Gestation is about 14 days long. The females do not have a marsupium, so the newborn crawl to the mammae and attach themselves to the nipples, which may number from 11 to 15. As they grow larger, the young begin traveling on the mother's back, sometimes curling their tails around hers. Reproductive season extends at least from March through August.

CONSERVATION STATUS
The species is not listed by the IUCN. Since it is found in both undisturbed and modified habitats, it probably is not facing major conservation problems.

SIGNIFICANCE TO HUMANS
Sometimes these opossums are kept as pets in rural communities. Some individuals have been found as stowaways in banana shipments to New York City. ◆

Gray slender mouse opossum
Marmosops incanus

SUBFAMILY
Didelphinae

TAXONOMY
Didelphis incana (Lund, 1840), Minas Gerais, Brazil.

OTHER COMMON NAMES
None known.

PHYSICAL CHARACTERISTICS
Length 4.7–7.9 in (12–20 cm); weight 1–1.9 oz (30–55 g). Dorsal fur is grayish brown to brown and underparts paler. The snout is long and pointed and the ears are large, naked, and pink to brown. There is a patch of black hair around each eye. The tail is long and naked.

DISTRIBUTION
Endemic to eastern Brazil from the state of Bahia to Parana.

HABITAT
Found in the Atlantic forest of the southeastern coast of Brazil, below 2,640 ft (800 m), in tropical moist forest and deeper inland in somewhat drier forest.

BEHAVIOR
Nocturnal and crepuscular. Solitary.

FEEDING ECOLOGY AND DIET
Feeds on insects, eggs, fruit, and small vertebrates.

REPRODUCTIVE BIOLOGY
Polygamous. A semelparous species in which individuals invest all their effort on a single reproductive event in their lives. Typically, the births occur in a single three-month period around the onset of the rainy season.

CONSERVATION STATUS
Considered Lower Risk/Near Threatened by the IUCN. The Atlantic forest of Brazil is one of the most endangered biomes of South America and, if deforestation trends continue, this and other species will face very serious extinction risks.

SIGNIFICANCE TO HUMANS
None known. ◆

Alston's woolly mouse opossum
Micoureus alstoni

SUBFAMILY
Didelphinae

TAXONOMY
Caluromys alstoni (J. A. Allen, 1900), Cartago, Costa Rica.

OTHER COMMON NAMES
French: Souris-opossum laineuse d'Alston; Spanish: Zorricí.

PHYSICAL CHARACTERISTICS
Length 6.7–7.9 in (17–20 cm); weight 2–5.3 oz (60–150 g). One of the largest mouse opossums. The dorsal hair is yellowish brown to reddish; underparts are paler. Distinct black mask over each eye. The tail is long, slender, and naked. There is no marsupium. The feet have strongly opposable thumbs.

DISTRIBUTION
Caribbean coast of Central America from Belize to Panama and into Colombia, and also in some Caribbean coastal islands.

HABITAT
Inhabits lowland tropical moist forest and cloud forest below 5,250 ft (1,600 m), and also areas of secondary forest.

BEHAVIOR
Nocturnal and solitary. Primarily moves in tree canopies and from branch to branch, but can also be found occasionally on the ground.

FEEDING ECOLOGY AND DIET
Polygamous. Feeds on insects, eggs, fruit, and small vertebrates. Attacks its prey by grasping it with its hands and apply-

▢ *Micoureus alstoni*
▮ *Philander opossum*

ing a series of quick bites all over the body of the prey. Then it is consumed from the head down. Legs and wings of insects are discarded.

REPRODUCTIVE BIOLOGY
Polygamous. Females build spherical nests with vegetation and debris. After a short gestation of less than two weeks, the young are born in a very undeveloped state. Births have been reported almost in every month of the year. Litter size varies from two to about 14 young.

CONSERVATION STATUS
The IUCN classified this species as Lower Risk/Near Threatened. Severe deforestation in Central America is likely having a strong negative impact on this and other species of the region, although it has been found in some protected areas in Costa Rica.

SIGNIFICANCE TO HUMANS
None known. ◆

Red-legged short-tailed opossum
Monodelphis brevicaudata

SUBFAMILY
Didelphinae

TAXONOMY
Didelphis brevicaudata (Erxleben, 1777), Surinam.

OTHER COMMON NAMES
German: Kurzschwanz-Spitzmausbeutelratte; Spanish: Colicorto de patas rojas.

Monodelphis kunsi
Monodelphis brevicaudata

PHYSICAL CHARACTERISTICS
Length 4.7–7 in (12–18 cm); weight 1.6–3.5 oz (45–100 g). Legs are relatively short as is the tail. Hair is relatively rigid, short, and dense. Color pattern is variable. The back can be dark gray, grizzled with pale hair tips. The sides are rich deep reddish and the change between the gray and red is a very sharp line. Underparts vary from pale, almost white, to light brown. Some individuals are reddish all over the back. The snout is conical, and the ears are rounded, dark, and naked. The tail is relatively short and hairy only at the base and dorsal surface of the basal one-third.

DISTRIBUTION
Venezuela, Guyana, Suriname, French Guiana, and Brazil north of the Amazon river. This species seems restricted to the northern half of the Amazon river basin.

HABITAT
Tropical moist forest in pristine and modified conditions. Can also be found in orchards and other agroecosystems, but usually upland and away from rivers and streams. It has been found only below 2,620 ft (800 m).

BEHAVIOR
This is among the few diurnal species in the family. It is also one of the least adapted to climbing and moving in trees; the majority of reports describe it as a terrestrial or strictly terrestrial species.

FEEDING ECOLOGY AND DIET
Feeds primarily on insects, rodents, birds, eggs, and some fruit. It hunts opportunistically, running and searching through the forest floor under the litter, under logs, and in hollow trees and fallen logs.

REPRODUCTIVE BIOLOGY
Polygamous. Breeding has been reported almost in every month of the year. Nests are built with plant matter inside or under fallen or hollow logs. The young are born after a gestation of about 15 days. Litters number between five and 12 young. As there is no marsupium, the young are attached to the female's nipples and are dragged under her. As they grow, they begin to travel clinging to their mother's back or sides.

CONSERVATION STATUS
This species is not listed by the IUCN. Although it has been found mainly in undisturbed forest, it seems able to survive also in modified forests and agroecosystems.

SIGNIFICANCE TO HUMANS
None known. ◆

Pygmy short-tailed opossum
Monodelphis kunsi

SUBFAMILY
Didelphinae

TAXONOMY
Monodelphis kunsi Pine, 1975, Beni, Bolivia.

OTHER COMMON NAMES
French: Opossum á queue courte de Kuns; German: Kuns-Spitzmausbeutelratte; Spanish: Colicorto pigmeo.

PHYSICAL CHARACTERISTICS
Length 2.4–3.1 in (6–8 cm); weight 0.6–1.2 oz (18–35 g). The back is uniform yellowish brown with paler underparts. The fur is short and dense. There is a gular gland in males, partially obscured by fur. The tail is short, slender, and almost completely covered with fine, sparse hair, except for its tip. The rostrum is conical and the eyes black. The ears are medium and naked. This is the smallest species in the genus *Monodelphis*.

DISTRIBUTION
Known only from a few localities in western Brazil and western Bolivia, apparently endemic to the upper Amazon river basin, below 2,100 ft (640 m).

HABITAT
Tropical moist forest, and maybe also some secondary vegetation.

BEHAVIOR
Probably solitary. It may also be nocturnal, but almost nothing is known about this species.

FEEDING ECOLOGY AND DIET
Presumably feeds on insects and other small animals.

REPRODUCTIVE BIOLOGY
Polygamous, but nothing else is known.

CONSERVATION STATUS
Classified by the IUCN as Endangered. Since this is a relatively rare species in nature, and western Amazonia is subject to major deforestation, the species faces serious extinction threats.

SIGNIFICANCE TO HUMANS
None known. ◆

Gray four-eyed opossum
Philander opossum

SUBFAMILY
Didelphinae

TAXONOMY
Didelphis opossum (Linnaeus, 1758), Paramaribo, Surinam.

OTHER COMMON NAMES
French: Opossum á quatre yeux; German: Vieraugenbeutelratte; Spanish: Tlacuache cuatro ojos, zorro cuatro ojos, comadreja cuatro ojos.

PHYSICAL CHARACTERISTICS
Length 8–13 in (20–33 cm); weight 7–24.7 oz (200–700 g). This relatively large opossum has dense and relatively short hair that varies from pale gray to dark gray dorsally and yellowish white ventrally. The cheeks and chin are also yellowish white, as are two conspicuous spots just above the eyes, which give this species its common name. The ears are large and naked, blackish with pink bases. There is a marsupial pouch that stains orange if the female has had young. The tail in most individuals is bicolored with the base dark and the final third to half white, naked, and scaly except the basal 0.8 in (2 cm), which are furred.

DISTRIBUTION
From eastern and southern Mexico south through Central America and into South America to Bolivia, Paraguay, northeastern Argentina, and southeastern Brazil.

HABITAT
Inhabits dense tropical moist forests, secondary areas, orchards, and other agricultural and modified ecosystems. It has been recorded from sea level up to about 5,400 ft (1,650 m). It is most abundant near streams, rivers, and other water bodies.

BEHAVIOR
When threatened, it hisses and gasps, snapping at the intruder. This is a mostly nocturnal species but sometimes may be active by day. Although most of the time it stays on the ground, it may also climb into trees. Like other didelphids, gray four-eyed opossums are solitary. When it is asleep, it rolls into a ball and the eyes are not visible, but the bright spots above the eyes give the appearance of an awake and alert opossum.

FEEDING ECOLOGY AND DIET
The four-eyed opossum is omnivorous. It feeds on many species of tropical and introduced fruits, corn, palm fruits, flower nectar, frogs, birds, rodents, and other small vertebrates, snails, insects, crustaceans, and carrion. Insects are most frequently eaten in the dry season. They tend to take fruits that are larger than 2 in (5 cm) in diameter and that are fleshy, juicy, with high contents of sugars or lipids, and low levels of nitrogen.

REPRODUCTIVE BIOLOGY
Polygamous. Breeds throughout the year or seasonally, depending on the region. Females have seven mammae. Although there may be more babies born than this number, a maximum of seven can be found in the pouch. Females may have more than one birth per year. They construct a nest of litter and other vegetation in tree limbs, in hollows in standing or fallen trees, or in dens underground. The nest is spherical and about 11.8 in (30 cm) in diameter. Litters of one to seven with an average of about five are reported. Females are sexually receptive when they are seven months old. Maximum longevity is about 2.5 years in the wild and 3.5 years in captivity.

CONSERVATION STATUS
This is generally an abundant species that can live in both pristine and disturbed conditions. It can also live in houses in rural areas. The IUCN has not listed it, and it does not seem to be threatened.

SIGNIFICANCE TO HUMANS
Sometimes considered a nuisance because it may raid henhouses. This species and other didelphids are reservoirs of *Trypanosoma cruzi*, the protozoan that causes Chagas' disease.

Common name / Scientific name/ Other common names	Physical characteristics	Habitat and behavior	Distribution	Diet	Conservation status
Black-shouldered opossum *Caluromysiops irrupta* French: Opossum àépaules noires; German: Bindenwollbeutelratten; Spanish: Cuica de hombros negros	Pearl gray dorsal fur with black patches extending from the shoulders down to each forefoot. From the shoulder patches, two parallel black lines run down the back to the rump. Tail is sparsely covered with hair and grayish black except for the final few inches (centimeters) which are white.	Nocturnal and solitary. Primarily arboreal species. Lives in undisturbed tropical moist forests.	Southeastern Peru and extreme western Brazil in the upper Amazon River basin.	Fruit, insects, eggs, and small vertebrates.	Vulnerable
Brown four-eyed opossum *Metachirus nudicaudatus* French: Brun opossum àyeux; German: Nacktschwanzbeutelratte; Spanish: Tlacuache cuatro ojos café, zorricí, cuica; Portuguese: Jupati	Uniform brown dorsum. Pale spots just above the eyes that give the appearance of a second set of eyes. Long, thin, tapered tail, which is naked and virtually unicolored. Length 8.3–11.8 in (21–30 cm).	Solitary and nocturnal. Mostly found in tropical moist forests. Rarely goes into trees, generally running along the forest floor. Sea level to bout 3,900 ft (1,200 m).	Extreme southern Mexico through Central America south to southern Brazil, Paraguay, and northern Argentina.	Arthropods, bird eggs, nestlings and other small vertebrates, fruit.	Not listed by IUCN
Southern opossum *Didelphis marsupialis* English: Common opossum; German: Nordopossum; Spanish: Tlacuache común, cuica, zorro; Portuguese: Gambá, saruê	Dark gray or blackish to pale gray with long, dense fur. The tail is long and naked, bicolor with the basal half black and the rest whitish. Length 12.6–19.7 in (32–50 cm).	Solitary and nocturnal. Terrestrial and arboreal. Movement in trees is aided by a prehensile tail. Females carry young in their pouch. Found in moist and dry tropical forests, cloud forests, semidesertic habitats, secondary vegetation, agricultural lands, and edges of towns and cities. Important as a reservoir of *Trypanosoma cruzi* which causes Chagas disease.	Central and eastern Mexico south to eastern Peru, northern Bolivia, and Brazil.	Omnivorous; feeds on fruit, insects and other invertebrates, small vertebrates, and carrion.	Not threatened
Northern gracile mouse opossum *Gracilinanus marica* Spanish: Chuchita costeña	Uniform reddish brown pelage on the back. Underparts paler. One large black patch over each eye. There is no pouch.	Solitary and nocturnal. Primarily arboreal. Lives in tropical and subtropical moist and dry forests and even in savannas, at an altitude of 4,920–8,530 ft (1,500–2,600 m).	Northern Colombia and western Venezuela.	Insects and other arthropods, eggs, and fruit.	Lower Risk/Near Threatened
Grayish mouse opossum *Marmosa canescens* French: Souris-opossum grisâtre; Spanish: Tlacuatzin; ratón tlacuache	Small opossum with grayish brown dorsal hair and a long, slender, tapered, prehensile, and naked tail. Two large black patches surround the eyes. Length 2.4–4.3 in (6–11 cm).	Solitary and nocturnal, mainly arboreal species. Makes spherical nests with vegetation among tree branches or in hollow trees, but has been found roosting inside hummingbird nests. Feeds primarily on insects but also on fruit, eggs, and small vertebrates. Found in tropical dry forest from sea level up to about 5,250 ft (1,600 m).	Endemic to western Mexico.	Fruit, arthropods, bird eggs and nestlings; also small animals such as snakes, bats, and lizards.	Data Deficient
Elegant fat-tailed opossum *Thylamys elegans* English: Chilean mouse opossum; French: Opossum àqueue adipeuse elegant; German: Elegantes Fettschwanzopossum; Spanish: Yaca, llaca	The hair is thick and dense, pale brown on the sides with a wide dark stripe along the back and paler underparts. The rostrum is short and conical and the ears medium sized and naked. The tail is almost completely naked and seasonally it is used to store fat reserves. Dark and rust-brown with white-tipped tail. Length 4.7–5.5 in (12–14 cm).	Solitary, semiarboreal, and nocturnal. This species lives in cloud forest, temperate southern rainforest, and scrub associated with forest edges. They build nests with plant matter and hairs among tree or shrub branches, or underground.	Southwestern coastal Peru and western Chile.	90% of their diet is composed of insects but they also eat fruit, eggs, and small vertebrates.	Not listed by IUCN

[continued]

Common name / Scientific name/ Other common names	Physical characteristics	Habitat and behavior	Distribution	Diet	Conservation status
Small fat-tailed opossum *Thylamys pusilla* English: White-bellied fat-tailed mouse opossum; French: Petit opossum à queue adipeuse; German: Kleines Fettschwanzopossum; Spanish: Comadrejita, marmosa común	Gray to brown on the back and white underparts. Tail relatively long, almost naked. It can store reserves and appear fatter seasonally. Length 4–5.5 in (10–14 cm).	Solitary, semiarboreal, and nocturnal. Found in the Andean piedmont and further above, from desert areas to submontane habitats.	Southern Brazil and Bolivia, Paraguay, and northern and central Argentina.	Mostly insects and other arthropods, but also fruit, eggs, and small vertebrates.	Not listed by IUCN
Slaty slender mouse opossum *Marmosops invictus* English: Slaty mouse opossum; French: Opossum schisteux de souris; Spanish: Marmosa de montaña	Dorsal hairs reddish brown to dark brown and paler underparts. Tail uniformly dark brown. Black patches on eyes. Length 4.7–5.9 in (12–15 cm).	Solitary, semiarboreal, nocturnal species. Found in rainforest at intermediate altitudes.	Eastern Panama, Colombia, and Venezuela.	Mostly insectivorous but also eats fruit, eggs, and small vertebrates.	Lower Risk/Near Threatened
Southern short-tailed opossum *Monodelphis dimidiata* English: Yellow-sided opossum, eastern short-tailed opossum; Spanish: Colicorto pampeano	Unicolored gray-brown on the back, with a short tail that is nearly naked. The snout is elongated and conical, and the ears are relatively short. Length 4.3–7.9 in (11–20 cm).	Solitary, terrestrial, diurnal. Found in pampas grasslands, induced grasslands, and wetlands. Males and females reproduce only once in their lifetime.	Southern Brazil, Uruguay, and Argentina.	Mostly insectivorous and not too carnivorous. Probably feeds also on fruit and eggs.	Lower Risk/Near Threatened
Black four-eyed opossum *Philander andersoni* Spanish: Comadreja cuatro ojos negra	Grayish black sides and a black stripe down the center of the dorsum. Two bright pale spots above the eyes that resemble a second pair of eyes. Tail as long as the head and body, naked, and bicolor: basal half dark, the rest white. Length 9–11 in (23–28 cm).	Solitary and nocturnal. Primarily terrestrial but also climbs trees. Lives in lowland tropical moist forest. Females have a pouch to protect young.	Northern and western Amazon basin, from Venezuela and eastern Colombia south through western Brazil, Ecuador, and Peru.	Omnivorous; feeds on insects, small vertebrates, eggs, and fruit.	Not listed by IUCN
Orange mouse opossum *Marmosa xerophila* English: Dryland mouse opossum; French: Souris-opossum orange; Spanish: Marmosa del desierto, comadrejita de los desiertos	The dorsum and flanks are orange-yellow and underparts white. Large, naked ears. Black patches on the eyes. Long, naked, semiprehensile tail. 3.5–5.5 in (9–14 cm).	Solitary and nocturnal. Mostly arboreal but readily takes to the ground. Inhabits desert and semidesert areas.	Northeastern Colombia and northwestern Venezuela.	Primarily insectivorous. Also feeds on birds' eggs and fruit.	Endangered

Resources

Books

Benton, M. J. *The Rise of the Mammals.* New York: Crescent Books, 1991.

Collins, L. *Monotremes and Marsupials.* Washington, DC: Smithsonian Institution Press, 1973.

Eisenberg, J. F. *Mammals of the Neotropics.* Vol.1, *The Northern Neotropics.* Chicago: University of Chicago Press, 1990.

Eisenberg, J. F., and K. H. Redford. *Mammals of the Neotropics.* Vol. 3, *The Central Neotropics.* Chicago: University of Chicago Press, 1999.

Hunsaker II, D. *The Biology of Marsupials.* New York: Academic Press, 1977.

Nowak, R. M. *Walker's Mammals of the World.* 6th ed., Vol. 1, Baltimore: John Hopkins University Press, 1999.

Redford, K. H., and J. F. Eisenberg. *Mammals of the Neotropics.* Vol. 2, *The Southern Cone.* Chicago: University of Chicago Press, 1992.

Wilson, D. E., and D. M. Reeder. *Mammal Species of the World.* Washington: Smithsonian Institution Press, 1993.

Periodicals

Alonso-Mejía, A., and R. A. Medellín. "*Marmosa mexicana.*" *Mammalian Species* 421 (1992): 1–4.

Caceres, N. C. "Food Habits and Seed Dispersal by the White-eared Opossum, *Didelphis albiventris,* in Southern Brazil." *Studies on Neotropical Fauna and Environment* 37 (2002): 97–104.

Castro, I., H. Zarza, and R. A. Medellín. "*Philander* opossum." *Mammalian Species* 638 (2000): 1–8.

Lemos, B., G. Marroig, and R. Cerqueira. "Evolutionary Rates and Stabilizing Selection in Large-bodied Opossum Skulls (Didelphimorphia: Didelphidae)." *Journal of Zoology* 255 (2001): 181–189.

Marshall, L. G. "*Chironectes minimus.*" *Mammalian Species* 109 (1978): 1–6.

———. "*Gironia venusta.*" *Mammalian Species* 107 (1978): 1–3.

———. "*Lestodelphys halli.*" *Mammalian Species* 81 (1977): 1–3.

McManus, J. J. "*Didelphis virginiana.*" *Mammalian Species* 40 (1974): 1–6.

Medellín, R. A. "Seed Dispersal of *Cecropia obtusifolia* by Two Species of Opossums." *Biotropica* 26 (1994): 400–407.

Rodrigo A. Medellín, PhD

Paucituberculata
Shrew opossums
(Caenolestidae)

Class Mammalia

Order Paucituberculata

Family Caenolestidae

Number of families 1

Thumbnail description
Small, shrewlike animals with small eyes, shaggy fur, and a long tail; females lack a pouch

Size
Head-body 3.5–5 in (9–13 cm); tail 2.5–5 in (6.5–13 cm); weight 0.7–1.5 oz (20–40 g)

Number of genera, species
2 genera; 5 species

Habitat
Temperate rainforest and alpine scrub bordering high altitude paramo meadow

Conservation status
Vulnerable: 1 species

Distribution
Western South America, including parts of Ecuador, Colombia, Venezuela, Peru, Chile, and Argentina

Evolution and systematics

The order Paucituberculata is represented by just a single living family, the Caenolestidae, with just two genera. The handful of living species is all that remains of what appears to have been a once-abundant marsupial dynasty during the Oligocene epoch, approximately 25 million years ago. There are seven extinct families, some of which were described from fossils before the living caenolestids were discovered, including the Groberiidae. Some of the earliest fossils, members of the family Polydolopidae, date back to the Palaeocene, more than 60 million years ago.

The decline of the paucituberculates began in the Oligocene, and it gathered pace in the Miocene, when the continents of North and South America were briefly joined. There was an influx of placental mammals from the north into what had been for many millions of years a bastion of marsupial diversity. The newcomers included rodents and primates whose descendants have since thrived at the expense of many ousted marsupials.

A number of similarities with other American marsupials has led some authorities to consider the shrew opossums to be no more than a subgroup of the order Didelphimorphia. However, molecular evidence supports the classification used here—it may be that at one time the Paucituberculata were as diverse as the extant Australian order Diprotodontia.

Until recently, there were thought to be three living caenolestid genera. The third, *Lestoros*, contained the species, *L. inca*, and is now considered part of the larger genus, *Caenolestes.*

Physical characteristics

Shrew opossums are all rather similar looking. The largest specimens (usually males) are no more than 10 in (25 cm) long, half of which is tail, except in the Chilean shrew opossum, whose tail is relatively short. The face is long and tapering, with a pointed snout, long whiskers, and very small eyes. The ears are quite large and project well beyond the animal's fur. The overall appearance is comparable to a rat or overgrown shrew.

Distribution

The living shrew opossum species all derive from western South America, from the coast to the high Andes.

Habitat

The preferred habit of most shrew opossums is densely vegetated and humid–temperate rainforest, montane woodland, or the lush high altitude meadows of the Andean paramo. Such habitats offer cover and an abundance of suitable nest sites. They also support a rich invertebrate fauna and thus present good feeding opportunities.

Behavior

The shrew opossums' small size and the remote nature of their habitats mean they have little contact with humans; consequently, details of their ecology and behavior are not well known. They are all largely nocturnal or crepuscular, and spend the day resting in hollow logs or holes around the roots of trees. They appear to live alone, and travel around the home range on regular runways. They are not territorial, and more than one individual use the paths. They are proficient climbers, and will scramble into the branches of shrubs, using the tail as a prop and a counterbalance, but not for grasping.

Feeding ecology and diet

The diet is predominantly insects and other invertebrates such as earthworms, which the shrew opossums find by rummaging in the surface litter and investigating likely nooks and crannies along their regular runways. Like placental shrews, they hunt mainly by smell; they also have sharp hearing, but their eyesight is relatively poor. Various species have also been reported feeding opportunistically on fruit, scavenging the flesh of dead animals, and killing and eating the young of other mammals such as rats found in nests.

No torpid or hibernating specimens have been recorded, but there is good evidence that shrew opossums can become torpid in times of food shortage. The Chilean shrew opossum builds up reserves of fat in its tail in late summer, which help it survive the cold Andean winter.

Reproductive biology

Very little is known about caenolestid reproduction, including mating system. Normal litter sizes can be guessed at from the number of teats—four in members of the genus, *Caenolestes*, five or seven in the *Rhyncholestes* species. Females do not have a pouch, so nursing young must latch on firmly to the mother's teat in order to survive. Once milk is flowing, the teat swells in the infant's mouth so that it cannot easily become detached. When the young become too big to be carried beneath the mother's body, she presumably leaves them in a nest while she goes out to feed, returning periodically to suckle them.

Conservation status

The Chilean shrew opossum (*Rhyncholestes raphanurus*) is listed as Vulnerable by the IUCN. Information regarding shrew opossum population density is very limited, since few specimens have ever been collected and field sightings are not well known. However, the temperate rainforest in which the Chilean shrew opossum lives is shrinking quickly, and other species in the same habitat have undergone a significant decline. *Caenolestes* shrew opossums are probably more secure.

Significance to humans

None known.

1. Silky shrew opossum (*Caenolestes fuliginosus*); 2. Chilean shrew opossum (*Rhyncholestes raphanurus*). (Illustration by Brian Cressman)

Species accounts

Silky shrew opossum
Caenolestes fuliginosus

TAXONOMY
Hyracodon fuliginosus (Tomes, 1863), Ecuador.

OTHER COMMON NAMES
English: Ecuador shrew opossum; French: Caenolestidé d'Ecuador; German: Ekuador-Opossumaus; Spanish: Ratón musarana de los Andes.

PHYSICAL CHARACTERISTICS
Head and body 3.5–5.5 in (9–13 cm) long; tail 3.5–5.5 in (9–13 cm) long; fur brown, soft, and shaggy, toes bear small sharp claws.

DISTRIBUTION
Alpine regions of northern Colombia, Ecuador, and western Venezuela.

HABITAT
Cool, humid thickets in alpine forests and meadows.

BEHAVIOR
Nocturnal, terrestrial, fast-moving; an excellent climber.

FEEDING ECOLOGY AND DIET
Mostly insect larvae, will also take young vertebrate prey and carrion.

REPRODUCTIVE BIOLOGY
Not well known, but young are probably born between June and September in litters of two to four.

CONSERVATION STATUS
Not listed by the IUCN.

SIGNIFICANCE TO HUMANS
None known. ◆

Chilean shrew opossum
Rhyncholestes raphanurus

TAXONOMY
Rhyncholestes raphanurus Osgood, 1924, Chiloe Island, Biobio, Chile.

OTHER COMMON NAMES
French: Caenolestidé du Chili; German: Chile-Opossumaus; Spanish: Comadrejita trompuda.

PHYSICAL CHARACTERISTICS
Head and body 4–5 in (10–13 cm) long; tail 2.5–3.5 in (6–9 cm); may fatten prior to onset of winter. Female has five or seven teats.

DISTRIBUTION
South central Chile and Chiloe Island.

HABITAT
Temperate rainforest.

BEHAVIOR
Nocturnal, terrestrial, burrows through surface litter; may enter torpor in winter.

FEEDING ECOLOGY AND DIET
Insects, worms, and other invertebrates caught by rummaging in topsoil and leaf litter.

REPRODUCTIVE BIOLOGY
Not well understood, but females are apparently capable of breeding at any time of year.

CONSERVATION STATUS
Listed as Vulnerable by IUCN.

SIGNIFICANCE TO HUMANS
None known.

Caenolestes fuliginosus
Rhyncholestes raphanurus

Resources

Books

Aplin, K. P., and M. Archer. "Recent Advances in Marsupial Systematics with a New Syncretic Classification." In *Possums and Opossums: Studies in Evolution*. Sydney, Australia: Surrey Beatty and Sons & Royal Zoological Society of New South Wales, 1987.

Feldhamer, G. A., L. C. Drickamer, S. H. Vessey, and J. F. Merritt. *Mammalogy: Adaptation, Diversity, and Ecology*. Boston: WCB McGraw-Hill, 1999.

Macdonald, D. *The New Encyclopedia of Mammals*. Oxford University Press, 2001.

Nowak, R. "Order Paucituberculata." In *Walker's Mammals of the World*. Vol. I, 6th ed. Baltimore and London: Johns Hopkins University Press, 1999.

Amy-Jane Beer, PhD

Significance to humans

The monito del monte is the subject of some extreme superstitions. Some Andean people regard it as a bearer of ill fortune. Being nimble and inquisitive, and interested in stored food such as fruit, monitos inevitably enter houses from time to time. They do little real harm, but in some places a "colo-colo" indoors is believed to be such bad luck that the only way to avert disaster is to move out and destroy the house completely. In actual fact, these animals probably do more good than harm by feeding on a variety of insect pests.

Resources

Books

Aplin, K. P., and M. Archer. "Recent Advances in Marsupial Systematics with a New Syncretic Classification." In *Possums and Opossums: Studies in Evolution.* Sydney: Surrey Beatty and Sons & Royal Zoological Society of New South Wales, 1987.

Feldhamer, G. A., L. C. Drickamer, S. H. Vessey, and J. F. Merritt. *Mammalogy: Adaptation, Diversity, and Ecology.* Boston: WCB McGraw-Hill, 1999.

Nowak, R. "Order Microbiotheria." In *Walker's Mammals of the World.* Vol. I, 6th ed. Baltimore and London: Johns Hopkins University Press, 1999.

Amy-Jane Beer, PhD

Dasyuromorphia

(Australasian carnivorous marsupials)

Class Mammalia

Order Dasyuromorphia

Number of families 3

Number of genera, species 23 genera; 71 species

Photo: The yellow-footed antechinus (*Antechinus flavipes*) forages in a rockpile in Warwick, Queensland, Australia. (Photo by B. G. Thomson/Photo Researchers, Inc. Reproduced by permission.)

Evolution and systematics

The order Dasyuromorphia includes three families of carnivorous marsupials in the superfamily Dasyuroidea: the Dasyuridae (dasyures), the Myrmecobiidae (numbat), and the Thylacinidae (thylacines). The dasyurids and thylacinids are more closely related to each other than they are to the numbat. The Australian marsupial radiation produced a number of other species of carnivorous marsupials in the otherwise herbivorous order Diprotodontia. These include two genera (*Thylacoleo* and *Wakaleo*) and seven species of large, up to 220 lb (100 kg), predatory marsupial lions of the family Thylacoleonidae, which are most closely related to koalas and wombats (superfamily Vombatoidea), and an omnivorous (partly flesh-eating) giant rat-kangaroo, in the subfamily Propleopinae, family Hypsiprymnodontidae, superfamily Macropodoidea.

The earliest known carnivorous marsupial in Australia, *Djarthia murgonensis*, comes from the early Eocene (55 million years ago [mya]). The taxonomic affiliation of this and two other early carnivorous marsupials is not certain, as key anatomical features used to clearly identify them to family level or even to separate them from the South American marsupial fauna are lacking in the fossils found to date. The Australian dasyuromorphian and American marsupial taxa are quite distinct but are allied in the possession of many incisors (polyprotodonty) which distinguish them from the herbivorous Diprotodontia. Dasyuromorphians originated in the late Oligocene. The early radiation comprised the very conservative or "primitive" thylacinids. Ranging in size from small dog-sized, 70.5–176.4 oz (2–5 kg) to slightly larger than 65 lb (30 kg), thylacines dominated the Australian carnivorous marsupial fauna until the late Miocene, after which they steadily declined to two species present in the Pleistocene and only one persisting until historic times. Dasyurids first appeared in the fossil record in the early to middle Miocene but were rare until the late Miocene when they diversified and replaced the thylacine as the dominant marsupial carnivore fauna. Most of the Pleistocene fossil dasyurids are from still-living taxa, although none of the living groups occur earlier than the Pliocene. Dasyurids are considered to be highly specialized or "derived" dasyuromorphians in terms of their morphology. The numbats are represented by only one living species which appeared in the fossil record as recently as the Pleistocene. The numbat is a highly specialised dasyuromorphian, with features of the skull, teeth, and tongue adapted for termite feeding.

Physical characteristics

Dasyuromorphians are quadrupedal (move on four legs), with four toes on the front feet, four or five toes (including

The Tasmanian devil (*Sarcophilus laniarius*) has teeth capable of crushing bone. (Photo by © Erwin & Peggy Bauer. Bruce Coleman, Inc. Reproduced by permission.)

in dasyures. All molar teeth (four in thylacines and dasyures; five in numbats) are similar in form, in contrast with the differentiation of slicing and grinding functions into separate teeth in the placental carnivores. Thylacine and dasyure molars each retain meat-slicing (carnassial) cusps and function, and grinding surfaces. This marsupial feature may be a consequence of reproductive mode. Permanent teat-attachment during tooth development appears to suppress the eruption of the deciduous teeth, which remain vestigial, leading to early eruption of the permanent dentition. Each of these permanent molars must, in turn, function as slicing and grinding/crushing teeth when they first erupt, and then either specialized slicing or crushing teeth when they achieve their final position in the mature jaw. Tooth structure ranges from simple, cuspless molars in the termite-eating numbat, to the more complex slicing/crushing molars of the other two families. The degree of carnassiality (or meat-slicing function) grades with diet. Highly carnivorous taxa, such as thylacines and devils, have well-developed carnassial cusps, reduced crushing surfaces, and molar orientation is more antero-posterior. This is particularly the case in the two rear molars, which are biomechanically positioned in adults for maximal slicing function, in a position halfway along the jaw bone comparable to the carnassial tooth in placental carnivores. As the diet becomes more insectivorous, the crushing surfaces (the rear part of each molar tooth) become larger

A mulgara (*Dasycercus cristicauda*) searches grass tops for food near Alice Springs, Northern Territory, Australia. (Photo by Photo Researchers, Inc. Reproduced by permission.)

a clawless toe called a hallux) on the hind feet, long tails and long, pointed snouts, and are considerably uniform in body shape despite the large size variation, from 0.14 oz (4 g) to more than 65 lb (30 kg). Extreme exceptions in body form include the kultarr, which has elongated hind legs and bounds rather than runs, and the robust form and massive skull, teeth and jaw musculature of the specialist scavenger, the Tasmanian devil. Unlike placental carnivores, the fleshy foot pad of thylacines and dasyures extends to the heel and wrist joint, which may contact the ground when stationary or moving slowly, although most species are digitigrade (run on the toes) when moving fast. Arboreal (tree-dwelling) species tend to have broader feet and a better-developed, more dexterous hallux. Some dasyurid species store fat in the tail. None are prehensile and in some such as the thylacine, the tail is semi-rigid. Tail length is generally shorter than body length, except in the numbat and in the long-tailed dunnart.

The dentition is polyprotodont, meaning many incisors (four upper, three lower), which distinguishes this group of marsupials from the Diprotodontia or herbivorous marsupials. Premolars number three in thylacines and numbats, two

A spotted-tailed quoll (*Dasyurus maculatus*) gets a drink in eastern Australia. (Photo by E. & P. Bauer. Bruce Coleman, Inc. Reproduced by permission.)

Distribution

The order Dasyuromorphia comprises the Australian radiation of the polyprotodont marsupials and is restricted to Australia, New Guinea, Tasmania, and some smaller close-by islands. At times during evolutionary history, including most recently the Pleistocene (2 million to 10,000 years ago), land bridges connected Australia and New Guinea, and Australia and Tasmania, and there was opportunity for interchange between the faunas of these major land masses. This historical pattern of connectivity between land masses is reflected in the distributions and genetic relationships of species. Two species of dasyurids in the genus *Sminthopsis*—*S. archeri* and *S. virginiae*—have ranges that cross Torres Strait from northern Australia to New Guinea, and one of the New Guinean quolls, the bronze quoll (*Dasyurus spartacus*), is very closely related to the chuditch (*D. geoffroii*) from the Australian mainland. All of the Tasmanian dasyurid species are currently, historically, or prehistorically (depending on the timing of mainland extinctions) represented in mainland populations (seven species). Where genetic studies have been carried out, mainland and Tasmanian populations represent different evolutionary significant units, a consequence of over 10,000 years of evolutionary separation. It is notable, however, that these faunal interchanges were restricted largely to savanna and woodland species. The cold, dry conditions that prevailed during the Pleistocene precluded the development of significant rainforest corridors. Rainforest *Antechinus* of Cape York and the sister clade *Murexia* from New Guinea did not make it across the land bridge, although some small dasyurids of wet habitats (*Antechinus minimus*—closed, wet heath, grass or sedgelands; *A. swainsonii*—wet forest and heath) do occur on either side of Bass Strait.

Habitat

Carnivorous marsupials occupy habitats from deserts to mountain tops to rainforests and show an accordingly wide range of behavioral, morphological, and physiological adap-

at the expense of the carnassial cusps (sharp, pointy bits) and the molar teeth become wider and more triangular in shape. A feature at least of the larger dasyuromorphians is continual or over-eruption of the canine teeth throughout life. This may be a consequence, again, of early eruption of the permanent teeth at an age when the juvenile animal is less than half of its adult size. Over-eruption, while probably a response to tooth wear and occlusal patterns, has the effect that the canine teeth continue to get bigger as the individual grows.

Coat color in most species ranges from sandy to reddish to grayish brown, sometimes with a lighter colored underbelly. Black is rare (only devils and one of the two color phases of the eastern quoll) and only eight taxa have distinct markings: stripes on the back in the thylacine, numbat and three groups of New Guinean dasyurids, a facial stripe in the numbat, spots in the quolls and white chest, shoulder and rump markings in the devil. Fur length is mostly short to slightly wispy, although tail fur can vary from short, to a bushy tip, to completely fluffy or bushy.

An eastern quoll (*Dasyurus viverrinus*) nest in Australia. (Photo by Animals Animals ©J. & C. Sohns. Reproduced by permission.)

The chuditch (*Dasyurus geoffroii*), also known as a western quoll, is nocturnal. (Photo by Michael Morcombe. Bruce Coleman, Inc. Reproduced by permission.)

tations. Australia's small desert dasyurids are remarkable in their ability to move very long distances in response to local rainfall and fire events, thereby utilizing different vegetative growth stages and short-term flushes in food availability. Planigales live in deep soil cracks and have a suitably flattened body shape. Arboreal and scansorial (above-ground) species, including the antechinuses, are more likely to have a well-developed and dexterous clawless hallux on the inside of the hind foot, which assists with grasping branches or rocks. Very capable climbers such as spotted-tailed quolls (*Dasyurus maculatus*), also possess fleshy ridges on their foot pads and sharp claws. Spotted-tailed quolls can climb large, straight trees to kill possum prey in their daytime tree-hollow refuges, move from tree to tree through the canopy, and climb head-first down trees.

Dasyurids live at extremes of temperature, have a low basal metabolic rate like other marsupials, experience a fluctuating food supply, and most species are small and so lose and gain heat easily. Carnivorous marsupials cannot sweat but resort to licking, panting, and lying flat out on the substrate to keep cool. Strategies to conserve body heat and reduce energy expenditure are diverse. Surface-dwelling desert species such as ningaui have a spherical body shape, which maximises heat

conservation. Antechinuses and Tasmanian devils increase fur thickness in winter and the fur color of some species (fat-tailed dunnart, *Sminthopsis crassicaudata*) becomes darker towards temperate regions. Most dasyurids use protected locations in hollow logs and trees, underground burrows, soil crevices, caves, and tussock grasses to rest during the day or between foraging bouts. Many species line their nest with insulative dead vegetation which they harvest and carry in their mouth, and some huddle in groups comprising adults, or mother and offspring. Huddling in communal nests reduces energy expenditure in dunnarts by 20%. Torpor, in which the body temperature is voluntarily reduced to between 52–82°F (11–28°C) for periods of several hours, is used by half of dasyurids and the numbat, species ranging from 0.17 oz (5 g) to 35 oz (1,000 g) in body size. Torpor is employed as a daily or occasional strategy to conserve energy while resting, under the stresses of cold or food restriction. The females of some species even go into torpor when with young. The slow marsupial development rate may allow this without adverse effects on the babies. Perhaps as a consequence of their relatively low basal metabolic rate, common to all marsupials, carnivorous marsupials have a pronounced ability to increase metabolic rate when cold that exceeds that of some placental mammals.

Another interesting question to ask is what determines how species live together. As most carnivorous marsupials are generalised predators that can take a wide size range of invertebrate and vertebrate prey, structurally complex habitats that offer a variety of different places to forage are an important contributor to local species diversity.

Rather counter-intuitively, it is the arid zone habitats, as well as forest and heath, that are more structurally complex and offer the most opportunities for multiple species to partition resources. This may explain why there are so many species of desert dasyurids. Although the scansorial or above-ground foraging niche, that is important in forest, is lacking in deserts, a wealth of opportunities to segregate are provided by cracks in clay soils, sandy, sand dune and rocky substrates, and structurally complex vegetation such as hummock grasslands.

In forests, the arboreal or above-ground foraging niche contributes to species coexistence. A number of antechinuses are scansorial, meaning that they scramble around both on the ground and a number of metres up into shrubs and trees, even denning in tree hollows. Among the larger marsupial carnivores of Tasmania, spotted-tailed quolls separate from their close competitors, eastern quolls (*Dasyurus viverrinus*) and devils (*Sarcophilus laniarus*), on their greater arboreal use of habitat.

Behavior

Australian mammals are almost all nocturnal or crepuscular and the carnivorous marsupials are no exception. Most are nocturnally active, although some diurnal foraging and basking activity has been recorded in a number of species for which detailed field observations are available, including in antechinuses and thylacines. In some populations, spotted-tailed quolls exploit opportunities to prey upon nocturnal possum prey asleep in tree hollows, and are almost arhythmic in their activity. Three species are substantially or completely diurnal, the numbat, the speckled dasyure from New Guinea, and the southern dibbler (*Parantechinus apicalis*) from Western Australia.

Well developed auditory and olfactory communication could be expected in these primarily nocturnal mammals, although carnivorous marsupials use a variety of visual displays as well. Vocalizations range from hisses, growls, squeaks, barks to screeches in the low (0.1 kHz) to ultrasonic (most less than 12 kHz) frequency range. Contact calls between mother and young start while young are still living in the pouch and consist of squeaks and wheezes which are returned by the mother at a lower pitch. Aggressive calls range from hisses, to growls and screeches. The screech of spotted-tailed quolls has been described as a blast from a circular saw and that of the Tasmanian devil sufficient to "raise the devil," a trait which may have contributed to its name. Female planigales and quolls in estrus emit male-attracting soft clucking calls when receptive to mating. Tasmanian devil females, and to a lesser extent males, crouch mouth to open mouth and give continuous soft barks, each of which ends in a whine. Unwelcome suitors are repelled aggressively.

Scent is an important vehicle for the dissemination of social and reproductive information between individuals, such as male dominance status and the estrous state of females. In carnivores, the information-laden metabolic breakdown products of reproductive hormones are excreted mainly via the feces, although urine of male antechinus also contains sex-specific compounds. Dasyurids have well-developed paracloacal glands from which a pungent, viscous, yellowish liquid is exuded during cloacal dragging form of scent marking. Male Tasmanian devils cloacal drag frequently in the presence of oestrous females, and both sexes mark frequently during non-breeding social interactions. This behavior is well established even in advanced pouch young before the glands have begun to produce scent. Sternal skin glands and chest rubbing are widespread primarily among males of the smaller dasyurids (antechinuses, dunnarts, phascogales), an activity which increases during the breeding season under the control of male reproductive hormones (testosterone). Female quolls and devils produce prodigious quantities of reddish oil in the pouch, the quality and quantity of which is an indicator of estrous state, but which probably also serves to prepare the pouch for occupancy by the young. That this oil also has a function in communicating reproductive state is suggested by the intense interest that males show in sniffing the female's pouch compared with her cloaca.

Visual signals include extensive repertoires of postures, which the use of light-amplifying equipment enables humans to observe. Tasmanian devils use in excess of 20 different postures in social interactions. Visual displays that accentuate body size and weaponry (threat displays) or reproductive readiness usually prelude potentially dangerous aggressive or reproductive interactions. Open mouthed threat displays that show the teeth are common.

The Tasmanian devil (*Sarcophilus laniarius*) has an open-mouthed threat display. (Photo by E. & P. Bauer. Bruce Coleman, Inc. Reproduced by permission.)

Feeding ecology and diet

Competition for food contributes to differences in foraging niche among all size ranges of carnivorous marsupials that live together, with the exception of the termite-eating numbat but including, historically, the thylacine. Competition appears to be more prevalent in mesic forests and heathlands than in arid environments, where droughts, floods, and unpredictable food supplies may often reduce populations to low levels. Competition from larger antechinus species excludes smaller species from the highest quality habitat, with the result that the smaller species eats smaller prey for which they probably have to work harder. In Tasmania, food competition among the larger carnivores has resulted in separation between species on prey size. If species consume different-sized prey, competition for the same prey items is reduced. Competition has been sufficiently intense over a long enough time scale to drive evolutionary changes in canine tooth size among the quolls. Canine tooth strength, which determines the size of prey that can be killed, have become evenly spaced. Even spacing in prey size among species minimizes competition.

Reproductive biology

Short life spans, leading in their extreme form to single breeding followed by death in the first year of life (semelparity), are a defining feature of carnivorous marsupials. Among mammals, semelparity has arisen only in dasyurids and didelphids, in which groups it has evolved at least six times, including in medium-sized species over 2.2 lb (1 kg) in body weight (northern quoll, *Dasyurus hallucatus*). All carnivorous marsupials, including thylacines and possibly numbats,

A spotted-tailed quoll (*Dasyurus maculatus*) displays defensive behavior. (Photo by © Erwin & Peggy Bauer. Bruce Coleman, Inc. Reproduced by permission.)

are short-lived, however, compared to similar-sized placental mammals. This entire taxon seems to be evolutionarily disposed towards early senescence.

Semelparity among dasyurids is obligate in most antechinuses (*Antechinus* spp.), in both *Phascogale* species and in the little red kaluta (*Dasykaluta rosamondae*), but faculatative in the dibbler and the northern quoll. In the larger antechinuses, complete male die-off occurs immediately after mating but females may live to breed in a second year. The adaptive explanation developed on antechinus to explain die-off postulates that as a consequence of small body size, slow growth and the consequent long period of time required to raise young to independence, and tightly seasonal environments in which food is limiting, males and females are unlikely to be able to gain sufficient energy to breed and then survive to a second year. Their best option may be to put all their energy into one big reproductive effort in the first year, even if it kills them. In antechinuses, a consequence of the need to wean young in a tightly seasonal climate at the time of year when food supplies are maximal is that mating occurs in winter when food resources are scarce. Males are unable to store sufficient fat to see them through the intense mating rut and overcome this energy deficit by using elevated levels of stress hormones to promote the use of protein in muscle tissue as an energy resource. This only becomes possible through complete destruction of the ability to produce more sperm before mating starts (sperm production would interfere with synthesis of stress hormones but also removes any possibility of breeding in a second year) and the shutdown of a negative feedback system that prevents sustained levels of damaging

stress hormones in most mammals. Death results from a multitudinous cascade of events related to dramatic loss of body condition and immunosuppression from prolonged elevation of stress hormone levels. Males become anaemic and support huge numbers of ectoparasites, which exacerbates their problems. The proximal cause of death is usually gastrointestinal ulcers or an outbreak of a normally benign disease.

This model is supported by the geographic distribution of semelparous species, which are generally restricted to tightly seasonal environments where reproductive opportunities are limited to a narrow window each year. Iteroparous or multiple-breeding species, such as some dunnarts, include most species from unpredictable arid environments where putting all the eggs in one reproductive basket would be a risky strategy indeed. The lack of sustained elevated stress hormone levels, and opportunities, that come with larger body size and tail fat stores, for sufficient fat storage to tide over the mating period in the northern quoll suggests that a universal model of male die-off remains obscure.

Very small quantities of some of the largest sperm produced by any mammal are other unique features of reproduction in carnivorous marsupials. Complete failure of sperm production prior to the mating period in antechinuses and limited sperm storage leave comparatively small amounts of sperm available for a very intense, once in a lifetime rut; in the brown antechinus (*Antechinus stuartii*) as few as eight ejaculations. Dasyurid sperm have an unusual form of motility which may compensate for this apparent disadvantage.

Copulation in semelparous species is an intense affair with intromission lasting as long as seven to 12 hours. Given the limited sperm supplies, it must be assumed that sperm is used carefully and ejaculation is timed to maximise the chance of fertilization. Most of the time and energy devoted to mating probably serves the dual functions of stimulating the female and mate guarding, both of which increase the male's chance of siring the young. Physical stimulation provided by the thrusting male during copulation improves sperm transport up the female reproductive tract, and occupying the female in copulo for a substantial proportion of the limited time in which she is in estrus physically prevents other males from mating with her. Antechinus, which mate in communal leks, are able to turn 180° while in copulo and fight off other males. Tasmanian devils indulge in ferocious mate guarding, keeping the female a prisoner in the den for days at a time without food or water, until her estrus is finished or the female's desire to escape reaches such an intensity that she succeeds in fighting off the male. Both male and female are likely to be injured in this process.

The female is by no means a passive spectator in the mating business. Female Tasmanian devils actively assess different males and solicit copulations from the male of their choice, avoiding or fighting off the others. Long periods of behavioral estrus allow time for females to mate with a number of males, resisting mate guarding attempts and fighting each one off in turn.

Multiple paternity has been found in all of the small number of species in which mating systems have been investigated using genetic paternity markers, and suggest high levels of promiscuity among females. Four fathers per litter is not uncommon in devils (four teats) and up to seven fathers have been recorded in the agile antechinus (six to 10 teats), with 50% of litters having three or more. Prolonged copulation, larger than average testes size (correlated with greater sperm production), strong male-biased sexual size dimorphism (males larger than females), intense mate guarding, long periods of behavioral estrus, prolonged periods of sperm storage within the female reproductive tract prior to ovulation, and higher population densities are associated with the likelihood of sperm competition (between the sperm of different males within the female reproductive tract) and possibly the intensity of female choice. Semelparous species such as antechinuses exhibit all of the above traits. Iteroparous species from less predictable, arid environments, such as some dunnarts, do not.

Carnivorous marsupials, at least the dasyurids, have one more trick up their metaphorical sleeves. Together with bats, dasyurids are unique among mammals in having sperm storage facilities in the female reproductive tract. (Sperm storage is common in birds and insects.) Ovulation occurs up to 12 days after behavioral estrus and the sperm of several males are stored, possibly right next to each other, in tiny crypts in the oviducts. The sperm is reactivated and released at ovulation. This puts quite a different spin on how sperm competition might operate.

Parental care in all carnivorous marsupials for which it is known is restricted to maternal care of the young during permanent attachment to the teat and lactation. Once young have permanently vacated the pouch, they are deposited in a vegetation-lined nest in a den (underground burrow, cave, or hollow log). Apart from records of chuditch (western quolls) moving young to new dens on their back, there are no other records of females escorting young outside the den. Juvenile chuditch gradually explore further and further from the mother's den as they grow and teach themselves to forage and hunt. As weaning from lactation approaches, both mother and young start to spend nights apart in different dens. The frequency of these separations increases until the male young disperse. Once the males move away from the mother's home range they move rapidly over long distances.

Among many mammals, and it appears most carnivorous marsupials, it is usually the male offspring that disperse from their mother's home range, females staying close to home for life. Females thus exact a longer term cost to the mother, although the difference in the number of young produced by both good and poor quality females will not be great. If, however, there are major differences among individual males in reproductive success, it may be advantageous for females to invest more heavily in male offspring. Strongly male-biased sex ratios occur in some species of dasyurids, including in agile antechinuses. The mechanism of sperm storage offers possibilities for sex-based sperm selection.

Conservation

The pattern of endangerment among the carnivorous marsupials is consistent with that for Australian mammals in gen-

The fat-tailed pseudantechinus (*Pseudantechinus macdonnellensis*) is a carnivorous marsupial. (Photo by B. G. Thomson/Photo Researchers, Inc. Reproduced by permission.)

eral. Australia accounts for 50% of the world's recent mammalian extinctions, the causes of which are multiple and confounded. Causes include habitat destruction, fragmentation, and degradation resulting from land clearance, altered fire regimes, and grazing by introduced livestock and rabbits, and vulnerability to predation by introduced foxes. Dasyurids track the overall mammalian pattern, in which extinctions and declines have been greatest in arid areas and in medium-sized terrestrial species. An exception to this is the thylacine, which has the distinction of being the only large carnivore globally to become extinct in recent times. The loss of the thylacine represents not just a species but also an entire genus and a family extirpated.

Significance to humans

Many species of native marsupials had a place in the dreamtime histories of aboriginal peoples. Aboriginal histories have indeed revealed ecological information on species that became extinct prior to European documentation. However, except for the very distinctive animals, aboriginal peoples did not necessarily distinguish between species of the smaller Australian mammals. The larger, more distinctive species were held in reverence, with dreaming stories, totemic status, and ritual treatment that sometimes precluded consumption. Aboriginal peoples probably hunted and ate all species of carnivorous marsupials depending on abundance.

Commercial exploitation of carnivorous marsupials for furs or skins has not been recorded, apart from early collecting for museums, despite the beautiful coat patterns of some of the larger species. Perhaps the lack of intensely cold climates and winter pelts contributed to this lack of economic interest. Some species may confer economic benefits in their dietary proclivity for agricultural and forest insect pests and their abilities in dispatching rodents. The diet of eastern quolls living on farmland is dominated at times of the year by pest pasture grubs (corbie grubs and wire worms).

Economic impacts of carnivorous marsupials are restricted to the larger species. The predatory abilities of carnivorous marsupials in tackling large prey (relative to their body size) means that any species larger than 4.8 oz (150 g) can take on domestic poultry. Inadequately housed poultry may be targeted by phascogales, quolls, and devils, particularly at night and especially by young animals and females feeding young. Lambs in the first 24 hours of life are vulnerable to Tas-manian devils. Lambs from multiple births and certain breeds of sheep are more vulnerable. Fencing is not difficult (strong mesh wire, no holes, footings 6 in [15 cm] below ground), but education and attitudinal change is a major hindrance. Thylacines would have killed sheep of all sizes. Persecution on individual properties can have a significant impact on local populations of devils in particular and certainly did on thylacines.

Resources

Books

Archer, M., T. Flannery, S. Hand, and J. Long. *Prehistoric Mammals of Australia and New Guinea: One Hundred Million Years of Evolution.* Sydney: UNSW Press, 2002.

Cockburn, A. "Living slow and dying young—senescence in marsupials." In *Marsupial Biology: Recent Research, New Perspectives.* Sydney: UNSW Press, 1997.

Dickman, R. R. "Distributional ecology of dasyurid marsupials." In *Predators with Pouches: The Biology of Carnivorous Marsupials.* Melbourne: CSIRO Publishing, 2003.

Dixon, J. M. "Thylacinidae." In *Fauna of Australia,* Canberra: Australian Government Publishing Service, 1989.

Flannery, T. *Mammals of New Guinea.* Sydney: Reed Books, 1995.

Friend, J. A. "Myrmecobiidae." In *Fauna of Australia,* pp. 583-590. Canberra: Australian Government Publishing Service, 1989.

Krajewski, C., and M. Westerman. "Molecular systematics of Dasyuromorphia." In *Predators with Pouches: The Biology of Carnivorous Marsupials.* Melbourne: CSIRO Publishing, 2003.

Maxwell, S., A. A. Burbidge, and K. Morris. "The Action Plan for Australian Marsupials and Monotremes. IUCN/SSC Australasian Marsupial and Monotreme Specialist Group, Wildlife Australia, Environment Australia, Canberra ACT Australia." In *Wildlife Australia, Endangered Species Program, Project Number 500 Report.* Canberra: Australian Government Publishing Service, 1996.

Morton, S. R., C. R. Dickman, and T. P. Fletcher. "Dasyuridae." In *Fauna of Australia.* Canberra: Australian Government Publishing Service, 1989.

Strahan, R., ed. *The Mammals of Australia.* Sydney: Reed Books, 1995.

Taggart, D. A., G. A. Shimmin, C. R. Dickman, and W. G. Breed. "Reproductive Biology of Carnivorous Marsupials: Clues to the Likelihood of Sperm Competition." In *Predators with Pouches: The Biology of Carnivorous Marsupials.* Melbourne: CSIRO Publishing, 2003.

Toftegaard, C. L., and A. J. Bradley. "Chemical communication in dasyurid marsupials." In *Predators with Pouches: The Biology of Carnivorous Marsupials.* Melbourne: CSIRO Publishing, 2003.

Ward, S. J. "Biased sex ratios in litters of carnivorous marsupials; Why, when & how?" In *Predators with Pouches: The Biology of Carnivorous Marsupials.* Melbourne: CSIRO Publishing, 2003.

Wilson, B. A., C. R. Dickman, and T. P. Fletcher. "Dasyurid dilemmas: Problems and solutions for conserving Australia's small carnivorous marsupials." In *Predators with Pouches: The Biology of Carnivorous Marsupials.* Melbourne: CSIRO Publishing, 2003.

Wroe, S. "Australian marsupial carnivores: an overview of recent advances in palaeontology." In *Predators with Pouches: The Biology of Carnivorous Marsupials.* Melbourne: CSIRO Publishing, 2003.

Periodicals

Dickman, C. R. "An experimental study of competition between two species of dasyurid marsupials." *Ecological Monographs* 56 (1986): 221–241.

Firestone, K. B., M. S. Elphinstone, W. B. Sherwin, and B. A. Houlden. "Phylogeographical population structure of tiger quolls *Dasyurus maculatus* (Dasyuridae: Marsupialia), an endangered carnivorous marsupial." *Molecular Ecology* 8 (1999): 1613–1625.

Fisher, D. O., and C. R. Dickman. "Body size–prey size relationships in insectivorous marsupials: tests of three hypotheses." *Ecology* 74 (1993): 1871–1883.

Jones, M. E. "Character displacement in Australian dasyurid carnivores: size relationships and prey size patterns." *Ecology* 78 (1997): 2569–2587.

Jones, M. E., and L. A. Barmuta. "Diet overlap and abundance of sympatric dasyurid carnivores: a hypothesis of competition?" *Journal of Animal Ecology* 67 (1998): 410–421.

Kraaijeveld-Smit, F. J. L., S. J. Ward, and P. D. Temple-Smith. "Multiple paternity in a field population of a small carnivorous marsupial, the agile antechinus, *Antechinus agilis.*" *Behavioral Ecology and Sociobiology* 52 (2002): 84–91.

Righetti, J., B. J. Fox, and D. B. Croft. "Behavioural mechanisms of competition in small dasyurid marsupials." *Australian Journal of Zoology* 48 (2000): 561–576.

Soderquist, T. R., and M. Serena. "Juvenile behaviour and dispersal of chuditch (*Dasyurus geoffroii*) (Marsupialia: Dasyuridae)." *Australian Journal of Zoology* 48 (2000): 551–560.

Taggart, D. A., and P. D. Temple-Smith. "An unusual mode of progressive motility in spermatozoa from the dasyurid marsupial, *Antechinus stuartii.*" *Reproduction, Fertilization and Development* 2 (1990): 107–114.

Resources

————. "Transport and storage of spermatozoa in the female reproductive tract of the brown marsupial mouse, *Antechinus stuartii* (Dasyuridae)." *Journal of Reproduction and Fertility* 93 (1991): 97–110.

————. "Comparative studies of epididymal morphology and sperm distribution in dasyurid marsupials during the breeding season." *Journal of Zoology* 232 (1994): 365–381.

Menna Jones, PhD

Marsupial mice and cats, Tasmanian devil
(Dasyuridae)

Class Mammalia

Order Dasyuormorphia

Family Dasyuridae

Thumbnail description
A large family of quadrupedal, predatory insectivores and carnivores, ranges in size from minute to medium

Size
1.8–25.7 in (46–652 mm); 0.07 oz–28.7 lb (2 g–13 kg)

Number of genera, species
17 genera; 69 species

Habitat
Occupy all terrestrial habitats in Australia and Papua New Guinea

Conservation status
Endangered: 6 species; Vulnerable: 9 species; Lower Risk: 4 species; Data Deficient: 9 species

Distribution
All of Australia, Tasmania, and Papua New Guinea, including close offshore islands

Evolution and systematics

First appearing in the fossil record in the early to middle Miocene, dasyurids were rare (only two species known) until the late Miocene, when they increased steadily in diversity to replace the thylacinids as the largest group of Australian carnivorous marsupials. Dasyurids comprise three extant subfamilies and one extinct subfamily (the earliest form), which was a sister group to the living subfamilies, and are most closely related to the thylacinids. Molecular data indicate that all four radiations of dasyurids took place in the late mid-Miocene, perhaps in response to climatic drying. Most of the species present in the Pleistocene were of living taxa, representatives of which occurred no earlier than the Pliocene. It is suggested that the dasyurids are highly specialized among dasyuromorphians in their morphology rather than primitive. The living fauna currently comprises 69 described species in seventeen genera (fifty-three restricted to Australia and islands, fourteen in New Guinea and islands). The number of species will almost certainly increase with taxonomic revisions, particularly with the recognition of morphologically cryptic but genetically distinct species. There is genetic and morphological differentiation at the subspecific level in some species.

Physical characteristics

Dasyurids are nearly all quadrupedal with long tails, long pointed snouts, four toes on the front feet, and four to five toes on the hind feet. If the fifth toe (inner rear) is present,

An eastern quoll (*Dasyurus viverrinus*) with young. (Photo by Tom McHugh/Photo Researchers, Inc. Reproduced by permission.)

A Tasmanian devil (*Sarcophilus laniarius*) feeds in Tasmania. (Photo by Animals Animals ©Don Brown. Reproduced by permission.)

it is a clawless hallux. The footpads, which may be placed on the ground when standing or moving slowly, extend to the heel and wrist joints. There are four upper and three lower incisors, two premolars, and four molars, each of which is similar in form and has distinct cusps with both slicing (carnassial) and grinding surfaces (less so in highly carnivorous forms such as the Tasmanian devil, *Sarcophilus laniarius*). Across the considerable size range (from the 0.14 oz [4 g] *Planigale ingrami*, the world's smallest marsupial, to the 28.6 lb [13 kg] devil), body shape is remarkably uniform. Arboreal species tend to have broader hind feet than terrestrial species, and a more prominent, dexterous hallux. One sandy desert species has fine bristles on the footpads. The most extreme variations in morphology include *Antechinomys laniger*, which has elongated hind legs and a bounding gait, and the heavily built, specialist scavenger, the Tasmanian devil, with massive skull, teeth, and jaw musculature. Coat color is mostly uniform shades of gray, sandy to reddish to dark brown, or black, sometimes with a lighter underbelly. Striking markings are the province of the larger dasyurids, with white spots on quolls and white markings in the devil, although six New Guinean dasyures (three *Murexia*, one *Myoictis*, and at least one *Phascolosorex*) have dark dorsal stripes, and two diurnal species have speckled (speckled dasyure, *Neophascogale lorentzii*) or grizzled (dibbler, *Parantechinus* spp.) gray fur, all of which may serve as camouflage. Some desert dasyurids have very sparse fine tail hair, the smaller quolls have soft, fluffy tails, while others are finely tufted at the tip, or sport a highly visible black bushy brush that contrasts with the pale body fur. Tail fur may function as flags in communication. Some species, notably the fat-tailed pseudantechinus (*Pseudantechinus macdonnellensis*), the fat-tailed dunnart (*Sminthopsis crassicaudata*), and the Tasmanian devil, store quantities of fat in their tail when environmental conditions are good, leading to a distinctly parsnip-shaped tail. The pouch is either well developed (though still quite shallow compared with diprotodont marsupials) as well

as backwards facing, or it is almost absent (raised lateral ridges of skin) and downward facing. Teat number varies from four to 12 and may vary within species. Male Tasmanian devils have a shallow scrotal pouch.

Distribution

Dasyurids occur in virtually every terrestrial environment at all altitudes in mainland Australia and some offshore islands, mainland New Guinea, and some islands between Australia and New Guinea. In Australia, species diversity of the smaller species is higher in arid regions than in the more mesic coastal and sub-coastal areas of the east, southeast, and extreme southwest, while the converse is true for the larger dasyurids (more than 17 oz [500 g]), quolls, and devils). Species richness of small dasyurids reaches its highest density in the spinifex hummock grasslands of arid central Australia (average 5.3, maximum eight species) and is correlated with structural complexity of the habitat that allows niche separation. Population density, on the other hand, reaches highest levels in coastal forest and heath among the smaller species, and is uniformly low in the arid zone. With the disappearance of the chuditch (*Dasyurus geoffroii*), there are no larger dasyurids in the arid zone. The island of Tasmania supports the largest assemblage of larger dasyurids (three species, historically four) following the historic and prehistoric extinction of three of these species from mainland Australia. Dasyurid distribution within New Guinea is poorly documented.

Habitat

Every type of terrestrial habitat in Australia and New Guinea is occupied by dasyurids. Habitat preferences of individual species are strongly associated with food supply, and with either protection from predators or suitable habitat

A Tasmanian devil (*Sarcophilus laniarius*) orphaned juvenile being raised in the Bonorong Wildlife Park in Tasmania, Australia. (Photo by Animals Animals ©Steven David Miller. Reproduced by permission.)

of this plan saw the recovery of this species from Endangered to Vulnerable in 1996. Chuditch may soon be removed from threatened species lists as well, although it is likely to retain the status of Lower Risk/Conservation Dependent, referring to the requirement in perpetuity for fox control. No recovery plans have yet been adopted for the smaller species.

Significance to humans

Apart from an occasional food source, the smaller species of daysurids seem not to have had great significance for aboriginal peoples. The larger, more distinctive species like quolls, were frequently totemic species and had dreaming histories and individual names that persisted long after they became extinct in a region.

1. Long-tailed planigale (*Planigale ingrami*); 2. Long-tailed dunnart (*Sminthopsis longicaudata*); 3. Brush-tailed phascogale (*Phascogale tapoatafa*); 4. Pilbara ningaui (*Ningaui timealeyi*); 5. Kultarr (*Antechinomys laniger*). (Illustration by Emily Damstra)

1. Speckled dasyure (*Neophascogale lorentzii*); 2. Southern dibbler (*Parantechinus apicalis*); 3. Mulgara (*Dasycrous cristicauda*); 4. Brown antechinus (*Antechinus stuartii*); 5. Tasmanian devil (*Sarcophilus laniarius*); 6. Chuditch (*Dasyurus geoffroii*). (Illustration by Emily Damstra)

Species accounts

Mulgara
Dasycercus cristicauda

SUBFAMILY
Dasyurinae

TAXONOMY
Dasycercus cristicauda (Krefft, 1867), South Australia, Australia, probably Lake Alexandrina. Two subspecies described.

OTHER COMMON NAMES
English: Crest-tailed marsupial mouse.

PHYSICAL CHARACTERISTICS
Length 4.9–8.7 in (125–220 mm). Light brown above, pale below with crest of long black fur distal two-thirds of short tail; short, rounded ears; stores fat in the base of the tail.

DISTRIBUTION
Inland central and western Australia.

HABITAT
Found in arid, sandy regions.

BEHAVIOR
Lives solitarily in burrows it digs in flats between or on lower slopes of sand dunes.

FEEDING ECOLOGY AND DIET
Generalized insectivore, takes small vertebrates.

REPRODUCTIVE BIOLOGY
Quite long-lived for a dasyurid. Breed for up to six years and produce up to eight young. Probably promiscuous.

CONSERVATION STATUS
Vulnerable. Decline possibly a result of predation by introduced foxes and cats, and changes in fire regimes. Susceptible to decline because of larger body size and restricted habitat.

SIGNIFICANCE TO HUMANS
None known. ◆

Chuditch
Dasyurus geoffroii

SUBFAMILY
Dasyurinae

TAXONOMY
Dasyurus geoffroii (Gould, 1841), Liverpool Plains, New South Wales, Australia.

OTHER COMMON NAMES
English: Western quoll.

PHYSICAL CHARACTERISTICS
Brown above, light below, with conspicuous white spots on body, and bushy tail black on distal half.

DISTRIBUTION
Formerly western two-thirds of inland Australia. Now restricted to nine localities (including reintroduction sites) in the extreme southwest.

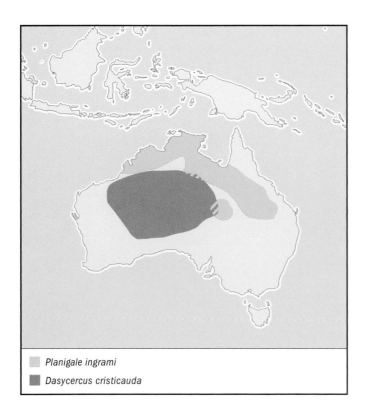

Planigale ingrami
Dasycercus cristicauda

Dasyurus geoffroii
Ningaui timealeyi

HABITAT
Open, dry eucalypt forests, woodlands, and shrublands. Formerly deserts.

BEHAVIOR
Solitary. Females maintain an exclusive core range that overlaps with several males.

FEEDING ECOLOGY AND DIET
Primarily insects, also birds, mammals, and reptiles. Can kill vertebrate prey larger than body size.

REPRODUCTIVE BIOLOGY
Up to six young. Longevity in the wild rarely more than three years. Probably promiscuous.

CONSERVATION STATUS
Vulnerable. Was Endangered until 1996. Introduced red foxes are the primary cause of decline. Current populations require ongoing fox baiting for their survival.

SIGNIFICANCE TO HUMANS
A nuisance as a predator of poultry. ◆

Speckled dasyure
Neophascogale lorentzi

SUBFAMILY
Dasyurinae

TAXONOMY
Neophascogale lorentzi (Jentink, 1911), Irian Jaya, Indonesia.

OTHER COMMON NAMES
None known.

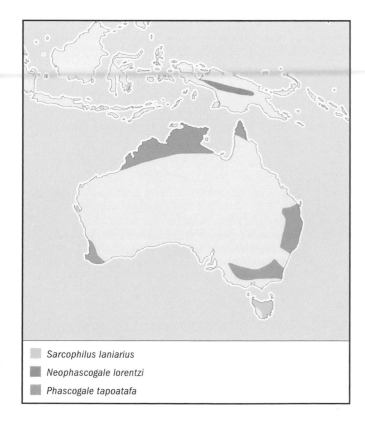

Sarcophilus laniarius

Neophascogale lorentzi

Phascogale tapoatafa

PHYSICAL CHARACTERISTICS
Length 6.5–8.7 in (166–220 mm). Silvery speckled gray to reddish coat with long, white-tipped tail; long, pointed snout, small eyes, and very long claws.

DISTRIBUTION
High mountain forests of western New Guinea (over 6,500 ft; 2,000 m).

HABITAT
Wet montane moss forest.

BEHAVIOR
Diurnal.

FEEDING ECOLOGY AND DIET
Large insects, probably small vertebrates.

REPRODUCTIVE BIOLOGY
Nothing is known, but probably promiscuous.

CONSERVATION STATUS
Data Deficient.

SIGNIFICANCE TO HUMANS
None known. ◆

Southern dibbler
Parantechinus apicalis

SUBFAMILY
Dasyurinae

TAXONOMY
Parantechinus apicalis (Gray, 1842), Southwestern Western Australia, Australia.

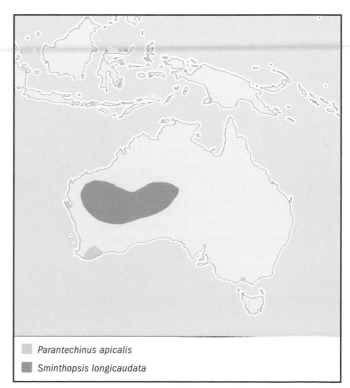

Parantechinus apicalis

Sminthopsis longicaudata

OTHER COMMON NAMES
English: Freckled antechninus, speckled marsupial mouse.

PHYSICAL CHARACTERISTICS
Length 5.5–5.7 in (140–145 mm). Coarse brownish gray fur, speckled with white, and pale below; tapered hairy tail; white ring around eye.

DISTRIBUTION
A few locations in extreme southwestern Western Australia and offshore islands.

HABITAT
Shrubland.

BEHAVIOR
Nothing is known.

FEEDING ECOLOGY AND DIET
Large insects, small reptiles.

REPRODUCTIVE BIOLOGY
Facultatively semelparous; male die-off in some years in island population but not on mainland. Probably promiscuous.

CONSERVATION STATUS
Endangered. Restricted habitat associations may play a role.

SIGNIFICANCE TO HUMANS
None known. ◆

Tasmanian devil
Sarcophilus laniarius

SUBFAMILY
Dasyurinae

TAXONOMY
Sarcophilus laniarius (Owen, 1838), Wellington caves (Pleistocene), Australia.

OTHER COMMON NAMES
None known.

PHYSICAL CHARACTERISTICS
Length 22.4–25.7 in (570–652 mm). Black all over with coarse long fur on medium-length tail; white markings most common on chest as well as on shoulders and rump; robust build with massive head, sloping hindquarters, and very short legs; fat storage in tail.

DISTRIBUTION
The island of Tasmania, Australia.

HABITAT
Open forests and woodlands.

BEHAVIOR
Social, but can be solitary; aggregates at carcasses. Promiscuous mating system.

FEEDING ECOLOGY AND DIET
Predator and specialized scavenger. Medium-sized mammals such as wallabies and possums.

REPRODUCTIVE BIOLOGY
Lives up to five or six years in the wild. Up to four young from ages two to six. Probably promiscuous.

CONSERVATION STATUS
Lower Risk. Common in suitable habitat.

SIGNIFICANCE TO HUMANS
Predates poultry and weak lambs. ◆

Brown antechinus
Antechinus stuartii

SUBFAMILY
Dasyurinae

TAXONOMY
Antechinus stuartii (Macleay, 1841), Manly, New South Wales, Australia. Two subspecies described.

OTHER COMMON NAMES
English: Macleay's marsupial mouse, Stuart's antechinus.

PHYSICAL CHARACTERISTICS
Length 2.9–5.5 in (74–140 mm). Uniform grayish brown; paler below, with thin, hairy tail almost body length; broad head with pale fur around eye.

DISTRIBUTION
Australian east coast and hinterland from southeast Queensland to southern New South Wales.

HABITAT
Wet to dry forests with dense ground cover and numerous logs.

BEHAVIOR
Nocturnal, but may be active during day if food scarce. Terrestrial; partly arboreal if sparse groundcover or larger terrestrial competitor present.

Antechinus stuartii

Antechinomys laniger

FEEDING ECOLOGY AND DIET
Small to large insects, beetles, and spiders.

REPRODUCTIVE BIOLOGY
Tightly synchronized mating season (two weeks). Semelparous with abrupt male die-off immediately after mating season. Probably promiscuous.

CONSERVATION STATUS
Lower Risk.

SIGNIFICANCE TO HUMANS
None known. ◆

Brush-tailed phascogale
Phascogale tapoatafa

SUBFAMILY
Dasyurinae

TAXONOMY
Phascogale tapoatafa (F. Meyer, 1793), Sydney, New South Wales, Australia.

OTHER COMMON NAMES
English: Tuan, common wambenger, black-tailed phascogale.

PHYSICAL CHARACTERISTICS
Length 5.8–10.3 in (148–261 mm). Uniform grizzled gray above, cream to white below, with large, naked ears, and conspicuous black brush tail with hairs up to 2.1 in (55 mm) long.

DISTRIBUTION
Mesic coastal and hinterland areas of southeastern and southwestern Australia (*P. t. tapoatafa*). Monsoonal northern Australia (*P. t. pirata*).

HABITAT
Dry eucalypt forest and woodlands with open under-story in temperate and tropical Australia.

BEHAVIOR
One of most arboreal of the dasyurids; very agile hunter aboveground. Nesting and mating in tree hollows.

FEEDING ECOLOGY AND DIET
Nectar, large insects, spiders, and small vertebrates. Forages by tearing away bark and probing into crevices.

REPRODUCTIVE BIOLOGY
Semelparous with complete male die-off after a synchronized three-week mating season. Probably promiscuous.

CONSERVATION STATUS
Lower Risk (Least Concern: *P. t. tapoatafa* and Near Threatened: *P. t. pirata*).

SIGNIFICANCE TO HUMANS
Occasionally takes penned poultry. ◆

Kultarr
Antechinomys laniger

SUBFAMILY
Sminthopsinae

TAXONOMY
Antechinomys laniger (Gould, 1856), interior New South Wales, Australia. Two subspecies described.

OTHER COMMON NAMES
English: Jerboa, marsupial mouse.

PHYSICAL CHARACTERISTICS
Length 2.8–3.9 in (70–100 mm). Grizzled fawn-gray above, white below, with very large ears, large eyes, and a long tail with black brushed tip; hind-foot greatly elongated with only four toes.

DISTRIBUTION
Broad band across central and southern arid Australia.

HABITAT
Stony and sandy desert plains and acacia scrubland where small bushes constitute the principal vegetation.

BEHAVIOR
Bounding gait; terrestrial; shelters beneath tussocks and in cracks in the soil, burrows, and logs. Nocturnal.

FEEDING ECOLOGY AND DIET
Large insects and spiders.

REPRODUCTIVE BIOLOGY
Iteroparous (breeds in multiple years) with long breeding season. Pouch a crescent-like fold over anterior part of mammary glands. Six or eight teats. Probably promiscuous.

CONSERVATION STATUS
Data Deficient. Uncommon over most of range. Populations appear to fluctuate seasonally. Not directly impacted by humans.

SIGNIFICANCE TO HUMANS
None known. ◆

Pilbara ningaui
Ningaui timealeyi

SUBFAMILY
Sminthopsinae

TAXONOMY
Ningaui timealeyi (Archer, 1975), Western Australia, Australia.

OTHER COMMON NAMES
English: Ealey's ningaui.

PHYSICAL CHARACTERISTICS
Length 1.8–2.2 in (46–57 mm). Minute predatory dasyruid with grizzled gray, bristly fur, and furred, long tail.

DISTRIBUTION
Restricted distribution on Hamersley Plateau, Western Australia.

HABITAT
Semi-arid grasslands. Prefers drainage lines where large hummocks of spinifex, scattered shrubs, and mallee trees grow.

BEHAVIOR
Mostly nocturnal.

FEEDING ECOLOGY AND DIET
Desert centipedes and cockroaches that may be larger than itself.

REPRODUCTIVE BIOLOGY
Probably promiscuous. Potentially long breeding season (six months) to allow for annual rains. Pouch a simple furless depression in belly. Semelparous. Four to six young are independent at 0.07 oz (2 g) body weight.

CONSERVATION STATUS
Lower Risk. Common. May survive only in moister pockets of habitat in dry seasons, repopulating after rains.

SIGNIFICANCE TO HUMANS
None known. ◆

Long-tailed dunnart
Sminthopsis longicaudata

SUBFAMILY
Sminthopsinae

TAXONOMY
Sminthopsis longicaudata (Spencer, 1909), Western Australia, Australia.

OTHER COMMON NAMES
English: Long-tailed marsupial mouse.

PHYSICAL CHARACTERISTICS
Length 3.1–3.9 in (80–100 mm). Gray above, white below, white feet; scaly tail with short hairs more than twice length of head and body.

DISTRIBUTION
Arid interior of western and central Australia.

HABITAT
Rugged scree, boulders, and rocky plateau, sparsely vegetated with shrubs and spinifex hummocks.

BEHAVIOR
Active and capable climber with striated footpads.

FEEDING ECOLOGY AND DIET
Small to large insects and spiders.

REPRODUCTIVE BIOLOGY
Probably promiscuous. Long breeding season, winter to spring.

CONSERVATION STATUS
Lower Risk. One of the most rarely recorded dasyurids, but with a broad, undisturbed range. May be locally common.

SIGNIFICANCE TO HUMANS
None known. ◆

Long-tailed planigale
Planigale ingrami

SUBFAMILY
Sminthopsinae

TAXONOMY
Planigale ingrami (Thomas, 1906), Alexandria, Northern Territory, Australia.

OTHER COMMON NAMES
English: Northern planigale.

PHYSICAL CHARACTERISTICS
Length 2.2–2.6 in (55–65 mm). Smallest of the planigales and the smallest marsupial with very flat head and thin tail longer than head and body.

DISTRIBUTION
Northern Australia.

HABITAT
Seasonally flooded grasslands and savanna woodlands.

BEHAVIOR
Forages and rests in crevices in moist, contracting (cracking) soils, under rocks, and in tussocks. Planigales may have evolved the very flat head to occupy the niche of foraging in seasonally flooded cracking soils.

FEEDING ECOLOGY AND DIET
Rapacious appetite; insects, lizards, and even young mammals almost as large as itself. Larger insects are killed by persistent biting.

REPRODUCTIVE BIOLOGY
Probably promiscuous. Breeding throughout year but concentrated in late summer. Four to eight young.

CONSERVATION STATUS
Lower Risk. Difficult to study on account of minute size. No accurate assessment of populations.

SIGNIFICANCE TO HUMANS
None known.

Common name / Scientific name/ Other common names	Physical characteristics	Habitat and behavior	Distribution	Diet	Conservation status
Dusky antechinus *Antechinus swainsonii;* English: Dusky marsupial; Spanish: Antechino sombrio	Pale pinkish fawn through gray to coppery brown. Underparts creamy or white. Short, dense, and rather coarse hair and long-haired tail. Head and body length 3.0–6.9 in (7.5–17.5 cm), tail length 2.6–6.1 in (6.5–15.5 cm). Typical male 1.7–3.2 oz (48–90 g), female 1.1–1.9 oz (31–55 g).	Found mostly in dense, moist forest. Monestrous, breeding pattern of three months followed by a synchronized male die-off. One litter produced per year.	Southeast Queensland, east New South Wales, east and southeast Victoria, coastal southeast Australia, and Tasmania.	Mostly invertebrates the size of small insects to domestic sparrows, but can supplement diet with fruit, such as blackberries.	Not threatened
Yellow-footed antechinus *Antechinus flavipes* Spanish: Antechino de patas amarillas	Pale pinkish fawn through gray to coppery brown. Underparts creamy or white. Short, dense and rather coarse hair, short, hooked claws. Head and body length 3.0–6.9 in (7.5–17.5 cm), tail length 2.6–6.1 in (6.5–15.5 cm).	Found in wide variety of forest and brushland habitat with sufficient cover. Very active, climbs well. Monestrous, three month breeding period, one litter produced per year.	Cape York Peninsula (Queensland) to Victoria and southeast South Australia, southwest Western Australia.	Mostly invertebrates.	Not threatened
Little red kaluta *Dasykaluta rosamondae* Spanish: Kalutica rojiza	Russet to coppery brown, rough fur. Head and body length 3.5–4.3 in (9–11 cm), tail length 2.2–2.8 in (5.5–7 cm), weight 0.7–1.4 oz (20–40 g).	Found in spinifex grassland. Mating occurs in September, followed by synchronized male die-off.	Pilbara region of northwestern Western Australia.	Insects and small vertebrates.	Not threatened
Northern quoll *Dasyurus hallucatus* English: Little northern cat; Spanish: Quoll norte'	Upperparts predominantly gray or darker brown. Tip and ventral surface of tail are dark brown or black. Head and body length 9.4–13.8 in (24–35 cm), tail length 8.3–12.2 in (21–31 cm).	Found in woodland and rocky areas. Monestrous, winter breeders, one litter produced a year. Cats maintained in pairs except when female has young.	Australia, in north Northern Territory, north and northeast Queensland, and north Western Australia.	Aggressive carnivores. Diet consists of mammals such as large rock rats, common rock rats, and sandstone antechinus, as well as reptiles, worms, ants, termites, grasshoppers, beetles, figs, and other soft fruits.	Lower Risk/ Near Threatened
Spotted-tailed quoll *Dasyurus maculatus* English: Tiger quoll; Spanish: Quoll de cola moteada	Upperparts grayish or dark brown, white spots or blotches on back and sides. Spots usually extend well onto the tail. Head and body length 15.7–29.9 in (40–76 cm), tail length 19.7–22.0 in (50–56 cm).	Found in dry forest and open country. Nocturnal hunter. Lengthy courtship. Monestrous, winter breeders, six to eight young produced per year.	Australia, in east Queensland, east New South Wales, east and south Victoria, southeast South Australia, and Tasmania. Formerly found in South Australia.	Predatory animal, but will also eat vegetable matter. May consume mammals as large as wallabies.	Vulnerable
Eastern quoll *Dasyurus viverrinus* Spanish: Quoll oriental	Upperparts mostly grayish, or olive brown to dark rufous brown, underparts paler yellowish or white. Prominent white spots or blotches on back and sides. Head and body length 13.8–17.7 in (35–45 cm), tail length 8.3–11.8 in (21–30 cm).	Variety of habitats, including rainforest, heathland, alpine areas, and scrub. Prefers dry grassland, forest mosaics bounded by agricultural land, drier forest, and open country. Shelters in rock piles or hollow logs. Five to six young per litter. Maximum known lifespan is six years, ten months.	Probably survives only in Tasmania; formerly South Australia, New South Wales, and Victoria.	Predatory animal, but will also consume vegetable matter. Insects and rodents also compose diet.	Lower Risk/ Near Threatened
New Guinean quoll *Dasyurus albopunctatus* Spanish: Quoll de Nueva Guinea	Upperparts grayish or olive brown to dark rufous brown. Coat is coarse with little underfur. Head and body length 9.4–13.8 in (24–35 cm), tail length 8.3–12.2 in (21–31 cm).	Found in dense, moist forest in a variety of conditions up to an altitude of 11,480 ft (3,500 m) in New Guinea. Primarily terrestrial, can climb well, nocturnal.	New Guinea.	High proportion of insects.	Vulnerable
Short-furred dasyure *Murexia longicaudata* French: Phascogalin's de Nouvelle-Guin'e; German: Neuguinea Beutelmäuse	Dull grayish brown upperparts, white underparts, long, sparsely haired tail, few long hairs at tip. Head and body length 4.1–11.2 in (10.5–28.5 cm) and tail length 5.7–9.4 in (14.5–24 cm).	Found in all lowland and midmountain forests of New Guinea, from sea level to 6,230 ft (1,900 m).	New Guinea, Aru Islands.	Insectivorous and carnivorous.	Not threatened
Broad-striped dasyure *Murexia rothschildi*	Dark, grayish brown upperparts. Broad, black dorsal stripe is present. Underparts are light brown, fur is short and dense. Head and body length 4.1–11.2 in (10.5–28.5 cm), tail length 5.7–9.4 in (14.5–24 cm).	Known from only eight specimens, this species has an altitudinal range of 3,280–6,560 ft (1,000–2,000 m).	Southeast New Guinea.	Presumed to be insectivorous and carnivorous.	Data Deficient

[continued]

Common name / Scientific name/ Other common names	Physical characteristics	Habitat and behavior	Distribution	Diet	Conservation status
Three-striped dasyure *Myoictis melas*	Upperparts richly variegated chestnut mixed with black and yellow, head is dark rusty red. Most colorful of all marsupials Long tapered tail and slender snout. Head and body length 6.7–9.8 in (17–25 cm), tail length 5.9–9.1 in (15–23 cm).	Occurs in most rainforests of the lowland and mid-mountains of New Guinea. Nocturnal and scansorial in habit.	New Guinea; Salawati Island and Aru Islands (Indonesia).	Plants and insects, mainly ants and termites.	Not threatened
Wongai ningaui *Ningaui ridei* Spanish: Ningaui vongai	Upperparts dark brown to black, underparts usually yellowish, sides of face salmon to brown. Thin tail. Length known only from two young samples: head and body length 1.9–2.1 in (4.9–5.3 cm), tail length 1.9–2.0 in (4.8–5 cm).	Found in dry grassland and savanna, mainly under dry conditions. Nocturnal. Uses dry, raspy sound to communicate.	Australia, in Western Australia to New South Wales and Victoria.	Insects and small invertebrates.	Not threatened
Sandstone dibbler *Parantechinus bilarni* English: Northern dibbler	Upperparts grizzled brown, underparts pale gray, cinnamon patches behind large ears. Head and body length 2.2–3.9 in (5.7–10 cm), tail length 3.2–4.5 in (8.2–11.5 cm), and weight 0.42–1.6 oz (12–44 g). Tail is long and thin.	Found in rugged, rocky country covered with eucalyptus forest, perennial grasses. Mating occurs late June to early July, four to five young constitute a litter.	Northern Territory, Australia.	Mostly insects.	Not threatened
Red-tailed phascogale *Phascogale calura* English: Red-tailed wamberger; Spanish: Fascogale de cola roja	Grayish upper parts, white underparts. Head and body length 3.7–4.8 in (9.3–12.2 cm), tail length 4.7–1.8 in (11.9–4.5 cm), and weight 1.3–2.4 oz (38–68 g). Rear half covered by long, silky, black hairs.	Found in heavy, humid forest and more sparsely timbered arid regions. Nests consist of leaves and twigs in the forks or holes of trees, some are built on ground. Nocturnal.	Inland southwest Western Australia, formerly in Northern Territory. South Australia, northwest Victoria, southwest New South Wales. Most likely extinct, except in Western Australia wheat belt.	Small mammals, birds, lizards, and insects.	Endangered
Narrow-striped marsupial shrew *Phascolosorex dorsalis* Spanish: Murasà marsupial rayada	Grizzled gray-brown coloration with chestnut red underneath. Head and body length 2.6–5.3 in (6.7–13.4 cm), tail length 2.4–4.3 in (6–11 cm). Thin black stripe runs from head to tail.	Occur in mountain forests at high altitudes from 3,970 to 10,170 ft (1,210–3,100 m). Nocturnal and scansorial (climbing) in habit.	West and east interior of New Guinea.	Mainly insects.	Not threatened
Pygmy planigale *Planigale maculata*	Upperparts pale tawny olive, darker tawny, or brownish gray; underside olive buff, fuscous, or light tan. Head and body length 2.0–3.9 in (5–10 cm), tail length 1.8–3.5 in (4.5–9 cm), average weight 0.54 oz (15.3 g) for males, 0.38 oz (10.9 g) for females.	Can be found in savanna woodland and grassland, and reportedly in rainforests. Shelter consists of rocky areas, clumps of grass, bases of trees, or hollow logs. Nests are saucer-shaped, composed of dry grass. Most nocturnal, primarily terrestrial.	East Queensland, north-east New South Wales, and north Northern Territory, Australia.	Insects, spiders, and small mammals.	Not threatened
Gile's planigale *Planigale gilesi* Spanish: Planigale de Gile	Upperparts pale tawny olive, darker tawny, or brownish gray, underparts are olive buff, fuscous, or light tan. Head and body length 2.0–3.9 in (5–10 cm), tail length 1.8–3.5 in (4.5–9 mm), weight 0.18 oz (5 g).	Can be found mainly in savanna woodland and grass-land. Seemingly nocturnal, but active throughout periods of the day. Avid predator.	Northeast South Australia, northwest New South Wales, and southwest Queensland, Australia.	Insects, spiders, small lizards, and small mammals.	Not threatened
Fat-tailed pseudantechinus *Pseudantechinus macdonnellensis* Spanish: Pseudantechino de cola ancha	Upperparts are grayish brown, chestnut patches behind ears, underparts are grayish white. Head and body length 3.7–4.1 in (9.5–10.5 cm), tail length 3.0–3.3 in (7.5–8.5 cm), and weight 0.71–1.6 oz (20–45 g).	Can be found mainly on rocky hills, breakaways and in termite mounds. Predominantly nocturnal. Females produce one litter annually. Mating occurs in winter and spring.	North Western Australia, Northern Territory, and central deserts in Australia.	Mainly insects.	Not threatened
Fat-tailed dunnart *Sminthopsis crassicaudata* Spanish: Dunart de cola gorda	Soft, fine, dense fur, buffy to grayish in color, underparts are white or grayish white. Feet usually white, tail is brownish or grayish. Some species have a median facial stripe. Head and body length 3.3 in (8.3 cm), weight 0.35–0.53 oz (10–15 g).	Can be found mainly in dry country, but sometimes in moist areas. Dig burrows or construct nest of grasses and leaves. Mainly terrestrial, but some are agile climbers. Nocturnal.	South Australia, south-west Queensland, south-east Northern Territory, south Western Australia, west New South Wales, and west Victoria.	Mainly insectivorous, but also eats small vertebrates, such as lizards and mice.	Not threatened

[continued]

Common name / Scientific name/ Other common names	Physical characteristics	Habitat and behavior	Distribution	Diet	Conservation status
Slender-tailed dunnart *Sminthopsis murina* German: Kleine Schmalfussbeutelmaus; Spanish: Dunart de cola delgada	Back and sides buffy to grayish, underparts white or grayish white. Feet usually white, tail is brownish or grayish. Head and body length 2.8–4.7 in (7–12 cm), and tail length 2.2–5.1 in (5.5–13 cm).	Can be found in moist forest or savanna. Terrestrial and nocturnal. Can be up to eight young in a litter.	Southwest Western Australia, southeast South Australia, Victoria, New South Wales, and east Queensland.	Mainly insectivorous, but also eats small vertebrates. May also jump high to catch moths.	Not threatened
Sandhill dunnart *Sminthopsis psammophila*	Back and sides buffy to grayish, underparts white or grayish white. Tail accumulates fat when food is scarce. Head and body length 2.8–4.7 in (7–12 cm), tail length 2.2–5.1 in (5.5–13 cm).	Can be found in moist forest or savanna, but also arid grassland and desert. Terrestrial and nocturnal.	Australia, in southwest Northern Territory (vicinity of Ayer's Rock) and Eyre Peninsula in South Australia.	Mainly insectivorous, but small vertebrates also eaten.	Endangered
White-footed dunnart *Sminthopsis leucopus* Spanish: Dunart de patas blancas	Back and sides buffy to grayish, underparts white or grayish white. Head and body length 4.4 in (11.2 cm), weight 1.1 oz (30 g).	Can be found in moist forest or savanna. Terrestrial and nocturnal.	South and southeast Victoria, Tasmania, New South Wales, and Queensland, Australia.	Insects and small vertebrates.	Data Deficient

Resources

Books

Archer, M., T. Flannery, S. Hand, and J. Long. (2002). *Prehistoric Mammals of Australia and New Guinea: One Hundred Million Years of Evolution.* Sydney: UNSW Press, 2002.

Dickman, R. R. "Distributional Ecology of Dasyurid Marsupials." In *Predators with Pouches: The Biology of Carnivorous Marsupials,* edited by M. E. Jones, C. R. Dickman, and M. Archer. Melbourne: CSIRO Publishing: 2003.

Flannery, T. *Mammals of New Guinea.* Sydney: Reed Books, 1995.

Geiser, F. "Thermal Biology and Energetics of Carnivorous Marsupials." In *Predators with Pouches: The Biology of Carnivorous Marsupials,* edited by M. E. Jones, C. R. Dickman, and M. Archer. Melbourne: CSIRO Books, 2003.

Krajewski, C., and M. Westerman. "Molecular Systematics of Dasyuromorphia." In *Predators with Pouches: The Biology of Carnivorous Marsupials,* edited by M. E. Jones, C. R. Dickman, and M. Archer. Melbourne: CSIRO Publishing, 2003.

McAllan, B. "Timing of Reproduction in Carnivorous Marsupials." In *Predators with Pouches: The Biology of Carnivorous Marsupials,* edited by M. E. Jones, C. R. Dickman, and M. Archer. Melbourne: CSIRO Publishing, 2003.

Morris, K., B. Johnson, P. Orell, A. Wayne, and G. Gaikorst. "Recovery of the Threatened Chuditch (*Dasyurus geoffroii* Gould, 1841: A Case Study." In *Predators with Pouches: The Biology of Carnivorous Marsupials,* edited by M. E. Jones, C.

R. Dickman, and M. Archer. Melbourne: CSIRO Publishing, 2003.

Morton, S. R., C. R. Dickman, and T. P. Fletcher. "Dasyuridae." In *Fauna of Australia,* edited by D. W. Walton, and B. J. Richardson. Canberra: Australian Government Publishing Service, 1989.

Strahan, R., ed. *The Mammals of Australia.* Sydney: Australian Museum, Reed Books, 1995.

Wilson, B. A., C. R. Dickman, and T. P. Fletcher. "Dasyurid Dilemmas: Problems and Solutions for Conserving Australia's Small Carnivorous Marsupials." In *Predators with Pouches: The Biology of Carnivorous Marsupials,* edited by M. E. Jones, C. R. Dickman, and M. Archer. Melbourne: CSIRO Publishing, 2003.

Wroe, S. "Australian Marsupial Carnivores: An Overview of Recent Advances in Paleontology." In *Predators with Pouches: The Biology of Carnivorous Marsupials,* edited by M. E. Jones, C. R. Dickman, and M. Archer. Melbourne: CSIRO Publishing, 2003.

Periodicals

Fisher, D. O., C. R. Dickman. (1993). "Body Size-Prey Size Relationships in Insectivorous Marsupials: Tests of Three Hypotheses." *Ecology* 74 (1993): 1871–1883.

Jones, M. E., and L. A. Barmuta. "Diet Overlap and Abundance of Sympatric Dasyurid Carnivores: A Hypothesis of Competition?" *Journal of Animal Ecology* 67 (1998): 410–421.

Menna Jones, PhD

Numbat
(Myrmecobiidae)

Class Mammalia
Order Dasyuomorphia
Family Myrmecobiidae

Thumbnail description
Medium-sized reddish brown specialized termite-feeder with five to six striking white stripes across lower back, and a very long tail with long, erect hairs; long, thin tongue can protrude well beyond end of snout

Size
7.9–10.8 in (200–274 mm); 0.66–1.5 lb (0.3–0.7 kg)

Number of genera, species
1 genus; 1 species

Habitat
Forest, woodland, and spinifex

Conservation status
Vulnerable

Distribution
Extreme southwestern Australia; formerly broad band across western, southern half of Australia

Evolution and systematics

The evolutionary history of this family is poorly known. There is only one known species, the living numbat (*Myrmecobius fasciatus*), which is represented in a few Pleistocene cave deposits in Western Australia and western New South Wales. Myrmecobiids appear to be a sister group of a combined thylacinid-dasyurid group within the Dasyuromorphia.

The taxonomy for this species is *Myrmecobius fasciatus* (Waterhouse, 1836), Mt. Kokeby, Western Australia, Australia.

Physical characteristics

Numbats are one of the more beautiful and strikingly marked Australian mammals. Morphologically similar to the dasyurids, they are quadrupedal, place the heel of the hind foot on the ground when standing, and the snout is elongated and sharply pointed. Unique features of numbats include a very long tail, almost equal to the head and body length, and ears that are furred, erect, and quite narrow. Numbats also have more teeth than dasyurids; with five structurally simple molar teeth lacking defined cusps. Males and females are a similar size (1–1.1 lb; 0.45–0.5 kg). The medium-length soft

fur is reddish brown in color, darker towards the rump and paler below, with five to six transverse white stripes across the lower back and a white-bordered black stripe running from the snout to the base of the ear. Tail hairs are long and often erect. The female has four teats surrounded by crimped hair on the lower abdomen but there is no visible pouch. The very long, thin tongue can protrude several inches/centimeters beyond the end of the snout when feeding.

Distribution

At the time of European settlement (late eighteenth century to early nineteenth century), numbats were distributed in a broad band across the southern half of central and Western Australia, the eastern and northern limits of their range represented by western New South Wales and southwestern Northern Territory, respectively. By 1985, numbats had disappeared from all but two small locations in the southwest of Western Australia. A program of feral red fox (*Vulpes vulpes*) control, reintroduction, and translocations has resulted in nine wild and two free-ranging fenced populations.

Numbat (*Myrmecobius fasciatus*). (Illustration by Marguette Dongvillo)

mer, when they divide their activity into two periods: dawn until midday and then late afternoon. When not active, numbats sleep in hollow logs or trees, or underground burrows that they have dug themselves. They make a nest in a den with grass or shredded bark, and they regularly use more than one den. Numbats are solitary except when females are rearing young and occupy home ranges from which other individuals of the same sex are excluded. Young disperse in December and have been recorded moving in excess of 9 mi (15 km).

Feeding ecology and diet

Numbats are highly specialized with a diet that consists almost entirely of termites, although some ants are taken incidentally. Numbats sniff out underground termite galleries and expose termites by digging small holes and turning over sticks and branches. They have extremely sharp claws that they use for digging but the forelimbs are not especially strong. The long, slender tongue is inserted deep into the winding termite galleries and withdrawn rapidly, insects adhering to saliva on the tongue. Numbats have very large salivary glands to supply the prodigious quantities of saliva required for this mode of feeding. The molar teeth are simple in structure with three almost equal cusps and the number can vary in individuals, and also from side to side in the same individual, suggesting that the molars receive light use. Unlike other mammalian anteaters, numbats show no obvious specializations for termite-eating in the stomach.

Reproductive biology

Breeding is probably promiscuous and is seasonal with most young born in summer, after a 14-day gestation. Males as well as females show an annual cycle of fertility. The female usually carries the full complement of four young that, in the ab-

Habitat

The key to numbat presence is an abundance of termites, their primary food. The second prerequisite seems to be adequate ground-level cover, in the form of thickets of dense vegetation or hollow logs, which provides a refuge from predators. Primary predators would have been diurnal raptors, but foxes are now the major force driving extinction of populations. Hollow logs, which provide complete protection from larger predators, are probably more important in the presence of foxes. Numbats formerly occupied a variety of vegetation types, from open forest to woodland to hummock grasslands in the arid zone, although most sites had eucalypt trees.

Behavior

The numbat stands out among Australian mammals in being exclusively diurnal, probably as a consequence of their termite diet. Seasonal patterns in daily activity correspond closely to the abundance of termites in galleries close to the surface. Numbats are active in the warmer parts of the day, from midmorning until late afternoon, except in the hottest part of sum-

Foraging on the forest floor, the numbat (*Myrmecobius fasciatus*) unearths insects. (Photo by H. & J. Beste/Nature Focus, Australian Museum. Reproduced by permission.)

Tasmanian wolves
(Thylacinidae)

Class Mammalia
Order Dasyuromorphia
Family Thylacinidae

Thumbnail description
Medium-to-large carnivore, characterized by a long, narrow snout, sloping hindquarters that taper to a long, semi-rigid tail, with broad back stripes from shoulders to tail base

Size
4.9–6.4 ft (1.5–2 m); 33–77 lb (15–35 kg)

Number of genera, species
1 genus; 1 species

Habitat
Forest and woodlands

Conservation status
Extinct

Distribution
Island of Tasmania, Australia; subfossil on Australian continent

Evolution and systematics

At least 14 species of Tasmanian wolves from six genera are known from the fossil record, including *Thylacinus*, the last species (*T. cynocephalus*), which persisted until historical times. Thylacinids originated in the late Oligocene, reached their greatest diversity, with coexisting species, in the Miocene, and then declined steadily with only two species, including the giant *T. potens*, living in the Pleistocene. Thylacinids ranged in size from small carnivores (4.4–11 lb; 2–5 kg) to slightly larger than the thylacine (66 lb; 30 kg). Thylacinids are morphologically conservative among the Dasyuromorphia, including *T. cynocephalus*, which were little derived from the late Oligocene thylacinids, and are most closely related to the dasyurids, although they are convergent with the extinct South American marsupial borhyaenids.

The taxonomy for this species is *Thylacinus cynocephalus* (Harris, 1808), Tasmania, Australia.

Physical characteristics

Tasmanian wolves are superficially dog-shaped. They walk on four legs, although the legs are shorter than most canids. The head is doglike with a long, narrow snout, medium-sized (3 in; 80 mm) erect ears, and a strong jaw. The hindquarters slope and taper to a long, semi-rigid tail. The footpads extend to the heel and wrist joints. The re-

cently extinct *T. cynocephalus* was sexually size dimorphic: females approximately 33 lb (15 kg), males up to 66 lb (30 kg). Body hair is short (to 0.6 in; 15 mm) and sandy brown in color, with 15–20 brown stripes across the back, extending from behind the shoulders to the base of the tail. The fe-

Tasmanian wolf (*Thylacinus cynocephalus*). (Illustration by Wendy Baker)

Tasmanian wolves had a large gape. (Illustration by Wendy Baker)

male pouch opens slightly posteriorly and contains four teats. Males also have a small pouch-like depression around the scrotum. There are four upper and three lower incisor teeth, one set of canines, and three sets of premolars. Each of the four molars is similar in form, with major slicing (carnassial) and minor grinding surfaces.

Distribution

Records from diaries and bounty payments in the nineteenth century indicate a historic range that incorporated the entire island of Tasmania, although the wolves were very scarce in the southwest and western regions, except on the coastal strip. This distribution is similar to the current range of the Tasmanian devil (*Sarcophilus laniarius*) and correlates with mean annual rainfall and associated vegetation. The thylacine reached its highest population densities in the low-to-moderate rainfall zones of the north, center, and east of the state, and thylacines occurred at all altitudes. Thylacines have been extinct on the Australian continent for not less than 2,000 years; subfossil and prehistoric distribution was broad. There are fossil records from New Guinea.

Habitat

Historic reports indicate a broad range of reasonably open habitats: grassy woodlands, coastal and alpine scrub, and open forests. Thylacines seem to have avoided the dense, wet rainforest of western and southwestern Tasmania. Their habitat preferences completely overlap with those of the devil and are consistent with the distribution of dense populations of prey. Tasmanian wolves are reported to have used dense vegetation and rocky outcrops during the day (probably for dens), hunt-

ing in adjacent open grassy woodlands and forests at night. The number of subfossil remains found in caves in Tasmania attest to the use of larger caves as lairs.

Behavior

Sightings of thylacines were usually of solitary animals. Occasional sightings of adult-sized animals together cannot be construed as evidence of pair-bonding, and there is no evidence to support territorial defense of home ranges. Tasmanian wolves were mostly nocturnal but were occasionally observed active during the day. Vocalizations included a coughing bark and a sigh emitted while hunting, and a warning hiss, a low growl, and an undulating screech that were thought to be antagonistic. These vocalizations are not dissimilar in structure to the vocal repertoire of the Tasmanian devil.

Feeding ecology and diet

Unfortunately, early research interest in the thylacine concerned classical anatomy and the species became extinct without any serious study of its ecology. What is known of diet, hunting, and killing behaviors has been gleaned from historical anecdotes or reconstructed from comparison of skeletal remains with its living relatives. Thylacines are reported to have taken a wide variety of prey, including wombats, macropods, possums, bandicoots, small mammals, and birds, suggesting they were generalist predators of prey between less than 2.2 lb (1 kg) and probably not much more than 66 lb (30 kg). Tasmanian wolves had a long, thin snout relative to all other mammalian carnivores, marsupial or placental, most like that of a fox. This translates to a relatively weak bite force at the canine teeth. Museum-collection skulls also have very low rates

The pouch of the Tasmanian wolf faced backwards. When the pouch young were large, the pouch bulged downward from the animal. (Illustration by Wendy Baker)

The Tasmanian wolf (*Thylacinus cynocephalus*) is believed to be extinct. This is one of very few known photos of the species alive and free-living. The date of the photo is unknown. (Photo by Photo Researchers, Inc. Reproduced by permission.)

of breakage of the canines. Combined with the dietary records, and in contrast to prey sizes taken by devils and the larger quolls (up to three times their body weight), this combination of features suggests that thylacines did not routinely kill very heavy-bodied prey or prey much larger than themselves (33–66 lb; 15–30 kg). While they are recorded killing kangaroos, it is unlikely that they regularly killed healthy large males (up to 155 lb; 70 kg) or the larger megafauna such as diprotodonts. A similar ovoid cross-sectional shape of the canine teeth to the living larger dasyurid carnivores suggests that thylacines probably killed their prey using a generalized crushing bite used in killing. Tasmanian wolves were probably not swift runners, which is indicated by comparison of their leg bone ratios with other marsupial and placental carnivores. Unlikely to be capable of sustained, fast pursuit, thylacines probably hunted using a combination of stealth, short pursuit, and ambush. Putting all of these pieces of information together, it is likely that the thylacine filled a niche more similar to a medium-sized canid such as a coyote than to a wolf.

Reproductive biology

Little is known of reproduction in the Tasmanian wolves and they were bred only once in captivity, although females with pouch young were trapped and kept in zoos. Breeding appears to have been timed, as with other dasyurid carnivores, so that young became independent in spring when food supply is maximal. Gestation was probably less than one month, pouch life thought to be around four months. Neither the period of maternal care, post-pouch-vacation, nor the mating system is known.

Conservation status

Thylacines are classified by IUCN criteria as Extinct. The story of the decline and extinction of the thylacine is a sad tale of a deliberate strategy of persecution and a convenient scapegoat. Eighteenth-century settlers, experiencing signifiant sheep losses, employed "tiger men" to destroy Tasmanian wolves on their properties and successfully lobbied the government to instigate a bounty. While there is no doubt that thylacines killed sheep, it is thought that poaching and feral dogs were responsible for the majority of missing and dead sheep. The intense pressure placed on populations (2,184 bounty payments in 22 years as well as unrecorded deaths) of this probably never-abundant top predator would have driven thylacines to very low densities. Thylacines suddenly became very scarce in the first decade

The Tasmanian wolf (*Thylacinus cynocephalus*) lived in open forests and grasslands of Australia. Shown here is a model. (Photo by Creation Jacana/Photo Researchers, Inc. Reproduced by permission.)

of the twentieth century, with bounty payments falling from 100 to 150 per year to none between 1905 and 1910, and populations never recovered. The bounty scheme was scrapped in 1912 and the species given official protection on July 14, 1936. The last confirmed living animal died in the Hobart Zoo on September 7, 1936, and the last confirmed killing of a wild Tasmanian wolf was in 1930. There have been and continue to be sightings that appear credible but no thylacine has turned up.

Significance to humans

Tasmanian wolves were deliberately killed and eaten by aboriginal peoples both on the Australian mainland and in Tasmania. Mainland aboriginal rock art depicts speared thylacines as well as females feeding young. Practices varied from tribe to tribe. George Augustus Robinson, an early colonist, recorded consumption of thylacines by some tribes, but others seemed to revere the species, building shelters to cover the body after skinning it and keeping the skull.

Resources

Books

Archer, M., T. Flannery, S. Hand, and J. Long. *Prehistoric Mammals of Australia and New Guinea: One Hundred Million Years of Evolution.* Sydney: UNSW Press, 2002.

Dixon, J. M. "Thylacinidae." In *Fauna of Australia*, edited by D. W. Walton, and B. J. Richardson. Canberra: Australian Government Publishing Service, 1989.

Guiler, E., and P. Godard. *Tasmanian Tiger, A Lesson to be Learnt.* Perth, Western Australia: Abrolhos Publishing, 1998.

Guiler, E. R. *Thylacine: The Tragedy of the Tasmanian Tiger.* Melbourne: Oxford University Press, 1985.

Jones, M. E. "Predators, Pouches and Partitioning: Ecomorphology and Guild Structure of Marsupial and Placental Carnivores." In *Predators with Pouches: The Biology of Carnivorous Marsupials*, edited by M. E. Jones, C. R. Dickman, and M. Archer. Melbourne: CSIRO Books, 2003.

Krajewski, C., and M. Westerman. "Molecular Systematics of Dasyuromorphia." In *Predators with Pouches: The Biology of Carnivorous Marsupials*, edited by M. E. Jones, C. R. Dickman, and M. Archer. Melbourne: CSIRO Publishing, 2003.

Paddle, R. *The Last Tasmanian Tiger. The History and Extinction of the Thylacine.* Cambridge: Cambridge University Press, 2000.

Strahan, R. *The Mammals of Australia.* Sydney: Australian Museum Reed Books, 1995.

Wroe, S. "Australian Marsupial Carnivores: An Overview of Recent Advances in Palaeontology." In *Predators with Pouches: The Biology of Carnivorous Marsupials*, edited by M. E. Jones, C. R. Dickman, and M. Archer. Melbourne: CSIRO Publishing, 2003.

Periodicals

Jones, M. E., and D. M. Stoddart. "Reconstruction of the Predatory Behaviour of the Extinct Marsupial Thylacine." *Journal of Zoology* 246 (1998): 239–246.

Other

The Thylacine Museum. <http://naturalworlds.org/thylacine/>.

Menna Jones, PhD

For further reading

Alcock, J. *Animal Behavior*. New York: Sinauer, 2001.

Alderton, D. *Rodents of the World*. New York: Facts on File, 1996.

Alterman, L., G. A. Doyle, and M. K. Izard, eds. *Creatures of the Dark: The Nocturnal Prosimians*. New York: Plenum Press, 1995.

Altringham, J. D. *Bats: Biology and Behaviour*. New York: Oxford University Press, 2001.

Anderson, D. F., and S. Eberhardt. *Understanding Flight*. New York: McGraw-Hill, 2001.

Anderson, S., and J. K. Jones Jr., eds. *Orders and Families of Recent Mammals of the World*. John Wiley & Sons, New York, 1984.

Apps, P. *Smithers' Mammals of Southern Africa*. Cape Town: Struik Publishers, 2000.

Attenborough, D. *The Life of Mammals*. London: BBC, 2002.

Au, W. W. L. *The Sonar of Dolphins*. New York: Springer Verlag, 1993.

Austin, C. R., and R. V. Short, eds. *Reproduction in Mammals*. 4 vols. Cambridge: Cambridge University Press, 1972.

Avise, J. C. *Molecular Markers, Natural History and Evolution*. London: Chapman & Hall, 1994.

Barber, P. *Vampires, Burial, and Death: Folklore and Reality*. New Haven: Yale University Press, 1998.

Barnett, S. A. *The Story of Rats*. Crows Nest, Australia: Allen & Unwin, 2001.

Baskin, L., and K. Danell. *Ecology of Ungulates. A Handbook of Species in Eastern Europe, Northern and Central Asia*. Heidelberg: Springer-Verlag, 2003.

Bates, P. J. J., and D. L. Harrison. *Bats of the Indian Subcontinent*. Sevenoaks, U. K.: Harrison Zoological Museum, 1997.

Bekoff, M., C. Allen, and G. M. Burghardt, eds. *The Cognitive Animal*. Cambridge: MIT Press, 2002.

Bennett, N. C., and C. G. Faulkes. *African Mole-rats: Ecology and Eusociality*. Cambridge: Cambridge University Press, 2000.

Benton, M. J. *The Rise of the Mammals*. New York: Crescent Books, 1991.

Berta, A., and L. Sumich. *Marine Mammals: Evolutionary Biology*. San Diego: Academic Press, 1999.

Bonaccorso, F. J. *Bats of Papua New Guinea*. Washington, DC: Conservation International, 1998.

Bonnichsen, R, and K. L. Turnmire, eds. *Ice Age People of North America*. Corvallis: Oregon State University Press. 1999.

Bright, P. and P. Morris. *Dormice* London: The Mammal Society, 1992.

Broome, D., ed. *Coping with Challenge*. Berlin: Dahlem University Press, 2001.

Buchmann, S. L., and G. P. Nabhan. *The Forgotten Pollinators*. Washington, DC: Island Press, 1997.

Burnie, D., and D. E. Wilson, eds. *Animal*. Washington, DC: Smithsonian Institution, 2001.

Caro, T., ed. *Behavioral Ecology and Conservation Biology*. Oxford: Oxford University Press, 1998.

Carroll, R. L. *Vertebrate Paleontology and Evolution*. New York: W. H. Freeman and Co., 1998.

Cavalli-Sforza, L. L., P. Menozzi, and A. Piazza. *The History and Geography of Human Genes*. Princeton: Princeton University Press, 1994.

Chivers, R. E., and P. Lange. *The Digestive System in Mammals: Food, Form and Function*. New York: Cambridge University Press, 1994.

Clutton-Brock, J. *A Natural History of Domesticated Mammals*. 2nd ed. Cambridge: Cambridge University Press, 1999.

Conley, V. A. *The War Against the Beavers*. Minneapolis: University of Minnesota Press, 2003.

Cowlishaw, G., and R. Dunbar. *Primate Conservation Biology*. Chicago: University of Chicago Press, 2000.

For further reading

Craighead, L. *Bears of the World* Blaine, WA: Voyager Press, 2000.

Crichton., E. G. and P. H. Krutzsch, eds. *Reproductive Biology of Bats.* New York: Academic Press, 2000.

Croft, D. B., and U. Gansloßer, eds. *Comparison of Marsupial and Placental Behavior.* Fürth, Germany: Filander, 1996.

Darwin, C. *The Autobiography of Charles Darwin 1809–1882 with original omissions restored.* Edited by Nora Barlow. London: Collins, 1958.

Darwin, C. *On The Origin of Species by Means of Natural Selection, or The Preservation of Favoured Races in the Struggle for Life.* London: John Murray, 1859.

Darwin, C. *The Zoology of the Voyage of HMS* Beagle *under the Command of Captain Robert FitzRoy RN During the Years 1832-1836.* London: Elder & Co., 1840.

Dawson, T. J. *Kangaroos: The Biology of the Largest Marsupials.* Kensington, Australia: University of New South Wales Press/Ithaca, 2002.

Duncan, P. *Horses and Grasses.* New York: Springer-Verlag Inc., 1991.

Easteal, S., C. Collett, and D. Betty. *The Mammalian Molecular Clock.* Austin, TX: R. G. Landes, 1995.

Eisenberg, J. F. *Mammals of the Neotropics.* Vol. 1, *The Northern Neotropics.* Chicago: University of Chicago Press, 1989.

Eisenberg, J. F., and K. H. Redford. *Mammals of the Neotropics.* Vol. 3, *The Central Neotropics.* Chicago: University of Chicago Press, 1999.

Ellis, R. *Aquagenesis.* New York: Viking, 2001.

Estes, R. D. *The Behavior Guide to African Mammals.* Berkeley: The University of California Press, 1991.

Estes, R. D. *The Safari Companion: A Guide to Watching African Mammals.* White River Junction, VT: Chelsea Green, 1999.

Evans, P. G. H., and J. A. Raga, eds. *Marine Mammals: Biology and Conservation.* New York: Kluwer Academic/Plenum, 2001.

Ewer, R. F. *The Carnivores.* Ithaca, NY: Comstock Publishing, 1998.

Feldhamer, G. A., L. C. Drickamer, A. H. Vessey, and J. F. Merritt. *Mammalogy. Adaptations, Diversity, and Ecology.* Boston: McGraw Hill, 1999.

Fenton, M. B. *Bats.* Rev. ed. New York: Facts On File Inc., 2001.

Findley, J. S. *Bats: A Community Perspective.* Cambridge: Cambridge University Press, 1993.

Flannery, T. F. *Mammals of New Guinea.* Ithaca: Cornell University Press, 1995.

Flannery, T. F. *Possums of the World: A Monograph of the Phalangeroidea.* Sydney: GEO Productions, 1994.

Fleagle, J. G. *Primate Adaptation and Evolution.* New York: Academic Press, 1999.

Frisancho, A. R. *Human Adaptation and Accommodation.* Ann Arbor: University of Michigan Press, 1993.

Garbutt, N. *Mammals of Madagascar.* New Haven: Yale University Press, 1999.

Geist, V. *Deer of the World: Their Evolution, Behavior, and Ecology.* Mechanicsburg, PA: Stackpole Books, 1998.

Geist, V. *Life Strategies, Human Evolution, Environmental Design.* New York: Springer Verlag, 1978.

Gillespie, J. H. *The Causes of Molecular Evolution.* Oxford: Oxford University Press, 1992.

Gittleman, J. L., ed. *Carnivore Behavior, Ecology and Evolution.* 2 vols. Chicago: University of Chicago Press, 1996.

Gittleman, J. L., S. M. Funk, D. Macdonald, and R. K. Wayne, eds. *Carnivore Conservation.* Cambridge: Cambridge University Press, 2001.

Givnish, T. I. and K. Sytsma. *Molecular Evolution and Adaptive Radiations.* Cambridge: Cambridge University Press, 1997.

Goldingay, R. L., and J. H. Scheibe, eds. *Biology of Gliding Mammals.* Fürth, Germany: Filander Verlag, 2000.

Goodman, S. M., and J. P. Benstead, eds. *The Natural History of Madagascar.* Chicago: The University of Chicago Press, 2003.

Gosling, L. M., and W. J. Sutherland, eds. *Behaviour and Conservation.* Cambridge: Cambridge University Press, 2000.

Gould, E., and G. McKay, eds. *Encyclopedia of Mammals.* 2nd ed. San Diego: Academic Press, 1998.

Groves, C. P. *Primate Taxonomy.* Washington, DC: Smithsonian Institute, 2001.

Guthrie, D. R. *Frozen Fauna of the Mammoth Steppe.* Chicago: University of Chicago Press. 1990.

Hall, L., and G. Richards. *Flying Foxes: Fruit and Blossom Bats of Australia.* Malabar, FL: Krieger Publishing Company, 2000.

Hancocks, D. *A Different Nature. The Paradoxical World of Zoos and Their Uncertain Future.* Berkeley: University of California Press, 2001.

Hartwig, W. C., ed. *The Primate Fossil Record.* New York: Cambridge University Press, 2002.

Hildebrand, M. *Analysis of Vertebrate Structure.* 4th ed. New York: John Wiley & Sons, 1994.

Hillis, D. M., and C. Moritz. *Molecular Systematics.* Sunderland, MA: Sinauer Associates, 1990.

Hoelzel, A. R., ed. *Marine Mammal Biology: An Evolutionary Approach.* Oxford: Blackwell Science, 2002.

Hunter, M. L., and A Sulzer. *Fundamentals of Conservation Biology.* Oxford, U. K.: Blackwell Science, Inc., 2001.

Jefferson, T. A., S. Leatherwood, and M. A. Webber, eds. *Marine Mammals of the World.* Heidelberg: Springer-Verlag, 1993.

Jensen, P., ed. *The Ethology of Domestic Animals: An Introductory Text.* Oxon, MD: CABI Publishing, 2002.

Jones, M. E., C. R. Dickman, and M. Archer. *Predators with Pouches: The Biology of Carnivorous Marsupials.* Melbourne: CSIRO Books, 2003.

Kardong, K. V. *Vertebrates: Comparative Anatomy, Function, Evolution.* Dubuque, IA: William C. Brown Publishers, 1995.

King, C. M. *The Handbook of New Zealand Mammals.* Auckland: Oxford University Press, 1990.

Kingdon, J. *The Kingdon Field Guide to African Mammals.* London: Academic Press, 1997.

Kingdon, J., D. Happold, and T. Butynski, eds. *The Mammals of Africa: A Comprehensive Synthesis.* London: Academic Press, 2003.

Kinzey, W. G., ed. *New World Primates: Ecology, Evolution, and Behavior.* New York: Aldine de Gruyter, 1997.

Kosco, M. *Mammalian Reproduction.* Eglin, PA: Allegheny Press, 2000.

Krebs, J. R., and N. B. Davies. *An Introduction to Behavioural Ecology.* 3rd ed. Oxford: Blackwell Scientific Publications, 1993.

Kunz, T. H., and M. B. Fenton, eds. *Bat Ecology.* Chicago: University of Chicago Press, 2003.

Lacey, E. A., J. L. Patton, and G. N. Cameron, eds. *Life Underground: The Biology of Subterranean Rodents.* Chicago: University of Chicago Press, 2000.

Lott, D. F. *American Bison: A Natural History.* Berkeley: University of California Press, 2002.

Macdonald, D. W. *European Mammals: Evolution and Behavior.* London: Collins, 1995.

Macdonald, D. W. *The New Encyclopedia of Mammals.* Oxford: Oxford University Press, 2001.

Macdonald, D. W. *The Velvet Claw: A Natural History of the Carnivores.* London: BBC Books, 1992.

Macdonald, D. W., and P. Barrett. *Mammals of Britain and Europe.* London: Collins, 1993.

Martin, R. E. *A Manual of Mammalogy: With Keys to Families of the World.* 3rd ed. Boston: McGraw-Hill, 2001.

Matsuzawa, T., ed. *Primate Origins of Human Cognition and Behavior.* Tokyo: Springer-Verlag, 2001.

Mayr, E. *What Evolution Is.* New York: Basic Books, 2001.

McCracken, G. F., A. Zubaid, and T. H. Kunz, eds. *Functional and Evolutionary Ecology of Bats.* Oxford: Oxford University Press, 2003.

McGrew, W. C., L. F. Marchant, and T. Nishida, eds. *Great Ape Societies.* Cambridge: Cambridge University Press, 1996.

Meffe, G. K., and C. R. Carroll. *Principles of Conservation Biology.* Sunderland, MA: Sinauer Associates, Inc., 1997.

Menkhorst, P. W. *A Field Guide to the Mammals of Australia.* Melbourne: Oxford University Press, 2001.

Mills, G., and M. Harvey. *African Predators.* Cape Town: Struik Publishers, 2001.

Mills, G., and L. Hes. *Complete Book of Southern African Mammals.* Cape Town: Struik, 1997.

Mitchell-Jones, A. J., et al. *The Atlas of European Mammals.* London: Poyser Natural History/Academic Press, 1999.

Neuweiler, G. *Biology of Bats.* Oxford: Oxford University Press, 2000.

Norton, B. G., et al. *Ethics on the Ark.* Washington, DC: Smithsonian Institution Press, 1995.

Nowak, R. M. *Walker's Bats of the World.* Baltimore: The Johns Hopkins University Press, 1994.

Nowak, R. M. *Walker's Mammals of the World.* 6th ed. Baltimore: Johns Hopkins University Press, 1999.

Nowak, R. M. *Walker's Primates of the World.* Baltimore: The Johns Hopkins University Press, 1999.

Payne, K. *Silent Thunder: The Hidden Voice of Elephants.* Phoenix: Wiedenfeld and Nicholson, 1999.

Pearce, J. D. *Animal Learning and Cognition.* New York: Lawrence Erlbaum, 1997.

Pereira, M. E., and L. A. Fairbanks, eds. *Juvenile Primates: Life History, Development, and Behavior.* New York: Oxford University Press, 1993.

Perrin, W. F., B. Würsig, and J. G. M. Thewissen. *Encyclopedia of Marine Mammals.* San Diego: Academic Press, 2002.

Popper, A. N., and R. R. Fay, eds. *Hearing by Bats.* New York: Springer-Verlag, 1995.

Pough, F. H., C. M. Janis, and J. B. Heiser. *Vertebrate Life.* 6th ed. Upper Saddle River, NJ: Prentice Hall, 2002.

Premack, D., and A. J. Premack. *Original Intelligence: The Architecture of the Human Mind.* New York: McGraw-Hill/Contemporary Books, 2002.

Price, E. O. *Animal Domestication and Behavior.* Cambridge, MA: CAB International, 2002.

Racey, P. A., and S. M. Swift, eds. *Ecology, Evolution and Behaviour of Bats.* Oxford: Clarendon Press, 1995.

Redford, K. H., and J. F. Eisenberg. *Mammals of the Neotropics.* Vol. 2, *The Southern Cone.* Chicago: University of Chicago Press, 1992.

Reeve, N. *Hedgehogs.* London: Poyser Natural History, 1994.

Reeves, R., B. Stewart, P. Clapham, and J. Powell. *Sea Mammals of the World.* London: A&C Black, 2002.

Reynolds, J. E. III, and D. K. Odell. *Manatees and Dugongs.* New York: Facts On File, 1991.

Reynolds, J. E. III, and S. A. Rommel, eds. *Biology of Marine Mammals.* Washington, DC: Smithsonian Institution Press, 1999.

Rice, D. W. *Marine Mammals of the World.* Lawrence, KS: Allen Press, 1998.

Ridgway, S. H., and R. Harrison, eds. *Handbook of Marine Mammals.* 6 vols. New York: Academic Press, 1985-1999.

Riedman, M. *The Pinnipeds.* Berkeley: University of California Press, 1990.

Rijksen, H., and E. Meijaard. *Our Vanishing Relative: The Status of Wild Orang-utans at the Close of the Twentieth Century.* Dordrecht: Kluwer Academic Publishers, 1999.

Robbins, C. T. *Wildlife Feeding and Nutrition.* San Diego: Academic Press, 1992.

Robbins, M. M., P. Sicotte, and K. J. Stewart, eds. *Mountain Gorillas: Three Decades of Research at Karisoke.* Cambridge: Cambridge University Press, 2001.

Roberts, W. A. *Principles of Animal Cognition.* New York: McGraw-Hill, 1998.

Schaller, G. B. *Wildlife of the Tibetan Steppe.* Chicago: University of Chicago Press, 1998.

Seebeck, J. H., P. R. Brown, R. L. Wallis, and C. M. Kemper, eds. *Bandicoots and Bilbies.* Chipping Norton, Australia: Surrey Beatty & Sons, 1990.

Shepherdson, D. J., J. D. Mellen, and M. Hutchins. *Second Nature: Environmental Enrichment for Captive Animals.* Washington, DC: Smithsonian Institution Press, 1998.

Sherman, P. W., J. U. M. Jarvis, and R. D. Alexander, eds. *The Biology of the Naked Mole-rat.* Princeton: Princeton University Press, 1991.

Shettleworth, S. J. *Cognition, Evolution, and Behavior.* Oxford: Oxford University Press, 1998.

Shoshani, J., ed. *Elephants.* London: Simon & Schuster, 1992.

Skinner, R., and R. H. N. Smithers. *The Mammals of the Southern African Subregion.* 2nd ed. Pretoria, South Africa: University of Pretoria, 1998.

Sowls, L. K. *The Peccaries.* College Station: Texas A&M Press, 1997.

Steele, M. A. and J. Koprowski. *North American Tree Squirrels.* Washington, DC: Smithsonian Institution Press, 2001.

Sunquist, M. and F. Sunquist. *Wild Cats of the World* Chicago: University of Chicago Press, 2002.

Sussman, R. W. *Primate Ecology and Social Structure.* 3 vols. Needham Heights, MA: Pearson Custom Publishing, 1999.

Szalay, F. S., M. J. Novacek, and M. C. McKenna, eds. *Mammalian Phylogeny.* New York: Springer-Verlag, 1992.

Thomas, J. A., C. A. Moss, and M. A. Vater, eds. *Echolocation in Bats and Dolphins.* Chicago: University of Chicago Press, 2003.

Thompson, H. V., and C. M. King, eds. *The European Rabbit: The History and Biology of a Successful Colonizer.* Oxford: Oxford University Press, 1994.

Tomasello, M., and J. Calli. *Primate Cognition.* Chicago: University of Chicago Press, 1997.

Twiss, J. R. Jr., and R. R. Reeves, eds. *Conservation and Management of Marine Mammals.* Washington, DC: Smithsonian Institution Press, 1999.

Van Soest, P. J. *Nutritional Ecology of the Ruminant.* 2nd ed. Ithaca, NY: Cornell University Press, 1994.

Vaughan, T., J. Ryan, and N. Czaplewski. *Mammalogy.* 4th ed. Philadelphia: Saunders College Publishing, 1999.

Vrba, E. S., G. H. Denton, T. C. Partridge, and L. H. Burckle, eds. *Paleoclimate and Evolution, with Emphasis on Human Origins.* New Haven: Yale University Press, 1995.

Vrba, E. S., and G. G. Schaller, eds. *Antelopes, Deer and Relatives: Fossil Record, Behavioral Ecology, Systematics and Conservation.* New Haven: Yale University Press, 2000.

Wallis, Janice, ed. *Primate Conservation: The Role of Zoological Parks.* New York: American Society of Primatologists, 1997.

Weibel, E. R., C. R. Taylor, and L. Bolis. *Principles of Animal Design.* New York: Cambridge University Press, 1998.

Wells, R. T., and P. A. Pridmore. *Wombats.* Sydney: Surrey Beatty & Sons, 1998.

Whitehead, G. K. *The Whitehead Encyclopedia of Deer.* Stillwater, MN: Voyager Press, 1993.

Wilson, D. E., and D. M. Reeder, eds. *Mammal Species of the World: a Taxonomic and Geographic Reference.* 2nd ed. Washington, DC: Smithsonian Institution Press, 1993.

Wilson, D. E., and S. Ruff, eds. *The Smithsonian Book of North American Mammals*. Washington, DC: Smithsonian Institution Press, 1999.

Wilson, E. O. *The Diversity of Life*. Cambridge: Harvard University Press, 1992.

Wójcik, J. M., and M. Wolsan, eds. *Evolution of Shrews*. Bialowieza, Poland: Mammal Research Institute, Polish Academy of Sciences, 1998.

Woodford, J. *The Secret Life of Wombats*. Melbourne: Text Publishing, 2001.

Wrangham, R. W., W. C. McGrew, F. B. M. de Waal, and P. G. Heltne, eds. *Chimpanzee Cultures*. Cambridge: Harvard University Press, 1994.

Wynne, C. D. L. *Animal Cognition*. Basingstoke, U. K.: Palgrave, 2001.

FOR FURTHER READING

Organizations

African Wildlife Foundation
1400 16th Street, NW, Suite 120
Washington, DC 20036 USA
Phone: (202) 939-3333
Fax: (202) 939-3332
E-mail: africanwildlife@awf.org
<http://www.awf.org/>

The American Society of Mammalogists
<http://www.mammalsociety.org/>

American Zoo and Aquarium Association
8403 Colesville Road, Suite 710
Silver Spring, MD 20910 USA.
Phone: (301) 562-0777
Fax: (301) 562-0888
<http://www.aza.org/>

Australian Conservation Foundation Inc.
340 Gore Street
Fitzroy, Victoria 3065 Australia
Phone: (3) 9416 1166
<http://www.acfonline.org.au>

The Australian Mammal Society
<http://www.australianmammals.org.au/>

Australian Regional Association of Zoological Parks and
Aquaria
PO Box 20
Mosman, NSW 2088
Australia
Phone: 61 (2) 9978-4797
Fax: 61 (2) 9978-4761
<http://www.arazpa.org>

Bat Conservation International
P.O. Box 162603
Austin, TX 78716 USA
Phone: (512) 327-9721
Fax: (512) 327-9724
<http://www.batcon.org/>

Center for Ecosystem Survival
699 Mississippi Street, Suite 106
San Francisco, 94107 USA
Phone: (415) 648-3392
Fax: (415) 648-3392

E-mail: info@savenature.org
<http://www.savenature.org/>

Conservation International
1919 M Street NW, Ste. 600
Washington, DC 20036
Phone: (202) 912-1000
<http://www.conservation.org>

The European Association for Aquatic Mammals
E-mail: info@eaam.org
<http://www.eaam.org/>

European Association of Zoos and Aquaria
PO Box 20164
1000 HD Amsterdam
The Netherlands
<http://www.eaza.net>

IUCN-The World Conservation Union
Rue Mauverney 28
Gland 1196 Switzerland
Phone: ++41(22) 999-0000
Fax: ++41(22) 999-0002
E-mail: mail@iucn.org
<http://www.iucn.org/>

The Mammal Society
2B, Inworth Street
London SW11 3EP United Kingdom
Phone: 020 7350 2200
Fax: 020 7350 2211
<http://www.abdn.ac.uk/mammal/>

Mammals Trust UK
15 Cloisters House
8 Battersea Park Road
London SW8 4BG United Kingdom
Phone: (+44) 020 7498 5262
Fax: (+44) 020 7498 4459
E-mail: enquiries@mtuk.org
<http://www.mtuk.org/>

The Marine Mammal Center
Marin Headlands
1065 Fort Cronkhite
Sausalito, CA 94965 USA
Phone: (415) 289-7325

Fax: (415) 289-7333
<http://www.marinemammalcenter.org/>

National Marine Mammal Laboratory
7600 Sand Point Way N.E. F/AKC3
Seattle, WA 98115-6349 USA
Phone: (206) 526-4045
Fax: (206) 526-6615
<http://nmml.afsc.noaa.gov/>

National Wildlife Federation
11100 Wildlife Center Drive
Reston, VA 20190-5362 USA
Phone: (703) 438-6000
<http://www.nwf.org/>

The Organization for Bat Conservation
39221 Woodward Avenue
Bloomfield Hills, MI 48303 USA
Phone: (248) 645-3232
E-mail: obcbats@aol.com
<http://www.batconservation.org/>

Scripps Institution of Oceanography
University of California-San Diego
9500 Gilman Drive
La Jolla, CA 92093 USA
<http://sio.ucsd.edu/gt;

Seal Conservation Society
7 Millin Bay Road
Tara, Portaferry
County Down BT22 1QD
United Kingdom

Phone: +44-(0)28-4272-8600
Fax: +44-(0)28-4272-8600
E-mail: info@pinnipeds.org
<http://www.pinnipeds.org>

The Society for Marine Mammalogy
<http://www.marinemammalogy.org/>

The Wildlife Conservation Society
2300 Southern Boulevard
Bronx, New York 10460
Phone: (718) 220-5100

Woods Hole Oceanographic Institution
Information Office
Co-op Building, MS #16
Woods Hole, MA 02543 USA
Phone: (508) 548-1400
Fax: (508) 457-2034
E-mail: information@whoi.edu
<http://www.whoi.edu/>

World Association of Zoos and Aquariums
PO Box 23
Liebefeld-Bern CH-3097
Switzerland
<http://www.waza.org>

World Wildlife Fund
1250 24th Street N.W.
Washington, DC 20037-1193 USA
Phone: (202) 293-4800
Fax: (202) 293-9211
<http://www.panda.org/>

ORGANIZATIONS

• • • • •

Contributors to the first edition

The following individuals contributed chapters to the original edition of Grzimek's Animal Life Encyclopedia, *which was edited by Dr. Bernhard Grzimek, Professor, Justus Liebig University of Giessen, Germany; Director, Frankfurt Zoological Garden, Germany; and Trustee, Tanzanian National Parks, Tanzania.*

Dr. Michael Abs
Curator, Ruhr University
Bochum, Germany

Dr. Salim Ali
Bombay Natural History Society
Bombay, India

Dr. Rudolph Altevogt
Professor, Zoological Institute,
University of Münster
Münster, Germany

Dr. Renate Angermann
Curator, Institute of Zoology,
Humboldt University
Berlin, Germany

Edward A. Armstrong
Cambridge University
Cambridge, England

Dr. Peter Ax
Professor, Second Zoological Institute
and Museum, University of Göttingen
Göttingen, Germany

Dr. Franz Bachmaier
Zoological Collection of the State of
Bavaria
Munich, Germany

Dr. Pedru Banarescu
Academy of the Roumanian Socialist
Republic, Trajan Savulescu Institute of
Biology
Bucharest, Romania

Dr. A. G. Bannikow
Professor, Institute of Veterinary
Medicine
Moscow, Russia

Dr. Hilde Baumgärtner
Zoological Collection of the State of
Bavaria
Munich, Germany

C. W. Benson
Department of Zoology, Cambridge
University
Cambridge, England

Dr. Andrew Berger
Chairman, Department of Zoology,
University of Hawaii
Honolulu, Hawaii, U.S.A.

Dr. J. Berlioz
National Museum of Natural
History
Paris, France

Dr. Rudolf Berndt
Director, Institute for Population
Ecology, Hiligoland Ornithological
Station
Braunschweig, Germany

Dieter Blume
Instructor of Biology, Freiherr-vom-
Stein School
Gladenbach, Germany

Dr. Maximilian Boecker
Zoological Research Institute and A.
Koenig Museum
Bonn, Germany

Dr. Carl-Heinz Brandes
Curator and Director, The Aquarium,
Overseas Museum
Bremen, Germany

Dr. Donald G. Broadley
Curator, Umtali Museum
Mutare, Zimbabwe

Dr. Heinz Brüll
Director; Game, Forest, and Fields
Research Station
Hartenholm, Germany

Dr. Herbert Bruns
Director, Institute of Zoology and the
Protection of Life
Schlangenbad, Germany

Hans Bub
Heligoland Ornithological Station
Wilhelmshaven, Germany

A. H. Chisholm
Sydney, Australia

Herbert Thomas Condon
Curator of Birds, South Australian
Museum
Adelaide, Australia

Dr. Eberhard Curio
Director, Laboratory of Ethology,
Ruhr University
Bochum, Germany

Dr. Serge Daan
Laboratory of Animal Physiology,
University of Amsterdam
Amsterdam, The Netherlands

Dr. Heinrich Dathe
Professor and Director, Animal Park
and Zoological Research Station,
German Academy of Sciences
Berlin, Germany

Dr. Wolfgang Dierl
Zoological Collection of the State of
Bavaria
Munich, Germany

Dr. Fritz Dieterlen
Zoological Research Institute and A.
Koenig Museum
Bonn, Germany

Dr. Rolf Dircksen
Professor, Pedagogical Institute
Bielefeld, Germany

Josef Donner
Instructor of Biology
Katzelsdorf, Austria

Dr. Jean Dorst
Professor, National Museum of
Natural History
Paris, France

Dr. Gerti Dücker
Professor and Chief Curator,
Zoological Institute, University of
Münster
Münster, Germany

Dr. Michael Dzwillo
Zoological Institute and Museum,
University of Hamburg
Hamburg, Germany

Dr. Irenäus Eibl-Eibesfeldt
Professor and Director, Institute of
Human Ethology, Max Planck
Institute for Behavioral Physiology
Percha/Starnberg, Germany

Dr. Martin Eisentraut
Professor and Director, Zoological
Research Institute and A. Koenig
Museum
Bonn, Germany

Dr. Eberhard Ernst
Swiss Tropical Institute
Basel, Switzerland

R. D. Etchecopar
Director, National Museum of
Natural History
Paris, France

Dr. R. A. Falla
Director, Dominion Museum
Wellington, New Zealand

Dr. Hubert Fechter
Curator, Lower Animals, Zoological
Collection of the State of Bavaria
Munich, Germany

Dr. Walter Fiedler
Docent, University of Vienna, and
Director, Schönbrunn Zoo
Vienna, Austria

Wolfgang Fischer
Inspector of Animals, Animal Park
Berlin, Germany

Dr. C. A. Fleming
Geological Survey Department of
Scientific and Industrial Research
Lower Hutt, New Zealand

Dr. Hans Frädrich
Zoological Garden
Berlin, Germany

Dr. Hans-Albrecht Freye
Professor and Director, Biological
Institute of the Medical School
Halle a.d.S., Germany

Günther E. Freytag
Former Director, Reptile and
Amphibian Collection, Museum of
Cultural History in Magdeburg
Berlin, Germany

Dr. Herbert Friedmann
Director, Los Angeles County
Museum of Natural History
Los Angeles, California, U.S.A.

Dr. H. Friedrich
Professor, Overseas Museum
Bremen, Germany

Dr. Jan Frijlink
Zoological Laboratory, University of
Amsterdam
Amsterdam, The Netherlands

Dr. H. C. Karl Von Frisch
Professor Emeritus and former
Director, Zoological Institute,
University of Munich
Munich, Germany

Dr. H. J. Frith
C.S.I.R.O. Research Institute
Canberra, Australia

Dr. Ion E. Fuhn
Academy of the Roumanian Socialist
Republic, Trajan Savulescu Institute of
Biology
Bucharest, Romania

Dr. Carl Gans
Professor, Department of Biology,
State University of New York at
Buffalo
Buffalo, New York, U.S.A.

Dr. Rudolf Geigy
Professor and Director, Swiss Tropical
Institute
Basel, Switzerland

Dr. Jacques Gery
St. Genies, France

Dr. Wolfgang Gewalt
Director, Animal Park
Duisburg, Germany

Dr. H. C. Viktor Goerttler
Professor Emeritus, University of Jena
Jena, Germany

Dr. Friedrich Goethe
Director, Institute of Ornithology,
Heligoland Ornithological Station
Wilhelmshaven, Germany

Dr. Ulrich F. Gruber
Herpetological Section, Zoological
Research Institute and A. Koenig
Museum
Bonn, Germany

Dr. H. R. Haefelfinger
Museum of Natural History
Basel, Switzerland

Dr. Theodor Haltenorth
Director, Mammalogy, Zoological
Collection of the State of Bavaria
Munich, Germany

Barbara Harrisson
Sarawak Museum, Kuching, Borneo
Ithaca, New York, U.S.A.

Dr. Francois Haverschmidt
President, High Court (retired)
Paramaribo, Suriname

Dr. Heinz Heck
Director, Catskill Game Farm
Catskill, New York, U.S.A.

Dr. Lutz Heck
Professor (retired), and Director,
Zoological Garden, Berlin
Wiesbaden, Germany

Dr. H. C. Heini Hediger
Director, Zoological Garden
Zurich, Switzerland

Dr. Dietrich Heinemann
Director, Zoological Garden, Münster
Dörnigheim, Germany

Dr. Helmut Hemmer
Institute for Physiological Zoology,
University of Mainz
Mainz, Germany

Dr. W. G. Heptner
Professor, Zoological Museum,
University of Moscow
Moscow, Russia

Dr. Konrad Herter
Professor Emeritus and Director
(retired), Zoological Institute, Free
University of Berlin
Berlin, Germany

Dr. Hans Rudolf Heusser
Zoological Museum, University of
Zurich
Zurich, Switzerland

Dr. Emil Otto Höhn
Associate Professor of Physiology,
University of Alberta
Edmonton, Canada

Dr. W. Hohorst
Professor and Director, Parasitological
Institute, Farbwerke Hoechst A.G.
Frankfurt-Höchst, Germany

Dr. Folkhart Hückinghaus
Director, Senckenbergische Anatomy,
University of Frankfurt a.M.
Frankfurt a.M., Germany

Francois Hüe
National Museum of Natural History
Paris, France

Dr. K. Immelmann
Professor, Zoological Institute,
Technical University of Braunschweig
Braunschweig, Germany

Dr. Junichiro Itani
Kyoto University
Kyoto, Japan

Dr. Richard F. Johnston
Professor of Zoology, University of
Kansas
Lawrence, Kansas, U.S.A.

Otto Jost
Oberstudienrat, Freiherr-vom-Stein
Gymnasium
Fulda, Germany

Dr. Paul Kähsbauer
Curator, Fishes, Museum of Natural
History
Vienna, Austria

Dr. Ludwig Karbe
Zoological State Institute and
Museum
Hamburg, Germany

Dr. N. N. Kartaschew
Docent, Department of Biology,
Lomonossow State University
Moscow, Russia

Dr. Werner Kästle
Oberstudienrat, Gisela Gymnasium
Munich, Germany

Dr. Reinhard Kaufmann
Field Station of the Tropical Institute,
Justus Liebig University, Giessen,
Germany
Santa Marta, Colombia

Dr. Masao Kawai
Primate Research Institute, Kyoto
University
Kyoto, Japan

Dr. Ernst F. Kilian
Professor, Giessen University and
Catedratico Universidad Australia,
Valdivia-Chile
Giessen, Germany

Dr. Ragnar Kinzelbach
Institute for General Zoology,
University of Mainz
Mainz, Germany

Dr. Heinrich Kirchner
Landwirtschaftsrat (retired)
Bad Oldesloe, Germany

Dr. Rosl Kirchshofer
Zoological Garden, University of
Frankfurt a.M.
Frankfurt a.M., Germany

Dr. Wolfgang Klausewitz
Curator, Senckenberg Nature
Museum and Research Institute
Frankfurt a.M., Germany

Dr. Konrad Klemmer
Curator, Senckenberg Nature
Museum and Research Institute
Frankfurt a.M., Germany

Dr. Erich Klinghammer
Laboratory of Ethology, Purdue
University
Lafayette, Indiana, U.S.A.

Dr. Heinz-Georg Klös
Professor and Director, Zoological
Garden
Berlin, Germany

Ursula Klös
Zoological Garden
Berlin, Germany

Dr. Otto Koehler
Professor Emeritus, Zoological
Institute, University of Freiburg
Freiburg i. BR., Germany

Dr. Kurt Kolar
Institute of Ethology, Austrian
Academy of Sciences
Vienna, Austria

Dr. Claus König
State Ornithological Station of Baden-
Württemberg
Ludwigsburg, Germany

Dr. Adriaan Kortlandt
Zoological Laboratory, University of
Amsterdam
Amsterdam, The Netherlands

Dr. Helmut Kraft
Professor and Scientific Councillor,
Medical Animal Clinic, University of
Munich
Munich, Germany

Dr. Friedrich Schaller
Professor and Chairman, First
Zoological Institute, University of
Vienna
Vienna, Austria

Dr. George B. Schaller
Serengeti Research Institute, Michael
Grzimek Laboratory
Seronera, Tanzania

Dr. Georg Scheer
Chief Curator and Director,
Zoological Institute, State Museum of
Hesse
Darmstadt, Germany

Dr. Christoph Scherpner
Zoological Garden
Frankfurt a.M., Germany

Dr. Herbert Schifter
Bird Collection, Museum of Natural
History
Vienna, Austria

Dr. Marco Schnitter
Zoological Museum, Zurich
University
Zurich, Switzerland

Dr. Kurt Schubert
Federal Fisheries Research Institute
Hamburg, Germany

Eugen Schuhmacher
Director, Animals Films, I.U.C.N.
Munich, Germany

Dr. Thomas Schultze-Westrum
Zoological Institute, University of
Munich
Munich, Germany

Dr. Ernst Schüt
Professor and Director (retired), State
Museum of Natural History
Stuttgart, Germany

Dr. Lester L. Short , Jr.
Associate Curator, American Museum
of Natural History
New York, New York, U.S.A.

Dr. Helmut Sick
National Museum
Rio de Janeiro, Brazil

Dr. Alexander F. Skutch
Professor of Ornithology, University
of Costa Rica
San Isidro del General, Costa Rica

Dr. Everhard J. Slijper
Professor, Zoological Laboratory,
University of Amsterdam
Amsterdam, The Netherlands

Bertram E. Smythies
Curator (retired), Division of Forestry
Management, Sarawak-Malaysia
Estepona, Spain

Dr. Kenneth E. Stager
Chief Curator, Los Angeles County
Museum of Natural History
Los Angeles, California, U.S.A.

Dr. H. C. Georg H. W. Stein
Professor, Curator of Mammals,
Institute of Zoology and Zoological
Museum, Humboldt University
Berlin, Germany

Dr. Joachim Steinbacher
Curator, Nature Museum and
Senckenberg Research Institute
Frankfurt a.M., Germany

Dr. Bernard Stonehouse
Canterbury University
Christchurch, New Zealand

Dr. Richard Zur Strassen
Curator, Nature Museum and
Senckenberg Research Institute
Frandfurt a.M., Germany

Dr. Adelheid Studer-Thiersch
Zoological Garden
Basel, Switzerland

Dr. Ernst Sutter
Museum of Natural History
Basel, Switzerland

Dr. Fritz Terofal
Director, Fish Collection, Zoological
Collection of the State of Bavaria
Munich, Germany

Dr. G. F. Van Tets
Wildlife Research
Canberra, Australia

Ellen Thaler-Kottek
Institute of Zoology, University of
Innsbruck
Innsbruck, Austria

Dr. Erich Thenius
Professor and Director, Institute of
Paleontolgy, University of Vienna
Vienna, Austria

Dr. Niko Tinbergen
Professor of Animal Behavior,
Department of Zoology, Oxford
University
Oxford, England

Alexander Tsurikov
Lecturer, University of Munich
Munich, Germany

Dr. Wolfgang Villwock
Zoological Institute and Museum,
University of Hamburg
Hamburg, Germany

Zdenek Vogel
Director, Suchdol Herpetological
Station
Prague, Czechoslovakia

Dieter Vogt
Schorndorf, Germany

Dr. Jiri Volf
Zoological Garden
Prague, Czechoslovakia

Otto Wadewitz
Leipzig, Germany

Dr. Helmut O. Wagner
Director (retired), Overseas Museum,
Bremen
Mexico City, Mexico

Dr. Fritz Walther
Professor, Texas A & M University
College Station, Texas, U.S.A.

John Warham
Zoology Department, Canterbury
University
Christchurch, New Zealand

Dr. Sherwood L. Washburn
University of California at Berkeley
Berkeley, California, U.S.A.

CONTRIBUTORS TO THE FIRST EDITION

Eberhard Wawra
First Zoological Institute, University
of Vienna
Vienna, Austria

Dr. Ingrid Weigel
Zoological Collection of the State of
Bavaria
Munich, Germany

Dr. B. Weischer
Institute of Nematode Research,
Federal Biological Institute
Münster/Westfalen, Germany

Herbert Wendt
Author, Natural History
Baden-Baden, Germany

Dr. Heinz Wermuth
Chief Curator, State Nature Museum,
Stuttgart
Ludwigsburg, Germany

Dr. Wolfgang Von Westernhagen
Preetz/Holstein, Germany

Dr. Alexander Wetmore
United States National Museum,
Smithsonian Institution
Washington, D.C., U.S.A.

Dr. Dietrich E. Wilcke
Röttgen, Germany

Dr. Helmut Wilkens
Professor and Director, Institute of
Anatomy, School of Veterinary
Medicine
Hannover, Germany

Dr. Michael L. Wolfe
Utah, U.S.A.

Hans Edmund Wolters
Zoological Research Institute and A.
Koenig Museum
Bonn, Germany

Dr. Arnfrid Wünschmann
Research Associate, Zoological Garden
Berlin, Germany

Dr. Walter Wüst
Instructor, Wilhelms Gymnasium
Munich, Germany

Dr. Heinz Wundt
Zoological Collection of the State of
Bavaria
Munich, Germany

Dr. Claus-Dieter Zander
Zoological Institute and Museum,
University of Hamburg
Hamburg, Germany

Dr. Fritz Zumpt
Director, Entomology and
Parasitology, South African Institute
for Medical Research
Johannesburg, South Africa

Dr. Richard L. Zusi
Curator of Birds, United States
National Museum, Smithsonian
Institution
Washington, D.C., U.S.A.

Glossary

Adaptive radiation—Diversification of a species or single ancestral type into several forms that are each adaptively specialized to a specific niche.

Agonistic—Behavioral patterns that are aggressive in context.

Allopatric—Occurring in separate, nonoverlapping geographic areas.

Alpha breeder—The reproductively dominant member of a social unit.

Altricial—An adjective referring to a mammal that is born with little, if any, hair, is unable to feed itself, and initially has poor sensory and thermoregulatory abilities.

Amphibious—Refers to the ability of an animal to move both through water and on land.

Austral—May refer to "southern regions," typically meaning Southern Hemisphere. May also refer to the geographical region included within the Transition, Upper Austral, and Lower Austral Life Zones as defined by C. Hart Merriam in 1892–1898. These zones are often characterized by specific plant and animal communities and were originally defined by temperature gradients especially in the mountains of southwestern North America.

Bergmann's rule—Within a species or among closely related species of mammals, those individuals in colder environments often are larger in body size. Bergmann's rule is a generalization that reflects the ability of endothermic animals to more easily retain body heat (in cold climates) if they have a high body surface to body volume ratio, and to more easily dissipate excess body heat (in hot environments) if they have a low body surface to body volume ratio.

Bioacoustics—The study of biological sounds such as the sounds produced by bats or other mammals.

Biogeographic region—One of several major divisions of the earth defined by a distinctive assemblage of animals and plants. Sometimes referred to as "zoogeographic regions or realms" (for animals) or "phytogeographic regions or realms" (for plants). Such terminology dates from the late nineteenth century and varies considerably. Major biogeographic regions each have a somewhat distinctive flora and fauna. Those generally recognized include Nearctic, Neotropical, Palearctic, Ethiopian, Oriental, and Australian.

Blow—Cloud of vapor and sea water exhaled by cetaceans.

Boreal—Often used as an adjective meaning "northern"; also may refer to the northern climatic zone immediately south of the Arctic; may also include the Arctic, Hudsonian, and Canadian Life Zones described by C. Hart Merriam.

Brachiating ancestor—Ancestor that swung around by the arms.

Breaching—A whale behavior—leaping above the water's surface, then falling back into the water, landing on its back or side.

Cephalopod—Member of the group of mollusks such as squid and octopus.

Cladistic—Evolutionary relationships suggested as "tree" branches to indicate lines of common ancestry.

Cline—A gradient in a measurable characteristic, such as size and color, showing geographic differentiation. Various patterns of geographic variation are reflected as clines or clinal variation, and have been described as "ecogeographic rules."

Cloaca—A common opening for the digestive, urinary, and reproductive tracts found in monotreme mammals.

Colony—A group of mammals living in close proximity, interacting, and usually aiding in early warning of the presence of predators and in group defense.

Commensal—A relationship between species in which one benefits and the other is neither benefited nor harmed.

Congeneric—Descriptive of two or more species that belong to the same genus.

Conspecific—Descriptive of two or more individuals or populations that belong to the same species.

Contact call—Simple vocalization used to maintain communication or physical proximity among members of a social unit.

Convergent evolution—When two evolutionarily unrelated groups of organisms develop similar characteristics due to adaptation to similar aspects of their environment or niche.

Coprophagy—Reingestion of feces to obtain nutrients that were not ingested the first time through the digestive system.

Cosmopolitan—Adjective describing the distribution pattern of an animal found around the world in suitable habitats.

Crepuscular—Active at dawn and at dusk.

Critically Endangered—A technical category used by IUCN for a species that is at an extremely high risk of extinction in the wild in the immediate future.

Cryptic—Hidden or concealed; i.e., well-camouflaged patterning.

Dental formula—A method for describing the number of each type of tooth found in an animal's mouth: incisors (I), canines (C), premolars (P), and molars (M). The formula gives the number of each tooth found in an upper and lower quadrant of the mouth, and the total is multiplied by two for the total number of teeth. For example, the formula for humans is: I2/2 C1/1 P2/2 M3/3 (total, 16, times two is 32 teeth).

Dimorphic—Occurring in two distinct forms (e.g., in reference to the differences in size between males and females of a species).

Disjunct—A distribution pattern characterized by populations that are geographically separated from one another.

Diurnal—Active during the day.

DNA-DNA hybridization—A technique whereby the genetic similarity of different animal groups is determined based on the extent to which short stretches of their DNA, when mixed together in solution in the laboratory, are able to join with each other.

Dominance hierarchy—The social status of individuals in a group; each animal can usually dominate those animals below it in a hierarchy.

Dorso-ventrally—From back to front.

Duetting—Male and female singing and integrating their songs together.

Echolocation—A method of navigation used by some mammals (e.g., bats and marine mammals) to locate objects and investigate surroundings. The animals emit audible "clicks" and determine pathways by using the echo of the sound from structures in the area.

Ecotourism—Travel for the primary purpose of viewing nature. Ecotourism is now "big business" and is used as a non-consumptive but financially rewarding way to protect important areas for conservation.

Ectothermic—Using external energy and behavior to regulate body temperature. "Cold-blooded."

Endangered—A term used by IUCN and also under the Endangered Species Act of 1973 in the United States in reference to a species that is threatened with imminent extinction or extirpation over all or a significant portion of its range.

Endemic—Native to only one specific area.

Endothermic—Maintaining a constant body-temperature using metabolic energy. "Warm-blooded."

Eocene—Geological time period; subdivision of the Tertiary, from about 55.5 to 33.7 million years ago.

Ethology—The study of animal behavior.

Exotic—Not native.

Extant—Still in existence; not destroyed, lost, or extinct.

Extinct—Refers to a species that no longer survives anywhere.

Extirpated—Referring to a local extinction of a species that can still be found elsewhere.

Feral—A population of domesticated animal that lives in the wild.

Flehmen—Lip curling and head raising after sniffing a female's urine.

Forb—Any herb that is not a grass or grass-like.

Fossorial—Adapted for digging.

Frugivorous—Feeds on fruit.

Granivorous—Feeding on seeds.

Gravid—Pregnant.

Gregarious—Occuring in large groups.

Hibernation—A deep state of reduced metabolic activity and lowered body temperature that may last for weeks or months.

Holarctic—The Palearctic and Nearctic bigeographic regions combined.

Hybrid—The offspring resulting from a cross between two different species (or sometimes between distinctive subspecies).

Innate—An inherited characteristic.

Insectivorous—Technically refers to animals that eat insects; generally refers to animals that feed primarily on insects and other arthropods.

Introduced species—An animal or plant that has been introduced to an area where it normally does not occur.

Iteroparous—Breeds in multiple years.

Jacobson's organ—Olfactory organ found in the upper palate that first appeared in amphibians and is most developed in these and in reptiles, but is also found in some birds and mammals.

Kiva—A large chamber wholly or partly underground, and often used for religious ceremonies in Pueblo Indian villages.

Mandible—Technically an animal's lower jaw. The plural, mandibles, is used to refer to both the upper and lower jaw. The upper jaw is technically the maxilla, but often called the "upper mandible."

Marsupial—A mammal whose young complete their embryonic development outside of the mother's body, within a maternal pouch.

Matrilineal—Describing a social unit in which group members are descended from a single female.

Melon—The fat-filled forehead of aquatic mammals of the order Cetacea.

Metabolic rate—The rate of chemical processes in living organisms, resulting in energy expenditure and growth. Metabolic rate decreases when an animal is resting and increases during activity.

Migration—A two-way movement in some mammals, often dramatically seasonal. Typically latitudinal, though in some species is altitudinal or longitudinal. May be short-distance or long-distance.

Miocene—The geological time period that lasted from about 23.8 to 5.6 million years ago.

Molecular phylogenetics—The use of molecular (usually genetic) techniques to study evolutionary relationships between or among different groups of organisms.

Monestrous—Experiencing estrus just once each year or breeding season.

Monogamous—A breeding system in which a male and female mate only with one another.

Monophyletic—A group (or clade) that shares a common ancestor.

Monotypic—A taxonomic category that includes only one form (e.g., a genus that includes only one species; a species that includes no subspecies).

Montane—Of or inhabiting the biogeographic zone of relatively moist, cool upland slopes below timberline dominated by large coniferous trees.

Morphology—The form and structure of animals and plants.

Mutualism—Ecological relationship between two species in which both gain benefit.

Near Threatened—A category defined by the IUCN suggesting possible risk of extinction in the medium term (as opposed to long or short term) future.

Nearctic—The biogeographic region that includes temperate North America. faunal region.

Neotropical—The biogeographic region that includes South and Central America, the West Indies, and tropical Mexico.

New World—A general descriptive term encompassing the Nearctic and Neotropical biogeographic regions.

Niche—The role of an organism in its environment; multidimensional, with habitat and behavioral components.

Nocturnal—Active at night.

Old World—A general term that usually describes a species or group as being from Eurasia or Africa.

Oligocene—The geologic time period occurring from about 33.7 to 23.8 million years ago.

Omnivorous—Feeding on a broad range of foods, both plant and animal matter.

Palearctic—A biogeographic region that includes temperate Eurasia and Africa north of the Sahara.

Paleocene—Geological period, subdivision of the Tertiary, from 65 to 55.5 million years ago.

Pelage—Coat, skin, and hair.

Pelagic—An adjective used to indicate a relationship to the open sea.

Pestiferous—Troublesome or annoying; nuisance.

Phylogeny—A grouping of taxa based on evolutionary history.

Piscivorous—Fish-eating.

Placental—A mammal whose young complete their embryonic development within the mother's uterus, joined to her by a placenta.

Pleistocene—In general, the time of the great ice ages; geological period variously considered to include the last 1 to 1.8 million years.

Pliocene—The geological period preceding the Pleistocence; the last subdivision of what is known as the Tertiary; lasted from 5.5 to 1.8 million years ago.

Polyandry—A breeding system in which one female mates with two or more males.

Polygamy—A breeding system in which either or both male and female may have two or more mates.

Polygyny—A breeding system in which one male mates with two or more females.

Polyphyletic—A taxonomic group that is believed to have originated from more than one group of ancestors.

Post-gastric digestion—Refers to the type of fermentative digestion of vegetative matter found in tapirs and other animals by which microorganisms decompose food in a caecum. This is not as thorough a decomposition as occurs in ruminant digesters.

Precocial—An adjective used to describe animals that are born in an advanced state of development such that they generally can leave their birth area quickly and obtain their own food, although they are often led to food and guarded by a parent.

Proboscis—The prehensile trunk (a muscular hydrostat) found in tapirs, elephants, etc.

Quaternary—The geological period, from 1.8 million years ago to the present, usually including two subdivisions: the Pleistocene, and the Holocene.

Refugium (pl. refugia)—An area relatively unaltered during a time of climatic change, from which dispersion and speciation may occur after the climate readjusts.

Reproductive longevity—The length of an animal's life over which it is capable of reproduction.

Ruminant—An even-toed, hoofed mammal with a four-chambered stomach that eats rapidly to regurgitate its food and chew the cud later.

Scansorial—Specialized for climbing.

Seed dispersal—Refers to how tapirs and other animals transport viable seeds from their source to near or distant, suitable habitats where they can successfully germinate. Such dispersal may occur through the feces, through sputum, or as the seeds are attached and later released from fur, etc.

Semelparity—A short life span, in which a single instance of breeding is followed by death in the first year of life.

Sexual dimorphism—Male and female differ in morphology, such as size, feather size or shape, or bill size or shape.

Sibling species—Two or more species that are very closely related, presumably having differentiated from a common ancestor in the recent past; often difficult to distinguish, often interspecifically territorial.

Sonagram—A graphic representation of sound.

Speciation—The evolution of new species.

Spy-hopping—Positioning the body vertically in the water, with the head raised above the sea surface, sometimes while turning slowly.

Steppe—Arid land with vegetation that can thrive with very little moisture; found usually in regions of extreme temperature range.

Suspensory—Moving around or hanging by the arms.

Sympatric—Inhabiting the same range.

Systematist—A specialist in the classification of organisms; systematists strive to classify organisms on the basis of their evolutionary relationships.

Taxon (pl. taxa)—Any unit of scientific classification (e.g., species, genus, family, order).

Taxonomist—A specialist in the naming and classification of organisms. (See also Systematist. Taxonomy is the older science of naming things; identification of evolutionary relationships has not always been the goal of taxonomists. The modern science of systematics generally incorporates taxonomy with the search for evolutionary relationships.)

Taxonomy—The science of identifying, naming, and classifying organisms into groups.

Territoriality—Refers to an animal's defense of a certain portion of its habitat against other conspecifics. This is often undertaken by males in relation to one another and as a lure to females.

Territory—Any defended area. Territorial defense is typically male against male, female against female, and within a species or between sibling species. Area defended varies greatly among taxa, seasons, and habitats. A territory may include the entire home range, only the area immediately around a nest, or only a feeding area.

Tertiary—The geological period including most of the Cenozoic; from about 65 to 1.8 million years ago.

Thermoregulation—The ability to regulate body temperature; can be either behavioral or physiological.

Tribe—A unit of classification below the subfamily and above the genus.

Truncal erectness—Sitting, hanging, arm-swinging (brachiating), walking bipedally with the backbone held vertical.

Ungulate—A hoofed mammal.

Upper cone—The circle in which the arm can rotate when raised above the head.

Viable population—A population that is capable of maintaining itself over a period of time. One of the major conservation issues of the twenty-first century is determining what is a minimum viable population size. Population geneticists have generally come up with estimates of about 500 breeding pairs.

Vulnerable—A category defined by IUCN as a species that is not Critically Endangered or Endangered, but is still facing a threat of extinction.

GLOSSARY

Mammals species list

Monotremata [Order]

Tachyglossidae [Family]
Tachyglossus [Genus]
T. aculeatus [Species]
Zaglossus [Genus]
Z. bruijni [Species]

Ornithorhynchidae [Family]
Ornithorhynchus [Genus]
O. anatinus [Species]

Didelphimorphia [Order]

Didelphidae [Family]
Caluromys [Genus]
C. derbianus [Species]
C. lanatus
C. philander
Caluromysiops [Genus]
C. irrupta [Species]
Chironectes [Genus]
C. minimus [Species]
Didelphis [Genus]
D. albiventris [Species]
D. aurita
D. marsupialis
D. virginiana
Glironia [Genus]
G. venusta [Species]
Gracilinanus [Genus]
G. aceramarcae [Species]
G. agilis
G. dryas
G. emiliae
G. marica
G. microtarsus
Lestodelphys [Genus]
L. halli [Species]
Lutreolina [Genus]
L. crassicaudata [Species]
Marmosa [Genus]
M. andersoni [Species]
M. canescens
M. lepida

M. mexicana
M. murina
M. robinsoni
M. rubra
M. tyleriana
M. xerophila
Marmosops [Genus]
M. cracens [Species]
M. dorothea
M. fuscatus
M. handleyi
M. impavidus
M. incanus
M. invictus
M. noctivagus
M. parvidens
Metachirus [Genus]
M. nudicaudatus [Species]
Micoureus [Genus]
M. alstoni [Species]
M. constantiae
M. demerarae
M. regina
Monodelphis [Genus]
M. adusta [Species]
M. americana
M. brevicaudata
M. dimidiata
M. domestica
M. emiliae
M. iheringi
M. kunsi
M. maraxina
M. osgoodi
M. rubida
M. scalops
M. sorex
M. theresa
M. unistriata
Philander [Genus]
P. andersoni [Species]
P. opossum
Thylamys [Genus]
T. elegans [Species]

T. macrura
T. pallidior
T. pusilla
T. velutinus

Paucituberculata [Order]

Caenolestidae [Family]
Caenolestes [Genus]
C. caniventer [Species]
C. convelatus
C. fuliginosus
Lestoros [Genus]
L. inca [Species]
Rhyncholestes [Genus]
R. raphanurus [Species]

Microbiotheria [Order]

Microbiotheriidae [Family]
Dromiciops [Genus]
D. gliroides [Species]

Dasyuromorphia [Order]

Dasyuridae [Family]
Antechinus [Genus]
A. bellus [Species]
A. flavipes
A. godmani
A. leo
A. melanurus
A. minimus
A. naso
A. stuartii
A. swainsonii
A. wilhelmina
Dasycercus [Genus]
D. byrnei [Species]
D. cristicauda
Dasykaluta [Genus]
D. rosamondae [Species]
Dasyurus [Genus]
D. albopunctatus [Species]
D. geoffroii
D. hallucatus

D. maculatus
D. spartacus
D. viverrinus
Murexia [Genus]
  M. longicaudata [Species]
  M. rothschildi
Myoictis [Genus]
  M. melas [Species]
Neophascogale [Genus]
  N. lorentzi [Species]
Ningaui [Genus]
  N. ridei [Species]
  N. timealeyi
  N. yvonnae
Parantechinus [Genus]
  P. apicalis [Species]
  P. bilarni
Phascogale [Genus]
  P. calura [Species]
  P. tapoatafa
Phascolosorex [Genus]
  P. doriae [Species]
  P. dorsalis
Planigale [Genus]
  P. gilesi [Species]
  P. ingrami
  P. maculata
  P. novaeguineae
  P. tenuirostris
Pseudantechinus [Genus]
  P. macdonnellensis [Species]
  P. ningbing
  P. woolleyae
Sarcophilus [Genus]
  S. laniarius [Species]
Sminthopsis [Genus]
  S. aitkeni [Species]
  S. archeri
  S. butleri
  S. crassicaudata
  S. dolichura
  S. douglasi
  S. fuliginosus
  S. gilberti
  S. granulipes
  S. griseoventer
  S. hirtipes
  S. laniger
  S. leucopus
  S. longicaudata
  S. macroura
  S. murina
  S. ooldea
  S. psammophila
  S. virginiae
  S. youngsoni

Myrmecobiidae [Family]
Myrmecobius [Genus]
  M. fasciatus [Species]

Thylacinidae [Family]
Thylacinus [Genus]
  T. cynocephalus [Species]

Peramelemorphia [Order]

Peramelidae [Family]
Chaeropus [Genus]
  C. ecaudatus [Species]
Isoodon [Genus]
  I. auratus [Species]
  I. macrourus
  I. obesulus
Macrotis [Genus]
  M. lagotis [Species]
  M. leucura
Perameles [Genus]
  P. bougainville [Species]
  P. eremiana
  P. gunnii
  P. nasuta

Peroryctidae [Family]
Echymipera [Genus]
  E. clara [Species]
  E. davidi
  E. echinista
  E. kalubu
  E. rufescens
Microperoryctes [Genus]
  M. longicauda [Species]
  M. murina
  M. papuensis
Peroryctes [Genus]
  P. broadbenti [Species]
  P. raffrayana
Rhynchomeles [Genus]
  R. prattorum [Species]

Notoryctemorphia [Order]

Notoryctidae [Family]
Notoryctes [Genus]
  N. caurinus [Species]
  N. typhlops

Diprotodontia [Order]

Phascolarctidae [Family]
Phascolarctos [Genus]
  P. cinereus [Species]

Vombatidae [Family]
Lasiorhinus [Genus]
  L. krefftii [Species]
  L. latifrons
Vombatus [Genus]
  V. ursinus [Species]

Phalangeridae [Family]
Ailurops [Genus]
  A. ursinus [Species]

Phalanger [Genus]
  P. carmelitae [Species]
  P. lullulae
  P. matanim
  P. orientalis
  P. ornatus
  P. pelengensis
  P. rothschildi
  P. sericeus
  P. vestitus
Spilocuscus [Genus]
  S. maculatus [Speccics]
  S. rufoniger
Strigocuscus [Genus]
  S. celebensis [Species]
  S. gymnotis
Trichosurus [Genus]
  T. arnhemensis [Species]
  T. caninus
  T. vulpecula
Wyulda [Genus]
  W. squamicaudata [Species]

Hypsiprymnodontidae [Family]
Hypsiprymnodon [Genus]
  H. moschatus [Species]

Potoroidae [Family]
Aepyprymnus [Genus]
  A. rufescens [Species]
Bettongia [Genus]
  B. gaimardi [Species]
  B. lesueur
  B. penicillata
Caloprymnus [Genus]
  C. campestris [Species]
Potorous [Genus]
  P. longipes [Species]
  P. platyops
  P. tridactylus

Macropodidae [Family]
Dendrolagus [Genus]
  D. bennettianus [Species]
  D. dorianus
  D. goodfellowi
  D. inustus
  D. lumholtzi
  D. matschiei
  D. scottae
  D. spadix
  D. ursinus
Dorcopsis [Genus]
  D. atrata [Species]
  D. hageni
  D. luctuosa
  D. muelleri
Dorcopsulus [Genus]
  D. macleayi [Species]
  D. vanheurni
Lagorchestes [Genus]

L. asomatus [Species]
L. conspicillatus
L. hirsutus
L. leporides
Lagostrophus [Genus]
   L. fasciatus [Species]
Macropus [Genus]
   M. agilis [Species]
   M. antilopinus
   M. bernardus
   M. dorsalis
   M. eugenii
   M. fuliginosus
   M. giganteus
   M. greyi
   M. irma
   M. parma
   M. parryi
   M. robustus
   M. rufogriseus
   M. rufus
Onychogalea [Genus]
   O. fraenata [Species]
   O. lunata
   O. unguifera
Petrogale [Genus]
   P. assimilis [Species]
   P. brachyotis
   P. burbidgei
   P. concinna
   P. godmani
   P. inornata
   P. lateralis
   P. penicillata
   P. persephone
   P. rothschildi
   P. xanthopus
Setonix [Genus]
   S. brachyurus [Species]
Thylogale [Genus]
   T. billardierii [Species]
   T. brunii
   T. stigmatica
   T. thetis
Wallabia [Genus]
   W. bicolor [Species]

Burramyidae [Family]
Burramys [Genus]
   B. parvus [Species]
Cercartetus [Genus]
   C. caudatus [Species]
   C. concinnus
   C. lepidus
   C. nanus

Pseudocheiridae [Family]
Hemibelideus [Genus]
   H. lemuroides [Species]
Petauroides [Genus]
   P. volans [Species]

Petropseudes [Genus]
   P. dahli [Species]
Pseudocheirus [Genus]
   P. canescens [Species]
   P. caroli
   P. forbesi
   P. herbertensis
   P. mayeri
   P. peregrinus
   P. schlegeli
Pseudochirops [Genus]
   P. albertisii [Species]
   P. archeri
   P. corinnae
   P. cupreus

Petauridae [Family]
Dactylopsila [Genus]
   D. megalura [Species]
   D. palpator
   D. tatei
   D. trivirgata
Gymnobelideus [Genus]
   G. leadbeateri [Species]
Petaurus [Genus]
   P. abidi [Species]
   P. australis
   P. breviceps
   P. gracilis
   P. norfolcensis

Tarsipedidae [Family]
Tarsipes [Genus]
   T. rostratus [Species]

Acrobatidae [Family]
Acrobates [Genus]
   A. pygmaeus [Species]
Distoechurus [Genus]
   D. pennatus [Species]

Xenarthra [Order]

Megalonychidae [Family]
Choloepus [Genus]
   C. didactylus [Species]
   C. hoffmanni

Bradypodidae [Family]
Bradypus [Genus]
   B. torquatus [Species]
   B. tridactylus
   B. variegatus

Myrmecophagidae [Family]
Cyclopes [Genus]
   C. didactylus [Species]
Myrmecophaga [Genus]
   M. tridactyla [Species]
Tamandua [Genus]
   T. mexicana [Species]
   T. tetradactyla

Dasypodidae [Family]
Chlamyphorus [Genus]
   C. retusus [Species]
   C. truncatus
Cabassous [Genus]
   C. centralis [Species]
   C. chacoensis
   C. tatouay
   C. unicinctus
Chaetophractus [Genus]
   C. nationi [Species]
   C. vellerosus
   C. villosus
Dasypus [Genus]
   D. hybridus [Species]
   D. kappleri
   D. novemcinctus
   D. pilosus
   D. sabanicola
   D. septemcinctus
Euphractus [Genus]
   E. sexcinctus [Species]
Priodontes [Genus]
   P. maximus [Species]
Tolypeutes [Genus]
   T. matacus [Species]
   T. tricinctus
Zaedyus [Genus]
   Z. pichiy [Species]

Insectivora [Order]

Erinaceidae [Family]
Atelerix [Genus]
   A. albiventris [Species]
   A. algirus
   A. frontalis
   A. sclateri
Erinaceus [Genus]
   E. amurensis [Species]
   E. concolor
   E. europaeus
Hemiechinus [Genus]
   H. aethiopicus [Species]
   H. auritus
   H. collaris
   H. hypomelas
   H. micropus
   H. nudiventris
Mesechinus [Genus]
   M. dauuricus [Species]
   M. hughi
Echinosorex [Genus]
   E. gymnura [Species]
Hylomys [Genus]
   H. hainanensis [Species]
   H. sinensis
   H. suillus
Podogymnura [Genus]
   P. aureospinula [Species]
   P. truei

Chrysochloridae [Family]
  Amblysomus [Genus]
    A. gunningi [Species]
    A. hottentotus
    A. iris
    A. julianae
  Calcochloris [Genus]
    C. obtusirostris [Species]
  Chlorotalpa [Genus]
    C. arendsi [Species]
    C. duthieae
    C. leucorhina
    C. sclateri
    C. tytonis
  Chrysochloris [Genus]
    C. asiatica [Species]
    C. stuhlmanni
    C. visagiei
  Chrysospalax [Genus]
    C. trevelyani [Species]
    C. villosus
  Cryptochloris [Genus]
    C. wintoni [Species]
    C. zyli
  Eremitalpa [Genus]
    E. granti [Species]

Tenrecidae [Family]
  Echinops [Genus]
    E. telfairi [Species]
  Geogale [Genus]
    G. aurita [Species]
  Hemicentetes [Genus]
    H. semispinosus [Species]
  Limnogale [Genus]
    L. mergulus [Species]
  Microgale [Genus]
    M. brevicaudata [Species]
    M. cowani
    M. dobsoni
    M. dryas
    M. gracilis
    M. longicaudata
    M. parvula
    M. principula
    M. pulla
    M. pusilla
    M. talazaci
    M. thomasi
  Micropotamogale [Genus]
    M. lamottei [Species]
    M. ruwenzorii
  Oryzorictes [Genus]
    O. hova [Species]
    O. talpoides
    O. tetradactylus
  Potamogale [Genus]
    P. velox [Species]
  Setifer [Genus]
    S. setosus [Species]

  Tenrec [Genus]
    T. ecaudatus [Species]

Solenodontidae [Family]
  Solenodon [Genus]
    S. cubanus [Species]
    S. marcanoi
    S. paradoxus

Nesophontidae [Family]
  Nesophontes [Genus]
    N. edithae [Species]
    N. hypomicrus
    N. longirostris
    N. major
    N. micrus
    N. paramicrus
    N. submicrus
    N. zamicrus

Soricidae [Family]
  Anourosorex [Genus]
    A. squamipes [Species]
  Blarina [Genus]
    B. brevicauda [Species]
    B. carolinensis
    B. hylophaga
  Blarinella [Genus]
    B. quadraticauda [Species]
    B. wardi
  Chimarrogale [Genus]
    C. hantu [Species]
    C. himalayica
    C. phaeura
    C. platycephala
    C. styani
    C. sumatrana
  Congosorex [Genus]
    C. polli [Species]
  Crocidura [Genus]
    C. aleksandrisi [Species]
    C. allex
    C. andamanensis
    C. ansellorum
    C. arabica
    C. armenica
    C. attenuata
    C. attila
    C. baileyi
    C. batesi
    C. beatus
    C. beccarii
    C. bottegi
    C. bottegoides
    C. buettikoferi
    C. caliginea
    C. canariensis
    C. cinderella
    C. congobelgica
    C. cossyrensis
    C. crenata

C. crossei
C. cyanea
C. denti
C. desperata
C. dhofarensis
C. dolichura
C. douceti
C. dsinezumi
C. eisentrauti
C. elgonius
C. elongata
C. erica
C. fischeri
C. flavescens
C. floweri
C. foxi
C. fuliginosa
C. fulvastra
C. fumosa
C. fuscomurina
C. glassi
C. goliath
C. gracilipes
C. grandiceps
C. grandis
C. grassei
C. grayi
C. greenwoodi
C. gueldenstaedtii
C. harenna
C. hildegardeae
C. hirta
C. hispida
C. horsfieldii
C. jacksoni
C. jenkinsi
C. kivuana
C. lamottei
C. lanosa
C. lasiura
C. latona
C. lea
C. leucodon
C. levicula
C. littoralis
C. longipes
C. lucina
C. ludia
C. luna
C. lusitania
C. macarthuri
C. macmillani
C. macowi
C. malayana
C. manengubae
C. maquassiensis
C. mariquensis
C. maurisca
C. maxi
C. mindorus

C. minuta
C. miya
C. monax
C. monticola
C. montis
C. muricauda
C. mutesae
C. nana
C. nanilla
C. neglecta
C. negrina
C. nicobarica
C. nigeriae
C. nigricans
C. nigripes
C. nigrofusca
C. nimbae
C. niobe
C. obscurior
C. olivieri
C. orii
C. osorio
C. palawanensis
C. paradoxura
C. parvipes
C. pasha
C. pergrisea
C. phaeura
C. picea
C. pitmani
C. planiceps
C. poensis
C. polia
C. pullata
C. raineyi
C. religiosa
C. rhoditis
C. roosevelti
C. russula
C. selina
C. serezkyensis
C. sibirica
C. sicula
C. silacea
C. smithii
C. somalica
C. stenocephala
C. suaveolens
C. susiana
C. tansaniana
C. tarella
C. tarfayensis
C. telfordi
C. tenuis
C. thalia
C. theresae
C. thomensis
C. turba
C. ultima
C. usambarae

C. viaria
C. voi
C. whitakeri
C. wimmeri
C. xantippe
C. yankariensis
C. zaphiri
C. zarudnyi
C. zimmeri
C. zimmermanni
Cryptotis [Genus]
    C. avia [Species]
    C. endersi
    C. goldmani
    C. goodwini
    C. gracilis
    C. hondurensis
    C. magna
    C. meridensis
    C. mexicana
    C. montivaga
    C. nigrescens
    C. parva
    C. squamipes
    C. thomasi
Diplomesodon [Genus]
    D. pulchellum [Species]
Feroculus [Genus]
    F. feroculus [Species]
Megasorex [Genus]
    M. gigas [Species]
Myosorex [Genus]
    M. babaulti [Species]
    M. blarina
    M. cafer
    M. eisentrauti
    M. geata
    M. longicaudatus
    M. okuensis
    M. rumpii
    M. schalleri
    M. sclateri
    M. tenuis
    M. varius
Nectogale [Genus]
    N. elegans [Species]
Neomys [Genus]
    N. anomalus [Species]
    N. fodiens
    N. schelkovnikovi
Notiosorex [Genus]
    N. crawfordi [Species]
Paracrocidura [Genus]
    P. graueri [Species]
    P. maxima
    P. schoutedeni
Ruwenzorisorex [Genus]
    R. suncoides [Species]
Scutisorex [Genus]
    S. somereni [Species]

Solisorex [Genus]
    S. pearsoni [Species]
Sorex [Genus]
    S. alaskanus [Species]
    S. alpinus
    S. araneus
    S. arcticus
    S. arizonae
    S. asper
    S. bairdii
    S. bedfordiae
    S. bendirii
    S. bucbariensis
    S. caecutiens
    S. camtschatica
    S. cansulus
    S. cinereus
    S. coronatus
    S. cylindricauda
    S. daphaenodon
    S. dispar
    S. emarginatus
    S. excelsus
    S. fumeus
    S. gaspensis
    S. gracillimus
    S. granarius
    S. haydeni
    S. hosonoi
    S. hoyi
    S. hydrodromus
    S. isodon
    S. jacksoni
    S. kozlovi
    S. leucogaster
    S. longirostris
    S. lyelli
    S. macrodon
    S. merriami
    S. milleri
    S. minutissimus
    S. minutus
    S. mirabilis
    S. monticolus
    S. nanus
    S. oreopolus
    S. ornatus
    S. pacificus
    S. palustris
    S. planiceps
    S. portenkoi
    S. preblei
    S. raddei
    S. roboratus
    S. sadonis
    S. samniticus
    S. satunini
    S. saussurei
    S. sclateri
    S. shinto

S. sinalis
S. sonomae
S. stizodon
S. tenellus
S. thibetanus
S. trowbridgii
S. tundrensis
S. ugyunak
S. unguiculatus
S. vagrans
S. ventralis
S. veraepacis
S. volnuchini
Soriculus [Genus]
  S. caudatus [Species]
  S. fumidus
  S. hypsibius
  S. lamula
  S. leucops
  S. macrurus
  S. nigrescens
  S. parca
  S. salenskii
  S. smithii
Suncus [Genus]
  S. ater [Species]
  S. dayi
  S. etruscus
  S. fellowesgordoni
  S. hosei
  S. infinitesimus
  S. lixus
  S. madagascariensis
  S. malayanus
  S. mertensi
  S. montanus
  S. murinus
  S. remyi
  S. stoliczkanus
  S. varilla
  S. zeylanicus
Surdisorex [Genus]
  S. norae [Species]
  S. polulus
Sylvisorex [Genus]
  S. granti [Species]
  S. howelli
  S. isabellae
  S. johnstoni
  S. lunaris
  S. megalura
  S. morio
  S. ollula
  S. oriundus
  S. vulcanorum

Talpidae [Family]
Desmana [Genus]
  D. moschata [Species]
Galemys [Genus]
  G. pyrenaicus [Species]

Condylura [Genus]
  C. cristata [Species]
Euroscaptor [Genus]
  E. grandis [Species]
  E. klossi
  E. longirostris
  E. micrura
  E. mizura
  E. parvidens
Mogera [Genus]
  M. etigo [Species]
  M. insularis
  M. kobeae
  M. minor
  M. robusta
  M. tokudae
  M. wogura
Nesoscaptor [Genus]
  N. uchidai [Species]
Neurotrichus [Genus]
  N. gibbsii [Species]
Parascalops [Genus]
  P. breweri [Species]
Parascaptor [Genus]
  P. leucura [Species]
Scalopus [Genus]
  S. aquaticus [Species]
Scapanulus [Genus]
  S. oweni [Species]
Scapanus [Genus]
  S. latimanus [Species]
  S. orarius
  S. townsendii
Scaptochirus [Genus]
  S. moschatus [Species]
Scaptonyx [Genus]
  S. fusicaudus [Species]
Talpa [Genus]
  T. altaica [Species]
  T. caeca
  T. caucasica
  T. europaea
  T. levantis
  T. occidentalis
  T. romana
  T. stankovici
  T. streeti
Urotrichus [Genus]
  U. pilirostris [Species]
  U. talpoides
Uropsilus [Genus]
  U. andersoni [Species]
  U. gracilis
  U. investigator
  U. soricipes

Scandentia [Order]

Tupaiidae [Family]
Anathana [Genus]
  A. ellioti [Species]

Dendrogale [Genus]
  D. melanura [Species]
  D. murina
Ptilocercus [Genus]
  P. lowii [Species]
Tupaia [Genus]
  T. belangeri [Species]
  T. chrysogaster
  T. dorsalis
  T. glis
  T. gracilis
  T. javanica
  T. longipes
  T. minor
  T. montana
  T. nicobarica
  T. palawanensis
  T. picta
  T. splendidula
  T. tana
Urogale [Genus]
  U. everetti [Species]

Dermoptera [Order]

Cynocephalidae [Family]
Cynocephalus [Genus]
  C. variegatus [Species]
  C. volans

Chiroptera [Order]

Pteropodidae [Family]
Acerodon [Genus]
  A. celebensis [Species]
  A. humilis
  A. jubatus
  A. leucotis
  A. lucifer
  A. mackloti
Aethalops [Genus]
  A. alecto [Species]
Alionycteris [Genus]
  A. paucidentata [Species]
Aproteles [Genus]
  A. bulmerae [Species]
Balionycteris [Genus]
  B. maculata [Species]
Boneia [Genus]
  B. bidens [Species]
Casinycteris [Genus]
  C. argynnis [Species]
Chironax [Genus]
  C. melanocephalus [Species]
Cynopterus [Genus]
  C. brachyotis [Species]
  C. horsfieldi
  C. nusatenggara
  C. sphinx
  C. titthaecheileus
Dobsonia [Genus]

D. beauforti [Species]
D. chapmani
D. emersa
D. exoleta
D. inermis
D. minor
D. moluccensis
D. pannietensis
D. peroni
D. praedatrix
D. viridis
Dyacopterus [Genus]
D. spadiceus [Species]
Eidolon [Genus]
E. dupreanum [Species]
E. helvum
Eonycteris [Genus]
E. major [Species]
E. spelaea
Epomophorus [Genus]
E. angolensis [Species]
E. gambianus
E. grandis
E. labiatus
E. minimus
E. wahlbergi
Epomops [Genus]
E. buettikoferi [Species]
E. dobsoni
E. franqueti
Haplonycteris [Genus]
H. fischeri [Species]
Harpyionycteris [Genus]
H. celebensis [Species]
H. whiteheadi
Hypsignathus [Genus]
H. monstrosus [Species]
Latidens [Genus]
L. salimalii [Species]
Macroglossus [Genus]
M. minimus [Species]
M. sobrinus
Megaerops [Genus]
M. ecaudatus [Species]
M. kusnotoi
M. niphanae
M. wetmorei
Megaloglossus [Genus]
M. woermanni [Species]
Melonycteris [Genus]
M. aurantius [Species]
M. melanops
M. woodfordi
Micropteropus [Genus]
M. intermedius [Species]
M. pusillus
Myonycteris [Genus]
M. brachycephala [Species]
M. relicta
M. torquata

Nanonycteris [Genus]
N. veldkampi [Species]
Neopteryx [Genus]
N. frosti [Species]
Notopteris [Genus]
N. macdonaldi [Species]
Nyctimene [Genus]
N. aello [Species]
N. albiventer
N. celaeno
N. cephalotes
N. certans
N. cyclotis
N. draconilla
N. major
N. malaitensis
N. masalai
N. minutus
N. rabori
N. robinsoni
N. sanctacrucis
N. vizcaccia
Otopteropus [Genus]
O. cartilagonodus [Species]
Paranyctimene [Genus]
P. raptor [Species]
Penthetor [Genus]
P. lucasi [Species]
Plerotes [Genus]
P. anchietai [Species]
Ptenochirus [Genus]
P. jagori [Species]
P. minor
Pteralopex [Genus]
P. acrodonta [Species]
P. anceps
P. atrata
P. pulchra
Pteropus [Genus]
P. admiralitatum [Species]
P. aldabrensis
P. alecto
P. anetianus
P. argentatus
P. brunneus
P. caniceps
P. chrysoproctus
P. conspicillatus
P. dasymallus
P. faunulus
P. fundatus
P. giganteus
P. gilliardi
P. griseus
P. howensis
P. hypomelanus
P. insularis
P. leucopterus
P. livingstonei
P. lombocensis

P. lylei
P. macrotis
P. mahaganus
P. mariannus
P. mearnsi
P. melanopogon
P. melanotus
P. molossinus
P. neohibernicus
P. niger
P. nitendiensis
P. ocularis
P. ornatus
P. personatus
P. phaeocephalus
P. pilosus
P. pohlei
P. poliocephalus
P. pselaphon
P. pumilus
P. rayneri
P. rodricensis
P. rufus
P. samoensis
P. sanctacrucis
P. scapulatus
P. seychellensis
P. speciosus
P. subniger
P. temmincki
P. tokudae
P. tonganus
P. tuberculatus
P. vampyrus
P. vetulus
P. voeltzkowi
P. woodfordi
Rousettus [Genus]
R. aegyptiacus [Species]
R. amplexicaudatus
R. angolensis
R. celebensis
R. lanosus
R. leschenaulti
R. madagascariensis
R. obliviosus
R. spinalatus
Scotonycteris [Genus]
S. ophiodon [Species]
S. zenkeri
Sphaerias [Genus]
S. blanfordi [Species]
Styloctenium [Genus]
S. wallacei [Species]
Syconycteris [Genus]
S. australis [Species]
S. carolinae
S. hobbit
Thoopterus [Genus]
T. nigrescens [Species]

P. aphylla [Species]
P. poeyi
Phyllops [Genus]
  P. falcatus [Species]
Phyllostomus [Genus]
  P. discolor [Species]
  P. elongatus
  P. hastatus
  P. latifolius
Platalina [Genus]
  P. genovensium [Species]
Platyrrhinus [Genus]
  P. aurarius [Species]
  P. brachycephalus
  P. chocoensis
  P. dorsalis
  P. helleri
  P. infuscus
  P. lineatus
  P. recifinus
  P. umbratus
  P. vittatus
Pygoderma [Genus]
  P. bilabiatum [Species]
Rhinophylla [Genus]
  R. alethina [Species]
  R. fischerae
  R. pumilio
Scleronycteris [Genus]
  S. ega [Species]
Sphaeronycteris [Genus]
  S. toxophyllum [Species]
Stenoderma [Genus]
  S. rufum [Species]
Sturnira [Genus]
  S. aratathomasi [Species]
  S. bidens
  S. bogotensis
  S. erythromos
  S. lilium
  S. ludovici
  S. luisi
  S. magna
  S. mordax
  S. nana
  S. thomasi
  S. tildae
Tonatia [Genus]
  T. bidens [Species]
  T. brasiliense
  T. carrikeri
  T. evotis
  T. schulzi
  T. silvicola
Trachops [Genus]
  T. cirrhosus [Species]
Uroderma [Genus]
  U. bilobatum [Species]
  U. magnirostrum
Vampyressa [Genus]

V. bidens [Species]
V. brocki
V. melissa
V. nymphaea
V. pusilla
Vampyrodes [Genus]
  V. caraccioli [Species]
Vampyrum [Genus]
  V. spectrum [Species]

Mormoopidae [Family]
  Mormoops [Genus]
    M. blainvillii [Species]
    M. megalophylla
  Pteronotus [Genus]
    P. davyi [Species]
    P. gymnonotus
    P. macleayii
    P. parnellii
    P. personatus
    P. quadridens

Noctilionidae [Family]
  Noctilio [Genus]
    N. albiventris [Species]
    N. leporinus

Mystacinidae [Family]
  Mystacina [Genus]
    M. robusta [Species]
    M. tuberculata

Natalidae [Family]
  Natalus [Genus]
    N. lepidus [Species]
    N. micropus
    N. stramineus
    N. tumidifrons
    N. tumidirostris

Furipteridae [Family]
  Amorphochilus [Genus]
    A. schnablii [Species]
  Furipterus [Genus]
    F. horrens [Species]

Thyropteridae [Family]
  Thyroptera [Genus]
    T. discifera [Species]
    T. tricolor

Myzopodidae [Family]
  Myzopoda [Genus]
    M. aurita [Species]

Molossidae [Family]
  Chaerephon [Genus]
    C. aloysiisabaudiae [Species]
    C. ansorgei
    C. bemmeleni
    C. bivittata
    C. chapini
    C. gallagheri

C. jobensis
C. johorensis
C. major
C. nigeriae
C. plicata
C. pumila
C. russata
Cheiromeles [Genus]
  C. torquatus [Species]
Eumops [Genus]
  E. auripendulus [Species]
  E. bonariensis
  E. dabbenei
  E. glaucinus
  E. hansae
  E. maurus
  E. perotis
  E. underwoodi
Molossops [Genus]
  M. abrasus [Species]
  M. aequatorianus
  M. greenhalli
  M. mattogrossensis
  M. neglectus
  M. planirostris
  M. temminckii
Molossus [Genus]
  M. ater [Species]
  M. bondae
  M. molossus
  M. pretiosus
  M. sinaloae
Mops [Genus]
  M. brachypterus [Species]
  M. condylurus
  M. congicus
  M. demonstrator
  M. midas
  M. mops
  M. nanulus
  M. niangarae
  M. niveiventer
  M. petersoni
  M. sarasinorum
  M. spurrelli
  M. thersites
  M. trevori
Mormopterus [Genus]
  M. acetabulosus [Species]
  M. beccarii
  M. doriae
  M. jugularis
  M. kalinowskii
  M. minutus
  M. norfolkensis
  M. petrophilus
  M. phrudus
  M. planiceps
  M. setiger
Myopterus [Genus]

M. daubentonii [Species]
M. whitleyi
Nyctinomops [Genus]
  N. aurispinosus [Species]
  N. femorosaccus
  N. laticaudatus
  N. macrotis
Otomops [Genus]
  O. formosus [Species]
  O. martiensseni
  O. papuensis
  O. secundus
  O. wroughtoni
Promops [Genus]
  P. centralis [Species]
  P. nasutus
Tadarida [Genus]
  T. aegyptiaca [Species]
  T. australis
  T. brasiliensis
  T. espiritosantensis
  T. fulminans
  T. lobata
  T. teniotis
  T. ventralis

Vespertilionidae [Family]
Antrozous [Genus]
  A. dubiaquercus [Species]
  A. pallidus
Barbastella [Genus]
  B. barbastellus [Species]
  B. leucomelas
Chalinolobus [Genus]
  C. alboguttatus [Species]
  C. argentatus
  C. beatrix
  C. dwyeri
  C. egeria
  C. gleni
  C. gouldii
  C. kenyacola
  C. morio
  C. nigrogriseus
  C. picatus
  C. poensis
  C. superbus
  C. tuberculatus
  C. variegatus
Eptesicus [Genus]
  E. baverstocki [Species]
  E. bobrinskoi
  E. bottae
  E. brasiliensis
  E. brunneus
  E. capensis
  E. demissus
  E. diminutus
  E. douglasorum
  E. flavescens
  E. floweri

E. furinalis
E. fuscus
E. guadeloupensis
E. guineensis
E. hottentotus
E. innoxius
E. kobayashii
E. melckorum
E. nasutus
E. nilssoni
E. pachyotis
E. platyops
E. pumilus
E. regulus
E. rendalli
E. sagittula
E. serotinus
E. somalicus
E. tatei
E. tenuipinnis
E. vulturnus
Euderma [Genus]
  E. maculatum [Species]
Eudiscopus [Genus]
  E. denticulus [Species]
Glischropus [Genus]
  G. javanus [Species]
  G. tylopus
Harpiocephalus [Genus]
  H. harpia [Species]
Hesperoptenus [Genus]
  H. blanfordi [Species]
  H. doriae
  H. gaskelli
  H. tickelli
  H. tomesi
Histiotus [Genus]
  H. alienus [Species]
  H. macrotus
  H. montanus
  H. velatus
Ia [Genus]
  I. io [Species]
Idionycteris [Genus]
  I. phyllotis [Species]
Kerivoula [Genus]
  K. aerosa [Species]
  K. africana
  K. agnella
  K. argentata
  K. atrox
  K. cuprosa
  K. eriophora
  K. flora
  K. hardwickei
  K. intermedia
  K. jagori
  K. lanosa
  K. minuta
  K. muscina

K. myrella
K. papillosa
K. papuensis
K. pellucida
K. phalaena
K. picta
K. smithi
K. whiteheadi
Laephotis [Genus]
  L. angolensis [Species]
  L. botswanae
  L. namibensis
  L. wintoni
Lasionycteris [Genus]
  L. noctivagans [Species]
Lasiurus [Genus]
  L. borealis [Species]
  L. castaneus
  L. cinereus
  L. ega
  L. egregius
  L. intermedius
  L. seminolus
Mimetillus [Genus]
  M. moloneyi [Species]
Miniopterus [Genus]
  M. australis [Species]
  M. fraterculus
  M. fuscus
  M. inflatus
  M. magnater
  M. minor
  M. pusillus
  M. robustior
  M. schreibersi
  M. tristis
Murina [Genus]
  M. aenea [Species]
  M. aurata
  M. cyclotis
  M. florium
  M. fusca
  M. grisea
  M. huttoni
  M. leucogaster
  M. puta
  M. rozendaali
  M. silvatica
  M. suilla
  M. tenebrosa
  M. tubinaris
  M. ussuriensis
Myotis [Genus]
  M. abei [Species]
  M. adversus
  M. aelleni
  M. albescens
  M. altarium
  M. annectans
  M. atacamensis

M. auriculus
M. australis
M. austroriparius
M. bechsteini
M. blythii
M. bocagei
M. bombinus
M. brandti
M. californicus
M. capaccinii
M. chiloensis
M. chinensis
M. cobanensis
M. dasycneme
M. daubentoni
M. dominicensis
M. elegans
M. emarginatus
M. evotis
M. findleyi
M. formosus
M. fortidens
M. frater
M. goudoti
M. grisescens
M. hasseltii
M. horsfieldii
M. hosonoi
M. ikonnikovi
M. insularum
M. keaysi
M. keenii
M. leibii
M. lesueuri
M. levis
M. longipes
M. lucifugus
M. macrodactylus
M. macrotarsus
M. martiniquensis
M. milleri
M. montivagus
M. morrisi
M. muricola
M. myotis
M. mystacinus
M. nattereri
M. nesopolus
M. nigricans
M. oreias
M. oxyotus
M. ozensis
M. peninsularis
M. pequinius
M. planiceps
M. pruinosus
M. ricketti
M. ridleyi
M. riparius
M. rosseti

M. ruber
M. schaubi
M. scotti
M. seabrai
M. sicarius
M. siligorensis
M. simus
M. sodalis
M. stalkeri
M. thysanodes
M. tricolor
M. velifer
M. vivesi
M. volans
M. welwitschii
M. yesoensis
M. yumanensis
Nyctalus [Genus]
    N. aviator [Species]
    N. azoreum
    N. lasiopterus
    N. leisleri
    N. montanus
    N. noctula
Nycticeius [Genus]
    N. balstoni [Species]
    N. greyii
    N. humeralis
    N. rueppellii
    N. sanborni
    N. schlieffeni
Nyctophilus [Genus]
    N. arnhemensis [Species]
    N. geoffroyi
    N. gouldi
    N. heran
    N. microdon
    N. microtis
    N. timoriensis
    N. walkeri
Otonycteris [Genus]
    O. hemprichi [Species]
Pharotis [Genus]
    P. imogene [Species]
Philetor [Genus]
    P. brachypterus [Species]
Pipistrellus [Genus]
    P. aegyptius [Species]
    P. aero
    P. affinis
    P. anchietai
    P. anthonyi
    P. arabicus
    P. ariel
    P. babu
    P. bodenheimeri
    P. cadornae
    P. ceylonicus
    P. circumdatus
    P. coromandra

P. crassulus
P. cuprosus
P. dormeri
P. eisentrauti
P. endoi
P. hesperus
P. imbricatus
P. inexspectatus
P. javanicus
P. joffrei
P. kitcheneri
P. kuhlii
P. lophurus
P. macrotis
P. maderensis
P. mimus
P. minahassae
P. mordax
P. musciculus
P. nanulus
P. nanus
P. nathusii
P. paterculus
P. peguensis
P. permixtus
P. petersi
P. pipistrellus
P. pulveratus
P. rueppelli
P. rusticus
P. savii
P. societatis
P. stenopterus
P. sturdeei
P. subflavus
P. tasmaniensis
P. tenuis
Plecotus [Genus]
    P. auritus [Species]
    P. austriacus
    P. mexicanus
    P. rafinesquii
    P. taivanus
    P. teneriffae
    P. townsendii
Rhogeessa [Genus]
    R. alleni [Species]
    R. genowaysi
    R. gracilis
    R. minutilla
    R. mira
    R. parvula
    R. tumida
Scotoecus [Genus]
    S. albofuscus [Species]
    S. hirundo
    S. pallidus
Scotomanes [Genus]
    S. emarginatus [Species]
    S. ornatus

MAMMALS SPECIES LIST

Scotophilus [Genus]
   S. borbonicus [Species]
   S. celebensis
   S. dinganii
   S. heathi
   S. kuhlii
   S. leucogaster
   S. nigrita
   S. nux
   S. robustus
   S. viridis
Tomopeas [Genus]
   T. ravus [Species]
Tylonycteris [Genus]
   T. pachypus [Species]
   T. robustula
Vespertilio [Genus]
   V. murinus [Species]
   V. superans

Primates [Order]

Lorisidae [Family]
  Arctocebus [Genus]
    A. aureus [Species]
    A. calabarensis
  Loris [Genus]
    L. tardigradus [Species]
  Nycticebus [Genus]
    N. coucang [Species]
    N. pygmaeus
  Perodicticus [Genus]
    P. potto [Species]

Galagidae [Family]
  Euoticus [Genus]
    E. elegantulus [Species]
    E. pallidus
  Galago [Genus]
    G. alleni [Species]
    G. gallarum
    G. matschiei
    G. moholi
    G. senegalensis
  Galagoides [Genus]
    G. demidoff [Species]
    G. zanzibaricus
  Otolemur [Genus]
    O. crassicaudatus [Species]
    O. garnettii

Cheirogaleidae [Family]
  Allocebus [Genus]
    A. trichotis [Species]
  Cheirogaleus [Genus]
    C. major [Species]
    C. medius
  Microcebus [Genus]
    Microcebus coquereli [Species]
    Microcebus murinus
    Microcebus rufus

Phaner [Genus]
   P. furcifer [Species]

Lemuridae [Family]
  Eulemur [Genus]
    E. coronatus [Species]
    E. fulvus
    E. macaco
    E. mongoz
    E. rubriventer
  Hapalemur [Genus]
    H. aureus [Species]
    H. griseus
    H. simus
  Lemur [Genus]
    L. catta [Species]
  Varecia [Genus]
    V. variegata [Species]

Indriidae [Family]
  Avahi [Genus]
    A. laniger [Species]
  Indri [Genus]
    I. indri [Species]
  Propithecus [Genus]
    P. diadema [Species]
    P. tattersalli
    P. verreauxi

Lepilemuridae [Family]
  Lepilemur [Genus]
    L. dorsalis [Species]
    L. edwardsi
    L. leucopus
    L. microdon
    L. mustelinus
    L. ruficaudatus
    L. septentrionalis

Daubentoniidae [Family]
  Daubentonia [Genus]
    D. madagascariensis [Species]

Tarsiidae [Family]
  Tarsius [Genus]
    T. bancanus [Species]
    T. dianae
    T. pumilus
    T. spectrum
    T. syrichta

Cebidae [Family]
  Alouatta [Genus]
    A. belzebul [Species]
    A. caraya
    A. coibensis
    A. fusca
    A. palliata
    A. pigra
    A. sara
    A. seniculus
  Callicebus [Genus]

    C. brunneus [Species]
    C. caligatus
    C. cinerascens
    C. cupreus
    C. donacophilus
    C. dubius
    C. hoffmannsi
    C. modestus
    C. moloch
    C. oenanthe
    C. olallae
    C. personatus
    C. torquatus
  Cebus [Genus]
    C. albifrons [Species]
    C. apella
    C. capucinus
    C. olivaceus
  Saimiri [Genus]
    S. boliviensis [Species]
    S. oerstedii
    S. sciureus
    S. ustus
    S. vanzolinii

Callitrichidae [Family]
  Callimico [Genus]
    C. goeldii [Species]
  Callithrix [Genus]
    C. argentata [Species]
    C. aurita
    C. flaviceps
    C. geoffroyi
    C. humeralifer
    C. jacchus
    C. kuhlii
    C. penicillata
    C. pygmaea
  Leontopithecus [Genus]
    L. caissara [Species]
    L. chrysomela
    L. chrysopygus
    L. rosalia
  Saguinus [Genus]
    S. bicolor [Species]
    S. fuscicollis
    S. geoffroyi
    S. imperator
    S. inustus
    S. labiatus
    S. leucopus
    S. midas
    S. mystax
    S. nigricollis
    S. oedipus
    S. tripartitus

Aotidae [Family]
  Aotus [Genus]
    A. azarai [Species]
    A. brumbacki

A. hershkovitzi
A. infulatus
A. lemurinus
A. miconax
A. nancymaae
A. nigriceps
A. trivirgatus
A. vociferans

Pitheciidae [Family]
  Cacajao [Genus]
    C. calvus [Species]
    C. melanocephalus
  Chiropotes [Genus]
    C. albinasus [Species]
    C. satanas
  Pithecia [Genus]
    P. aequatorialis [Species]
    P. albicans
    P. irrorata
    P. monachus
    P. pithecia

Atelidae [Family]
  Ateles [Genus]
    A. belzebuth [Species]
    A. chamek
    A. fusciceps
    A. geoffroyi
    A. marginatus
    A. paniscus
  Brachyteles [Genus]
    B. arachnoides [Species]
  Lagothrix [Genus]
    L. flavicauda [Species]
    L. lagotricha

Cercopithecidae [Family]
  Allenopithecus [Genus]
    A. nigroviridis [Species]
  Cercocebus [Genus]
    C. agilis [Species]
    C. galeritus
    C. torquatus
  Cercopithecus [Genus]
    C. ascanius [Species]
    C. campbelli
    C. cephus
    C. diana
    C. dryas
    C. erythrogaster
    C. erythrotis
    C. hamlyni
    C. lhoesti
    C. mitis
    C. mona
    C. neglectus
    C. nictitans
    C. petaurista
    C. pogonias
    C. preussi
    C. sclateri

C. solatus
C. wolfi
Chlorocebus [Genus]
  C. aethiops [Species]
Colobus [Genus]
  C. angolensis [Species]
  C. guereza
  C. polykomos
  C. satanas
Erythrocebus [Genus]
  E. patas [Species]
Lophocebus [Genus]
  L. albigena [Species]
Macaca [Genus]
  M. arctoides [Species]
  M. assamensis
  M. cyclopis
  M. fascicularis
  M. fuscata
  M. maura
  M. mulatta
  M. nemestrina
  M. nigra
  M. ochreata
  M. radiata
  M. silenus
  M. sinica
  M. sylvanus
  M. thibetana
  M. tonkeana
Mandrillus [Genus]
  M. leucophaeus [Species]
  M. sphinx
  Miopithecus
  M. talapoin
Nasalis [Genus]
  N. concolor [Species]
  N. larvatus
Papio [Genus]
  P. hamadryas [Species]
Presbytis [Genus]
  P. comata [Species]
  P. femoralis
  P. frontata
  P. hosei
  P. melalophos
  P. potenziani
  P. rubicunda
  P. thomasi
Procolobus [Genus]
  P. badius [Species]
  P. pennantii
  P. preussi
  P. rufomitratus
  P. verus
Pygathrix [Genus]
  P. avunculus [Species]
  P. bieti
  P. brelichi
  P. nemaeus

P. roxellana
Semnopithecus [Genus]
  S. entellus [Species]
Theropithecus [Genus]
  T. gelada [Species]
Trachypithecus [Genus]
  T. auratus [Species]
  T. cristatus
  T. francoisi
  T. geei
  T. johnii
  T. obscurus
  T. phayrei
  T. pileatus
  T. vetulus

Hylobatidae [Family]
  Hylobates [Genus]
    H. agilis [Species]
    H. concolor
    H. gabriellae
    H. hoolock
    H. klossii
    H. lar
    H. leucogenys
    H. moloch
    H. muelleri
    H. pileatus
    H. syndactylus

Hominidae [Family]
  Gorilla [Genus]
    G. gorilla [Species]
  Homo [Genus]
    H. sapiens [Species]
  Pan [Genus]
    P. paniscus [Species]
    P. troglodytes
  Pongo [Genus]
    P. pygmaeus [Species]

Carnivora [Order]

Canidae [Family]
  Alopex [Genus]
    A. lagopus [Species]
    Atelocynus
    A. microtis
  Canis [Genus]
    C. adustus [Species]
    C. aureus
    C. latrans
    C. lupus
    C. mesomelas
    C. rufus
    C. simensis
  Cerdocyon [Genus]
    C. thous [Species]
  Chrysocyon [Genus]
    C. brachyurus [Species]
  Cuon [Genus]

C. alpinus [Species]
Dusicyon [Genus]
  D. australis [Species]
Lycaon [Genus]
  L. pictus [Species]
Nyctereutes [Genus]
  N. procyonoides [Species]
Otocyon [Genus]
  O. megalotis [Species]
Pseudalopex [Genus]
  P. culpaeus [Species]
  P. griseus
  P. gymnocercus
  P. sechurae
  P. vetulus
Speothos [Genus]
  S. venaticus [Species]
Urocyon [Genus]
  U. cinereoargenteus [Species]
  U. littoralis
Vulpes [Genus]
  V. bengalensis [Species]
  V. cana
  V. chama
  V. corsac
  V. ferrilata
  V. pallida
  V. rueppelli
  V. velox
  V. vulpes
  V. zerda

Ursidae [Family]
  Ailuropoda [Genus]
    A. melanoleuca [Species]
  Ailurus [Genus]
    A. fulgens [Species]
  Helarctos [Genus]
    H. malayanus [Species]
  Melursus [Genus]
    M. ursinus [Species]
  Tremarctos [Genus]
    T. ornatus [Species]
  Ursus [Genus]
    U. americanus [Species]
    U. arctos
    U. maritimus
    U. thibetanus

Procyonidae [Family]
  Bassaricyon [Genus]
    B. alleni [Species]
    B. beddardi
    B. gabbii
    B. lasius
    B. pauli
  Potos [Genus]
    P. flavus [Species]
  Bassariscus [Genus]
    B. astutus [Species]
    B. sumichrasti

Nasua [Genus]
  N. narica [Species]
  N. nasua
Nasuella [Genus]
  N. olivacea [Species]
Procyon [Genus]
  P. cancrivorus [Species]
  P. gloveralleni
  P. insularis
  P. lotor
  P. maynardi
  P. minor
  P. pygmaeus

Mustelidae [Family]
  Amblonyx [Genus]
    A. cinereus [Species]
  Aonyx [Genus]
    A. capensis [Species]
    A. congicus
  Arctonyx [Genus]
    A. collaris [Species]
  Conepatus [Genus]
    C. chinga [Species]
    C. humboldtii
    C. leuconotus
    C. mesoleucus
    C. semistriatus
  Eira [Genus]
    E. barbara [Species]
  Enhydra [Genus]
    E. lutris [Species]
  Galictis [Genus]
    G. cuja [Species]
    G. vittata
  Gulo [Genus]
    G. gulo [Species]
  Ictonyx [Genus]
    I. libyca [Species]
    I. striatus
  Lontra [Genus]
    L. canadensis [Species]
    L. felina
    L. longicaudis
    L. provocax
  Lutra [Genus]
    L. lutra [Species]
    L. maculicollis
    L. sumatrana
  Lutrogale [Genus]
    L. perspicillata [Species]
  Lyncodon [Genus]
    L. patagonicus [Species]
  Martes [Genus]
    M. americana [Species]
    M. flavigula
    M. foina
    M. gwatkinsii
    M. martes
    M. melampus
    M. pennanti

    M. zibellina
  Meles [Genus]
    M. meles [Species]
  Mellivora [Genus]
    M. capensis [Species]
  Melogale [Genus]
    M. everetti [Species]
    M. moschata
    M. orientalis
    M. personata
  Mephitis [Genus]
    M. macroura [Species]
    M. mephitis
  Mustela [Genus]
    M. africana [Species]
    M. altaica
    M. erminea
    M. eversmannii
    M. felipei
    M. frenata
    M. kathiah
    M. lutreola
    M. lutreolina
    M. nigripes
    M. nivalis
    M. nudipes
    M. putorius
    M. sibirica
    M. strigidorsa
    M. vison
  Mydaus [Genus]
    M. javanensis [Species]
    M. marchei
  Poecilogale [Genus]
    P. albinucha [Species]
  Pteronura [Genus]
    P. brasiliensis [Species]
  Spilogale [Genus]
    S. putorius [Species]
    S. pygmaea
  Taxidea [Genus]
    T. taxus [Species]
  Vormela [Genus]
    V. peregusna [Species]

Viverridae [Family]
  Arctictis [Genus]
    A. binturong [Species]
  Arctogalidia [Genus]
    A. trivirgata [Species]
  Chrotogale [Genus]
    C. owstoni [Species]
  Civettictis [Genus]
    C. civetta [Species]
  Cryptoprocta [Genus]
    C. ferox [Species]
  Cynogale [Genus]
    C. bennettii [Species]
  Diplogale [Genus]
    D. hosei [Species]
  Eupleres [Genus]

E. goudotii [Species]
Fossa [Genus]
   F. fossana [Species]
Genetta [Genus]
   G. abyssinica [Species]
   G. angolensis
   G. genetta
   G. johnstoni
   G. maculata
   G. servalina
   G. thierryi
   G. tigrina
   G. victoriae
Hemigalus [Genus]
   H. derbyanus [Species]
Nandinia [Genus]
   N. binotata [Species]
Macrogalidia [Genus]
   M. musschenbroekii [Species]
Paguma [Genus]
   P. larvata [Species]
Paradoxurus [Genus]
   P. hermaphroditus [Species]
   P. jerdoni
   P. zeylonensis
Osbornictis [Genus]
   O. piscivora [Species]
Poiana [Genus]
   P. richardsonii [Species]
Prionodon [Genus]
   P. linsang [Species]
   P. pardicolor
Viverra [Genus]
   V. civettina [Species]
   V. megaspila
   V. tangalunga
   V. zibetha
Viverricula [Genus]
   V. indica [Species]

Herpestidae [Family]
   Atilax [Genus]
      A. paludinosus [Species]
   Bdeogale [Genus]
      B. crassicauda [Species]
      B. jacksoni
      B. nigripes
   Crossarchus [Genus]
      C. alexandri [Species]
      C. ansorgei
      C. obscurus
   Cynictis [Genus]
      C. penicillata [Species]
   Dologale [Genus]
      D. dybowskii [Species]
   Galerella [Genus]
      G. flavescens [Species]
      G. pulverulenta
      G. sanguinea
      G. swalius
   Galidia [Genus]

G. elegans [Species]
Galidictis [Genus]
   G. fasciata [Species]
   G. grandidieri
Helogale [Genus]
   H. hirtula [Species]
   H. parvula
Herpestes [Genus]
   H. brachyurus [Species]
   H. edwardsii
   H. ichneumon
   H. javanicus
   H. naso
   H. palustris
   H. semitorquatus
   H. smithii
   H. urva
   H. vitticollis
Ichneumia [Genus]
   I. albicauda [Species]
Liberiictis [Genus]
   L. kuhni [Species]
Mungos [Genus]
   M. gambianus [Species]
   M. mungo
Mungotictis [Genus]
   M. decemlineata [Species]
Paracynictis [Genus]
   P. selousi [Species]
Rhynchogale [Genus]
   R. melleri [Species]
Salanoia [Genus]
   S. concolor [Species]
Suricata [Genus]
   S. suricatta [Species]

Hyaenidae [Family]
   Crocuta [Genus]
      C. crocuta [Species]
   Hyaena [Genus]
      H. hyaena [Species]
   Parahyaena [Genus]
      P. brunnea [Species]
   Proteles [Genus]
      P. cristatus [Species]

Felidae [Family]
   Acinonyx [Genus]
      A. jubatus [Species]
   Caracal [Genus]
      C. caracal [Species]
   Catopuma [Genus]
      C. badia [Species]
      C. temminckii
   Felis [Genus]
      F. bieti [Species]
      F. chaus
      F. margarita
      F. nigripes
      F. silvestris

Herpailurus [Genus]
   H. yaguarondi [Species]
Leopardus [Genus]
   L. pardalis [Species]
   L. tigrinus
   L. wiedii
Leptailurus [Genus]
   L. serval [Species]
Lynx [Genus]
   L. canadensis [Species]
   L. lynx
   L. pardinus
   L. rufus
Neofelis [Genus]
   N. nebulosa [Species]
Oncifelis [Genus]
   O. colocolo [Species]
   O. geoffroyi
   O. guigna
Oreailurus [Genus]
   O. jacobita [Species]
Otocolobus [Genus]
   O. manul [Species]
Panthera [Genus]
   P. leo [Species]
   P. onca
   P. pardus
   P. tigris
   Pardofelis
   P. marmorata
Prionailurus [Genus]
   P. bengalensis [Species]
   P. planiceps
   P. rubiginosus
   P. viverrinus
Profelis [Genus]
   P. aurata [Species]
Puma [Genus]
   P. concolor [Species]
Uncia [Genus]
   U. uncia [Species]

Otariidae [Family]
   Arctocephalus [Genus]
      A. australis [Species]
      A. forsteri
      A. galapagoensis
      A. gazella
      A. philippii
      A. pusillus
      A. townsendi
      A. tropicalis
   Callorhinus [Genus]
      C. ursinus [Species]
   Eumetopias [Genus]
      E. jubatus [Species]
   Neophoca [Genus]
      N. cinerea [Species]
   Otaria [Genus]
      O. byronia [Species]

Phocarctos [Genus]
  P. hookeri [Species]
Zalophus [Genus]
  Z. californianus [Species]

Odobenidae [Family]
Odobenus [Genus]
  O. rosmarus [Species]

Phocidae [Family]
Cystophora [Genus]
  C. cristata [Species]
Erignathus [Genus]
  E. barbatus [Species]
Halichoerus [Genus]
  H. grypus [Species]
Hydrurga [Genus]
  H. leptonyx [Species]
Leptonychotes [Genus]
  L. weddellii [Species]
Lobodon [Genus]
  L. carcinophagus [Species]
Mirounga [Genus]
  M. angustirostris [Species]
  M. leonina
Monachus [Genus]
  M. monachus [Species]
  M. schauinslandi
  M. tropicalis
Ommatophoca [Genus]
  O. rossii [Species]
Phoca [Genus]
  P. caspica [Species]
  P. fasciata
  P. groenlandica
  P. hispida
  P. largha
  P. sibirica
  P. vitulina

Cetacea [Order]

Platanistidae [Family]
Platanista [Genus]
  P. gangetica [Species]
  P. minor

Lipotidae [Family]
Lipotes [Genus]
  L. vexillifer [Species]

Pontoporiidae [Family]
Pontoporia [Genus]
  P. blainvillei [Species]

Iniidae [Family]
Inia [Genus]
  I. geoffrensis [Species]

Phocoenidae [Family]
Australophocaena [Genus]
  A. dioptrica [Species]
Neophocaena [Genus]

N. phocaenoides [Species]
Phocoena [Genus]
  P. phocoena [Species]
  P. sinus
  P. spinipinnis
Phocoenoides [Genus]
  P. dalli [Species]

Delphinidae [Family]
Cephalorhynchus [Genus]
  C. commersonii [Species]
  C. eutropia
  C. heavisidii
  C. hectori
Delphinus [Genus]
  D. delphis [Species]
Feresa [Genus]
  F. attenuata [Species]
Globicephala [Genus]
  G. macrorhynchus [Species]
  G. melas
Grampus [Genus]
  G. griseus [Species]
Lagenodelphis [Genus]
  L. hosei [Species]
Lagenorhynchus [Genus]
  L. acutus [Species]
  L. albirostris
  L. australis
  L. cruciger
  L. obliquidens
  L. obscurus
Lissodelphis [Genus]
  L. borealis [Species]
  L. peronii
Orcaella [Genus]
  O. brevirostris [Species]
Orcinus [Genus]
  O. orca [Species]
Peponocephala [Genus]
  P. electra [Species]
Pseudorca [Genus]
  P. crassidens [Species]
Sotalia [Genus]
  S. fluviatilis [Species]
Sousa [Genus]
  S. chinensis [Species]
  S. teuszii
Stenella [Genus]
  S. attenuata [Species]
  S. clymene
  S. coeruleoalba
  S. frontalis
  S. longirostris
Steno [Genus]
  S. bredanensis [Species]
Tursiops [Genus]
  T. truncatus [Species]

Ziphiidae [Family]
Berardius [Genus]

B. arnuxii [Species]
B. bairdii
Hyperoodon [Genus]
  H. ampullatus [Species]
  H. planifrons
Indopacetus [Genus]
  I. pacificus [Species]
Mesoplodon [Genus]
  M. bidens [Species]
  M. bowdoini
  M. carlhubbsi
  M. densirostris
  M. europaeus
  M. ginkgodens
  M. grayi
  M. hectori
  M. layardii
  M. mirus
  M. peruvianus
  M. stejnegeri
Tasmacetus [Genus]
  T. shepherdi [Species]
Ziphius [Genus]
  Z. cavirostris [Species]

Physeteridae [Family]
Kogia [Genus]
  K. breviceps [Species]
  K. simus
Physeter [Genus]
  P. catodon [Species]

Monodontidae [Family]
Delphinapterus [Genus]
  D. leucas [Species]
Monodon [Genus]
  M. monoceros [Species]

Eschrichtiidae [Family]
Eschrichtius [Genus]
  E. robustus [Species]

Neobalaenidae [Family]
Caperea [Genus]
  C. marginata [Species]

Balaenidae [Family]
Balaena [Genus]
  B. mysticetus [Species]
Eubalaena [Genus]
  E. australis [Species]
  E. glacialis

Balaenopteridae [Family]
Balaenoptera [Genus]
  B. acutorostrata [Species]
  B. borealis
  B. edeni
  B. musculus
  B. physalus
Megaptera [Genus]
  M. novaeangliae [Species]

Tubulidentata [Order]

Orycteropodidae [Family]
Orycteropus [Genus]
O. afer [Species]

Proboscidea [Order]

Elephantidae [Family]
Elephas [Genus]
E. maximus [Species]
Loxodonta [Genus]
L. africana [Species]
L. cyclotis

Hyracoidea [Order]

Procaviidae [Family]
Dendrohyrax [Genus]
D. arboreus [Species]
D. dorsalis
D. validus
Heterohyrax [Genus]
H. antineae [Species]
H. brucei
Procavia [Genus]
P. capensis [Species]

Sirenia [Order]

Dugongidae [Family]
Dugong [Genus]
D. dugon [Species]
Hydrodamalis [Genus]
H. gigas [Species]

Trichechidae [Family]
Trichechus [Genus]
T. inunguis [Species]
T. manatus
T. senegalensis

Perissodactyla [Order]

Equidae [Family]
Equus [Genus]
E. asinus [Species]
E. burchellii
E. caballus
E. grevyi
E. hemionus
E. kiang
E. onager
E. quagga
E. zebra

Tapiridae [Family]
Tapirus [Genus]
T. bairdii [Species]
T. indicus
T. pinchaque
T. terrestris

Rhinocerotidae [Family]
Ceratotherium [Genus]
C. simum [Species]

Dicerorhinus [Genus]
D. sumatrensis [Species]
Diceros [Genus]
D. bicornis [Species]
Rhinoceros [Genus]
R. sondaicus [Species]
R. unicornis

Artiodactyla [Order]

Suidae [Family]
Babyrousa [Genus]
B. babyrussa [Species]
Phacochoerus [Genus]
P. aethiopicus [Species]
P. africanus
Hylochoerus [Genus]
H. meinertzhageni [Species]
Potamochoerus [Genus]
P. larvatus [Species]
P. porcus
Sus [Genus]
S. barbatus [Species]
S. bucculentus
S. cebifrons
S. celebensis
S. heureni
S. philippensis
S. salvanius
S. scrofa
S. timoriensis
S. verrucosus

Tayassuidae [Family]
Catagonus [Genus]
C. wagneri [Species]
Pecari [Genus]
P. tajacu [Species]
Tayassu [Genus]
T. pecari [Species]

Hippopotamidae [Family]
Hexaprotodon [Genus]
H. liberiensis [Species]
H. madagascariensis
Hippopotamus [Genus]
H. amphibius [Species]
H. lemerlei

Camelidae [Family]
Camelus [Genus]
C. bactrianus [Species]
C. dromedarius
Lama [Genus]
L. glama [Species]
L. guanicoe
L. pacos
Vicugna [Genus]
V. vicugna [Species]

Tragulidae [Family]
Hyemoschus [Genus]

H. aquaticus [Species]
Moschiola [Genus]
M. meminna [Species]
Tragulus [Genus]
T. javanicus [Species]
T. napu

Cervidae [Family]
Alces [Genus]
A. alces [Species]
Axis [Genus]
A. axis [Species]
A. calamianensis
A. kuhlii
A. porcinus
Blastocerus [Genus]
B. dichotomus [Species]
Capreolus [Genus]
C. capreolus [Species]
C. pygargus
Cervus [Genus]
C. albirostris [Species]
C. alfredi
C. duvaucelii
C. elaphus
C. eldii
C. mariannus
C. nippon
C. schomburgki
C. timorensis
C. unicolor
Dama [Genus]
D. dama [Species]
D. mesopotamica
Elaphodus [Genus]
E. cephalophus [Species]
Elaphurus [Genus]
E. davidianus [Species]
Hippocamelus [Genus]
H. antisensis [Species]
H. bisulcus
Hydropotes [Genus]
H. inermis [Species]
Mazama [Genus]
M. americana [Species]
M. bricenii
M. chunyi
M. gouazoupira
M. nana
M. rufina
Moschus [Genus]
M. berezovskii [Species]
M. chrysogaster
M. fuscus
M. moschiferus
Muntiacus [Genus]
M. atherodes [Species]
M. crinifrons
M. feae
M. gongshanensis

M. muntjak
M. reevesi
Odocoileus [Genus]
   O. hemionus [Species]
   O. virginianus
Ozotoceros [Genus]
   O. bezoarticus [Species]
Pudu [Genus]
   P. mephistophiles [Species]
   P. puda
Rangifer [Genus]
   R. tarandus [Species]

Giraffidae [Family]
  Giraffa [Genus]
   G. camelopardalis [Species]
  Okapia [Genus]
   O. johnstoni [Species]

Antilocapridae [Family]
  Antilocapra [Genus]
   A. americana [Species]

Bovidae [Family]
  Addax [Genus]
   A. nasomaculatus [Species]
  Aepyceros [Genus]
   A. melampus [Species]
  Alcelaphus [Genus]
   A. buselaphus [Species]
  Ammodorcas [Genus]
   A. clarkei [Species]
  Ammotragus [Genus]
   A. lervia [Species]
  Antidorcas [Genus]
   A. marsupialis [Species]
  Antilope [Genus]
   A. cervicapra [Species]
  Bison [Genus]
   B. bison [Species]
   B. bonasus
  Bos [Genus]
   B. frontalis [Species]
   B. grunniens
   B. javanicus
   B. sauveli
   B. taurus
  Boselaphus [Genus]
   B. tragocamelus [Species]
  Bubalus [Genus]
   B. bubalis [Species]
   B. depressicornis
   B. mephistopheles
   B. mindorensis
   B. quarlesi
  Budorcas [Genus]
   B. taxicolor [Species]
  Capra [Genus]
   C. caucasica [Species]
   C. cylindricornis
   C. falconeri
   C. hircus

C. ibex
C. nubiana
C. pyrenaica
C. sibirica
C. walie
Cephalophus [Genus]
   C. adersi [Species]
   C. callipygus
   C. dorsalis
   C. harveyi
   C. jentinki
   C. leucogaster
   C. maxwellii
   C. monticola
   C. natalensis
   C. niger
   C. nigrifrons
   C. ogilbyi
   C. rubidus
   C. rufilatus
   C. silvicultor
   C. spadix
   C. weynsi
   C. zebra
Connochaetes [Genus]
   C. gnou [Species]
   C. taurinus
Damaliscus [Genus]
   D. hunteri [Species]
   D. lunatus
   D. pygargus
Dorcatragus [Genus]
   D. megalotis [Species]
Gazella [Genus]
   G. arabica [Species]
   G. bennettii
   G. bilkis
   G. cuvieri
   G. dama
   G. dorcas
   G. gazella
   G. granti
   G. leptoceros
   G. rufifrons
   G. rufina
   G. saudiya
   G. soemmerringii
   G. spekei
   G. subgutturosa
   G. thomsonii
Hemitragus [Genus]
   H. hylocrius [Species]
   H. jayakari
   H. jemlahicus
Hippotragus [Genus]
   H. equinus [Species]
   H. leucophaeus
   H. niger
Kobus [Genus]
   K. ellipsiprymnus [Species]

K. kob
K. leche
K. megaceros
K. vardonii
Litocranius [Genus]
   L. walleri [Species]
Madoqua [Genus]
   M. guentheri [Species]
   M. kirkii
   M. piacentinii
   M. saltiana
Naemorhedus [Genus]
   N. baileyi [Species]
   N. caudatus
   N. crispus
   N. goral
   N. sumatraensis
   N. swinhoei
Neotragus [Genus]
   N. batesi [Species]
   N. moschatus
   N. pygmaeus
Oreamnos [Genus]
   O. americanus [Species]
Oreotragus [Genus]
   O. oreotragus [Species]
Oryx [Genus]
   O. dammah [Species]
   O. gazella
   O. leucoryx
Ourebia [Genus]
   O. ourebi [Species]
Ovibos [Genus]
   O. moschatus [Species]
Ovis [Genus]
   O. ammon [Species]
   O. aries
   O. canadensis
   O. dalli
   O. nivicola
   O. vignei
Pantholops [Genus]
   P. hodgsonii [Species]
Pelea [Genus]
   P. capreolus [Species]
Procapra [Genus]
   P. gutturosa [Species]
   P. picticaudata
   P. przewalskii
Pseudois [Genus]
   P. nayaur [Species]
   P. schaeferi
Raphicerus [Genus]
   R. campestris [Species]
   R. melanotis
   R. sharpei
Redunca [Genus]
   R. arundinum [Species]
   R. fulvorufula
   R. redunca

Rupicapra [Genus]
   R. pyrenaica [Species]
   R. rupicapra
Saiga [Genus]
   S. tatarica [Species]
Sigmoceros [Genus]
   S. lichtensteinii [Species]
Sylvicapra [Genus]
   S. grimmia [Species]
Syncerus [Genus]
   S. caffer [Species]
Taurotragus [Genus]
   T. derbianus [Species]
   T. oryx
Tetracerus [Genus]
   T. quadricornis [Species]
Tragelaphus [Genus]
   T. angasii [Species]
   T. buxtoni
   T. eurycerus
   T. imberbis
   T. scriptus
   T. spekii
   T. strepsiceros

Pholidota [Order]

Manidae [Family]
   Manis [Genus]
      M. crassicaudata [Species]
      M. gigantea
      M. javanica
      M. pentadactyla
      M. temminckii
      M. tetradactyla
      M. tricuspis

Rodentia [Order]

Aplodontidae [Family]
   Aplodontia [Genus]
      A. rufa [Species]

Sciuridae [Family]
   Aeretes [Genus]
      A. melanopterus [Species]
   Aeromys [Genus]
      A. tephromelas [Species]
      A. thomasi
   Ammospermophilus [Genus]
      A. harrisii [Species]
      A. insularis
      A. interpres
      A. leucurus
      A. nelsoni
   Atlantoxerus [Genus]
      A. getulus [Species]
   Belomys [Genus]
      B. pearsonii [Species]
   Biswamoyopterus [Genus]
      B. biswasi [Species]

Callosciurus [Genus]
   C. adamsi [Species]
   C. albescens
   C. baluensis
   C. caniceps
   C. erythraeus
   C. finlaysonii
   C. inornatus
   C. melanogaster
   C. nigrovittatus
   C. notatus
   C. orestes
   C. phayrei
   C. prevostii
   C. pygerythrus
   C. quinquestriatus
Cynomys [Genus]
   C. gunnisoni [Species]
   C. leucurus
   C. ludovicianus
   C. mexicanus
   C. parvidens
Dremomys [Genus]
   D. everetti [Species]
   D. lokriah
   D. pernyi
   D. pyrrhomerus
   D. rufigenis
Epixerus [Genus]
   E. ebii [Species]
   E. wilsoni
Eupetaurus [Genus]
   E. cinereus [Species]
Exilisciurus [Genus]
   E. concinnus [Species]
   E. exilis
   E. whiteheadi
Funambulus [Genus]
   F. layardi [Species]
   F. palmarum
   F. pennantii
   F. sublineatus
   F. tristriatus
Funisciurus [Genus]
   F. anerythrus [Species]
   F. bayonii
   F. carruthersi
   F. congicus
   F. isabella
   F. lemniscatus
   F. leucogenys
   F. pyrropus
   F. substriatus
Glaucomys [Genus]
   G. sabrinus [Species]
   G. volans
Glyphotes [Genus]
   G. simus [Species]
Heliosciurus [Genus]
   H. gambianus [Species]

   H. mutabilis
   H. punctatus
   H. rufobrachium
   H. ruwenzorii
   H. undulatus
Hylopetes [Genus]
   H. alboniger [Species]
   H. baberi
   H. bartelsi
   H. fimbriatus
   H. lepidus
   H. nigripes
   H. phayrei
   H. sipora
   H. spadiceus
   H. winstoni
Hyosciurus [Genus]
   H. heinrichi [Species]
   H. ileile
Iomys [Genus]
   I. horsfieldi [Species]
   I. sipora
Lariscus [Genus]
   L. hosei [Species]
   L. insignis
   L. niobe
   L. obscurus
Marmota [Genus]
   M. baibacina [Species]
   M. bobak
   M. broweri
   M. caligata
   M. camtschatica
   M. caudata
   M. flaviventris
   M. himalayana
   M. marmota
   M. menzbieri
   M. monax
   M. olympus
   M. sibirica
   M. vancouverensis
Menetes [Genus]
   M. berdmorei [Species]
Microsciurus [Genus]
   M. alfari [Species]
   M. flaviventer
   M. mimulus
   M. santanderensis
Myosciurus [Genus]
   M. pumilio [Species]
Nannosciurus [Genus]
   N. melanotis [Species]
Paraxerus [Genus]
   P. alexandri [Species]
   P. boehmi
   P. cepapi
   P. cooperi
   P. flavovittis
   P. lucifer

P. ochraceus
P. palliatus
P. poensis
P. vexillarius
P. vincenti
Petaurillus [Genus]
P. emiliae [Species]
P. hosei
P. kinlochii
Petaurista [Genus]
P. alborufus [Species]
P. elegans
P. leucogenys
P. magnificus
P. nobilis
P. petaurista
P. philippensis
P. xanthotis
Petinomys [Genus]
P. crinitus [Species]
P. fuscocapillus
P. genibarbis
P. hageni
P. lugens
P. sagitta
P. setosus
P. vordermanni
Prosciurillus [Genus]
P. abstrusus [Species]
P. leucomus
P. murinus
P. weberi
Protoxerus [Genus]
P. aubinnii [Species]
P. stangeri
Pteromys [Genus]
P. momonga [Species]
P. volans
Pteromyscus [Genus]
P. pulverulentus [Species]
Ratufa [Genus]
R. affinis [Species]
R. bicolor
R. indica
R. macroura
Rheithrosciurus [Genus]
R. macrotis [Species]
Rhinosciurus [Genus]
R. laticaudatus [Species]
Rubrisciurus [Genus]
R. rubriventer [Species]
Sciurillus [Genus]
S. pusillus [Species]
Sciurotamias [Genus]
S. davidianus [Species]
S. forresti
Sciurus [Genus]
S. aberti [Species]
S. aestuans
S. alleni

S. anomalus
S. arizonensis
S. aureogaster
S. carolinensis
S. colliaei
S. deppei
S. flammifer
S. gilvigularis
S. granatensis
S. griseus
S. ignitus
S. igniventris
S. lis
S. nayaritensis
S. niger
S. oculatus
S. pucheranii
S. pyrrhinus
S. richmondi
S. sanborni
S. spadiceus
S. stramineus
S. variegatoides
S. vulgaris
S. yucatanensis
Spermophilopsis [Genus]
S. leptodactylus [Species]
Spermophilus [Genus]
S. adocetus [Species]
S. alashanicus
S. annulatus
S. armatus
S. atricapillus
S. beecheyi
S. beldingi
S. brunneus
S. canus
S. citellus
S. columbianus
S. dauricus
S. elegans
S. erythrogenys
S. franklinii
S. fulvus
S. lateralis
S. madrensis
S. major
S. mexicanus
S. mohavensis
S. mollis
S. musicus
S. parryii
S. perotensis
S. pygmaeus
S. relictus
S. richardsonii
S. saturatus
S. spilosoma
S. suslicus
S. tereticaudus

S. townsendii
S. tridecemlineatus
S. undulatus
S. variegatus
S. washingtoni
S. xanthoprymnus
Sundasciurus [Genus]
S. brookei [Species]
S. davensis
S. fraterculus
S. hippurus
S. hoogstraali
S. jentinki
S. juvencus
S. lowii
S. mindanensis
S. moellendorffi
S. philippinensis
S. rabori
S. samarensis
S. steerii
S. tenuis
Syntheosciurus [Genus]
S. brochus [Species]
Tamias [Genus]
T. alpinus [Species]
T. amoenus
T. bulleri
T. canipes
T. cinereicollis
T. dorsalis
T. durangae
T. merriami
T. minimus
T. obscurus
T. ochrogenys
T. palmeri
T. panamintinus
T. quadrimaculatus
T. quadrivittatus
T. ruficaudus
T. rufus
T. senex
T. sibiricus
T. siskiyou
T. sonomae
T. speciosus
T. striatus
T. townsendii
T. umbrinus
Tamiasciurus [Genus]
T. douglasii [Species]
T. hudsonicus
T. mearnsi
Tamiops [Genus]
T. macclellandi [Species]
T. maritimus
T. rodolphei
T. swinhoei
Trogopterus [Genus]

T. xanthipes [Species]
Xerus [Genus]
    X. erythropus [Species]
    X. inauris
    X. princeps
    X. rutilus

Castoridae [Family]
  Castor [Genus]
    C. canadensis [Species]
    C. fiber

Geomyidae [Family]
  Geomys [Genus]
    G. arenarius [Species]
    G. bursarius
    G. personatus
    G. pinetis
    G. tropicalis
  Orthogeomys [Genus]
    O. cavator [Species]
    O. cherriei
    O. cuniculus
    O. dariensis
    O. grandis
    O. heterodus
    O. hispidus
    O. lanius
    O. matagalpae
    O. thaeleri
    O. underwoodi
  Pappogeomys [Genus]
    P. alcorni [Species]
    P. bulleri
    P. castanops
    P. fumosus
    P. gymnurus
    P. merriami
    P. neglectus
    P. tylorhinus
    P. zinseri
  Thomomys [Genus]
    T. bottae [Species]
    T. bulbivorus
    T. clusius
    T. idahoensis
    T. mazama
    T. monticola
    T. talpoides
    T. townsendii
    T. umbrinus
  Zygogeomys [Genus]
    Z. trichopus [Species]

Heteromyidae [Family]
  Chaetodipus [Genus]
    C. arenarius [Species]
    C. artus
    C. baileyi
    C. californicus

C. fallax
C. formosus
C. goldmani
C. hispidus
C. intermedius
C. lineatus
C. nelsoni
C. penicillatus
C. pernix
C. spinatus
Dipodomys [Genus]
  D. agilis [Species]
  D. californicus
  D. compactus
  D. deserti
  D. elator
  D. elephantinus
  D. gravipes
  D. heermanni
  D. ingens
  D. insularis
  D. margaritae
  D. merriami
  D. microps
  D. nelsoni
  D. nitratoides
  D. ordii
  D. panamintinus
  D. phillipsii
  D. spectabilis
  D. stephensi
  D. venustus
Microdipodops [Genus]
  M. megacephalus [Species]
  M. pallidus
Heteromys [Genus]
  H. anomalus [Species]
  H. australis
  H. desmarestianus
  H. gaumeri
  H. goldmani
  H. nelsoni
  H. oresterus
Liomys [Genus]
  L. adspersus [Species]
  L. irroratus
  L. pictus
  L. salvini
  L. spectabilis
Perognathus [Genus]
  P. alticola [Species]
  P. amplus
  P. fasciatus
  P. flavescens
  P. flavus
  P. inornatus
  P. longimembris
  P. merriami
  P. parvus
  P. xanthanotus

Dipodidae [Family]
  Allactaga [Genus]
    A. balikunica [Species]
    A. bullata
    A. elater
    A. euphratica
    A. firouzi
    A. hotsoni
    A. major
    A. severtzovi
    A. sibirica
    A. tetradactyla
    A. vinogradovi
  Allactodipus [Genus]
    A. bobrinskii [Species]
  Cardiocranius [Genus]
    C. paradoxus [Species]
  Dipus [Genus]
    D. sagitta [Species]
  Eozapus [Genus]
    E. setchuanus [Species]
  Eremodipus [Genus]
    E. lichtensteini [Species]
  Euchoreutes [Genus]
    E. naso [Species]
  Jaculus [Genus]
    J. blanfordi [Species]
    J. jaculus
    J. orientalis
    J. turcmenicus
  Napaeozapus [Genus]
    N. insignis [Species]
  Paradipus [Genus]
    P. ctenodactylus [Species]
  Pygeretmus [Genus]
    P. platyurus [Species]
    P. pumilio
    P. shitkovi
  Salpingotus [Genus]
    S. crassicauda [Species]
    S. heptneri
    S. kozlovi
    S. michaelis
    S. pallidus
    S. thomasi
  Sicista [Genus]
    S. armenica [Species]
    S. betulina
    S. caucasica
    S. caudata
    S. concolor
    S. kazbegica
    S. kluchorica
    S. napaea
    S. pseudonapaea
    S. severtzovi
    S. strandi
    S. subtilis
    S. tianshanica
  Stylodipus [Genus]

MAMMALS SPECIES LIST

S. andrewsi [Species]
S. sungorus
S. telum
Zapus [Genus]
Z. hudsonius [Species]
Z. princeps
Z. trinotatus

Muridae [Family]
Abditomys [Genus]
A. latidens [Species]
Abrawayaomys [Genus]
A. ruschii [Species]
Acomys [Genus]
A. cahirinus [Species]
A. cilicicus
A. cinerasceus
A. ignitus
A. kempi
A. louisae
A. minous
A. mullah
A. nesiotes
A. percivali
A. russatus
A. spinosissimus
A. subspinosus
A. wilsoni
Aepeomys [Genus]
A. fuscatus [Species]
A. lugens
Aethomys [Genus]
A. bocagei [Species]
A. chrysophilus
A. granti
A. hindei
A. kaiseri
A. namaquensis
A. nyikae
A. silindensis
A. stannarius
A. thomasi
Akodon [Genus]
A. aerosus [Species]
A. affinis
A. albiventer
A. azarae
A. bogotensis
A. boliviensis
A. budini
A. cursor
A. dayi
A. dolores
A. fumeus
A. hershkovitzi
A. illuteus
A. iniscatus
A. juninensis
A. kempi
A. kofordi
A. lanosus

A. latebricola
A. lindberghi
A. longipilis
A. mansoensis
A. markhami
A. mimus
A. molinae
A. mollis
A. neocenus
A. nigrita
A. olivaceus
A. orophilus
A. puer
A. sanborni
A. sanctipaulensis
A. serrensis
A. siberiae
A. simulator
A. spegazzinii
A. subfuscus
A. surdus
A. sylvanus
A. toba
A. torques
A. urichi
A. varius
A. xanthorhinus
Allocricetulus [Genus]
A. curtatus [Species]
A. eversmanni
Alticola [Genus]
A. albicauda [Species]
A. argentatus
A. barakshin
A. lemminus
A. macrotis
A. montosa
A. roylei
A. semicanus
A. stoliczkanus
A. stracheyi
A. strelzowi
A. tuvinicus
Ammodillus [Genus]
A. imbellis [Species]
Andalgalomys [Genus]
A. olrogi [Species]
A. pearsoni
Andinomys [Genus]
A. edax [Species]
Anisomys [Genus]
A. imitator [Species]
Anonymomys [Genus]
A. mindorensis [Species]
Anotomys [Genus]
A. leander [Species]
Apodemus [Genus]
A. agrarius [Species]
A. alpicola
A. argenteus

A. arianus
A. chevrieri
A. draco
A. flavicollis
A. fulvipectus
A. gurkha
A. hermonensis
A. hyrcanicus
A. latronum
A. mystacinus
A. peninsulae
A. ponticus
A. rusiges
A. semotus
A. speciosus
A. sylvaticus
A. uralensis
A. wardi
Apomys [Genus]
A. abrae [Species]
A. datae
A. hylocoetes
A. insignis
A. littoralis
A. microdon
A. musculus
A. sacobianus
Arborimus [Genus]
A. albipes [Species]
A. longicaudus
A. pomo
Archboldomys [Genus]
A. luzonensis [Species]
Arvicanthis [Genus]
A. abyssinicus [Species]
A. blicki
A. nairobae
A. niloticus
A. somalicus
Arvicola [Genus]
A. sapidus [Species]
A. terrestris
Auliscomys [Genus]
A. boliviensis [Species]
A. micropus
A. pictus
A. sublimis
Baiomys [Genus]
B. musculus [Species]
B. taylori
Bandicota [Genus]
B. bengalensis [Species]
B. indica
B. savilei
Batomys [Genus]
B. dentatus [Species]
B. granti
B. salomonseni
Beamys [Genus]
B. hindei [Species]

J. vulpinus
Kadarsanomys [Genus]
  K. sodyi [Species]
Komodomys [Genus]
  K. rintjanus [Species]
Kunsia [Genus]
  K. fronto [Species]
  K. tomentosus
Lagurus [Genus]
  L. lagurus [Species]
Lamottemys [Genus]
  L. okuensis [Species]
Lasiopodomys [Genus]
  L. brandtii [Species]
  L. fuscus
  L. mandarinus
Leggadina [Genus]
  L. forresti [Species]
  L. lakedownensis
Leimacomys [Genus]
  L. buettneri [Species]
Lemmiscus [Genus]
  L. curtatus [Species]
Lemmus [Genus]
  L. amurensis [Species]
  L. lemmus
  L. sibiricus
Lemniscomys [Genus]
  L. barbarus [Species]
  L. bellieri
  L. griselda
  L. hoogstraali
  L. linulus
  L. macculus
  L. mittendorfi
  L. rosalia
  L. roseveari
  L. striatus
Lenomys [Genus]
  L. meyeri [Species]
Lenothrix [Genus]
  L. canus [Species]
Lenoxus [Genus]
  L. apicalis [Species]
Leopoldamys [Genus]
  L. edwardsi [Species]
  L. neilli
  L. sabanus
  L. siporanus
Leporillus [Genus]
  L. apicalis [Species]
  L. conditor
Leptomys [Genus]
  L. elegans [Species]
  L. ernstmayri
  L. signatus
Limnomys [Genus]
  L. sibuanus [Species]
Lophiomys [Genus]
  L. imhausi [Species]

Lophuromys [Genus]
  L. cinereus [Species]
  L. flavopunctatus
  L. luteogaster
  L. medicaudatus
  L. melanonyx
  L. nudicaudus
  L. rahmi
  L. sikapusi
  L. woosnami
Lorentzimys [Genus]
  L. nouhuysi [Species]
Macrotarsomys [Genus]
  M. bastardi [Species]
  M. ingens
Macruromys [Genus]
  M. elegans [Species]
  M. major
Malacomys [Genus]
  M. cansdalei [Species]
  M. edwardsi
  M. longipes
  M. lukolelae
  M. verschureni
Malacothrix [Genus]
  M. typica [Species]
Mallomys [Genus]
  M. aroaensis [Species]
  M. gunung
  M. istapantap
  M. rothschildi
Malpaisomys [Genus]
  M. insularis [Species]
Margaretamys [Genus]
  M. beccarii [Species]
  M. elegans
  M. parvus
Mastomys [Genus]
  M. angolensis [Species]
  M. coucha
  M. erythroleucus
  M. hildebrandtii
  M. natalensis
  M. pernanus
  M. shortridgei
  M. verheyeni
Maxomys [Genus]
  M. alticola [Species]
  M. baeodon
  M. bartelsii
  M. dollmani
  M. hellwaldii
  M. hylomyoides
  M. inas
  M. inflatus
  M. moi
  M. musschenbroekii
  M. ochraceiventer
  M. pagensis
  M. panglima

  M. rajah
  M. surifer
  M. wattsi
  M. whiteheadi
Mayermys [Genus]
  M. ellermani [Species]
Megadendromus [Genus]
  M. nikolausi [Species]
Megadontomys [Genus]
  M. cryophilus [Species]
  M. nelsoni
  M. thomasi
Megalomys [Genus]
  M. desmarestii [Species]
  M. luciae
Melanomys [Genus]
  M. caliginosus [Species]
  M. robustulus
  M. zunigae
Melasmothrix [Genus]
  M. naso [Species]
Melomys [Genus]
  M. aerosus [Species]
  M. bougainville
  M. burtoni
  M. capensis
  M. cervinipes
  M. fellowsi
  M. fraterculus
  M. gracilis
  M. lanosus
  M. leucogaster
  M. levipes
  M. lorentzii
  M. mollis
  M. moncktoni
  M. obiensis
  M. platyops
  M. rattoides
  M. rubex
  M. rubicola
  M. rufescens
  M. spechti
Meriones [Genus]
  M. arimalius [Species]
  M. chengi
  M. crassus
  M. dahli
  M. hurrianae
  M. libycus
  M. meridianus
  M. persicus
  M. rex
  M. sacramenti
  M. shawi
  M. tamariscinus
  M. tristrami
  M. unguiculatus
  M. vinogradovi
  M. zarudnyi

Mesembriomys [Genus]
    M. gouldii [Species]
    M. macrurus
Mesocricetus [Genus]
    M. auratus [Species]
    M. brandti
    M. newtoni
    M. raddei
Microdillus [Genus]
    M. peeli [Species]
Microhydromys [Genus]
    M. musseri [Species]
    M. richardsoni
Micromys [Genus]
    M. minutus [Species]
Microryzomys [Genus]
    M. altissimus [Species]
    M. minutus
Microtus [Genus]
    M. abbreviatus [Species]
    M. agrestis
    M. arvalis
    M. bavaricus
    M. breweri
    M. cabrerae
    M. californicus
    M. canicaudus
    M. chrotorrhinus
    M. daghestanicus
    M. duodecimcostatus
    M. evoronensis
    M. felteni
    M. fortis
    M. gerbei
    M. gregalis
    M. guatemalensis
    M. guentheri
    M. hyperboreus
    M. irani
    M. irene
    M. juldaschi
    M. kermanensis
    M. kirgisorum
    M. leucurus
    M. limnophilus
    M. longicaudus
    M. lusitanicus
    M. majori
    M. maximowiczii
    M. mexicanus
    M. middendorffi
    M. miurus
    M. mongolicus
    M. montanus
    M. montebelli
    M. mujanensis
    M. multiplex
    M. nasarovi
    M. oaxacensis
    M. obscurus

M. ochrogaster
M. oeconomus
M. oregoni
M. pennsylvanicus
M. pinetorum
M. quasiater
M. richardsoni
M. rossiaemeridionalis
M. sachalinensis
M. savii
M. schelkovnikovi
M. sikimensis
M. socialis
M. subterraneus
M. tatricus
M. thomasi
M. townsendii
M. transcaspicus
M. umbrosus
M. xanthognathus
Millardia [Genus]
    M. gleadowi [Species]
    M. kathleenae
    M. kondana
    M. meltada
Muriculus [Genus]
    M. imberbis [Species]
Mus [Genus]
    M. baoulei [Species]
    M. booduga
    M. bufo
    M. callewaerti
    M. caroli
    M. cervicolor
    M. cookii
    M. crociduroides
    M. famulus
    M. fernandoni
    M. goundae
    M. haussa
    M. indutus
    M. kasaicus
    M. macedonicus
    M. mahomet
    M. mattheyi
    M. mayori
    M. minutoides
    M. musculoides
    M. musculus
    M. neavei
    M. orangiae
    M. oubanguii
    M. pahari
    M. phillipsi
    M. platythrix
    M. saxicola
    M. setulosus
    M. setzeri
    M. shortridgei
    M. sorella

M. spicilegus
M. spretus
M. tenellus
M. terricolor
M. triton
M. vulcani
Mylomys [Genus]
    M. dybowskii [Species]
Myomys [Genus]
    M. albipes [Species]
    M. daltoni
    M. derooi
    M. fumatus
    M. ruppi
    M. verreauxii
    M. yemeni
Myopus [Genus]
    M. schisticolor [Species]
Myospalax [Genus]
    M. aspalax [Species]
    M. epsilanus
    M. fontanierii
    M. myospalax
    M. psilurus
    M. rothschildi
    M. smithii
Mystromys [Genus]
    M. albicaudatus [Species]
Nannospalax [Genus]
    N. ehrenbergi [Species]
    N. leucodon
    N. nehringi
Neacomys [Genus]
    N. guianae [Species]
    N. pictus
    N. spinosus
    N. tenuipes
Nectomys [Genus]
    N. palmipes [Species]
    N. parvipes
    N. squamipes
Nelsonia [Genus]
    N. goldmani [Species]
    N. neotomodon
Neofiber [Genus]
    N. alleni [Species]
Neohydromys [Genus]
    N. fuscus [Species]
Neotoma [Genus]
    N. albigula [Species]
    N. angustapalata
    N. anthonyi
    N. bryanti
    N. bunkeri
    N. chrysomelas
    N. cinerea
    N. devia
    N. floridana
    N. fuscipes
    N. goldmani

N. lepida
N. martinensis
N. mexicana
N. micropus
N. nelsoni
N. palatina
N. phenax
N. stephensi
N. varia
Neotomodon [Genus]
   N. alstoni [Species]
Neotomys [Genus]
   N. ebriosus [Species]
Nesomys [Genus]
   N. rufus [Species]
Nesokia [Genus]
   N. bunnii [Species]
   N. indica
Nesoryzomys [Genus]
   N. darwini [Species]
   N. fernandinae
   N. indefessus
   N. swarthi
Neusticomys [Genus]
   N. monticolus [Species]
   N. mussoi
   N. oyapocki
   N. peruviensis
   N. venezuelae
Niviventer [Genus]
   N. andersoni [Species]
   N. brahma
   N. confucianus
   N. coxingi
   N. cremoriventer
   N. culturatus
   N. eha
   N. excelsior
   N. fulvescens
   N. hinpoon
   N. langbianis
   N. lepturus
   N. niviventer
   N. rapit
   N. tenaster
Notiomys [Genus]
   N. edwardsii [Species]
Notomys [Genus]
   N. alexis [Species]
   N. amplus
   N. aquilo
   N. cervinus
   N. fuscus
   N. longicaudatus
   N. macrotis
   N. mitchellii
   N. mordax
Nyctomys [Genus]
   N. sumichrasti [Species]
Ochrotomys [Genus]

O. nuttalli [Species]
Oecomys [Genus]
   O. bicolor [Species]
   O. cleberi
   O. concolor
   O. flavicans
   O. mamorae
   O. paricola
   O. phaeotis
   O. rex
   O. roberti
   O. rutilus
   O. speciosus
   O. superans
   O. trinitatis
Oenomys [Genus]
   O. hypoxanthus [Species]
   O. ornatus
Oligoryzomys [Genus]
   O. andinus [Species]
   O. arenalis
   O. chacoensis
   O. delticola
   O. destructor
   O. eliurus
   O. flavescens
   O. fulvescens
   O. griseolus
   O. longicaudatus
   O. magellanicus
   O. microtis
   O. nigripes
   O. vegetus
   O. victus
Ondatra [Genus]
   O. zibethicus [Species]
Onychomys [Genus]
   O. arenicola [Species]
   O. leucogaster
   O. torridus
Oryzomys [Genus]
   O. albigularis [Species]
   O. alfaroi
   O. auriventer
   O. balneator
   O. bolivaris
   O. buccinatus
   O. capito
   O. chapmani
   O. couesi
   O. devius
   O. dimidiatus
   O. galapagoensis
   O. gorgasi
   O. hammondi
   O. intectus
   O. intermedius
   O. keaysi
   O. kelloggi
   O. lamia

O. legatus
O. levipes
O. macconnelli
O. melanotis
O. nelsoni
O. nitidus
O. oniscus
O. palustris
O. polius
O. ratticeps
O. rhabdops
O. rostratus
O. saturatior
O. subflavus
O. talamancae
O. xantheolus
O. yunganus
Osgoodomys [Genus]
   O. banderanus [Species]
Otomys [Genus]
   O. anchietae [Species]
   O. angoniensis
   O. denti
   O. irroratus
   O. laminatus
   O. maximus
   O. occidentalis
   O. saundersiae
   O. sloggetti
   O. tropicalis
   O. typus
   O. unisulcatus
Otonyctomys [Genus]
   O. hatti [Species]
Ototylomys [Genus]
   O. phyllotis [Species]
Oxymycterus [Genus]
   O. akodontius [Species]
   O. angularis
   O. delator
   O. hiska
   O. hispidus
   O. hucucha
   O. iheringi
   O. inca
   O. nasutus
   O. paramensis
   O. roberti
   O. rufus
Pachyuromys [Genus]
   P. duprasi [Species]
Palawanomys [Genus]
   P. furvus [Species]
Papagomys [Genus]
   P. armandvillei [Species]
   P. theodorverhoeveni
Parahydromys [Genus]
   P. asper [Species]
Paraleptomys [Genus]
   P. rufilatus [Species]

MAMMALS SPECIES LIST

P. wilhelmina
Parotomys [Genus]
   P. brantsii [Species]
   P. littledalei
Paruromys [Genus]
   P. dominator [Species]
   P. ursinus
Paulamys [Genus]
   P. naso [Species]
Pelomys [Genus]
   P. campanae [Species]
   P. fallax
   P. hopkinsi
   P. isseli
   P. minor
Peromyscus [Genus]
   P. attwateri [Species]
   P. aztecus
   P. boylii
   P. bullatus
   P. californicus
   P. caniceps
   P. crinitus
   P. dickeyi
   P. difficilis
   P. eremicus
   P. eva
   P. furvus
   P. gossypinus
   P. grandis
   P. gratus
   P. guardia
   P. guatemalensis
   P. gymnotis
   P. hooperi
   P. interparietalis
   P. leucopus
   P. levipes
   P. madrensis
   P. maniculatus
   P. mayensis
   P. megalops
   P. mekisturus
   P. melanocarpus
   P. melanophrys
   P. melanotis
   P. melanurus
   P. merriami
   P. mexicanus
   P. nasutus
   P. ochraventer
   P. oreas
   P. pectoralis
   P. pembertoni
   P. perfulvus
   P. polionotus
   P. polius
   P. pseudocrinitus
   P. sejugis
   P. simulus

P. sitkensis
P. slevini
P. spicilegus
P. stephani
P. stirtoni
P. truei
P. winkelmanni
P. yucatanicus
P. zarhynchus
Petromyscus [Genus]
   P. barbouri [Species]
   P. collinus
   P. monticularis
   P. shortridgei
Phaenomys [Genus]
   P. ferrugineus [Species]
Phaulomys [Genus]
   P. andersoni [Species]
   P. smithii
Phenacomys [Genus]
   P. intermedius [Species]
   P. ungava
Phloeomys [Genus]
   P. cumingi [Species]
   P. pallidus
Phyllotis [Genus]
   P. amicus [Species]
   P. andium
   P. bonaeriensis
   P. caprinus
   P. darwini
   P. definitus
   P. gerbillus
   P. haggardi
   P. magister
   P. osgoodi
   P. osilae
   P. wolffsohni
   P. xanthopygus
Pithecheir [Genus]
   P. melanurus [Species]
   P. parvus
Phodopus [Genus]
   P. campbelli [Species]
   P. roborovskii
   P. sungorus
Platacanthomys [Genus]
   P. lasiurus [Species]
Podomys [Genus]
   P. floridanus [Species]
Podoxymys [Genus]
   P. roraimae [Species]
Pogonomelomys [Genus]
   P. bruijni [Species]
   P. mayeri
   P. sevia
Pogonomys [Genus]
   P. championi [Species]
   P. loriae
   P. macrourus

P. sylvestris
Praomys [Genus]
   P. delectorum [Species]
   P. hartwigi
   P. jacksoni
   P. minor
   P. misonnei
   P. morio
   P. mutoni
   P. rostratus
   P. tullbergi
Prionomys [Genus]
   P. batesi [Species]
Proedromys [Genus]
   P. bedfordi [Species]
Prometheomys [Genus]
   P. schaposchnikowi [Species]
Psammomys [Genus]
   P. obesus [Species]
   P. vexillaris
Pseudohydromys [Genus]
   P. murinus [Species]
   P. occidentalis
Pseudomys [Genus]
   P. albocinereus [Species]
   P. apodemoides
   P. australis
   P. bolami
   P. chapmani
   P. delicatulus
   P. desertor
   P. fieldi
   P. fumeus
   P. fuscus
   P. glaucus
   P. gouldii
   P. gracilicaudatus
   P. hermannsburgensis
   P. higginsi
   P. johnsoni
   P. laborifex
   P. nanus
   P. novaehollandiae
   P. occidentalis
   P. oralis
   P. patrius
   P. pilligaensis
   P. praeconis
   P. shortridgei
Pseudoryzomys [Genus]
   P. simplex [Species]
Punomys [Genus]
   P. lemminus [Species]
Rattus [Genus]
   R. adustus [Species]
   R. annandalei
   R. argentiventer
   R. baluensis
   R. bontanus
   R. burrus

R. colletti
R. elaphinus
R. enganus
R. everetti
R. exulans
R. feliceus
R. foramineus
R. fuscipes
R. giluwensis
R. hainaldi
R. hoffmanni
R. hoogerwerfi
R. jobiensis
R. koopmani
R. korinchi
R. leucopus
R. losea
R. lugens
R. lutreolus
R. macleari
R. marmosurus
R. mindorensis
R. mollicomulus
R. montanus
R. mordax
R. morotaiensis
R. nativitatis
R. nitidus
R. norvegicus
R. novaeguineae
R. osgoodi
R. palmarum
R. pelurus
R. praetor
R. ranjiniae
R. rattus
R. sanila
R. sikkimensis
R. simalurensis
R. sordidus
R. steini
R. stoicus
R. tanezumi
R. tawitawiensis
R. timorensis
R. tiomanicus
R. tunneyi
R. turkestanicus
R. villosissimus
R. xanthurus
Reithrodon [Genus]
R. auritus [Species]
Reithrodontomys [Genus]
R. brevirostris [Species]
R. burti
R. chrysopsis
R. creper
R. darienensis
R. fulvescens
R. gracilis

R. hirsutus
R. humulis
R. megalotis
R. mexicanus
R. microdon
R. montanus
R. paradoxus
R. raviventris
R. rodriguezi
R. spectabilis
R. sumichrasti
R. tenuirostris
R. zacatecae
Rhabdomys [Genus]
R. pumilio [Species]
Rhagomys [Genus]
R. rufescens [Species]
Rheomys [Genus]
R. mexicanus [Species]
R. raptor
R. thomasi
R. underwoodi
Rhipidomys [Genus]
R. austrinus [Species]
R. caucensis
R. couesi
R. fulviventer
R. latimanus
R. leucodactylus
R. macconnelli
R. mastacalis
R. nitela
R. ochrogaster
R. scandens
R. venezuelae
R. venustus
R. wetzeli
Rhizomys [Genus]
R. pruinosus [Species]
R. sinensis
R. sumatrensis
Rhombomys [Genus]
R. opimus [Species]
Rhynchomys [Genus]
R. isarogensis [Species]
R. soricoides
Saccostomus [Genus]
S. campestris [Species]
S. mearnsi
Scapteromys [Genus]
S. tumidus [Species]
Scolomys [Genus]
S. melanops [Species]
S. ucayalensis
Scotinomys [Genus]
S. teguina [Species]
S. xerampelinus
Sekeetamys [Genus]
S. calurus [Species]
Sigmodon [Genus]

S. alleni [Species]
S. alstoni
S. arizonae
S. fulviventer
S. hispidus
S. inopinatus
S. leucotis
S. mascotensis
S. ochrognathus
Sigmodontomys [Genus]
S. alfari [Species]
S. aphrastus
Solomys [Genus]
S. ponceleti [Species]
S. salamonis
S. salebrosus
S. sapientis
S. spriggsarum
Spalax [Genus]
S. arenarius [Species]
S. giganteus
S. graecus
S. microphthalmus
S. zemni
Spelaeomys [Genus]
S. florensis [Species]
Srilankamys [Genus]
S. ohiensis [Species]
Stenocephalemys [Genus]
S. albocaudata [Species]
S. griseicauda
Steatomys [Genus]
S. caurinus [Species]
S. cuppedius
S. jacksoni
S. krebsii
S. parvus
S. pratensis
Stenomys [Genus]
S. ceramicus [Species]
S. niobe
S. richardsoni
S. vandeuseni
S. verecundus
Stochomys [Genus]
S. longicaudatus [Species]
Sundamys [Genus]
S. infraluteus [Species]
S. maxi
S. muelleri
Synaptomys [Genus]
S. borealis [Species]
S. cooperi
Tachyoryctes [Genus]
T. ankoliae [Species]
T. annectens
T. audax
T. daemon
T. macrocephalus
T. naivashae

T. rex
T. ruandae
T. ruddi
T. spalacinus
T. splendens
Taeromys [Genus]
   T. arcuatus [Species]
   T. callitrichus
   T. celebensis
   T. hamatus
   T. punicans
   T. taerae
Tarsomys [Genus]
   T. apoensis [Species]
   T. echinatus
Tateomys [Genus]
   T. macrocercus [Species]
   T. rhinogradoides
Tatera [Genus]
   T. afra [Species]
   T. boehmi
   T. brantsii
   T. guineae
   T. inclusa
   T. indica
   T. kempi
   T. leucogaster
   T. nigricauda
   T. phillipsi
   T. robusta
   T. valida
Taterillus [Genus]
   T. arenarius [Species]
   T. congicus
   T. emini
   T. gracilis
   T. harringtoni
   T. lacustris
   T. petteri
   T. pygargus
Tscherskia [Genus]
   T. triton [Species]
Thallomys [Genus]
   T. loringi [Species]
   T. nigricauda
   T. paedulcus
   T. shortridgei
Thalpomys [Genus]
   T. cerradensis [Species]
   T. lasiotis
Thamnomys [Genus]
   T. kempi [Species]
   T. venustus
Thomasomys [Genus]
   T. aureus [Species]
   T. baeops
   T. bombycinus
   T. cinereiventer
   T. cinereus
   T. daphne

T. eleusis
T. gracilis
T. hylophilus
T. incanus
T. ischyurus
T. kalinowskii
T. ladewi
T. laniger
T. monochromos
T. niveipes
T. notatus
T. oreas
T. paramorum
T. pyrrhonotus
T. rhoadsi
T. rosalinda
T. silvestris
T. taczanowskii
T. vestitus
Tokudaia [Genus]
   T. muenninki [Species]
   T. osimensis
Tryphomys [Genus]
   T. adustus [Species]
Tylomys [Genus]
   T. bullaris [Species]
   T. fulviventer
   T. mirae
   T. nudicaudus
   T. panamensis
   T. tumbalensis
   T. watsoni
Typhlomys [Genus]
   T. chapensis [Species]
   T. cinereus
Uranomys [Genus]
   U. ruddi [Species]
Uromys [Genus]
   U. anak [Species]
   U. caudimaculatus
   U. hadrourus
   U. imperator
   U. neobritanicus
   U. porculus
   U. rex
Vandeleuria [Genus]
   V. nolthenii [Species]
   V. oleracea
Vernaya [Genus]
   V. fulva [Species]
Volemys [Genus]
   V. clarkei [Species]
   V. kikuchii
   V. millicens
   V. musseri
Wiedomys [Genus]
   W. pyrrhorhinos [Species]
Wilfredomys [Genus]
   W. oenax [Species]
   W. pictipes

Xenomys [Genus]
   X. nelsoni [Species]
Xenuromys [Genus]
   X. barbatus [Species]
Xeromys [Genus]
   X. myoides [Species]
Zelotomys [Genus]
   Z. hildegardeae [Species]
   Z. woosnami
Zygodontomys [Genus]
   Z. brevicauda [Species]
   Z. brunneus
Zyzomys [Genus]
   Z. argurus [Species]
   Z. maini
   Z. palatilis
   Z. pedunculatus
   Z. woodwardi

Anomaluridae [Family]
   Anomalurus [Genus]
      A. beecrofti [Species]
      A. derbianus
      A. pelii
      A. pusillus
   Idiurus [Genus]
      I. macrotis [Species]
      I. zenkeri
   Zenkerella [Genus]
      Z. insignis [Species]

Pedetidae [Family]
   Pedetes [Genus]
      P. capensis [Species]

Ctenodactylidae [Family]
   Ctenodactylus [Genus]
      C. gundi [Species]
      C. vali
   Felovia [Genus]
      F. vae [Species]
   Massoutiera [Genus]
      M. mzabi [Species]
   Pectinator [Genus]
      P. spekei [Species]

Myoxidae [Family]
   Dryomys [Genus]
      D. laniger [Species]
      D. nitedula
      D. sichuanensis
   Eliomys [Genus]
      E. melanurus [Species]
      E. quercinus
   Glirulus [Genus]
      G. japonicus [Species]
   Graphiurus [Genus]
      G. christyi [Species]
      G. hueti
      G. lorraineus
      G. monardi
      G. ocularis

G. parvus
G. rupicola
Muscardinus [Genus]
   M. avellanarius [Species]
Myomimus [Genus]
   M. personatus [Species]
   M. roachi
   M. setzeri
Myoxus [Genus]
   M. glis [Species]
Selevinia [Genus]
   S. betpakdalaensis [Species]

Petromuridae [Family]
  Petromus [Genus]
   P. typicus [Species]

Thryonomyidae [Family]
  Thryonomys [Genus]
   T. gregorianus [Species]
   T. swinderianus

Bathyergidae [Family]
  Bathyergus [Genus]
   B. janetta [Species]
   B. suillus
  Cryptomys [Genus]
   C. bocagei [Species]
   C. damarensis
   C. foxi
   C. hottentotus
   C. mechowi
   C. ochraceocinereus
   C. zechi
  Georychus [Genus]
   G. capensis [Species]
  Heliophobius [Genus]
   H. argenteocinereus
  Heterocephalus [Genus]
   H. glaber [Species]

Hystricidae [Family]
  Atherurus [Genus]
   A. africanus [Species]
   A. macrourus
  Hystrix [Genus]
   H. africaeaustralis [Species]
   H. brachyura
   H. crassispinis
   H. cristata
   H. indica
   H. javanica
   H. pumila
   H. sumatrae
  Trichys [Genus]
   T. fasciculata [Species]

Erethizontidae [Family]
  Coendou [Genus]
   C. bicolor [Species]
   C. koopmani
   C. prehensilis

C. rothschildi
Echinoprocta [Genus]
   E. rufescens [Species]
Erethizon [Genus]
   E. dorsatum [Species]
Sphiggurus [Genus]
   S. insidiosus [Species]
   S. mexicanus
   S. pallidus
   S. spinosus
   S. vestitus
   S. villosus

Chinchillidae [Family]
  Chinchilla [Genus]
   C. brevicaudata [Species]
   C. lanigera
  Lagidium [Genus]
   L. peruanum [Species]
   L. viscacia
   L. wolffsohni
  Lagostomus [Genus]
   L. maximus [Species]

Dinomyidae [Family]
  Dinomys [Genus]
   D. branickii [Species]

Caviidae [Family]
  Cavia [Genus]
   C. aperea [Species]
   C. fulgida
   C. magna
   C. porcellus
   C. tschudii
  Dolichotis [Genus]
   D. patagonum [Species]
   D. salinicola
  Galea [Genus]
   G. flavidens [Species]
   G. spixii
  Kerodon [Genus]
   K. rupestris [Species]
  Microcavia [Genus]
   M. australis [Species]
   M. niata
   M. shiptoni

Hydrochaeridae [Family]
  Hydrochaeris [Genus]
   H. hydrochaeris [Species]

Dasyproctidae [Family]
  Dasyprocta [Genus]
   D. azarae [Species]
   D. coibae
   D. cristata
   D. fuliginosa
   D. guamara
   D. kalinowskii
   D. leporina
   D. mexicana

D. prymnolopha
D. punctata
D. ruatanica
Myoprocta [Genus]
   M. acouchy [Species]
   M. exilis

Agoutidae [Family]
  Agouti [Genus]
   A. paca [Species]
   A. taczanowskii

Ctenomyidae [Family]
  Ctenomys [Genus]
   C. argentinus [Species]
   C. australis
   C. azarae
   C. boliviensis
   C. bonettoi
   C. brasiliensis
   C. colburni
   C. conoveri
   C. dorsalis
   C. emilianus
   C. frater
   C. fulvus
   C. haigi
   C. knighti
   C. latro
   C. leucodon
   C. lewisi
   C. magellanicus
   C. maulinus
   C. mendocinus
   C. minutus
   C. nattereri
   C. occultus
   C. opimus
   C. pearsoni
   C. perrensis
   C. peruanus
   C. pontifex
   C. porteousi
   C. saltarius
   C. sericeus
   C. sociabilis
   C. steinbachi
   C. talarum
   C. torquatus
   C. tuconax
   C. tucumanus
   C. validus

Octodontidae [Family]
  Aconaemys [Genus]
   A. fuscus [Species]
   A. sagei
  Octodon [Genus]
   O. bridgesi [Species]
   O. degus
   O. lunatus

Octodontomys [Genus]
    O. gliroides [Species]
Octomys [Genus]
    O. mimax [Species]
Spalacopus [Genus]
    S. cyanus [Species]
Tympanoctomys [Genus]
    T. barrerae [Species]

Abrocomidae [Family]
    Abrocoma [Genus]
        A. bennetti [Species]
        A. boliviensis
        A. cinerea

Echimyidae [Family]
    Boromys [Genus]
        B. offella [Species]
        B. torrei
    Brotomys [Genus]
        B. contractus [Species]
        B. voratus
    Carterodon [Genus]
        C. sulcidens [Species]
    Clyomys [Genus]
        C. bishopi [Species]
        C. laticeps
    Chaetomys [Genus]
        C. subspinosus [Species]
    Dactylomys [Genus]
        D. boliviensis [Species]
        D. dactylinus
        D. peruanus
    Diplomys [Genus]
        D. caniceps [Species]
        D. labilis
        D. rufodorsalis
    Echimys [Genus]
        E. blainvillei [Species]
        E. braziliensis
        E. chrysurus
        E. dasythrix
        E. grandis
        E. lamarum
        E. macrurus
        E. nigrispinus
        E. pictus
        E. rhipidurus
        E. saturnus
        E. semivillosus
        E. thomasi
        E. unicolor
    Euryzygomatomys [Genus]
        E. spinosus [Species]
    Heteropsomys [Genus]
        H. antillensis [Species]
        H. insulans
    Hoplomys [Genus]
        H. gymnurus [Species]
    Isothrix [Genus]
        I. bistriata [Species]

        I. pagurus
    Kannabateomys [Genus]
        K. amblyonyx [Species]
    Lonchothrix [Genus]
        L. emiliae [Species]
    Makalata [Genus]
        M. armata [Species]
    Mesomys [Genus]
        M. didelphoides [Species]
        M. hispidus
        M. leniceps
        M. obscurus
        M. stimulax
    Olallamys [Genus]
        O. albicauda [Species]
        O. edax
    Proechimys [Genus]
        P. albispinus [Species]
        P. amphichoricus
        P. bolivianus
        P. brevicauda
        P. canicollis
        P. cayennensis
        P. chrysaeolus
        P. cuvieri
        P. decumanus
        P. dimidiatus
        P. goeldii
        P. gorgonae
        P. guairae
        P. gularis
        P. hendeei
        P. hoplomyoides
        P. iheringi
        P. longicaudatus
        P. magdalenae
        P. mincae
        P. myosuros
        P. oconnelli
        P. oris
        P. poliopus
        P. quadruplicatus
        P. semispinosus
        P. setosus
        P. simonsi
        P. steerei
        P. trinitatis
        P. urichi
        P. warreni
    Puertoricomys [Genus]
        P. corozalus [Species]
    Thrichomys [Genus]
        T. apereoides [Species]

Capromyidae [Family]
    Capromys [Genus]
        C. pilorides [Species]
    Geocapromys [Genus]
        G. brownii [Species]
        G. thoracatus
    Hexolobodon [Genus]

        H. phenax [Species]
    Isolobodon [Genus]
        I. montanus [Species]
        I. portoricensis
    Mesocapromys [Genus]
        M. angelcabrerai [Species]
        M. auritus
        M. nanus
        M. sanfelipensis
    Mysateles [Genus]
        M. garridoi [Species]
        M. gundlachi
        M. melanurus
        M. meridionalis
        M. prehensilis
    Plagiodontia [Genus]
        P. aedium [Species]
        P. araeum
        P. ipnaeum
    Rhizoplagiodontia [Genus]
        R. lemkei [Species]

Heptaxodontidae [Family]
    Amblyrhiza [Genus]
        A. inundata [Species]
    Clidomys [Genus]
        C. osborni [Species]
        C. parvus
    Elasmodontomys [Genus]
        E. obliquus [Species]
    Quemisia [Genus]
        Quemisia gravis [Species]

Myocastoridae [Family]
    Myocastor [Genus]
        M. coypus [Species]

Lagomorpha [Order]

Ochotonidae [Family]
    Ochotona [Genus]
        O. alpina [Species]
        O. cansus
        O. collaris
        O. curzoniae
        O. dauurica
        O. erythrotis
        O. forresti
        O. gaoligongensis
        O. gloveri
        O. himalayana
        O. hyperborea
        O. iliensis
        O. koslowi
        O. ladacensis
        O. macrotis
        O. muliensis
        O. nubrica
        O. pallasi
        O. princeps
        O. pusilla
        O. roylei

O. rufescens
O. rutila
O. thibetana
O. thomasi
Prolagus [Genus]
   P. sardus [Species]

Leporidae [Family]
  Brachylagus [Genus]
    B. idahoensis [Species]
  Bunolagus [Genus]
    B. monticularis [Species]
  Caprolagus [Genus]
    C. hispidus [Species]
  Lepus [Genus]
    L. alleni [Species]
    L. americanus
    L. arcticus
    L. brachyurus
    L. californicus
    L. callotis
    L. capensis
    L. castroviejoi
    L. comus
    L. coreanus
    L. corsicanus
    L. europaeus
    L. fagani
    L. flavigularis
    L. granatensis
    L. hainanus
    L. insularis

L. mandshuricus
L. nigricollis
L. oiostolus
L. othus
L. pequensis
L. saxatilis
L. sinensis
L. starcki
L. timidus
L. tolai
L. townsendii
L. victoriae
L. yarkandensis
Nesolagus [Genus]
  N. netscheri [Species]
Oryctolagus [Genus]
  O. cuniculus [Species]
Pentalagus [Genus]
  P. furnessi [Species]
Poelagus [Genus]
  P. marjorita [Species]
Pronolagus [Genus]
  P. crassicaudatus [Species]
  P. randensis
  P. rupestris
Romerolagus [Genus]
  R. diazi [Species]
Sylvilagus [Genus]
  S. aquaticus [Species]
  S. audubonii
  S. bachmani
  S. brasiliensis

S. cunicularius
S. dicei
S. floridanus
S. graysoni
S. insonus
S. mansuetus
S. nuttallii
S. palustris
S. transitionalis

Macroscelidea [Order]

Macroscelididae [Family]
  Elephantulus [Genus]
    E. brachyrhynchus [Species]
    E. edwardii
    E. fuscipes
    E. fuscus
    E. intufi
    E. myurus
    E. revoili
    E. rozeti
    E. rufescens
    E. rupestris
  Macroscelides [Genus]
    M. proboscideus [Species]
  Petrodromus [Genus]
    P. tetradactylus [Species]
  Rhynchocyon [Genus]
    R. chrysopygus [Species]
    R. cirnei
    R. petersi

• • • • •

A brief geologic history of animal life

A note about geologic time scales: A cursory look will reveal that the timing of various geological periods differs among textbooks. Is one right and the others wrong? Not necessarily. Scientists use different methods to estimate geological time—methods with a precision sometimes measured in tens of millions of years. There is, however, a general agreement on the magnitude and relative timing associated with modern time scales. The closer in geological time one comes to the present, the more accurate science can be—and sometimes the more disagreement there seems to be. The following account was compiled using the more widely accepted boundaries from a diverse selection of reputable scientific resources.

Geologic time scale

Era	Period	Epoch	Dates	Life forms
Proterozoic			2,500-544 mya*	First single-celled organisms, simple plants, and invertebrates (such as algae, amoebas, and jellyfish)
Paleozoic	Cambrian		544-490 mya	First crustaceans, mollusks, sponges, nautiloids, and annelids (worms)
	Ordovician		490-438 mya	Trilobites dominant. Also first fungi, jawless vertebrates, starfishes, sea scorpions, and urchins
	Silurian		438-408 mya	First terrestrial plants, sharks, and bony fishes
	Devonian		408-360 mya	First insects, arachnids (scorpions), and tetrapods
	Carboniferous	Mississippian	360-325 mya	Amphibians abundant. Also first spiders, land snails
		Pennsylvanian	325-286 mya	First reptiles and synapsids
	Permian		286-248 mya	Reptiles abundant. Extinction of trilobytes. Most modern insect orders
Mesozoic	Triassic		248-205 mya	Diversification of reptiles: turtles, crocodiles, therapsids (mammal-like reptiles), first dinosaurs, first flies
	Jurassic		205-145 mya	Insects abundant, dinosaurs dominant in later stage. First mammals, lizards, frogs, and birds
	Cretaceous		145-65 mya	First snakes and modern fish. Extinction of dinosaurs and ammonites, rise and fall of toothed birds
Cenozoic	Tertiary	Paleocene	65-55.5 mya	Diversification of mammals
		Eocene	55.5-33.7 mya	First horses, whales, monkeys, and leafminer insects
		Oligocene	33.7-23.8 mya	Diversification of birds. First anthropoids (higher primates)
		Miocene	23.8-5.6 mya	First hominids
		Pliocene	5.6-1.8 mya	First australopithecines
	Quaternary	Pleistocene	1.8 mya-8,000 ya	Mammoths, mastodons, and Neanderthals
		Holocene	8,000 ya-present	First modern humans

*Millions of years ago (mya)

· · · · ·

Index

Bold page numbers indicate the primary discussion of a topic; page numbers in italics indicate illustrations; "t" indicates a table.

INDEX

Bobcats, 12:*50*, 14:370, 14:*373*, 14:*378*,
14:*384*, 14:389
Bobrinski's jerboas, 16:212, 16:213, 16:216,
16:*218*, 16:*219*, 16:222
Bocages mole-rats, 16:349*t*
Bodenheimer's pipistrelles, 13:313
Body fat, 12:120
in hominids, 12:23–24
in neonates, 12:125
storage of, 12:19, 12:174
See also Physical characteristics
Body mass, 12:11–12, 12:24–25
Body size, 12:41
bats, 12:58–59
blue whales, 12:213
competition and, 12:117
domestication and, 12:174
field studies and, 12:200–202
Ice Ages and, 12:17–25
life histories and, 12:97–98
nocturnal mammals and, 12:115
reproduction and, 12:15
sexual dimorphism and, 12:99
tropics and, 12:21
See also Physical characteristics
Body temperature
body size and, 12:21
sperm and, 12:91, 12:103
subterranean mammals, 12:73–74
See also Thermoregulation
Bogdanowicz, W., 13:387
Bohor reedbucks, 16:27, 16:29, 16:30, 16:34
Boinski, S., 14:104
Bolivian chinchilla rats, 16:443, 16:444,
16:445, 16:*446*, 16:*447*, 16:448
Bolivian hemorrhagic fever, 16:270
Bolivian red howler monkeys, 14:167*t*
Bolivian squirrel monkeys, 14:101, 14:102,
14:103, 14:*107*, 14:*108*
Bolivian tuco-tucos, 16:430*t*
Bones, 12:41
bats, 12:58–59
growth of, 12:10
See also Physical characteristics
Bongos, 15:138, 15:*141*, 16:4, 16:5, 16:11
Bonn Convention for the Conservation of
Migratory Animals
Brazilian free-tailed bats, 13:493
free-tailed bats, 13:487
Bonnet macaques, 14:194
Bonobos, 14:*236*, 14:*239*–240
behavior, 14:228, 14:231–232
conservation status, 14:235
distribution, 14:227
evolution, 12:33, 14:225
feeding ecology, 14:232–233
habitats, 14:227
humans and, 14:235
language, 12:162
physical characteristics, 14:226–227
reproduction, 14:235
taxonomy, 14:225
Bonteboks, 16:27, 16:29, 16:34, 16:*35*,
16:*39*–40
Boocercus euryceros, 15:141
Boodies, 13:74, 13:77, 13:79–*80*
Boraker, D. K., 16:436
Border collies, 14:288–289
Borean long-tailed porcupines. *See* Long-
tailed porcupines

Bornean bay cats. *See* Bay cats
Bornean gibbons. *See* Mueller's gibbons
Bornean orangutans, 12:*156*, 12:222, 14:225,
14:*236*, 14:*237*
Bornean smooth-tailed tree shrews, 13:292,
13:*293*, 13:*294*–295
Bornean tree shrews, 13:292, 13:297*t*
Bornean yellow muntjacs, 15:346, 15:*348*,
15:*349*, 15:352
Borneo short-tailed porcupines. *See* Thick-
spined porcupines
Bos spp., 16:11, 16:13, 16:14
Bos domestica. See Bali cattle
Bos frontalis. See Gayals
Bos gaurus. See Gaurs
Bos grunniens. See Yaks
Bos indicus. See Brahma cattle
Bos javanicus. See Bantengs
Bos mutus. See Yaks
Bos primigenius. See Aurochs
Bos sauveli. See Koupreys
Bos taurus. See Aurochs; Domestic cattle
Boselaphini, 16:1, 16:11
Boselaphus tragocamelus. See Nilgais
Bothma, J. P., 15:171
Botos, 15:5, 15:*8*, **15:27–31**, 15:*28*, 15:*29*,
15:*30*
Botta's pocket gophers. *See* Valley pocket
gophers
Bottleheads. *See* Northern bottlenosed whales
Bottleneck species, 12:221–222
Bottlenosed dolphins, 12:*27*, 12:*67*, 12:68,
15:2
behavior, 12:*143*, 15:44, 15:45, 15:46
chemoreception, 12:84
conservation status, 15:49–50
distribution, 15:5
echolocation, 12:87
feeding ecology, 15:47
habitats, 15:44
humans and, 15:10–11
language, 12:*161*
physical characteristics, 12:67–68
reproduction, 15:48
See also Common bottlenosed dolphins
Bottlenosed whales, 15:7
See also Northern bottlenosed whales;
Southern bottlenosed whales
Bovidae, **16:1–106**
Antilopinae, 16:1, **16:45–58**, 16:*49*,
16:56*t*–57*t*
behavior, 12:145, 12:146–147, 16:5–6
Bovinae, 16:1, **16:11–25**, 16:*16*, 16:*17*,
16:24*t*–25*t*
conservation status, 16:8
digestive systems of, 12:14
distribution, 16:4
evolution, 15:137, 15:138, 15:265, 15:266,
15:267, 16:1–2
feeding ecology, 15:142, 15:271, 16:6–7
habitats, 16:4–5
Hippotraginae, 16:1, **16:27–43**, 16:*35*,
16:41*t*–42*t*
horns, 12:40
humans and, 16:8–9
Neotraginae, 16:1, **16:59–72**, 16:*65*,
16:71*t*
physical characteristics, 15:142, 15:267,
15:268, 16:2–4
reproduction, 15:143, 16:7–8

taxonomy, 16:1–2
See also Caprinae; Duikers
Bovinae, 16:1, **16:11–25**, 16:*16*, 16:*17*,
16:24*t*–25*t*
Bovini, 16:1, 16:11
Bowdoin's beaked whales. *See* Andrew's
beaked whales
Bowhead whales, **15:107–118**, 15:*114*,
15:*115*–116
behavior, 15:109–110
conservation status, 15:9–10, 15:111–112
distribution, 15:5, 15:*107*, 15:109
evolution, 12:*65*, 15:107
feeding ecology, 15:110
habitats, 15:109
humans and, 15:112
migrations, 12:87, 12:170
physical characteristics, 15:107, 15:109
reproduction, 15:110–111
species of, 15:*115*–118
taxonomy, 15:107
Brachydelphis spp., 15:23
Brachylagus spp. *See* Pygmy rabbits
Brachylagus idahoensis. See Pygmy rabbits
Brachyphylla cavernarum. See Antillean fruit-
eating bats
Brachyphyllinae, 13:413, 13:415
Brachyteles spp. *See* Muriquis
Brachyteles arachnoides. See Southern muriquis
Brachyteles hypoxanthus. See Northern muriquis
Brachyuromy betsileonensis. See Malagasy reed rats
Bradbury, J. W., 13:342
Bradypodidae. *See* Three-toed tree sloths
Bradypus spp. *See* Three-toed tree sloths
Bradypus infuscatus. See Three-toed tree sloths
Bradypus pygmaeus. See Monk sloths
Bradypus torquatus. See Maned sloths
Bradypus tridactylus. See Pale-throated three-
toed sloths
Bradypus variegatus. See Brown-throated
three-toed sloths
Brahma cattle, 16:*4*
Braincase, 12:8, 12:9
Brains, 12:6, 12:36, 12:49
enlargement of, 12:9
evolution of, 12:19
learning and, 12:141
life spans and, 12:97–98
Neanderthals, 12:24
placentation and, 12:94
size of, 12:149–150
See also Physical characteristics
Brandt's hamsters, 16:247*t*
Brandt's hedgehogs, 13:213*t*
Branick's giant rats. *See* Pacaranas
Braude, S., 12:71
Brazilian agoutis. *See* Red-rumped agoutis
Brazilian free-tailed bats, 13:*484*, 13:*485*,
13:*486*, 13:*489*, 13:*492*, 13:493
behavior, 13:312, 13:484–485
distribution, 13:310
feeding ecology, 13:485, 13:486
humans and, 13:487–488
maternal recognition, 12:85
reproduction, 13:315, 13:486–487
Brazilian guinea pigs, 16:*392*
Brazilian shrew mice, 16:*271*, 16:273, 16:274
Brazilian spiny tree rats, 16:451, 16:458*t*
Brazilian tapirs. *See* Lowland tapirs
Brazilian three-banded armadillos, 13:192*t*

C

Cratogeomys zinseri. See Zinser's pocket gophers
Creek rats, 16:*254,* 16:*255,* 16:258–259
Creodonta, 12:11
Crepuscular activity cycle, 12:55
   See also Behavior
Crest-tailed marsupial mice. *See* Mulgaras
Crested agoutis, 16:409, 16:415*t*
Crested free-tailed bats. *See* Lesser-crested mastiff bats
Crested gibbons, 14:207, 14:210, 14:215, 14:*216,* 14:*220,* 14:221
Crested rats, 16:*287,* 16:*289,* 16:293
Crevice bats. *See* Peruvian crevice-dwelling bats
Cricetinae. *See* Hamsters
Cricetomyinae, 16:282
Cricetomys gambianus. See Gambian rats
Cricetulus spp. *See* Rat-like hamsters
Cricetulus barabensis. See Striped dwarf hamsters
Cricetulus migratorius. See Gray dwarf hamsters
Cricetus spp. *See* Black-bellied hamsters
Cricetus cricetus. See Black-bellied hamsters
Crocidura spp., 13:265, 13:266–267, 13:268
Crocidura attenuata. See Gray shrews
Crocidura canariensis. See Canary shrews
Crocidura dsinezumi. See Dsinezumi shrews
Crocidura fuliginosa. See Southeast Asian shrews
Crocidura horsfieldii. See Horsfield's shrews
Crocidura leucodon. See Bicolored shrews
Crocidura monticola. See Sunda shrews
Crocidura negrina. See Negros shrews
Crocidura russula. See Common European white-toothed shrews
Crocidura suaveolens. See Lesser white-toothed shrews
Crocidurinae. *See* White-toothed shrews
Crocuta crocuta. See Spotted hyenas
Cross foxes. *See* Red foxes
Cross River Allen's bushbabies, 14:32*t*
Crossarchus spp., 14:347
Crossarchus obscurus. See Western cusimanses
Crowned gibbons. *See* Pileated gibbons
Crowned lemurs, 14:*9,* 14:*51,* 14:*54,* 14:57, 14:59–60
Cryptic golden moles, 13:215
Cryptochloris spp., 13:215
Cryptochloris wintoni. See De Winton's golden moles
Cryptochloris zyli. See Van Zyl's golden moles
Cryptomys spp., 16:339, 16:342–343, 16:344, 16:345
   See also African mole-rats
Cryptomys amatus. See Zambian mole-rats
Cryptomys anselli. See Zambian mole-rats
Cryptomys bocagei. See Bocages mole-rats
Cryptomys damarensis. See Damaraland mole-rats; Kalahari mole-rats
Cryptomys darlingi. See Mashona mole-rats
Cryptomys foxi. See Nigerian mole-rats
Cryptomys hottentotus. See Common mole-rats
Cryptomys hottentotus hottentotus. See Common mole-rats
Cryptomys hottentotus pretoriae. See Highveld mole-rats
Cryptomys mechowi. See Giant Zambian mole-rats

Cryptomys ochraceocinereus. See Ochre mole-rats
Cryptomys zechi. See Togo mole-rats
Cryptoprocta spp., 14:347, 14:348
Cryptoprocta ferox. See Fossa
Cryptotis spp., 13:198, 13:248, 13:250
Cryptotis meridensis. See Mérida small-eared shrews
Cryptotis parva. See American least shrews
Ctenodactylidae. *See* Gundis
Ctenodactylus spp., 16:311, 16:312, 16:313
Ctenodactylus gundi. See North African gundis
Ctenodactylus vali. See Desert gundis
Ctenomyidae. *See* Tuco-tucos
Ctenomys spp. *See* Tuco-tucos
Ctenomys australis. See Southern tuco-tucos
Ctenomys azarae. See Azara's tuco-tucos
Ctenomys boliviensis. See Bolivian tuco-tucos
Ctenomys colburni. See Colburn's tuco-tucos
Ctenomys conoveri, 16:425
Ctenomys emilianus. See Emily's tuco-tucos
Ctenomys latro. See Mottled tuco-tucos
Ctenomys magellanicus. See Magellanic tuco-tucos
Ctenomys mattereri, 16:427
Ctenomys mendocinus, 16:427
Ctenomys nattereri. See Natterer's tuco-tucos
Ctenomys opimus, 16:427
Ctenomys pearsoni. See Pearson's tuco-tucos
Ctenomys perrensis. See Goya tuco-tucos
Ctenomys pundti, 16:425
Ctenomys rionegrensis. See Rio Negro tuco-tucos
Ctenomys saltarius. See Salta tuco-tucos
Ctenomys sociabilis. See Social tuco-tucos
Ctenomys talarum. See Talas tuco-tucos
Ctenomys torquatus. See Collared tuco-tucos
Cuban flower bats, 13:435*t*
Cuban funnel-eared bats. *See* Small-footed funnel-eared bats
Cuban hutias, 16:461, 16:*462,* 16:463, 16:*464,* 16:*465*
Cuban solenodons, 13:237, 13:*238*
Cuis, 16:391–392, 16:*394,* 16:*395,* 16:396
Culpeos, 14:265, 14:284*t*
Culture, 12:157–159
   See also Behavior
Cuon alpinus. See Dholes
Cursorial locomotion. *See* Running
Cuscomys spp., 16:443
Cuscomys ashaninka. See Ashaninka rats
Cuscomys oblativa, 16:443
Cuscuses, 12:137, 13:31, 13:32, 13:34, 13:35, 13:39, **13:57–67,** 13:*62,* 13:66*t*
Cusimanses, 14:349, 14:350
Cuvier's beaked whales, 15:*60,* 15:61, 15:62, 15:*63,* 15:*65,* 15:66
Cyclopes spp., 13:151
Cyclopes didactylus. See Silky anteaters
Cynictis spp., 14:347
Cynictis penicillata. See Yellow mongooses
Cynocephalidae. *See* Colugos
Cynocephalus spp. *See* Colugos
Cynocephalus variegatus. See Malayan colugos
Cynocephalus volans. See Philippine colugos
Cynodontia, 12:10, 12:11
Cynogale bennettii. See Otter civets
Cynomys spp. *See* Prairie dogs
Cynomys gunnisoni. See Gunnison's prairie dogs

Cynomys ludovicianus. See Black-tailed prairie dogs
Cynomys mexicanus, 16:147
Cynopterus spp. *See* Short-nosed fruit bats
Cynopterus brachyotis. See Lesser short-nosed fruit bats
Cynopterus sphinx. See Indian fruit bats
Cyomys ludovicianus. See Black-tailed prairie dogs
Cyprus spiny mice, 16:261*t*
Cystophora cristata. See Hooded seals
Cyttarops spp., 13:355

D

Dactylomys dactylinus. See Amazon bamboo rats
Dactylomys peruanus, 16:451
Dactylopsila spp., 13:127, 13:128
Dactylopsila megalura. See Great-tailed trioks
Dactylopsila palpator. See Long-fingered trioks
Dactylopsila tatei. See Tate's trioks
Dactylopsila trivirgata. See Striped possums
Dactylopsilinae, 13:125, 13:126
Dagestan turs. *See* East Caucasian turs
Dairy cattle. *See* Domestic cattle
Dairy goats. *See* Domestic goats
d'Albertis's ringtail possums, 13:122*t*
   See also Ringtail possums
Dall's porpoises, 15:33–*34,* 15:35, 15:36, 15:39*t*
Dall's sheep, 12:*103,* 12:178
Dalmatians, 14:289, 14:292
Dalpiazinidae, 15:13
Dama dama. See Fallow deer
Dama gazelles, 16:48, 16:57*t*
Dama wallabies. *See* Tammar wallabies
Damaliscus hunteri. See Hunter's hartebeests
Damaliscus lunatus. See Topis
Damaliscus lunatus jimela, 16:33
Damaliscus lunatus korrigum. See Korrigums
Damaliscus lunatus lunatus. See Tsessebes
Damaliscus pygargus. See Blesboks; Bonteboks
Damara mole-rats. *See* Damaraland mole-rats
Damaraland dikdiks. *See* Kirk's dikdiks
Damaraland mole-rats, 16:124, 16:*342,* 16:344–345, 16:*346,* 16:348
Darién pocket gophers, 16:196*t*
Dark Annamite muntjacs. *See* Truong Son muntjacs
Dark-handed gibbons. *See* Agile gibbons
Dark kangaroo mice, 16:*201,* 16:209*t*
Darwin, Charles
   on animal weapons, 15:268
   on armadillos, 13:181
   on Bovidae, 16:2
   on *Calomys laucha,* 16:267
   on domestic pigs, 15:149
   on mind and behavior, 12:149, 12:150
   on ungulates, 15:143
   See also Evolution
Darwin's foxes, 14:273
Dassie rats, 16:121, **16:329–332,** 16:*330*
Dasycercus cristicauda. See Mulgaras
Dasykaluta rosamondae. See Little red kalutas
Dasypodidae. *See* Armadillos
Dasypodinae, 13:182
Dasyprocta spp., 16:407, 16:*408*
Dasyprocta azarae. See Azara's agoutis

INDEX

Moles *(continued)*
    feeding ecology, 12:75–76, 13:198,
       13:281–282
    habitats, 13:280
    humans and, 13:200, 13:282
    physical characteristics, 13:195–196,
       13:279–280
    reproduction, 13:199, 13:282
    species of, 13:284–287, 13:287t–288t
    taxonomy, 13:193, 13:194, 13:279
    See also Marsupial moles; specific types of
      moles
Moloch gibbons, 14:215, 14:217, 14:219,
    14:220
    behavior, 14:210, 14:211
    conservation status, 14:11, 14:215
    distribution, 14:208
    evolution, 14:207
    feeding ecology, 14:213
Molossidae. *See* Free-tailed bats; Mastiff bats
Molossops spp., 13:483
Molossops greenhalli. See Greenhalli's dog-faced
    bats
Molossus ater. See Greater house bats
Molossus molossus. See Pallas's mastiff bats
Molossus pretiosus. See Miller's mastiff bats
Molsher, Robyn, 12:185
Molts, 12:38
Moluccan cuscuses, 13:66t
Monachinae, 14:417
Monachus monachus. See Mediterranean monk
    seals
Monachus schauinslandi. See Hawaiian monk
    seals
Monachus tropicalis. See West Indian monk
    seals
Mongolian gazelles, 12:134, 16:48, 16:49,
    16:53–54
Mongolian hamsters, 16:239, 16:247t
Mongolian pikas. *See* Pallas's pikas
Mongolian wild horses. *See* Przewalski's
    horses
Mongoose lemurs, 14:6, 14:8, 14:48, 14:50,
    14:53, 14:56–57
Mongooses, **14:347–358**, 14:353
    behavior, 12:114, 12:115, 12:146–147,
      14:259, 14:349–351
    in biological control, 12:190–191
    conservation status, 14:262, 14:352
    distribution, 12:136, 14:348–349
    evolution, 14:347
    feeding ecology, 14:260, 14:351–352
    habitats, 14:349
    humans and, 14:352
    physical characteristics, 14:347–348
    reproduction, 14:261, 14:352
    species of, 14:354–356, 14:357t
    taxonomy, 14:256, 14:347
Monitos del monte, 12:132, **12:273–275,**
    12:274
Monk sakis, 14:143, 14:144–145, 14:147,
    14:153t
Monk seals, 12:138, 14:256, 14:418
    Hawaiian, 12:138, 14:420, 14:422, 14:425,
      14:429, 14:431–432
    Mediterranean, 14:262, 14:418, 14:422,
      14:435t
    West Indian, 14:422, 14:435t
Monk sloths, 13:166, 13:168
Monkeys, 14:3, 14:8

Allen's swamp, 14:194, 14:196, 14:197
    behavior, 12:148
    coloration in, 12:38
    encephalization quotient, 12:149
    learning set, 12:152
    lesser white-nosed, 14:192
    memory, 12:152–154
    numbers and, 12:155
    theory of mind, 12:159–160
    tool use, 12:157
    in zoos, 12:207
    See also specific types of monkeys
Monodelphis spp. *See* Short-tailed opossums
Monodelphis americana, 12:251
Monodelphis brevicaudata. See Red-legged
    short-tailed opossums
Monodelphis dimidiata. See Southern short-
    tailed opossums
Monodelphis domestica, 12:251
Monodelphis iheringi, 12:251
Monodelphis kunsi. See Pygmy short-tailed
    opossums
Monodelphis osgoodi, 12:250
Monodelphis rubida, 12:250
Monodelphis scallops, 12:250
Monodelphis sorex, 12:250
Monodelphis unistriata, 12:250
Monodon monoceros. See Narwhals
Monodontidae, **15:81–91**, 15:89
Monogamy, 12:98, 12:107, 14:6–7
    Canidae, 14:271
    Carnivora, 14:261
    gibbons, 14:213, 14:214
    See also Behavior
Monophyllus redmani, 13:415
Monotremata. *See* Monotremes
Monotrematum sudamericanum. See South
    American monotremes
Monotremes, **12:227–234**, 15:133
    behavior, 12:230–231
    conservation status, 12:233–234
    development, 12:106
    distribution, 12:136–137, 12:230
    evolution, 12:11, 12:33, 12:227–228
    feeding ecology, 12:231
    habitats, 12:230
    humans and, 12:234
    physical characteristics, 12:38, 12:228–230
    reproduction, 12:12, 12:51, 12:89–93,
      12:108, 12:231–233
    skeletons, 12:41
Montane bamboo rats, 16:450–451, 16:452,
    16:455, 16:457
Montane guinea pigs, 12:181, 16:399t
Montane tree shrews, 13:297t
Montane woolly flying squirrels, 16:136
Montezuma (Aztec chief), 12:203
Moon-toothed degus, 16:433, 16:440t
Moonrats, 13:194, 13:197, 13:198, 13:199,
    13:203, 13:204, 13:208, 13:212
Moose, 15:133, 15:145–146, 15:263, 15:269,
    15:387
    behavior, 15:383, 15:394
    conservation status, 15:384, 15:394
    distribution, 12:132, 15:391, 15:394
    evolution, 15:379, 15:380, 15:381
    feeding ecology, 15:394
    habitats, 15:383, 15:394
    humans and, 15:384–385
    migrations, 12:164–165

    physical characteristics, 15:140, 15:381,
      15:394
    reproduction, 15:384, 15:394
    taxonomy, 15:394
Mops midas. See Midas free-tailed bats
Mops niangarae, 13:489
Mops spurelli. See Spurelli's free-tailed bats
Moraes, P. L. R., 14:164
Moreno glacier, 12:20
Morenocetus parvus, 15:107
Morganucodon spp., 12:11
Mormoopidae. *See* Moustached bats
Mormoops spp. *See* Ghost-faced bats
Mormoops blainvillii. See Antillean ghost-faced
    bats
Mormoops megalophylla. See Ghost-faced bats
Mormopterus spp., 13:483
Mormopterus acetabulosus. See Natal free-tailed
    bats
Mormopterus beccarii. See Beccari's mastiff bats
Mormopterus phrudus, 13:487
Moschidae. *See* Musk deer
Moschinae. *See* Musk deer
Moschus berezovskii, 15:333
Moschus chrysogaster. See Himalayan musk
    deer
Moschus fuscus, 15:333
Moschus moschiferus. See Siberian musk deer
Moss, Cynthia, 15:167
Moss-forest ringtails, 13:114, 13:115, 13:122t
Mottle-faced tamarins, 14:132t
Mottled tuco-tucos, 16:427, 16:430t
Mouflons, 12:178, 15:147, 16:87, 16:91
Mt. Graham red squirrels, 16:167
Mountain beavers, 12:131, 16:122, 16:123,
    16:126–127, **16:131–134**, 16:132, 16:133
Mountain brushtail possums, 13:59, 13:60,
    13:66t
Mountain cavies, 16:389, 16:390, 16:391,
    16:392, 16:393, 16:394, 16:395, 16:399t
Mountain cottontails, 16:481, 16:516t
Mountain cows. *See* Central American tapirs
Mountain deer, 15:384
Mountain gazelles, 12:139, 16:48, 16:49,
    16:54–55
Mountain goats, 12:21, 12:131, 12:165,
    15:134, 15:138, 16:5, 16:87, 16:92,
    16:93–94, 16:96, 16:99
Mountain gorillas, 12:218, 14:225, 14:227
    See also Eastern gorillas
Mountain hares, 12:129, 16:481, 16:506,
    16:510, 16:512–513
Mountain lions. *See* Pumas
Mountain marmots. *See* Hoary marmots
Mountain nyalas, 16:11, 16:25t
Mountain pacas, 16:422, 16:423–424
Mountain pocket gophers, 16:185, 16:186,
    16:195t
Mountain pygmy possums, 13:35, 13:109,
    13:110, 13:111
    behavior, 13:37, 13:106–107
    conservation status, 13:39, 13:108
    evolution, 13:105
    feeding ecology, 13:38, 13:107
    habitats, 13:34, 13:106
    reproduction, 13:38
Mountain reedbucks, 16:27, 16:29, 16:30,
    16:32
Mountain sheep, 12:165–167, 15:138, 16:4,
    16:95

INDEX

example, "silent mutations" (notably in third base positions) and mutations within non-coding regions of genes (introns) are inherently likely to accumulate faster because they do not lead to changes in amino acid sequences and are hence not subject to natural selection. In sum, it is now widely recognized that the concept of the "molecular clock" must be used with caution and that there may be quite marked differences between lineages in the rates of molecular change. Methods have therefore been developed to identify differences in rates of change between lineages and to apply the notion of "local clocks".

It should be noted that molecular data cannot directly yield information on elapsed time and that phylogenetic trees produced with such data always require calibration using information from the fossil record. Once a tree that is characterized by relatively uniform rates of change has been calibrated with at least one date from the fossil record, it is possible to convert genetic distances into time differences. However, conversion of genetic distances into time differences requires that genetic change should be linearly related to time. This generally seems to be the case once a global correction has been made for repeated mutation at a given site. Unfortunately, even if rates of molecular change along lineages are approximately linear (as is required for reliable application of a clock model), calibration dates derived from paleontological evidence introduce an additional source of error. The problem is that the fossil record can only yield a minimum date for the time of emergence of a particular lineage, by taking the age of the earliest known member of that lineage. The lineage may have existed for some considerable period of time prior to the earliest known fossil representative. Clearly, the size of the gap between the actual date of emergence of a lineage and the age of the earliest known representative of that lineage will vary according to the quality of the fossil record. If the fossil record is relatively well documented, as is probably the case with large-bodied hoofed mammals, the earliest known fossil representative may be quite close to the time of origin. In other cases, however, use of the age of the earliest known fossil to calibrate a phylogenetic tree may lead to considerable underestimation of dates of divergence. For instance, the earliest known undoubted primates are about 55 million years old, but statistical modeling indicates that we have so far discovered less than 5% of extinct fossil primate species. Correction for the numerous gaps in the primate fossil record indicates that the common ancestor of living primates existed about 85 million years ago (mya), rather than 60–65 mya as is commonly assumed. Molecular evolutionary phylogenetic trees have also been accurately determined for the common chimpanzee, pygmy chimpanzee, gorilla, and orangutan.

Overall molecular trees for mammals

Large-scale combined studies of nDNA and mtDNA have yielded phylogenetic trees for mammals that generally fit the conclusions derived from traditional morphological comparisons, but also show some differences in detail. For instance, molecular data have generally confirmed that the monotremes branched away first in the mammalian tree and that there was a subsequent division between marsupials and placentals. Interestingly, however, comparisons of complete mtDNA sequences have suggested that the monotremes and marsupials form a group separate from the placentals (Marsupiontia). As this aberrant result conflicts with other molecular evidence as well as with a well-established body of morphological evidence, it probably reflects an artifact of some kind. Indeed, it is noteworthy that the main points of conflict between different molecular trees involve relatively deep branches in the mammalian tree, which are precisely the branches that have posed the greatest challenges in morphological studies. Nevertheless, there is a gathering consensus from broad-based molecular studies that there are four major groups of placental mammals (Afrotheria, Xenarthra, Euarchontoglires and Laurasiatheria). As there are a number of consistent novel features of these groups, some modifications of conclusions based on morphological studies are undoubtedly required. For instance, the existence of the assemblage "Afrotheria" had not been identified from morphological comparisons and was first revealed by molecular studies. Moreover, it would seem that the order Insectivora is not only an artificial grouping of relatively primitive mammals (as has long been expected) but in fact includes widely separate lineages that belong either in Afrotheria (tenrecs, golden moles) or in Laurasiatheria (hedgehogs, shrews, and moles).

Overall phylogenetic trees for mammals based on molecular data have been calibrated in a variety of ways, and a fairly consistent picture has emerged. This conflicts with the long-accepted interpretation, according to which the evolutionary radiation of modern mammals did not begin until the dinosaurs died out at the end of the Cretaceous, 65 mya. Instead, it would seem that the four major groups of placental mammals began to diverge over 100 mya and that many (if not all) modern orders of placental mammals had become established by the end of the Cretaceous. For instance, numerous lines of evidence indicate that primates diverged from other placental mammals about 90 mya. This revised interpretation of the timing of mammalian evolution is significant not only because it indicates that dinosaurs and early relatives of modern placental mammals were contemporaries, but also because it suggests that continental drift may have played a major part in the early evolution of mammals. Contrary to the long-accepted interpretation that the evolutionary radiation of modern mammals began after the end of the Cretaceous, the new interpretation based on molecular data indicates that the early evolution of both placental mammals and marsupials took place at a time when the southern supercontinent Gondwana was undergoing active subdivision. As one outcome of this process, it seems that the endemic group Afrotheria became isolated in Africa.

Resources

Books

Avise, J. C. *Molecular Markers, Natural History and Evolution.* London: Chapman & Hall, 1994.

Easteal, Simon, C. Collett, and D. Betty. *The Mammalian Molecular Clock.* Austin: Texas, R.G. Landes, 1995.

Gillespie, J. H. *The Causes of Molecular Evolution.* Oxford: Oxford University Press, 1992.

Givnish, T. I., and K. Sytsma. *Molecular Evolution and Adaptive Radiations.* Cambridge: Cambridge University Press, 1997.

Hennig, W. *Phylogenetic Systematics. (Reprint of 1966 edition with a foreword by Donn E. Rosen, Gareth Nelson, and Colin Patterson)* Urbana: University of Illinois Press, 1979.

Hillis, David M., and Craig Moritz. *Molecular Systematics.* Sunderland: MA, Sinauer Associates, 1990.

Kimura, M. *The Neutral Theory of Molecular Evolution.* Cambridge: Cambridge University Press, 1983.

Lewin, Roger. *Patterns in Evolution: The New Molecular View.* San Francisco: W. H. Freeman, 1997.

Li, W.-H. *Molecular Evolution.* Sunderland: MA, Sinauer Associates, 1997.

Li, W.-H., and D. Graur. *Fundamentals of Molecular Evolution.* Sunderland: MA, Sinauer Associates, 1991.

Nei, M. *Molecular Evolutionary Genetics.* New York: Columbia University Press, 1987.

Ohno, S. *Evolution by Gene Duplication.* Berlin: Springer Verlag, 1970.

Scheffler, I. E. *Mitochondria.* New York: Wiley-Liss, 1999.

Periodicals

Allard, Mark W., R. L. Honeycutt, and M. J. Novacek. "Advances in higher level mammalian relationships." *Cladistics* 15 (1999): 213–219.

Anderson, S., M. H. L. de Bruijn, A. R. Coulson, I. C. Eperon, F. Sanger, and I. G. Young. "Complete sequence of bovine mitochondrial DNA: Conserved features of the mammalian mitochondrial genome." *Journal of Molecular Biology* 156 (1982): 683–717.

Arnason, Ulfur, Anette Gullberg, S. Gratarsdottir, B. Ursing, and A. Janke. "The mitochondrial genome of the sperm whale and a new molecular reference for estimating eutherian divergence dates." *Journal of Molecular Evolution* 50 (2000): 569–578.

Bromham, L., M. J. Phillips, and D. Penny. "Growing up with dinosaurs; molecular dates and the mammalian radiation." *Trends in Ecological Evolution* 14 (1999): 113–118.

Cao, Y., J. Adachi, Axel Janke, Svante Pääbo, and M. Hasegawa. "Phylogenetic relationships among eutherian orders estimated from inferred sequences of mitochondrial proteins: instability of a tree based on a single gene." *Journal of Molecular Evolution* 39 (1994): 519–527.

Gatesy, Y., C. Hayashi, M. A. Cronin, and P. Arctander. "Evidence from milk casein genes that cetaceans are close relatives of hippopotamid artiodactyls." *Molecular Biology and Evolution* 13 (1996): 954–963.

Gray, M. W. "Origin and evolution of mitochondrial DNA." *Annual Review of Cell Biology* 5 (1989): 25–50.

Hedges, S. Blair, Patrick H. Parker, Charles G. Sibley, and S. Kumar. "Continental breakup and the ordinal diversification of birds and mammals." *Nature* 381 (1996): 226–229.

Janke, Axel, X. Xu, and U. Arnason. "The complete mitochondrial genome of the wallaroo (*Macropus robustus*) and the phylogenetic relationship among Monotremata, Marsupialia and Eutheria." *Proceedings of the National Academy of Sciences, USA* 94 (1997): 1276-1281.

Kumar, Sudhir, and S. B. Hedges. "A molecular timescale for vertebrate evolution." *Nature* 392 (1998): 917–920.

Li, W.-H., M. Gouy, P. M. Sharp, C. O'hUigin, and Y.-W. Yang. "Molecular phylogeny of Rodentia, Lagomorpha, Primates, Artiodactyla, and Carnivora and molecular clocks." *Proceedings of the National Academy of Sciences, USA* 87 (1990): 6703–6707.

Li, W.-H., M. Tanimura, and P. M. Sharp. "An evaluation of the molecular clock hypothesis using mammalian DNA sequences." *Journal of Molecular Evolution* 25 (1987): 330–432.

Madsen, Ole, et al. "Parallel adaptive radiations in two major clades of placental mammals." *Nature* 409 (2001): 610–614.

Miyamoto, M. M. "A congruence study of molecular and morphological data for eutherian mammals." *Molecular Phylological Evolution* 6 (1996): 373-390.

Murphy, William J., et al. "Molecular phylogenetics and the origins of placental mammals." *Nature* 409 (2001): 614–618.

Murphy, William J., et al. "Resolution of the early placental mammal radiation using Bayesian phylogenetics." *Science* 294 (2001): 2348–2351.

Nikaido, M., A. P. Rooney, and N. Okada. "Phylogenetic relationships among certartiodactyls based on insertions of short and long interspersed elements: Hippopotamuses are the closest extant relatives of whales." *Proceedings of the National Academy of Sciences, USA* 96 (1999): 10261–10266.

Rambaut, A., and L. Bromham. "Estimating dates of divergence from molecular sequences." *Molecular Biological Evolution* 15 (1998): 442–448.

Schmitz, Jürgen, Martina Ohme, and H. Zischler. "The complete mitochondrial genome of *Tupaia belangeri* and the phylogenetic affiliation of Scandentia to other eutherian orders." *Molecular Biological Evolution* 17 (2000): 1334–1343.

Shedlock, A. M., and N. Okada. "SINE insertions: powerful tools for molecular systematics." *BioEssays* 22 (2000): 148–160.

Shimamura, Mitsuru, et al. "Molecular evidence from retroposons that whales form a clade within even-toed ungulates." *Nature* 388 (1997): 666–670.

Springer, Mark S., et al. "Endemic African mammals shake the phylogenetic tree." *Nature* 388 (1997): 61–64.

Tavaré, Simon, Charles R. Marshall, Oliver Will, Christophe Soligo, and R. D. Martin. "Using the fossil record to estimate the age of the last common ancestor of extant primates." *Nature* 416 (2002): 726–729.

A barbary ape (*Macaca sylvanus*) juvenile in a female's arms. (Photo by Animals Animals ©J. & P. Wegner. Reproduced by permission.)

ber of neurons in the neocortex it must be folded to fit within the skull of a mammal. For example, the surface area of the human neocortex is about 1.5 ft² (0.14 m²). With this area, it could not be simply laid over the deeper parts of the brain; folding produces gyri and sulci (folds and grooves, respectively), which gives the eutherian brain a convoluted appearance. Small mammals do not usually have convolutions, but they are almost always found once a species has reached a particular body size. Some researchers believe that the convolutions simply serve to increase the number of neurons in the neocortex, while others propose that the primary purpose of the convolutions is to increase surface area for heat dissipation. The brain produces a large amount of metabolic heat and must be cooled. Increasing the surface area provides more area for heat transfer (radiation) to occur; i.e., the convolutions produce a "radiator" for the brain.

The neocortex may be more developed or less developed depending on the mammalian species. In echolocating bats, for example, it comprises less than 50% of the brain surface because most of the bat brain is devoted to the auditory centers. Specific regions of the neocortex are specialized for par-

A Bengal tiger's (*Panthera tigris*) stripes make it stand out on a beach, but keep the tiger well camouflaged as it hunts from thickets, long grasses, or shrubs along riverbanks. (Photo by Jeff Lepore/Photo Researchers, Inc. Reproduced by permission.)

picked up by the animal and reingested. This reingestation of feces is called coprophagy. The soft pellet then goes through the digestive process a second time and the end product is a hard fecal pellet devoid of nutrients. Many owners of pet rabbits are familiar with the hard pellet, often called a "raisin." The softer pellet is usually consumed at night (when coprophagy goes unobserved by the pet owner) and is called the "midnight pellet." Coprophagy is efficient; voles are able to extract 67–75% of the energy contained in their food.

Nervous system and sensory organs

Mammals have relatively larger brains than other vertebrates. From monotremes to marsupials to eutherians, the mammal brain increases in size and complexity, primarily by the expansion of the neopallium. The neopallium (or neocortex) is a mantle of gray matter that first appeared as a small region between the olfactory bulb and the larger archipallium. The neopallium in mammals has expanded over the primitive parts of the vertebrate brain, dominating it as the cerebral cortex. The cerebral cortex is a thin laminar structure consisting of six sheets of neurons. In order to increase the num-

The bison's heavy coat helps it to survive winter blizzards in Wyoming, USA. (Photo by © Jeff Vanuga/Corbis. Reproduced by permission.)

Mammal hearing is highly developed. In mammals, the articular and quadrate bones of the reptilian jaw were modified to become the malleus and incus bones, which, along with the stapes, form the auditory ossicles in the mammal skull. The auditory ossicles conduct vibrations to the inner ear. Another mammal modification is the evolution of a pinna, an external flap that directs sound waves into the ear canal (external acoustic meatus). Many mammal species have mobil pinnae that enable them to pinpoint the location of the sound source. Pinnae are most elaborate in the insectivorous bats, but completely lacking in most marine and subterranean species.

The mammal eye is based on the reptilian eye. Many mammals are able to see very well in low-light conditions because of a reflective mirror-like layer (tapetum lucidum) in the choroid coat beneath the retina. The tapetum lucidum produces "eyeshine," such as seen in the eyes of deer staring into automobile headlights. Vision is improved as light is reflected back across the retina so that photoreceptors can interact with the light multiple times. Some mammals have an abundance of cones in the retina providing for color vision. This is especially true of the anthropoid primates, but this is also found in other mammals, e.g., Old World fruit bats. Aquatic species usually have a nictitating membrane that covers the eye, providing protection in the underwater environment.

Thermoregulation

Mammals produce their own body heat (endothermy) as opposed to absorbing energy from the outside environment. This metabolic heat is produced mainly in their mitochondria. Internal organs such as the heart, kidney, and brain are larger in mammals than reptiles and the corresponding increase in mitochondrial membrane surface area adds to their

ticular functions. For example, the occipital region is a visual center, the temporal region is involved with hearing, and the parietal lobe interprets touch. A structure found only in the eutherian brain is the corpus callosum, a concentration of nerve fibers that connect the two cerebral hemispheres and serve as a communication conduit between them.

Brain structure accounts for mammals' great ability to learn from their experiences. Their brain structure, combined with other neural characteristics, also accounts for mammals' acute sensory abilities. For example, mammalian smell is very acute. In some mammals, it is the most developed sense. Mammals have an elongated palate and, consequently, the nasal cavity is elongated as well. A structure in the palate of many mammals, the vomeronasal organ, detects smells from food. The development of turbinal bones covered by sensory mucosa in the nasal cavities has allowed more efficient detection of odors. Even so, some mammals have a poorer sense of smell than others, e.g., insectivorous bats, higher primates, and whales. In fact, dolphins and porpoises completely lack the olfactory apparatus. The receptors for taste are located on the tongue. Taste, interpreted in the brain in conjunction with olfactory stimuli, helps mammals identify whether food is safe to eat or not.

An agile bobcat (*Lynx rufus*) leaps across rocks. (Photo by Hans Reinhard. Bruce Coleman, Inc. Reproduced by permission.)

A Galápagos sea lion (*Zalophus californianus wollebaecki*) showing its streamlined shape enabling dives to catch fish. Photo by Animals Animals ©T. De Roi, OSF. Reproduced by permission.)

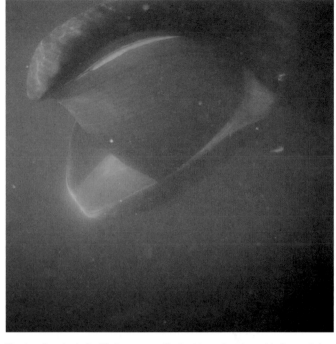

The bowhead whale (*Balaena mysticetus*) has developed baleen plates to filter the tiny organisms on which it feeds. (Photo by Glenn Williams/Ursus. Reproduced by permission.)

worldwide are beginning to yield Burgess-quality fossils, with perhaps many more waiting to be discovered.

Since that explosion of new forms some 530 mya, however, few new marine animals have evolved. Analysis of the evolution of marine animals suggests that a sufficient variety of life forms in an environment suppresses further innovation. About 530 mya, during the Cambrian period, after a long period in which animals were essentially jellyfishes or worms, marine animal life exploded into a variety of fundamentally new body types. Arthropods turned up inside external skeletons, mollusks put on their calcareous shells, and seven other new and different body plans appeared; an additional one showed up shortly thereafter. But since then, there's been nothing new in terms of basic body types, which form the basis of the top-level classification of the animal kingdom called phyla.

Research presented at a 1994 meeting of the Geological Society of America lends support to the idea that once evolution fills the world with sufficient variety, further innovation may be for naught. There are only so many ways marine animals can feed themselves—preying on others or scavenging debris, for example. And there are only so many places to

do it: on the sea floor, beneath it, or some distance above it. When all the nooks and crannies of this "ecospace" are filled, latecomers never get a foot in the door.

Challenges

Because water is so dense (up to 800 times denser than air), it can easily support an animal's body, eliminating the need for weight bearing skeletons like terrestrial animals. Water is also more viscous than air, and this coupled with the high density has resulted in aquatic animals adapting a very streamlined shape, particularly the carnivores. This makes them very fast and powerful swimmers, enabling them to catch their prey.

Many of the adaptations of aquatic organisms have to do with maintaining suitable conditions inside their bodies. The living "machinery" inside most organisms is rather sensitive and can only operate within a narrow range of conditions. Therefore, aquatic organisms have devised ways to keep their internal environments within this range no matter what external conditions are like.

Thermoregulation

Most aquatic animals are ectotherms, or poikilotherms, or what is often referred to as "cold-blooded." As the temperature of the surrounding water rises and falls, so does their body temperature and, consequently, their metabolic rate. Many become quite sluggish in unusually cold water. This "slowing down" caused by cold water is a disadvantage for active swimmers. Some large fish, such as certain tunas and sharks, can maintain body temperatures that are considerably

The fluke of the humpback whale (*Megaptera novaeangliae*) has evolved to be wide with scalloped edges, which enables the mammal to reach great heights when breaching. (Photo by John K. B. Ford/Ursus. Reproduced by permission.)

The Antarctic fur seal (*Arctocephalus gazella*) and her pup have a thick layer of blubber to keep them warm. (Photo by Animals Animals ©Johnny Johnson. Reproduced by permission.)

warmer than the surrounding water. They do this by retaining the heat produced in their large and active muscles. This allows them to remain active even in cold water.

Aquatic mammals are able to keep their body temperatures more or less constant regardless of water temperature. Marine mammals deposit most of their body fat into a thick layer of blubber that lies just underneath the skin. This blubber layer not only insulates them but also streamlines the body and functions as an energy reserve. The fusiform body shape and reduced limb size of many marine mammals and organisms decreases the amount of surface area exposed to the external environment. This helps conserve body heat. An interesting example of this body form adaptation can be seen in dolphins: those adapted to cooler, deeper water generally have larger bodies and smaller flippers than coastal dolphins, further reducing the surface area of their skin.

Arteries in the flippers, flukes, and dorsal fins of marine mammals are surrounded by veins. Thus, some heat from the blood traveling through the arteries is transferred to the venous blood rather than the outside environment. This countercurrent heat exchange also helps to conserve body heat.

Mammalian Sense of Smell

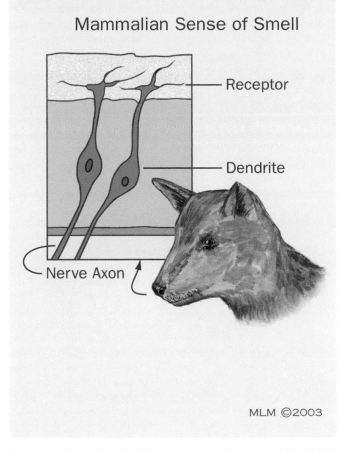

Receptor

Dendrite

Nerve Axon

MLM ©2003

The sense of smell displayed by many mammals owes much to the internal structure of the nasal passages. These are most highly developed in predators such as dogs. (Illustration by Michelle Meneghini)

both spray urine and kick feces at specific locations (middens) in their territories.

Many species of mammals also have glandular organs that contribute to their olfactory signatures. These organs typically include sebaceous and/or sudoriferous glands that synthesize odoriferous molecules. Behavior that transfers the glandular product(s) to other surfaces, sometimes to other animals, is called "scent marking." An example is the chinning behavior of male rabbits, which serves to place the products of exocrine glands located on the chin on the surfaces being marked. Scent glandular organs often are visually conspicuous, enhancing their role in advertisement. The behavior of mammals rubbing scent glands on surfaces makes the glands even more conspicuous, as in male white-tailed deer marking twigs with scent from their tear ducts during rut. In some mammals, scent glandular organs are associated with specialized hairs called osmotrechia, which are typically quite different from body hairs, being larger in diameter, sometimes longer, and often with a different scale structure; osmetrichia hold and transfer odoriferous molecules.

Although the wing sacs of some sheath-tailed bats (family Emballonuridae) have been referred to as glands, closer ex-

amination reveals that they lack glandular tissue. Rather, the wing sacs are fermentation chambers to which the bats (adult males) add various ingredients to enhance their personal scent. Greater sac-winged bats (*Saccopteryx bilineata*) put saliva, urine, and products of glands located near the anus into the mix in the sac, where fermentation produces the distinctive odors. Using their wing sacs, males can mark objects ranging from females in their group to their roosting sites.

Acoustics

The importance of sounds (acoustics) to mammals should be obvious. As in vision, binaural cues are timing differences between the arrival of sounds at one ear before the other, and they assist in the localization of sources of sounds. Humans use acoustical information to recognize the voices of family and friends or to locate an accident from the wail of an emergency vehicle's siren. In odontocetes, the ability to use binaural hearing is improved by an evolutionary shifting of the bones of the skull so that the hearing anatomy of the skull is asymmetrical. This makes odontocetes particularly sensitive to the direction of an incoming sound.

The western tarsier (*Tarsius bancanus*) has a heightened sense of hearing to help it avoid danger and capture prey. (Photo by Fletcher & Baylis/Photo Researchers, Inc. Reproduced by permission.)

The American bison (*Bison bison*) uses its vomero-nasal gland located on the anterior palate in the roof of its mouth to sense females in estrus. (Photo by © Layne Kennedy/Corbis. Reproduced by permission.)

The auditory system of most mammals consists of the following five main components:

- the pinnae, an external structure that acts as a sound collector

- the ear drum, or tympanum, that converts vibrations in air (sounds) to mechanical vibrations

- an amplifying system, the auditory ossicles (malleus, incus, and stapes) or bones of the middle ear

- a transducer (the oval window), where mechanical vibrations are converted to vibrations in fluid in the inner ear

- sites for converting vibrations in fluid to electrical stimuli (hair cells attached to the basilar membrane in the cochlea)

Through these components, electronic representations of the sounds are generated and transmitted to the brain via the auditory nerve.

Fossorial mammals, those that live most of their lives underground, may lack pinnae (which would only collect dirt). Many, but not all, aquatic mammals also lack pinnae (which would collect water). In fact, as a mammal progresses from amphibious (otters, seals, walrus, and sea lions) to totally aquatic (whales and dolphins), the pinnae go from small and valvular to absent. In fossorial mammals, considerable fusion of the auditory osscicles has reduced sensitivity to high-frequency sounds and emphasized the importance of low-frequency ones. In odontocetes, the lower jaw probably serves to conduct sounds to the middle ear and into the rest of the auditory system. Because water is a denser medium than air, it transmits sound more effectively (sound velocity in water is 4.5 times faster than in air), meaning that the auditory systems of odontocetes, even

without pinnae, are no less sensitive than those of humans. In fact, the effective communication distance for all marine mammals is much greater than for any terrestrial animal because of the density of water. In contrast, the effective communication distance for fossorial mammals would be very small, being limited by the reflective tunnel/burrow environment.

Sounds used by mammals can be of very different pitch or frequency, depending upon the species and situation. African elephants are sensitive to sounds at frequencies below 40 Hz; blue whales produce sounds as low as 20 Hz. These are referred to as "infrasounds," because they are below the range of human hearing (arbitrarily, 40 Hz). Other mammals, notably many bats, most carnivores (Felidae, Canidae, Mustelidae, Viverridae), and dolphins (Delphinidae), use sounds that are well above the range of human hearing (these are ultrasounds, theoretically >20,000 Hz). Humans hear best at frequencies from about 100–5,000 Hz, while some bats and dolphins hear very well at more than 200,000 Hz. In general, low-frequency sounds carry much farther (propogate) than high-frequency ones, and sounds greater than 20,000 Hz are rapidly eroded by the atmosphere (attenuated).

Touch

Mammals use their sense of touch in different ways. Tactile interactions are important for intraspecific communication, well known to a human who has benefited from the comfort of a hug. Often, touch plays an important role in female mammals recognizing their infants. Seal pups often reunite with their mothers by exchanges of vocalizations that

The giraffes' 18-foot (5.5-meter) height and excellent vision make it easy for them to spot predators from a distance. (Photo by David M. Maylen, III. Reproduced by permission.)

mand is, for example, generally typical of primates, most of which show parental carriage of the infant. Because infants that suckle on demand are usually slow growing but quite active precocial singeletons, the milk tends to be low in protein and fat but relatively high in sugar. This is the case with human milk, which is evidently naturally adapted for suckling on demand.

Eventually, provision of milk by the mother comes to an end and the offspring are weaned. In altricial mammal species, there is usually a fairly constant lactation period, and weaning tends to occur within a few weeks after birth. In precocial mammals, the lactation period can be quite variable and it may last months or even years. In fact, there is some indication that for certain hoofed mammals and primates there is a feedback relationship between the frequency of suckling and the mother's resumption of fertility, driven by the level of maternal nutrition. If food availability is low, the mother produces more dilute milk, which results in an increased suckling frequency. A higher frequency of suckling can suppress maternal fertility and also lead to an extension of the lactation period.

After weaning, the developing offspring must forage for food independently in order to meet its nutrient requirements for growth and maintenance. Eventually, it will attain sexual maturity and enter the breeding population. Here, too, altricial mammal species tend to have fairly standard ages for the attainment of sexual maturity, whereas precocial mammals can show more flexibility, according to prevailing environmental conditions. Despite such variability, the age of sexual maturity is an important milestone for comparisons among mammal species. It should be noted, incidentally, that sexual maturity may or may not coincide with the attainment of the adult condition in other respects, for example in body size and/or skeletal and dental maturity. In some cases, individuals may continue to grow for some time after achieving sexual maturity. It is noteworthy, however, that mammals (like birds) differ from reptiles, amphibians, and fish in showing a target body size. In each species, individuals tend to cease growing at a fairly standard size.

Life histories

All of the basic parameters that can be identified in the life cycle of any given mammal species, such as gestation period, litter size, lactation period, time taken to reach sexual maturity, and life span (longevity), contribute to its overall life history. Numerous lines of evidence suggest that these individual components of the life history of a species together constitute an adaptive complex that has been shaped by natural selection. For instance, one fundamental finding is that species that are subject to relatively heavy mortality under natural conditions tend to breed earlier. Because early breeding typically translates into a higher reproductive turnover, it can be concluded that heavy mortality promotes rapid breeding. In such comparisons between species, it has become commonplace to refer to "reproductive strategies." However, this is an anthropomorphic term and it is perhaps better to use a more neutral term like "life-history pattern."

In examining life-history patterns across species, it is essential to take account of the scaling influence of body size. It is only to be expected that large-bodied species will breed more slowly than small-bodied species, as it is generally likely that the time taken to grow to maturity will increase with increasing body size. However, it is also possible for mammals to show divergent life-history patterns even when body size is the same. David Western has aptly referred to these two kinds of difference as "first-order strategies" and "second-order strategies." For any given animal population, increase in population size is typically geometric until the population reaches a particular level (carrying capacity) determined by limiting factors (e.g. food supply, predation) in its environment. During the geometric phase of population growth, population increase depends on the intrinsic rate of increase (r) that is permitted by the reproductive parameters of the species. Because it is very difficult and time-consuming to obtain field data on the intrinsic rate of increase for natural populations, especially for large-bodied species, comparisons are often based on the maximum possible intrinsic rate of increase (r_{max}) that is permitted by standard reproductive values. A number of basic parameters such as age at sexual maturity, gestation period, litter size, interbirth interval, and longevity are used to calculate the r_{max} value for each species. As expected, for mammals r_{max} generally declines with increasing body size. In addition, there are distinctions between groups in the value found at any given body size. Primates, for example, generally have markedly lower values of r_{max} than other mammals. The two parameters that have the greatest influence in the calculation of r_{max} values are litter size and age at sexual maturity. Hence, the distinction between altricial and precocial mammals, in which litter size is a crucial feature, is clearly connected to a divergence in life-history patterns.

Various attempts have been made to develop general theories to explain the evolution of life-history patterns in mammals and other organisms. One basic problem that is encountered is that differences found between species do not fit well with the patterns that are observed within species. For instance, within a mammal species a particularly long gestation period is typically associated with a decrease in the duration of postnatal growth. In comparisons between species, however, it is found that species with long gestation periods tend to have extended periods of postnatal growth as well. It is therefore unclear how natural selection can shape life-history patterns. One model that has been suggested is "r- and K-selection." In this, it is proposed that species living in unpredictable environments with occasional catastrophic mortality will be subject to selection to increase r_{max} (r-selection). By contrast, species that live in predictable environments with moderate mortality will be subject to selection to increase competitiveness at carrying capacity (K-selection). An alternative model is that of "bet-hedging," in which it is suggested that high mortality among juveniles will favor slow breeding whereas high mortality among adults will favor rapid breeding. In fact, neither of these models fits all of the facts, so a convincing overall model remains elusive.

One point that deserves special mention is a potential link between life span (longevity) and relative brain size. Various authors have reported that mammals with relatively large

Fossa (*Cryptoprocta ferox*) breeding in the canopy at dawn in Madagascar. (Photo by Harald Schütz. Reproduced by permission.)

brains tend to have particularly long life spans. Although this proposed link is controversial, especially because of claims that it may be based on a secondary correlation, there is certainly enough evidence to indicate that some kind of connection exists. Hence, slow-breeding, long-lived mammals may also have relatively large brains as part of their overall life-history patterns.

Mating systems

Mammals exhibit a wide variety of mating systems, which can be basically divided into promiscuous, monogamous, polygynous, and multi-male. In mammals that live in gregarious social groups, the mating system is commonly (but not always) reflected by the composition of those groups, whereas in dispersed mammals patterns of mating must be determined from observations of interactions between separately ranging, "solitary" individuals. In all cases, however, it must be remembered that social systems and mating systems do not necessarily coincide. Even with mammals that are seemingly monogamous, genetic tests of paternity are quite likely to produce surprises just as they have already done for several bird species.

Monogamy, which is the predominant pattern of social organization and widely assumed to be the dominant mating

system among birds, is relatively rare among mammals. It is somewhat more common in carnivores and primates than in other mammals, but even in those groups it is found in only a minority of species. It seems likely that promiscuous mating was present in ancestral mammals in association with their likely nocturnal, dispersed habits, as it is in various relatively primitive nocturnal mammals today that lack any obvious social networks (e.g. many marsupials, insectivores, carnivores, and rodents). Another common mating pattern among mammals is polygyny, in which a single male has exclusive or almost exclusive mating access to a number of females. Polygyny is commonly found, for example, in hoofed mammals, pinnipeds, and elephants. By contrast, it is relatively rare to find multi-male systems in which several adult males are present in a social network or group competing for mating access to females, although it is widespread among higher primates. A key issue here is the potential occurrence of competition between sperm from different males. In mating systems in which a single male has clear priority of access to one or more females (monogamous and polygynous systems), the probability of sperm competition is presumably low, whereas in promiscuous and multi-male systems there is likely to be a high incidence of sperm competition. This expectation has been confirmed by studies of the relative size of the testes in mammals. Species with promiscuous or multi-male mating systems generally have significantly larger testes, rel-

Ecology

Ecology, the study of an organism's relationship to its surroundings, consists of several distinct areas, all of which have their own specific approaches and methods. Ecophysiology deals with physiological mechanisms, evolutionary ecology is concerned with life history-related, fitness-relevant, and population-genetic aspects, behavioral ecology looks at how the animal deals with its surroundings, and community ecology asks how groups of species can live together. This text will approach how mammals are able to adapt to extreme conditions and will also consider spatial and temporal distribution, predator-prey relationships, and relationships between species forming similar niches.

Mammals are endotherms. This means that they have to put a great deal of energy into regulating their body temperatures. In a cold environment, there are several strategies to deal with those challenges: mammals, contrary to reptiles, are often capable of developing a rather thick isolatory tissue (subcutaneous fat) plus thick fur. This option is not open to reptiles due to the fact that they need to retain the high thermal conductancy of their integument in order to heat themselves by exposure to sun rays. Small mammals, however, have this capability in a lesser extent than larger ones. An additional option for larger species is migration.

Another mammalian adaptation is known as non-shivering thermogenesis (NST), the burning of so-called brown fat, a special tissue rich in mitochondria and often deposited around the neck or between the shoulder blades. The most effective way of dealing with the challenge of a cold environment is torpor, the reduction of one's body temperature and basal metabolic rate, in some species to around or even slightly below 32°F (0°C). For example, the Arctic ground squirrel (*Spermophilus parryii*) goes down to a startling body temperature of 28°F (-2°C). Daily torpor, or larger periods of hibernation, can be found in members of at least five placental and two marsupial orders. The largest species found with "real torpor," lowering their body temperatures by at least 50°F (10°C), are badgers (both the American and the Eurasian species—in the latter case it was found in an individual of 27 lb [13 kg] body weight). Bears also become dormant in winter, but their body temperature is lowered only by about 41°F (5°C), and their physiological mechanisms are different from those of the smaller species.

How do mammals deal with desert conditions? Deserts are not only characterized by extreme temperatures (hot and cold, many small species thus also exhibit daily torpor), but also by arid conditions and a low biodiversity. One strategy for smaller mammals to deal with this low productivity again is heterothermy, to reduce basal metabolism and thus economize on one's energy demand. Large mammals such as camels can also store heat quite effectively in their large bodies; camels can increase their body temperatures under heat-stress up to 106°F (41°C) during the day, and lower it to around 93°F (34°C) at night. This can save about 12,000 kJ and 1.3 gal (5 l) of sweat. Water balance in many species of desert-dwelling mammals is improved by a counter-current system in their nasal conchae, and by recycling water in the kidneys. Kangaroo rats (*Dipodomys*) are among those species that regularly live without the need to drink open water; they are able to survive on the water content of their food (seeds), and from oxidation. Grazing mammals can improve their water balance by feeding at night, because grass species are rather rich in tissue fluid at that time. Carnivores extract water from vertebrate prey—even fennecs can keep their water balance simply by feeding on mice. The effectiveness of desert mammals is increased by behavioral adaptations, such as regularly retreating into burrows or shady areas where it's not only cooler but also more humid.

An even more challenging habitat for mammals is extreme high altitude. While desert means cold plus dry plus food scarcity, high altitude means desert plus low oxygen. Thus, physiological adaptations found in mammals at high altitudes include all those just discussed, plus specific ones to improve gas exchange (lung tissue and blood capillaries becoming more intricate), better oxygen transport in the blood, and better oxygen-dissociative capabilities from hemoglobin to body tissues. Also, smaller mammals in arid or mountainous habitats often retreat into underground burrows.

Living underground continually, or at least for a large part of the animal's daily activities, is as challenging an environment as the one they might escape from. Members of at least 11 families (marsupials, insectivores, and rodents) have adopted a subterranean lifestyle with two different methods of digging: hand digging, as performed by moles and the marsupial mole, mostly in loose soils, or tooth digging, performed by rodents in hard substrates. The environment in an underground tunnel has several specific

Migrating herds of blue wildebeest (*Connochaetes taurinus*), Paradise Plains, Masai Mara. (Photo by Animals Animals ©Rick Edwards. Reproduced by permission.)

characteristics. It is often hot, rather humid, and carries more carbon dioxide and less oxygen than fresh air does. Most subterranean species thus are rather small (the largest reaching a few kilograms only), sparsely furred, have a low resting metabolic rate, and have a more effective cardiac and respiratory system (for example, myoglobin density and capillarization in muscle tissue is higher, hemoglobin has a higher oxygen affinity, hearts are bigger and more effective, and the animals are more tolerant of hypoxia than related, non-burrowing species). Another way of dealing with high carbon dioxide concentrations seems to be excreting bicarbonates via urine, which is achieved by a high concentration of calcium-ion (Ca^{2+}) and magnesium-ion (Mg^{2+}) in the urine. Subterranean mammals also encounter a unique sensory environment. Their preferred mode of communication, even in small species, is low-frequency acoustics, as they are rather insensitive to high frequencies and vibratory communication. Some mole rats (*Cryptomys hottentotus*, *Spalax ehrenbergi*) are also capable of magnetic field orientation.

There are, especially for smaller mammals, other ecological factors at least as important (if not even more so) that also determine which activity patterns (being diurnal, nocturnal or crepuscular, being active in short or long bouts, etc.) are most adaptive under given situations. One of these factors is predation. Even larger species, such as kangaroos, tend be active at times when their predators are less likely to attack. In small mammals, predator-prey relationships are perhaps even more decisive. This holds true not only for prey species but also for the predators. Small mammalian predators such as weasels, mongooses, or small dasyurids are potential prey to other raptors and larger mammals themselves. On the other hand, being of small body size means that the energy demands and constraints are particularly severe. Thus, they have to term their activity patterns much more carefully than larger species. Being potential prey puts a heavy ecological load on all smaller mammals. Being active at times of low predator activity (dusk and dawn, when most diurnal raptors are no longer active and many owls and mammals not yet active) is one possibility of escape. Being active in a synchronized way provides safety in numbers (the dilution effect), and predation stress can explain ecologically the often dramatic suddenness in the onset of activity. It is also interesting to note that the onset of activity, in most species, is more fixed by internal factors—termination, however, is more variable.

Besides predation, inter- and possibly intra-specific competition also must be considered as influencing activity. Being active at different times of day or using different parts of a habitat at different times can raise the possibility of niche separation, as has been shown in communities of *Gerbillus* as well as heteromyid species. The behaviorally or ecologically

• • • • •

Distribution and biogeography

Introduction

Mammals are distributed in virtually every part of the globe. The only extensive areas of land from which they are absent are the Antarctic ice caps. Even in the Antarctic, seals occur on the coastal ice and may haul out on shore. A few species live at high Arctic latitudes; polar bears (*Ursus maritimus*) have been recorded as far as 88°N and ring seals (*Phoca hispida*) have reached the vicinity of the North Pole. Mammals are found on all the remaining continents, on all except the most remote islands, and in all of Earth's oceans and seas. Marine mammals are known to reach depths of 3,280 ft (1,000 m) while land mammals are found from below sea level to elevations above 21,500 ft (6,500 m). They are distributed in all biomes, including tundra, deserts, grasslands, and forests. Species in a wide range of families have adapted to an aquatic lifestyle in swamps, lakes, and rivers. Their distribution extends below the surface of the earth in the case of fossorial or burrowing mammals, and above it through adoption of an arboreal mode of life or by means of flight, in the case of bats. This essay focuses on modern zoogeography (animal distribution) and human effects on zoogeography in modern and historical times.

Mammals are classified into 26 orders, 136 families, and more than 1,150 genera. The number of living and recently extinct species exceeds 4,800. This figure fluctuates as new species are discovered and the status of certain forms is revised. Advances in techniques of molecular genetic analysis have enabled taxonomists to assign taxa with increasing certainty to either specific or subspecific status. During the 1990s intensive fieldwork in forested mountains along the border between Laos and Vietnam revealed the existence of several new large mammals. These consist of a new species and genus of bovid, the saola (*Pseudoryx nghetinhensis*), and three species of muntjac, a type of small deer. These also included a new genus, *Megamuntiacus*.

Fewer than 100 mammal species are either known or believed to have become extinct during the last 500 years. Some of these extinct species remain virtually unknown. For example the red gazelle (*Gazella rufina*) of North Africa is known only from three specimens obtained in Algeria toward the end of the nineteenth century; no living specimen has ever been described, and nothing is known of its ecology. Even more vague is the Jamaican monkey (*Xenothrix mcgregori*), which is known only from a sub-fossil jaw bone.

The order Tubulidentata contains only one species, the aardvark (*Orycteropus afer*), which is restricted to Africa. The Monotremata (monotremes) and four orders of marsupials are confined to the Australian region. Two orders of marsupials are found only in a relatively small area of South America. The two largest orders are Rodentia (rodents), with over 2,000 species and Chiroptera (bats), with almost 1,000. Both of these occur naturally on all continents except Antarctica, and they are the only orders to have reached many oceanic islands. Artiodactyla (even-toed ungulates) and Carnivora occur on all continents except Antarctica and Australia, though representatives of both have been introduced to Australia. The Cetacea (whales and dolphins) and Pinnipedia (seals and sea lions) have a worldwide distribution.

Similar wide variations exist at family and species level. No mammal species is truly cosmopolitan, that is, it occurs in every region and every habitat, though a few species have an extensive distribution covering several continents. The gray wolf (*Canis lupus*) has one of the widest distributions of any terrestrial mammal and is found across North America, Europe, the Middle East, central and northern Asia, and India. The common leopard (*Panthera pardus*) also has an extensive range through Africa, Arabia, Turkey, the Caucasus and much of Asia, north to the Russian Far East. In the New World, the mountain lion or cougar (*Puma concolor*) is distributed from Canada to southern Chile. At the other extreme, some species have a distribution that encompasses only a few square miles.

Some mammals show a discontinuous distribution. Thus the mountain hare (*Lepus timidus*) has its main distribution across the polar and boreal zones of Eurasia, but a disjunct population is still found at high elevations in the Alps, a relict from the last Ice Age when the mountains provided a refuge above the ice sheet. That of the lion (*Panthera leo*) is a relatively recent phenomenon, and a consequence of human activity. It is now found widely in eastern and southern Africa and in a small area of western India. Well into the historical period the distribution also included North Africa, the Middle East, southeast Europe, and Iran but habitat loss and hunting removed it from the intervening areas. In the late Pleistocene the lion reached even North America, where it

Bison (*Bison bison*) among the hot springs of Yellowstone National Park, Wyoming, USA. (Photo by E & P Bauer. Bruce Coleman, Inc. Reproduced by permission.)

attained a large size, as did other Eurasian immigrants such as the mammoth and moose.

The diversity and richness of the mammal fauna present at any one locality is also subject to great variation. Some places contain representatives of a few orders or groups and have only a few species in total while others have an abundance of species from across a wide range of groups. The reasons for these differences are a complex combination of evolutionary history, degree of isolation, primary productivity, and habitat complexity. A latitudinal gradient is a general pattern, with the number of species increasing from the poles towards the tropics. A similar altitudinal gradient is also commonly observed, with the number of mammal species declining with increasing elevation, though in neither of these cases is the correlation uniform. There is also a generally positive correlation between species diversity and habitat complexity. Smaller islands tend to have impoverished mammal faunas, which is partly a function of the species-area relationship by which smaller areas tend to contain fewer species of all groups.

Ancestral forms of mammals evolved at a time when the continents were still connected in the single land mass of Pangaea. This split into two large continents, Laurasia in the Northern Hemisphere and Gondwana in the Southern Hemi-

sphere. Mammals continued to evolve as these then slowly separated into a series of plates that drifted apart to form the present continents. When plates remained isolated for long periods of geological history as in the case of Madagascar, Australia, and South America, unique mammal faunas evolved that were very different from those elsewhere. In other cases, plates gradually collided or were connected intermittently when lower sea levels exposed land connections, allowing mammals to mingle and disperse. Fluctuations in sea levels alternately exposed and flooded land bridges between regions, thus permitting and preventing dispersal of mammal groups between the two areas at different times. On occasions, a chain of islands formed, rather than a full land bridge, allowing some species to cross, but preventing others from doing so. The result of these movements is that each region has a mammal fauna composed of elements that originated there and others that arrived subsequently and at different times from somewhere else. Dispersing species interact with the existing fauna and may replace or modify the existing species.

In many cases, groups of mammals became extinct in their center of origin but survived elsewhere. Equids first evolved in North America and dispersed into Asia, Europe, and Africa and also later into South America. Wild equids survive in Asia and Africa but disappeared from the Americas around 10,000

An Australian sea lion (*Neophoca cinerea*) juvenile acknowledges an adult. (Photo by Hans Reinhard/OKAPIA/Photo Researchers, Inc. Reproduced by permission.)

or food, and provide those resources to one's potential social or mating companions.

The diversity of mammalian social systems

Before approaching explanatory questions by means of Tinbergen's questions again, a brief attempt at categorization of social systems: in order to categorize the diversity of mammalian social systems, there are several variables that need to be described for each species. One is the degree of sociality. We find at least three types of social organization here: first are the solitary individuals that do not regularly have any social contact with conspecifics outside the narrow timespan of reproduction. Individuals of solitary species are commonly found alone in periods of both activity and inactivity. Examples are several species of shrews, small mustelids, and probably some other small carnivores. Next are the individuals of species with a dispersed social system that are also mostly found alone during their period of activity. They do, however, have a network of non-aggressive social relationships with neighbors (often closely related individuals) and may form sleeping groups in periods of inactivity. Examples are many prosimian species, several small possums, some wallabies and rat-kangaroos, but also brown bears, female northern white rhinos, female roe deer, and possibly many other species of ungulates formerly classified as solitary. Finally, gregarious or "social" species are those mostly found in groups, such as larger canids, zebras, or savanna-living bovids.

The second variable to consider is territorial defense. A territory is some area that is actively defended at least against members of the owner's age/sex class, where males at least do not tolerate other fully adult and reproductively active males. Territories thus cannot be "automatically" assumed as a

A Madagascar hedgehog (*Microgale longicaudata*) demonstrates typical prehensile tail behavior. (Photo by Harald Schütz. Reproduced by permission.)

species' characteristic trait, from the fact that some individuals are solitary. Solitary species may well live in undefended, overlapping home-ranges, or even avoid each other actively without defending a territory, as can be seen in females of smaller cats as well as domestic cats in suburban areas (there are, however, also social feral cats). On the other hand, active defense of territories can also be found in truly social species, such as the European badger, the chimpanzee, larger canids, or the spotted hyena.

The third variable to describe mammalian social systems concerns the degree of overlap in the home range. This is, of course, something that can only be found in species with a dispersed or gregarious system. We can roughly distinguish four types here:

- Pairs are found, when one male and one female overlap in their range. This does not necessarily mean that they are found together, such as in gibbon pairs. So-called solitary ranging pairs such as tupaias, red fox, or some prosimians are a common type of mammalian social organization. Pair-living also is not necessarily connected with a monoga-

Bison (*Bison bison*) cluster closely when fleeing, a behavior that greatly reduces the opportunity for wolves to take down a single bison. (Photo by Erwin and Peggy Bauer. Bruce Coleman, Inc. Reproduced by permission.)

mous reproductive system, because extra-pair copulations are not uncommon.

- Polygynous systems are those with one male and several females' ranges overlapping. This system is often called a "harem," or "uni-male group." Again, from looking purely at numbers of animals in the group,

A leopard (*Panthera pardus*) practices fighting with its mother. (Photo by Fritz Pölking. Bruce Coleman, Inc. Reproduced by permission.)

we cannot fully describe the structure. "Harems" may be kept together solely by the male's herding behavior, such as in hamadryas baboons, or they may stay together even in the male's absence, such as in plains or mountain zebra, even though the mares are not related to each other. Or, they may consist of a matriline, a clan of closely related females, such as in patas monkeys, forest guenons, or Eurasian wild boar.

- Polyandrous systems are those in which two or more males overlap with one female. This is found in some large canids, e.g. the African hunting dog, but is generally more common in birds than mammals.

- Multi-male/multi-female systems where more than one adult of both sexes overlap are typical for many diurnal primates, large bovids, lions, or small diurnal mongooses. In these, but also in polyandrous (rarely in polygynous) systems we cannot automatically assume that all adult members are reproductively active. Helpers, such as in canids or dwarf mongoose, can be fully adult but reproductively suppressed individuals. The degree of reproductive cooperation and suppression is thus the last variable to consider, again mostly for gregarious (or theoretically at least, disperse) species. There are very few truly eusocial species of

Washoe, a chimpanzee reared by Beatrice and Allen Gardner beginning in 1966, was taught American Sign Language (ASL). Washoe's acquisition and use of ASL were interpreted to demonstrate that a great ape could acquire a human language. The strong version of this interpretation was questioned from two perspectives. First, H. S. Terrace, who in the late 1970s trained a chimpanzee named Nim Chimpsky to learn ASL, questioned the referential nature of a chimpanzee's use of ASL. Terrace suggested that the primary basis for Nim's "linguistic" ability was not linguistic or referential in the way that human language is used to refer to objects, concepts, and experiences, but was more likely based on associative learning and imitation in goal-oriented situations. That is, Nim appeared to reply to questions posed by his teachers in a manner that was more imitative of the teacher's gestures than suggestive that he was generating linguistic utterances of his own that referred to objects, concepts, and experiences in his environment.

A second related question addressed syntax, or grammar. Chimpanzees who learned ASL (and there were more than Washoe and Nim involved in such studies) did not follow strict grammatical rules of word order required in human linguistic utterances, whether spoken or signed. Rather, the chimpanzees would repeat words or phrases, often varying the order of words as the utterances were repeated. Such behavior was consistent with Terrace's suggestion that the chimpanzees were not using signs to communicate. Beginning in 1973, Duane Rumbaugh trained a chimpanzee named Lana to produce sentences through the use of a computer-operated keyboard containing abstract symbols that represented verbs, nouns, adjectives, and adverbs. These symbols, called lexigrams, were composed of abstract geometric shapes combined to form unique configurations, each of which referred to a particular word or concept. Rumbaugh developed a language he called "Yerkish" named for the Yerkes Regional Primate Research Center where the Lana project began. Lana's symbols followed rules of grammar and the results of this project showed that Lana and other chimpanzees in the project could learn rules of syntax and use abstract visual symbols in a communicative manner. Chimpanzees Sherman and Austin later showed the ability to use this computer-based system to communicate information to each other.

Although for some investigators of animal cognition the question of ape language is still at issue, the ape language projects have expanded in content and have provided important discoveries about ape cognitive abilities. In such a project, Chantek, an orangutan reared by Lynn Miles beginning in 1977, learned ASL. In this project Chantek's acquisition of symbolic communication was studied in the context of overall cognitive development. This project is unique in its breadth, providing understanding of the development of symbolic communication as one feature of a suite of cognitive abilities.

In 1971, David Premack first reported results of a project in which he trained chimpanzee Sarah to communicate using arbitrary abstract symbols. These symbols were colored plastic shapes. Sarah learned concepts such as "same" and "different," as well as nouns, verbs, and adjectives. That these

Bottlenosed dolphins (*Tursiops truncatus*) have a complex language that may eventually help humans to communicate with them. (Photo by © Stuart Westmorland/Corbis. Reproduced by permission.)

symbols were representational to Sarah was demonstrated when she was provided with the symbol for apple, and asked to describe its physical features. Although the symbol for apple was a blue triangle, Sarah described the object presented to her as "red" and "round," referring to attributes of the object rather than of the symbol. She also showed the ability to reason analogically through her understanding of "same" and "different." The primary contribution of this language project was not Sarah's linguistic abilities, but how Premack used Sarah's representational capacity to show the breadth of cognitive flexibility and conceptual understanding available to a chimpanzee provided with Sarah's rich cognitive environment.

In a similar fashion, two current ape language projects use the animals' symbolic ability to uncover cognitive representational abilities that would be difficult to access without this means of symbolic communication. In 1995 Robert Shumaker began to work with orangutans Azy and Indah, who are learning an abstract symbolic communication system presented as lexigrams on a touch-sensitive video screen. When questions are posed, the orangutans indicate the appropriate symbol from an array of symbols by touching it. Shumaker is using the animals' symbolic abilities as a window into other related cognitive processes such as number comprehension. In a project begun in 1978 by Kiyoko Murofushi, Toshio Asano, and Tetsuro Matsuzawa, chimpanzee Ai continues to demonstrate her sophisticated cognitive abilities using a touch-sensitive video screen and a vocabulary of lexigrams and Arabic numerals. For example, Ai labels objects and their characteristics such as color and number; she clearly understands numbers conceptually, and counts using Arabic numerals from zero to nine; and she "spells" by constructing lexigrams from their components. Ai's infant Ayumu, born in 2000, is learning to use the touch-screen system, providing insight into the development of symbolic cognitive skills as he interacts with his mother and observes her using the system.

The symbolic ability of great apes in the context of a language-learning setting was most strongly demonstrated by Sue Savage-Rumbaugh who in 1979 extended the Lana project to another great ape species, the bonobo (*Pan paniscus*). Kanzi, a young bonobo who lived with his mother during her training with Rumbaugh's computer-based language, surprised researchers when he showed clear understanding of the task and of particular symbols simply as a result of having been passively exposed to the symbols while his mother was learning the task. Following this discovery, Savage-Rumbaugh focused on Kanzi's ability and motivation to use abstract communicative symbols. Kanzi not only uses the abstract symbolic language system, he also has a demonstrated understanding of spoken language, and Savage-Rumbaugh has reported that he appears to attempt to communicate vocally by imitating acoustic properties of human speech. Her interpretation of Kanzi's behavior is that bonobos appear to have not only the ability to acquire abstract human-derived symbols, but that they may have additional communicative skills that can provide insight into the evolution of human language.

Koko the gorilla has been using ASL under the tutelage of Francine Patterson since 1976. She is probably the most publicly recognizable language-trained ape. Indeed, Patterson has extended her project into conservation efforts in the United States by promoting Koko to the public. Further, Patterson translated a children's book about Koko into French to distribute in French-speaking Africa as a way to educate children about the cognitive and emotional capacities of gorillas and the importance of preserving them in the wild.

The importance of the ape language projects has not been in the demonstration of a human capacity in great apes, but rather in exposing the complex symbolic skills available to animals in a rich interactive environment. The cognitive skills shown by the nonhuman participants in these projects extend beyond the specific symbolic skills trained in individual projects. They have opened a window into the minds of animals that enriches our understanding of the animal mind, while also providing new theoretical perspectives on the evolution of human cognition and language.

Enculturation of apes

Much of the information we have about cognition in great apes has come from research projects involving intensive study of one or two apes, usually chimpanzees. In most of these research projects the apes have had extensive interaction with humans during their early development. These interactions not only take the form of explicit cognitive and behavioral tests, but they also include teaching and guidance. The apes acquire sophistication with human artifacts including computers and other electronic or mechanical objects.

The environment for these animals is complex, and not typical of the environment in which apes evolved. Although some of the ecological challenges met by animals living in the natural environment are missing, the early experience of these animals is enriched and challenged in different ways. Apes who experience this intensive experience with humans are called "enculturated" apes, those who at some level have been exposed to and integrated into certain human social/cultural experiences. Because they have not been exposed to their natural environment, some researchers question the value of the conclusions based on their cognitive abilities. Clearly, these animals have had enriched early experiences, including direct teaching of skills by human caretakers, and are not representative of chimpanzees developing with their mothers in the wild environment. This challenge has been met by the response that although the cognitive tasks provided to enculturated animals differ from those in the natural environment, they challenge the cognitive skills required for survival in the wild.

Further, the possible enhancement of cognitive skills provided by the enriched environment shows the extents and limits of cognitive abilities under circumstances conducive to high levels of performance. The results of these studies tell us what cognitive skills great apes are capable of; the extensive behavioral studies of great apes in their natural environment show us how they use these cognitive abilities to solve daily social and environmental challenges. Although an enculturated ape has had a different early environment, he or she continues to be an ape and to show cognitive skills available to an ape. Integration of data from field and laboratory continues to provide the richest understanding of great ape cognition.

Resources

Books

Bekoff, M., C. Allen, and G. M. Burghardt, eds. *The Cognitive Animal.* Cambridge, MA: MIT Press, 2002.

Candland, D. K. *Feral Children and Clever Animals.* New York: Oxford University Press, 1993.

Hauser, M. D. *Wild Minds.* New York: Henry Holt and Company, 2000.

Matsuzawa, T., ed. *Primate Origins of Human Cognition and Behavior.* Tokyo: Springer-Verlag, 2001.

of Libya, today it is assumed extinct. The second subspecies is the Somali wild ass (*E. a. somaliensis*). It was also domesticated in the area around the Persian Gulf, and is today nearly extinct. The descendants of both subspecies crossbred and quickly spread, thanks to military expeditions and lively mercantile bustle, through Palestine to western Asia and farther to the east through Morroco to the Pyrenean Peninsula, where they arrived in the second millenium B.C. It is known that during the first millenium, the Celts were breeding donkeys. The Asiatic ass (*Equus hemionus*) was for some time an object of interest but domestication did not succeed due to its uncontrollable nature.

Donkey breeds have never reached the number of varieties that horse breeds have. They differ in height (from 2 to 5 ft [0.6–1.5 m]), in weight (175 lb–990 lb [80–450 kg]), in color, and in the quality of hair. The donkey is still a common component of country life in the Mediterranean, the Balkans, south Asia, South America, and other subtropical and tropical areas. Feral donkey populations live for many generations in the southwest United States and in Australia.

Humans have used horses and donkeys to breed hybrid species. The most famous is the mule, the offspring of a female horse (mare) and a male donkey (jackass). A second hybrid is less known, the dunce hinny, whose parents are a female donkey (jenny) and a male horse (stallion). Hybrids inherit more traits from the mother. Another interesting interspecies hybrid is the zebroid. It is the offspring of some zebra species and a horse, or rarely a donkey.

Cattle (*Bos primigenius* f. *taurus*)

Cattle are the oldest domestic animals. Their importance lies in giving meat, milk, and working power. Leather, fat, hooves, and horns are also valuable, and dried excrements are used as fuel, building material, and fertilizer. Cattle were first used as draft and riding animals, allowing people to drastically change their way of life. Some African people, for example the Hottentots, use them for riding even today. Cattle are the main source of meat and milk and the second most numerous domestic animal in the world, after domestic fowl. Under the cattle category, we can include the descendants of the aurochs, and those of other domestic cattle such as the yak, gayal, Bali cattle, and buffalo. They come from areas with extreme climatic conditions (high in the mountains or from the tropics), where they are used as domestic cattle. The kouprey (*Bos sauveli*) from the forests of Cambodia occupies a special place among cattle, for it may be the last surviving form of the wild ox, which went through an early form of domestication and then ran wild again.

The aurochs (*Bos primigenius*) was a progenitor of domestic cattle. It occupied the forests of the whole temperate zone of the Old World from Europe to north Africa and west Asia to the China Sea at the end of the last glacial period. In this large area, it developed more subspecies. The wild ox had survived almost until the end of that glacial period in Asia and north Africa, and in the middle and west European forests, until the end of the Middle Ages. The last individual became extinct in Poland in 1627. The domestication of the wild ox

The goat (*Capra hircus*) has many uses to humans, such as a food and clothing source. (Photo by Hans Reinhard. Bruce Coleman, Inc. Reproduced by permission.)

began in 7000 B.C. and it is assumed that it started almost at the same time in several places—Greece, Macedonia, the Fertile Crescent (Mesopotamia, Egypt, Persia), and later (5000 B.C.) in the Indus Valley. It is possible (according to new genetic research) that an independent domestic center existed also in north Africa and could even be the oldest.

Why did people try to domesticate such massive and dangerous animals? They already bred sheep and goats at that time, which were sufficient as a source of food. It is possible that at first there were religious reasons. Wild cattle symbolized fertility and power for many cultures and they were significant sacrificial animals. Cattle derived from aurochs include some 450 breeds today and are divided into two basic groups. One is humpless cattle, which are the European descendants of the aurochs (*B. p. primigenius*, syn. *Bos taurus*). The second group includes cattle with a hump (zebu) whose progenitor is assumed to be the Indian subspecies of the aurochs (*B. p. namadicus*, syn. *B. indicus*).

The oldest known bone findings, which confirm the existence of humped zebu (thorny projections of neck vertebrae distinctly bifurcated) are almost 6,000 years old and come from Iraq. However, the domestication of the zebu probably occurred in the Indus Valley. The zebu adapted to subtropical and tropical climates and became resistant to tropical diseases. After domestication, it spread quickly to Malaysia, Indonesia, and China and to the west to Africa. Today it is one of the most plentiful cattle in the African tropics and subtropics and in the Indian subcontinent.

Domestic yak (*Bos mutus* f. *grunniens*)

The yak was domesticated in Tibet from 3000 to 1000 B.C., and its progenitor is the wild yak (*Bos mutus*). It is almost one

This domesticated dog has been trained to help hunters retrieve game. (Photo by St. Meyers/Okapia/Photo Researchers, Inc. Reproduced by permission.)

third smaller than its wild progenitor and it has markedly weaker horns or no horns. It moves without any problems at altitudes of 9,800–19,500 ft (2,990–5,940 m). It is an indispensible helper for the mountain inhabitants and the basic source of livelihood. The yak provides milk, meat, and wool. Dried excrement is used as fuel. The yak is a great working animal. It bears around 330 lb (150 kg) very easily, it serves as a riding and draft animal, and is used for plowing. Twelve million yaks live in the mountains areas of Tibet, China, Nepal, Mongolia, and southern Siberia. It is bred to a small extent in North America and in the Swiss Alps.

Gayal (*Bos gaurus* f. *frontalis*)

The gayal or mithan is the domesticated form of the wild gaur (*B. gaurus*). For a long time, it was not clear if the gayal and gaur were the same separated species or if the gayal had developed by crossbreeding of a gaur with bantengs or zebu. The gayal is noticeably smaller than the gaur, it has shorter conical horns, a markedly shorter skull, and a wider and flatter forehead. It has a large double dewlap on the chin and neck. It is most commonly brown and black but can also be spotted or white. The gayal is not a typical domestic animal; it usually lives in small groups in the jungle at the periphery of a village and comes back to the village only towards evening, lured by salt. The economic use of the gayal is insignificant. Sometimes is it used for field work or for its meat; its milk is not drinkable. The gayal was used by many people as a sacrificial animal or as currency. Sometimes the animals escape and run wild,

but the domestication influence is still seen in their calm temperament. Feral populations live in northern India.

Bali cattle (*Bos javanicus* f. *domestica*)

Bali cattle is the domestic form of wild banteng (*B. javanicus*), which lives in the forests of Java, Borneo, Malaysia, and Thailand today. Its domestication took place in Java around 1000 B.C. Wild bantengs were caught and tamed in the Middle Ages, in Bali, Sumatra, and Java until the eighteenth century. Bali cattle are smaller than banteng, the horns lack the characteristic curvature, and the skull is smaller. The external sexual signs are weaker than in the wild species. The domestic species grows more rapidly and matures earlier. Elegant Bali cattle have adapted to life in tropics better than the zebu. It was never bred for a specific purpose and because of that it has no major economic value. Approximately 1.5 million individuals are bred today. It is used as other domestic cattle for field work and riding. Milk utility is low but meat has excellent quality and taste. Bali cattle are crossbred with the taurine cattle, and with the zebu, but male descendants are infertile. Bali cattle often run wild, and feral populations live in savanna in the south of Sulawesi and in Australia.

Sheep (*Ovis ammon* f. *aries*)

Sheep and goats are assumed to be the oldest domestic livestock. Sheep are bred in different areas around the world from lowlands to mountains and from tropics to cold north moorlands. They are exceptionally acclimatized to extreme conditions. They are also very useful, providing meat, milk, tallow, wool, fur, leather, horn, lanolin, dung, and they carry loads in Tibet. No culture or religion forbids killing sheep or eating sheep meat.

The exact origin of the domestic sheep is not clear because all wild sheep in the genus *Ovis* are fertile when bred together and zoologists still do not agree about their taxonomy. Two groups of wild sheep are considered as having first undergone domestication. They differ in the size of the body and in their habitats. The arkhar or argali sheep (*O. ammon*) is the representative of the "mountain" group. It has six subspecies and it lives in central Asia in altitudes ranging from 16,500 to 19,500 ft (5,030–5,940 m). The Asiatic mouflon (*O. orientalis*) and the urial sheep (*O. vignei*) are members of the "steppe" group of wild sheep. They also have several subspecies and live in lower regions from west Asia to northwest India. The European mouflon (*O. musimon* syn. *O. ammon musimon*) is a special case. It comes from Corsica and Sardinia and it was assumed to be the progenitor of domestic sheep for a long time. However it is itself a feral form of the early Neolithic domestic sheep, which came with humans to Corsica 9,000 years ago. The question of domestic sheep progenitors is still debated. With certainty we can eliminate only the species that were not domesticated. These are the American bighorn sheep (*O. canadensis*), Dall's sheep (*O. dalli*), and the Siberian snow sheep (*O. nivicola*), which were not "in the right place at the right time." All other species are good candidates.

of mammalian invasions contains many of the basic principles needed to better comprehend the plethora of media stories on invasive mammals.

Many of the invasive mammal examples have been from islands, as scientists prefer to study the simplest possible finite systems before venturing forth to tackle larger continental problems. In other words, the island serves as a laboratory for studying invasions at their simplest. Even in Australia, many rabbit control solutions were first tested on small islands to perfect them before introducing them to the mainland continent.

The invasive mammal problems in some areas are likely to grow worse before they get better. But knowing that many of these invasive species are so successful in their new homes because they left behind the predators, parasites, diseases, and other natural control factors that kept populations under control in their native lands suggests one avenue of control. Namely searching the native lands of the pestiferous mammals for natural control factors that can be safely introduced elsewhere, like the myxomatosis virus introduced successfully into Australia for rabbit control.

However, if nothing else is learned, it is that great care must be taken so that the cures introduced are not worse than the original invasive mammal problems, as was the case with the introduction of the mongoose for rat control. Since humans created most of the invasive mammal problems, it might be reasonable to expect that humans can collectively atone by researching new solutions that minimize the possibility of ecological harm.

In the future, look for clever ecological manipulations of populations, like on the Channel Islands, as well as more molecular, biotechnology solutions like immunocontraception for rabbits. But rather than expecting the newest technological solution to be the ultimate answer, remember that organisms can and often do adapt, just as the rabbits in Australia built up immunity to the myxomatosis virus and the virus attenuated over time.

If nothing else, the half century of experience with rabbits in Australia points to the need for a control strategy integrating multiple techniques (like the myxomatosis virus plus ripping rabbit warrens) to achieve the best and longest-lasting results. Hopefully, the magic bullet approach of the pesticide era will be replaced with this more comprehensive integrated pest management approach.

However, the human side of the equation must never be forgotten when dealing with invasive mammals, as many of these animals in other non-pest contexts are highly valued. Hence, pestiferous feral cats, rabbits, wild horses, burros, pigs, and other mammals causing problems cannot be treated as the object of extermination like cockroaches or termites. Every mammal seems to be loved by some group, be it hunters and indigenous people who favor wild pigs or animal rights groups who champion freedom for minks. Right or wrong, good or bad, these varied human sensibilities need to be taken into account in designing any integrated pest management program to control invasive mammals. For example, in the western United States, capturing wild horses and letting people adopt them has replaced the old practice of herding the horses into canyons and shooting them. This type of solution may have more to do with politics or social science and consensus building than with biological or ecological principles, but ignoring the human species behind the invasive mammal problems is to invite failure.

Resources

Books

Crosby, Alfred. *Ecological Imperialism: The Biological Expansion of Europe, 900-1900.* Cambridge: Cambridge University Press, 1986.

Kirch, Patrick, and Terry Hunt. *Historical Ecology in the Pacific Islands: Prehistoric Environmental and Landscape Change.* New Haven, CT: Yale University Press, 1997.

Perrings, Charles, Mark Williamson, and Silvana Dalmazzone. *The Economics of Biological Invasions.* Northampton, MA: Edward Elgar Publishing, 2000.

Staples, George, and Robert Cowie. *Hawaii's Invasive Species: A Guide to the Invasive Alien Animals and Plants of the Hawaiian Islands.* Honolulu: Mutual Publishing and Bishop Museum Press, 2001.

U.S. Congress, Office of Technology Assessment. *Harmful Non-Indigenous Species in the United States, OTA-F-565.* Washington, DC: U.S. Government Printing Office, 1993.

Vitousek, Peter, et al. *Biological Diversity and Ecosystem Function on Islands.* Heidelberg, Germany: Springer-Verlag, 1995.

Periodicals

Flux, J. E. C. "Relative Effect of Cats, Myxomatosis, Traditional Control, or Competitors in Removing Rabbits From Islands." *New Zealand Journal of Zoology* 20 (1993): 13–18.

Molsher, Robyn. "Trappability of feral cats (*Felis catus*) in central New South Wales." *Wildlife Research* 28 (2001): 631–636.

Pimentel, David, et al. "Environmental and Economic Costs Associated With Nonindigenous Species in the United States." *Bioscience* 50 (2000): 53–65; 309–319.

Roemer, Gary W., et al. "Feral Pigs Facilitate Hyperpredation by Golden Eagles and Indirectly Cause the Decline of the Island Fox." *Animal Conservation* 4 (2000): 307–318.

Joel H. Grossman

Mammals and humans: Field techniques for studying mammals

There are two general reasons for studying mammals in the field. The first is to provide numbers that are needed for biodiversity measures or population management; the second is to provide natural history information that is needed to better understand species' requirements or their roles within the natural community. This chapter provides broad guidelines that would assist someone in selecting appropriate field techniques. References provided at the end of the chapter provide more details on both study design and specific techniques.

Biodiversity surveys

Biodiversity measures are based on the ability to accurately count the number of species within a given area and usually some measure of their relative or absolute abundance. Population management of both common and rare species relies on accurate measures of population numbers or at least a way to measure population trends. Most mammal populations or communities are too complex for every individual to be counted, therefore a sample is often taken of the population and the number is estimated based on that sample. Unfortunately, obtaining these estimates is not an easy task. More than other vertebrate groups, mammals occupy a wide array of habitats and possess a broad range of body sizes. These factors make them difficult to survey as a group, and survey techniques have to be tailored for a specific species, or suite of species. Planning a biodiversity survey is a two-step process: the first step is to determine the level of information needed to meet objectives, and the second step is to tailor a survey to fit the attributes of the species.

Three levels of information that can be obtained from a survey include a species list, a relative index of abundance for each species, and an absolute density for each species. Generally, there is increasing cost and complexity as the level of information increases. For some mammal species, it is prohibitively expensive to estimate absolute density because of the habits of the mammal or the habitats it occupies. Solitary bats, which live in trees under strips of bark or in crevices, are a good example of a suite of species whose density estimate is logistically difficult to obtain. When planning a survey, the first consideration should be how necessary the increased information is to the management or research objectives. The initial survey of a park would not start with a

density estimate of each mammal species, but rather a list of species found in the park. Often a mammal's relative density is adequate information to track changes in abundance within a park, and the saved money can be used for other conservation tasks such as patrolling. Within broad conservation plans for an area, mammal surveys should reflect a nested subset design. For example, following a complete species list for the area, some species from this list are monitored through an abundance index, and select animals from this group are targeted for detailed population and ecology studies that might include a density estimate.

If field technicians are working with a rare species or a harvested species, they might not be concerned with the higher levels of organization and start with a focal study on the target species. However, even under these circumstances, an index survey over a larger area may be more appropriate than a density estimate at one site. Project goals and financial logistics usually produce a compromise in how much information can be gathered. It is important the data collected are not stretched beyond their purpose, when compromises are made. Unfortunately, the scientific literature is full of indexes used to calculate densities and species lists used as indexes. Surveys are powerful tools in wildlife conservation and management, but when stretched beyond their ability they convey more confidence in the trends then the data warrant.

Species list

Species lists can be as complex as a complete mammal survey or as simple as presence/absence of a focal species. Species lists are composed through use of multiple means. Traditional field surveys that use traps, cameras, or transects can be supplemented by sociological techniques such as examination of local markets, discussions with local hunters, and inspecting kitchen remains in villages. When using traps or cameras, they are usually placed in a line to traverse as many habitat types as possible. The distance between the traps would be less than the smallest home range of the animal targeted. The larger the home range, the longer the traps should remain in one spot in order to account for the time it takes an animal to traverse its entire home range. For a rodent, traps might be in place for a few days; for a large predator, a few weeks would be more appropriate.

A hamadryas baboon (*Papio hamadryas*) at the Phoenix Zoo in Phoenix, Arizona. Zoos attempt to educate the public about animals and their habitats. (Photo by G. C. Kelley/Photo Researchers, Inc. Reproduced by permission.)

There are great advances being made in exotic mammal husbandry as of this writing, but much work is still needed. The reproduction and birth of captive elephants is becoming a major priority. Both the African (*Loxodonta africana*) and the Asian (*Elephas maximus*) populations are aging and without more success both captive populations could be in trouble. Efforts are currently underway by numerous institutions to produce captive born animals. The silky anteater (*Cyclopes didactylus*) is another animal that does poorly in captivity. The average life span in captivity is approximately thirty days. Their has never been a silky anteater born in captivity. The wild caught animals usually arrive very weak, dehydrated, malnourished, and heavily parasitized. Also, it is unclear if diets used for other species of anteaters in captivity are sufficient for silky anteaters.

Population management programs

Zoos worldwide have intensive conservation programs for mammals. Although the names and acronyms vary by the geographic region of zoo associations, their functions are very much the same. In North America, that plan is called the

Species Survival Plan (SSP), a copyright name and program implemented in 1981 by the American Zoo and Aquarium Association (AZA). The Species Survival Plan is defined as a cooperative breeding and conservation program designed to maintain a genetically viable and demographically stable population of a species in captivity and to organize zoo and aquarium-based efforts to preserve the species in captivity and in natural habitats. SSPs participate in a variety of other cooperative conservation activities, such as research, public education, reintroduction, and field projects. Currently, 108 SSPs covering 159 individual species are administered by the AZA.

Most SSP species are endangered or threatened in the wild, or "flagship species." These are well known animals that arouse strong feelings in the public for their protection and that of their habitat. For an animal to have an SSP there must be qualified professionals with time to dedicate to conservation. Each SSP has a coordinator and management committee. They use tools such as population management, scientific research, education, and reintroduction to formulate a master plan. This plan outlines the goals of the program, based on what is most appropriate and attainable based on the current captive population.

Animal husbandry

No day at the zoo is the same as the previous; each day brings new events and challenges. A rigid work schedule will not survive. Disruptions and chaos are inevitable, requiring flexibility and an open mind. The management of zoo business and animal husbandry is in itself constantly evolving. Internal and external ideas from both local and global political groups, scientists, the general public, and environmental emergencies interact and contribute to change, while the health and biological status of animals determines the overall degree of success.

Basic needs of mammals

The basic needs of a mammal are food and shelter. A zookeeper's primary responsibility is to satisfy these basic needs. However, their duties are usually much more complex, and with their daily responsibilities, a profound bond between human and the other mammal occurs. Early diagnosis of health problems in the zoo population can often be done by simply paying close attention to condition of the skin. The skin should be free of rough areas or exposed tissue. The fur should also be somewhat silky and shiny rather than dry and dull in appearance. The health maintenance of elephants and rhinos is one example. The skin of an elephant or rhinoceros is a good indicator of their overall health. It should be free of wounds and have a firm condition, a loose skin texture can be indicative of serious health problems. These animals are also very susceptible to foot problems. This is due to the fact that they do not walk as much in captivity as they do in the wild. Zookeepers provide all required health maintenance while also concerning themselves with the following subjects.

Enrichment

In addition to physical needs, a captive mammal has psychological needs that must be satisfied. "Enrichment" means the application of environmental stimuli in an attempt to cre-

A gorilla (*Gorilla gorilla*) at the Frankfurt Zoo, Germany. The gorilla's native habitiat has been reduced in size by human expansion. Survival of the species is threatened, and zoos educate their visitors about the need to protect animal environments. (Photo by Bildarchiv Okapia/Photo Researchers, Inc. Reproduced by permission.)

ate psychological and physiological events that improve the overall quality of life of an animal in captivity. Enrichment helps to overcome stereotypical and undesirable behavior, as well as encourages mental and physical exercise in captive mammals. It usually involves adding novel items to an animal enclosure or providing visual or olfactory cues to a novel item. The desired behavior is curiosity or investigation, which results in mental and physical stimulation or exercise. Zookeepers benefit from developing enrichment programs for the same reasons. They stimulate themselves to rethink and reduce the stagnancy of routines. In some facilities, visitors participate in enrichment as well. Many institutions have "enrichment days" in which the public participates in making enrichment items. The safety and suitability of each item for a specific animal must always be the primary concern. For example, the Minnesota Zoo uses fir trees as enrichment tool. The animals love to investigate and move around them, and tear branches apart. Snowmen are also a popular enrichment item with many of the animals at this zoo. The bison love to

bash theirs, but the gibbons aren't really sure what to do when snow shows up in their enclosure.

The types of enrichment and the applications are limitless. Some groups of mammals have been documented to have specific preferences. For example, exotic cats in captivity are drawn to spices such as cinnamon, nutmeg, and paprika sprinkled in their enclosures. Perfume sprayed on surfaces in their enclosure also arouses their curiosity and excitement. For other mammals, cardboard boxes and objects that they can roll are enjoyed. Enrichment can be used to elicit "natural" behaviors in captivity as well as improve overall physical health through exercise. Food is also frequently used as an enrichment tool because it solicits the natural hunting and foraging behaviors of animals. Food with interesting textures or new flavors, and food that is hidden in hard to reach places, all make good enrichment items. Many animals love "popsicles," blocks of ice with food or bone inside. One of the Minnesota Zoo tigers is reported to put her popsicles into the tiger pool to make the ice melt faster.

Health care

Most facilities employ staff veterinarians who specialize in exotic animal medicine. Zoological medicine has been identified as a distinct and identifiable specialty of veterinary medicine on the basis of academic programs in colleges of veterinary medicine in the United States from the 1960s. In 1972, the National Academy of Science committee on Veterinary Medical Research and Education first documented the need for veterinarians specially trained to manage the health of zoo animals. It is typical for each mammal within a zoo population to receive a routine health exam at least once a year if they have no signs of illness. Care is provided as often as needed if health problems are suspected. Routine physicals usually involve blood serum chemistry screening profiles, a complete red blood cell and white blood cell count, x rays,

The thick glass allows these children to get extremely close to a bear at the Seattle Zoo. (Photo by © Mug Shots/Corbis. Reproduced by permission.)

every U. S. state (and many national governments around the world) has a consumer advocate office that can help evaluate the non-profit organizations that solicit conservation contributions. Typically these consumer advocate offices can provide information on the percentage of contributions that go to support actual conservation activities (as opposed to paying staff personnel, for example). They can almost always warn of organizations that support outright frauds.

IT IS IMPORTANT TO EVALUATE THE PROGRAM ADVOCATED BY A CANDIDATE OR CONSERVATION ORGANIZATION

A key to evaluation is to determine whether an organization's stated objectives are meaningful and realistic. Here are four hypothetical statements of objectives that should be questioned:

- *Elephants are in terrible danger, and your contribution will save the lives of countless elephants in southern Africa.* The organization should offer some idea of how the promise will be fulfilled. Furthermore, words like "countless" should ring loud alarm bells.

- *If I am elected, I will protect lands in such a way as to conserve functioning ecosystems in which living organisms can interact in complex ways.* Every living system—from rice fields to rainforests to urban gardens to septic tanks—meets this criterion. This is a meaningless promise, since it will automatically be kept.

- *The goal of our policy is to preserve appropriate natural, aesthetic values for future generations.* Both authors of this essay are teachers. Part of our job is evaluation, and we don't give tests that we cannot grade. Thus we are wary of claims that cannot be checked. We like the idea of preserving values—but we wouldn't offer our votes or our dollars until we learned many more specifics.

- *The objectives of this program are to integrate economic and intrinsic wildlife values in a holistic program that recognizes human rights to sustainable development and national responsibilities for conservation of biodiversity.* This statement sounds great. It uses most of the favorite vocabulary words of the conservation community. However, we have no idea what the statement means—and we wrote it. We would certainly look for specific, measurable objectives before we were tempted to support such a program.

Resources

Books

Caro, Tim, ed. *Behavioral Ecology and Conservation Biology.* Oxford: Oxford University Press, 1998.

Carr, Archie. *Ulendo: Travels of a Naturalist In and Out of Africa.* New York: Knopf, 1974.

Ehrlich, Paul. *Human Natures: Genes, Cultures, and the Human Prospect.* New York: Penguin Books, 2002.

Gosling, L. Morris, and William J. Sutherland, eds. *Behaviour and Conservation.* Cambridge: Cambridge University Press, 2000.

Hunter, M. L., and A. Sulzer. *Fundamentals of Conservation Biology.* Oxford, UK: Blackwell Science, Inc., 2001.

Kleiman, Devra G., and Anthony B. Rylands, eds. *Lion Tamarins: Biology and Conservation.* Washington, DC: Smithsonian Institution Press, 2002.

Leopold, Aldo. *A Sand County Almanac.* Reissue ed., New York: Ballantine Books, 1990.

Meffe, G. K., and C. R. Carroll. *Principles of Conservation Biology.* Sunderland, MA: Sinauer Associates, Inc., 1997.

Primack, Richard B. *Essentials of Conservation Biology.* 3rd ed. Sunderland, Massachusetts: Sinauer Associates, 2002.

Wilson, Edward O. *The Diversity of Life.* Cambridge, MA: Harvard University Press, 1992.

Periodicals

Daszak, P., et al. "Anthropogenic Environmental Change and the Emergence of Infectious Diseases in Wildlife." *Acta Tropica* 78 (2001): 103–116.

Fischer, J., and D. B. Lindemayer. "An Assessment of the Published Results of Animal Relocations." *Biological Conservation* 96 (2000): 1–11.

Lindburg, D. G., et al. "Hormonal and Behavioral Relationships During Estrus in the Giant Panda." *Zoo Biology* 20 (2001): 537–543.

McShane, Thomas O. "The Devil in the Detail of Biodiversity Conservation." *Conservation Biology* 17 (2003): 1–3.

Myers, N., et al. "Biodiversity Hotspots for Conservation Priorities." *Nature* 403 (2000): 853.

Peng, Jianjun, et al. "Status and Conservation of the Giant Panda (*Ailuropoda melanoleuca*): A Review." *Folia Zoologica* 50 (2001): 81–88.

Rosser, Alison M., and Sue A. Mainka. "Overexploitation and Species Extinction." *Conservation Biology* 16 (2002): 584–586.

Organizations

Conservation International. 1919 M Street NW, Ste. 600, Washington, DC 20036. Phone: (202) 912-1000. Web site: <http://www.conservation.org>.

IUCN—The World Conservation Union. Rue Mauverney 28, Gland, 1196 Switzerland. Phone: ++41(22) 999-0000. Fax: ++41(22) 999-0002. E-mail: mail@iucn.org Web site: <http://www.iucn.org/>.

Wildlife Conservation Society. 2300 Southern Blvd., Bronx, NY 10460. Phone: (718) 220-5100. Web site: <http://wcs.org>.

World Wildlife Fund. 1250 24th Street N.W., Washington, DC 20037-1193 USA. Phone: (202) 293-4800. Fax: (202) 293-9211. Web site: <http://www.panda.org/>

Kimberley A. Phillips, PhD
Clarence L. Abercrombie, PhD

Resources

Flannery, T. F. *Mammals of New Guinea.* Sydney: Reed Books, 1995.

Griffiths, M. *Echidnas.* Oxford and New York: Pergamon Press, 1968.

Griffiths, M. *The Biology of the Monotremes.* New York: Academic Press, 1978.

Nicol, S. C., and N. A. Andersen. "Patterns of hibernation of echidnas in Tasnamia." In *Life in the Cold: Tenth International Hibernation Symposium,* edited by G. Heldmaier and M. Klingenspor. Berlin: Springer, 2000.

Rismiller, P. D. *The Echidna, Australia's Enigma.* Southport, Connecticut: Hugh Lauter Levin Associates, 1999.

Rismiller, P. D., and M. W. McKelvey. "Sex, torpor and activity in temperate climate echidnas." In *Adaptations to the Cold,* edited by F. Geiser, A. J. Hulbert, and S. C. Nicol. Armidale: University of New England Press, 1996.

Wells R. T. and Pledge, N. S. "Vertebrate Fossils." In *Natural History of the South East,* edited by M. J. Tyler, C. R. Twidale, J. K. Ling and J. W. Holmes. Adelaide: Royal Society of South Australia, 1983.

Periodicals

Abensperg-Traun, M. "A study of home range, movements and shelter use in adult and juvenile echidnas *Tachyglossus aculeatus* (Monotremata: Tachyglossidae) in Western Australian wheatbelt reserves." *Australian Mammalogy* 14 (1991): 13–21.

Dulhunty, J. A. "Potassium-Argon basalt dates and their significance in the Ilford-Mudgee-Gulgong region." *Proceedings of the Royal Society of New South Wales* 104 (1971): 9–44.

Flannery, T. F., and C. P. Groves. "A revision of the genus *Zaglossus* (Monotremata, Tachyglossidae) with description of new species and subspecies." *Mammalia* 62, no. 3 (1998): 367–396.

Gates, G. A. "Vision in the monotreme anteater (*Tachyglossus aculeatus*)." *Australian Zoologist* 20 (1978): 147–169.

Grigg, G. C., L. A. Beard, and M. L. Augee. "Hibernation in a monotreme, the echidna *Tachyglossus aculeatus.*" *Comparative Biochemistry and Physiology A* 92 (1989): 609–612.

Murray, P. F. "Late Cenozoic monotreme anteaters." *Australian Zoologist* 20 (1978): 29–55.

Pledge, N. S. "Giant echidnas in South Australia." *South Australian Naturalist* 2 (1980): 27–30.

Rismiller, P. D., and M. W. McKelvey. "Frequency of breeding and recruitment in the short-beaked echidna, *Tachyglossus aculeatus.*" *Journal of Mammalogy* 81, no. 1 (2000): 1–17.

Other

Echidna the Survivor. Documentary videotape. Australian Broadcasting Corporation, 1995.

Peggy Rismiller, PhD

Species accounts

Bare-tailed woolly opossum
Caluromys philander

SUBFAMILY
Caluromyinae

TAXONOMY
Didelphis philander (Linnaeus, 1758), America, restricted to Surinam; Ghana.

OTHER COMMON NAMES
French: Opossum laineux; German: Gelbe Wollbeutelratte; Spanish: Tlacuache lanudo, comadreja lanuda, cuica lanuda.

PHYSICAL CHARACTERISTICS
Length 7–10 in (18–28 cm); weight 6.0–15 oz (180–450 g). The back is nearly uniform cinnamon brown and the face is gray with brown bulging eyes, with a black fine line between the eyes. Ears are large, naked, pink, and membranous. More than half of the tail is furry.

DISTRIBUTION
Venezuela, Trinidad and Tobago, Guyana, Suriname, French Guiana, and Brazil.

Caluromys philander
Didelphis virginiana

HABITAT
Primary and secondary tropical lowland moist forest in both swampy and well-drained areas, from sea level up to about 2,000 ft (600 m). Rarely, it has also been found in plantations and other agroecosystems. Often found high in the canopy but also rarely seen on the ground or close to it.

BEHAVIOR
This is a solitary species. The only groups reported are those composed of a female and her suckling young attached to the mammae. Primarily arboreal and rarely abandons the shelter of the high and medium canopy. They are primarily nocturnal and crepuscular, decreasing their activity in periods of high levels of lunar illumination.

FEEDING ECOLOGY AND DIET
Feeds primarily on fruits but also takes some leaves, insects, bird eggs, and nestlings.

REPRODUCTIVE BIOLOGY
Polygamous. Females construct nests with plant matter inside hollow trees. After a brief gestation of less than 15 days, one to six young are born blind, naked, and with closed ears. The young crawl to the nipple area, where each attaches to one. There is no well-developed pouch, only lateral folds of skin. Reproduction may occur throughout the year but most frequently at the start of the rainy season.

CONSERVATION STATUS
Seems to be dependent on undisturbed tropical moist forest, although it has sometimes been found in secondary vegetation. Destruction of its habitat is the most serious threat. The IUCN lists the species as Lower Risk/Near Threatened.

SIGNIFICANCE TO HUMANS
It is sometimes kept as a pet, and otherwise the species is considered harmful to banana and citrus plantations. ◆

Water opossum
Chironectes minimus

SUBFAMILY
Didelphinae

TAXONOMY
Latra minima (Zimmerman, 1780), Cayenne, French Guiana.

OTHER COMMON NAMES
French: Opossum aquatique; German: Schwimmbeutler; Spanish: Tlacuache de agua, cuica de agua, yapok, zorro de agua, comadreja de agua.

PHYSICAL CHARACTERISTICS
Length 10.6–15.7 in (27–40 cm); weight 21.2–28 oz (600–790 g). Dense and silky hair with four to five broad dark brown bands across the back joined along the dorsal spine. Venter is silvery white and tail is almost naked and scaly. The eyes are large and black, and the face has another dark band across the eyes. The tail is slightly flattened laterally and bicolored, with the distant half whitish. The toes are clearly webbed to aid in swimming.

Lestodelphys halli

Chironectes minimus

Virginia opossum
Didelphis virginiana

SUBFAMILY
Didelphinae

TAXONOMY
Didelphis virginiana Kerr, 1792, Virginia, United States.

OTHER COMMON NAMES
French: Opossum de Virginie; German: Nordopossum; Spanish: Tlacuache común, tlacuache norteño.

PHYSICAL CHARACTERISTICS
Length 14–21.6 in (35–55 cm); weight 28–176 oz (800–5,000 g). Long hair and dense underfur; colors vary from uniform whitish to blackish brown. Ears are black and naked, and cheeks are white. Tail nearly naked, scaly, and bicolored: the basal half is black and the distal half whitish. The feet have opposable thumbs that render their footprints unmistakable.

DISTRIBUTION
Extreme southwestern Canada, the west coast and the eastern half of the United States, tropical and subtropical Mexico, and Central America south to Costa Rica.

HABITAT
Tropical and temperate forests, in wet and subhumid ecosystems. Found in a wide variety of human-disturbed habitats from logged and secondary forests to agricultural lands and even landfills and urban areas.

BEHAVIOR
Solitary and nocturnal. Roosts in hollow trees and branches, in leaf litter, crevices and caves under rocks, and in the soil. Feigns death when threatened, gaping its mouth wide and lying on its side while emitting an offensive musky odor.

FEEDING ECOLOGY AND DIET
Omnivorous. Eats fruit, insects, eggs, small vertebrates, spoiled fruit, carrion, and even trash. Forages at night opportunistically and avoids other medium and large-sized mammals such as raccoons and skunks. It is not a good hunter but rather eats items that are easily available.

REPRODUCTIVE BIOLOGY
Polygamous. The female builds a nest with leaf litter and vegetation that she transports with the help of the tail. The reproductive season extends from January through July and two birth peaks are reported. After a gestation of about 13 days, as many as 21 young are born undeveloped. Only the first to make it to the pouch are able to survive, as the female has only 13 teats and rarely are all occupied by a young one.

CONSERVATION STATUS
Not threatened. Inhabits both pristine and modified ecosystems. They have colonized towns and cities and they are not rare in New York City, Miami, or other large cities. The species does not face any immediate threats of extinction.

SIGNIFICANCE TO HUMANS
Sometimes consumed for food by indigenous or mixed human groups. The species has been pointed out as an agricultural pest or a disease reservoir. ◆

DISTRIBUTION
Southern Mexico, through Central America, Colombia, Venezuela, Guyana, Suriname, French Guiana, Ecuador, Peru, Brazil, Bolivia, Paraguay, and Argentina.

HABITAT
Rivers and streams in primary lowland tropical moist forest, generally from sea level up to about 3,300 ft (1,000 m), although it has been found at 6,230 ft (1,900 m).

BEHAVIOR
A solitary species that swims under water to avoid danger. It is primarily nocturnal, secretive, and silent.

FEEDING ECOLOGY AND DIET
Feeds mainly on fish, crustaceans, and mollusks. This is the most carnivorous of the New World opossums.

REPRODUCTIVE BIOLOGY
Polygamous. The female constructs a den on the bank of a river or stream where she builds a nest with vegetation. The young are born very undeveloped after a short gestation. The female keeps the young in a well-developed pouch that closes hermetically when she swims. The male also has a marsupium to protect the testicles.

CONSERVATION STATUS
Considered Lower Risk/Near Threatened by the IUCN, but the water opossum has disappeared from many areas in its historical distribution. Deforestation and water pollution are two factors that determine their local extinction.

SIGNIFICANCE TO HUMANS
None known. ◆

Microbiotheria
Monitos del monte
(*Microbiotheriidae*)

Class Mammalia

Order Microbiotheria

Family Microbiotheriidae

Number of families 1

Thumbnail description
Small, mouse-like, South American nocturnal marsupials

Size
Head and body length 3–5 in (8–13 cm); tail length 3.5–5.2 in (9–13 cm); weight 16–42g (0.5–1.4 oz)

Number of genera, species
1 genus; 1 species

Habitat
Occupies dense, humid forest

Conservation status
Vulnerable

Distribution
Chile and Argentina

Evolution and systematics

The mouse-like monito del monte is a small, inconspicuous species of marsupial from South America. Its unprepossessing appearance belies its huge zoological importance. This drab little mammal is a relic of a bygone era—all that remains of a once-prominent group that may have given rise to the entire diverse marsupial fauna that now lives on the other side of the world in Australia. The Microbiotheridae are therefore considered part of the otherwise exclusively Australian group known as the cohort Australidelphia.

Of the seven species of microbiothere that have been described, six members of the genus *Microbiotherium* are long extinct and are known only from fossil remains. This order of marsupials has only one living family, the Microbiotheriidae, with a single representative, *Dromiciops gliroides*, the diminutive monito del monte, also known in its native Chile as the *colocolo*. *Dromiciops* has several very primitive traits and is only distantly related to other South American marsupials (opossums). In fact, it has more in common with some Australian marsupials, in particular the carnivorous dasyurids,

from which it must have been separated for at least 40 million years. The monito del monte is thus thought to be the only surviving member of an early offshoot of the marsupial lineage that lived on the ancient supercontinent of Gondwana, to which Australia, Antarctica, and South America were once joined. A microbiotherian marsupial could have been the ancestor of the diverse marsupial fauna seen in Australia today. While the descendants of ancient micorbiotheres were doing well in Australia, the initially successful marsupial radiation in South America came to an end when placental mammals arrived from the north. Like most other South American marsupial orders, the little microbiotheres were mostly edged out—all but *Dromiciops*, which alone survives to this day.

The taxonomy of this species is *Dromiciops gliroides* (Thomas, 1894), Biobio, Chile, "Huite, NE Chiloe Island."

Physical characteristics

The monito del monte is small (head and body less than 5 in [13 cm] long) and superficially mouse-like, with a pointed

Monito del monte (*Dromiciops gliroides*). (Illustration by Michelle Meneghini)

face, small rounded ears, and large eyes. Its furry tail is prehensile, up to 5 in (16 cm) long and sometimes very thick, especially around the base where fat accumulates prior to hibernation. There is a patch of naked skin on the underside of the tail tip, which helps improves traction when the tail is used to grasp branches. The paws are similar to those of opossums, each having opposable toes adapted for gripping. The face is short and pointed, with very large eyes ringed with dark fur. The ears are oval and erect. The body is covered in soft brownish gray fur, which fades to buff-white on the belly and sometimes shows a pattern of swirls on the shoulders and back.

Distribution

The monito del monte is found only in the southern Andes mountains between the latitudes of 36° and 43°S in Chile, and just over the border into southwestern Argentina. This very limited distribution makes the Microbiotheria the least widespread of all living mammalian orders.

Habitat

This small, nimble marsupial lives for the most part off the ground, in trees and shrubs and especially in thickets of Chilean *Chusquea* bamboo. These grow best in the cool, humid forests of the Andes foothills. The species' common name means "mouse of the mountain."

Behavior

Monitos del monte are nocturnal and active most of the year in milder parts of their range. However, prolonged periods of cold winter weather or food shortage induce hibernation during which the animals survive on reserves of fat stored in the base of the tail. They spend most of their life aboveground, climbing with great skill using hands, feet, and tail to grasp branches and stems. They build intricate but sturdy nests of twigs and bamboo leaves woven into a ball. Preferred nest sites are tree holes or dense thickets, but rocky crevices and hollow fallen logs are also used, and nests are sometimes built suspended aboveground in trailing liana vines. The bamboo leaves with which the nests are made are waterproof, so the lining material of soft moss and grass remains dry and snug. Mosses are also sometime used to adorn the outside of the nest as well, helping to make it less conspicuous.

Individuals typically live in pairs or in small groups that usually comprise a mother and up to four young of the current year. Neighbors and family members communicate by sound; the most distinctive call is a long, trilling sound ending in a soft, hoarse cough.

Feeding ecology and diet

Monitos mainly eat insect grubs and pupae, also flies and small lizards that they collect from leaves and from cracks and crevices in tree bark. They grab prey with their nimble hands. They will also eat fruits, especially when preparing to hibernate in the fall.

Reproductive biology

The breeding season is in spring, and males and females mate in October soon after emerging from hibernation. The first litters appear in November, but continue to be born until January. Litters of between one and four young are normal; there are some records of five but, since the female only has four teats, the chances are the fifth will not survive. The young spend the first few weeks of life in their mother's pouch, attached to a teat. As soon as they are old enough to maintain their own body heat, the female leaves them in the nest while she goes out to feed. A little later, the young family accompanies the mother, at first riding on her back, and then learning to climb and feed themselves. Once independent, they often remain close to their mother for the rest of the year. They are sexually mature at one year old.

Conservation status

The monito del monte was once considered a common animal in parts of its range, but it is increasingly under pressure from habitat loss due to deforestation and development. The species' very restricted range means that even small losses can be significant. Having survived apparently little change for at least 20 million years, it would be a tragedy if the monito del monte was to become extinct as a result of human negligence—its loss would represent the demise of an entire order of animals. The species is listed as Vulnerable by the IUCN.

A fat-tailed dunnart (*Sminthopsis crassicaudata*) forages in Australia. (Photo by Animals Animals ©Hans & Judy Beste. Reproduced by permission.)

structure for hunting. Even the largest species require dense vegetation or crevices as refuge from mammalian and raptorial predators. Tasmanian devils cover many miles (kilometers) in a night's foraging and show a preference for habitats with an open understory or routes through dense vegetation.

Behavior

The majority of dasyurids for which spacing patterns have been studied occupies undefended home ranges that overlap with other individuals of both sexes. The females of two species of quolls maintain a core or major part of their home range exclusive to other females, but overlap with several males. The mechanism of territorial defense is not known. At the other extreme, among very small arid-zone dasyurids, drifting home ranges, transience, and high mobility are common. Very long-distance movements relative to the diminutive size of these animals have been recorded, including movements in excess of 0.6 mi (1 km) in 24 hours in the 1.05 oz (30 g) white-footed dunnart (*Sminthopsis leucopus*). This strategy is adaptive in environments where insect prey abundance is low and unpredictable.

Feeding ecology and diet

The huge range of body sizes in the Dasyuridae means that diet encompasses a broad range of invertebrate and vertebrate prey sizes. Prey size increases with body size. Dasyurids that are less than 5.2 oz (150 g) in body size are mostly insectivorous, although they may kill and eat small mammals, lizards, and frogs, and eat carrion of larger species if it is available. Carnivory (consumption of vertebrate prey) gradually replaces insectivory as body size increases. At approximately 2.2 lb (1 kg), dasyurids become too large to support themselves primarily on invertebrates, and carnivory takes over as the

principal component of the diet. Only the two largest species, the Tasmanian devil and the spotted-tailed quoll (*Dasyurus maculatus*), in which adult females and males exceed 4.4 lb (2 kg), are exclusively carnivorous. Tasmanian devils are specialized scavengers as well as being highly effective predators, although all species are likely to eat carrion if it is available. Several species have been recorded eating soft fruit or flowers seasonally, including antechinuses and eastern quolls.

Some species, both large and small, are renowned for their ferocity and take prey up to several times their body size. Prey is killed using generalized crushing bites towards the anterior end. The rear of the skull and the nape are often targeted in small vertebrates, and devils and spotted-tailed quolls go for the throat or chest of macropods.

Reproductive biology

The degree of reproductive synchrony and seasonality in dasyurids is associated with latitude and climatic predictability. Reproductive seasonality is known for approximately half of species with an information bias on a few temperate and

A red-tailed phascogale (*Phascogale calura*) eats a gecko (*Gymnodactylus* sp.). (Photo by Eric Lindgren/Photo Researchers, Inc. Reproduced by permission.)

Brush-tailed phascogales (*Phascogale tapoatafa*) prefer Australia's eucalyptus forests for foraging and nesting sites. (Photo by Michael Morcombe. Bruce Coleman, Inc. Reproduced by permission.)

Among the smaller species, larger (3.5–17.6 oz;100–500 g) body size and restricted habitat associations correlate strongly with endangerment. Habitat loss and fragmentation, altered fire regimes, and predation are the main threatening processes. The larger dasyurids have been more affected by human impacts than the smaller species. This is perhaps a consequence of their lower population densities and greater needs for space. They also are more likely to run into direct conflict with humans over livestock depredations, and are susceptible to non-target poisoning from fox baits and road mortality. The principal factor threatening the smaller quolls has been predation by red foxes, resulting in catastrophic declines and population extinctions across continental Australia everywhere fox populations are abundant. Tasmania has remained dingo- and fox-free until very recently (2000) and has functioned as a refuge for larger dasyurids, supporting healthy populations of three species.

A recovery plan implemented in 1992 for the chuditch (western quoll) has used captive breeding, reintroduction, and translocation of quolls to suitable areas of habitat within its former distributional range, as well as intense ongoing fox control through poison-baiting programs. The success

arid zone animals; very little is known of the New Guinean species. Most Australian dasyurids are seasonal breeders, and probably promiscuous. Reproduction is tightly synchronous (three to four weeks) in many temperate species, particularly the semelparous antechinuses and phascogales, but can extend over a number of months in arid zone animals. New Guinean dasyurids from wet tropical forests, for the two species of *Murexia* and one species of *Phascolosorex*, breed year-round. Changes in photoperiod seem to be the most important force driving timing of reproduction, which is consistent with aseasonal breeding in the wet tropical forests. In arid areas, rainfall events are important in defining the precise timing of reproduction within the broader seasonal window. This flexibility in arid-zone species enables reproduction to be synchronized with maximal food supply after rain, as breeding in predictably seasonal regions is timed so that young emerge in the late spring food flush.

Conservation status

Fourteen (24%) of the 58 smaller (less than 17.6 oz; 500 g) and two (40%) of the five larger (more than 17.6 oz; 500 g) Australian dasyurids are classified as Vulnerable, Endangered, or Data Deficient (IUCN criteria). This list does not include another five small and two larger species that are Lower Risk/Near Threatened. There is insufficient information available to assess the status of the New Guinean dasyurids.

A slender-tailed dunnart (*Sminthopsis murina*) scenting the air. (Photo by Animals Animals ©B. & B. Wells, OSF. Reproduced by permission.)

A numbat (*Myrmecobius fasciatus*) searches for termites. (Photo by Frans Lanting/Minden Pictures. Reproduced by permission.)

sence of a pouch, maintain attachment orally and by entwining the forelimbs in the crimped fur of the mammary region. Development is slow and young are carried for six to seven months, after which they are deposited in a nest. At this stage they are furred with visible stripes, but their eyes are not yet open. The young are suckled for another three months, until at least late October. During this time they gradually explore and forage within their mother's home range. The female may move them to another nest, particularly in response to disturbance, and does so by carrying small young on her back.

Conservation status

The decline of the numbat, from its formerly wide distribution at the time of European settlement, is documented. Populations disappeared gradually in an east-west progression, with the expansion in range of introduced foxes. The rate of disappearance accelerated after 1920 when fox populations suddenly exploded. By the 1960s, numbats persisted in only two locations: the Gibson Desert and the southwest of Western Australia. The desert population disappeared first, leaving only two populations to the southwest of Perth by 1985.

An experimental fox control program, initiated in the early 1980s, demonstrated that numbat populations increased when fox populations were suppressed by monthly poison baiting. Fox predation was confirmed as the primary factor in the decline of numbats. Since 1985, there has been a successful recovery program involving translocation of wild individuals, supplemented with the reintroduction of captive-bred numbats to suitable habitat in nature reserves within their former southwestern range. This program, combined with regular fox baiting, has increased wild populations to nine localities. An additional two populations live within large, fenced reserves in South Australia and New South Wales. Rates of increase in translocated populations vary with the levels of predation by (native) raptors, residual levels of foxes and feral cats, dispersal opportunities, and habitat type that are related to food supply. Numbats probably never occurred in high density, even though they were widespread. Populations in which wide dispersal is limited by fencing or surrounding farmland increase more rapidly. In 1994, numbats were upgraded from an Endangered listing to Vulnerable under IUCN Red List criteria.

Significance to humans

Numbat is an aboriginal name from South Australia. Central Australian aboriginal peoples knew the animal as

Numbats (*Myrmecobius fasciatus*) use scent and a long sticky tongue to locate their prey. The tongue can extend as far as 4 in (10 cm) from its mouth. (Photo by Bill Bachman/Photo Researchers, Inc. Reproduced by permission.)

Two juvenile numbats (*Myrmecobius fasciatus*) in Western Australia. (Photo by Animals Animals ©A. Wells, OSF. Reproduced by permission.)

"walpurti" and hunted it to eat. Individuals were tracked to burrows where they were dug up. No commercial exploitation of numbats is recorded, but up to 200 have been collected for museum specimens.

Resources

Books

Archer, M., T. Flannery, S. Hand, and J. Long. *Prehistoric Mammals of Australia and New Guinea: One Hundred Million Years of Evolution.* Sydney: UNSW Press, 2002.

Friend, J. A. "Myrmecobiidae." In *Fauna of Australia.* Canberra: Australian Government Publishing Service, 1989.

Friend, J. A., and N. D. Thomas. "Conservation of the Numbat (*Myrmecobius fasciatus*)." In *Predators with Pouches: The Biology of Carnivorous Marsupials,* edited by M. E. Jones, C. R. Dickman, and M. Archer. Melbourne: CSIRO Publishing, 2003.

Krajewski, C., and M. Westerman. "Molecular Systematics of Dasyuromorphia." In *Predators with Pouches: The Biology of Carnivorous Marsupials,* edited by M. E. Jones, C. R. Dickman, and M. Archer. Melbourne: CSIRO Publishing, 2003.

Strahan, R. *The Mammals of Australia.* Sydney: Australian Museum, Reed Books, 1995.

Menna Jones, PhD

CONTRIBUTORS TO THE FIRST EDITION

Dr. Helmut Kramer
Zoological Research Institute and A.
Koenig Museum
Bonn, Germany

Dr. Franz Krapp
Zoological Institute, University of
Freiburg
Freiburg, Switzerland

Dr. Otto Kraus
Professor, University of Hamburg,
and Director, Zoological Institute and
Museum
Hamburg, Germany

Dr. Hans Krieg
Professor and First Director (retired),
Scientific Collections of the State of
Bavaria
Munich, Germany

Dr. Heinrich Kühl
Federal Research Institute for
Fisheries, Cuxhaven Laboratory
Cuxhaven, Germany

Dr. Oskar Kuhn
Professor, formerly University
Halle/Saale
Munich, Germany

Dr. Hans Kumerloeve
First Director (retired), State
Scientific Museum, Vienna
Munich, Germany

Dr. Nagamichi Kuroda
Yamashina Ornithological Institute,
Shibuya-Ku
Tokyo, Japan

Dr. Fred Kurt
Zoological Museum of Zurich
University, Smithsonian Elephant
Survey
Colombo, Ceylon

Dr. Werner Ladiges
Professor and Chief Curator,
Zoological Institute and Museum,
University of Hamburg
Hamburg, Germany

Leslie Laidlaw
Department of Animal Sciences,
Purdue University
Lafayette, Indiana, U.S.A.

Dr. Ernst M. Lang
Director, Zoological Garden
Basel, Switzerland

Dr. Alfredo Langguth
Department of Zoology, Faculty of
Humanities and Sciences, University
of the Republic
Montevideo, Uruguay

Leo Lehtonen
Science Writer
Helsinki, Finland

Bernd Leisler
Second Zoological Institute,
University of Vienna
Vienna, Austria

Dr. Kurt Lillelund
Professor and Director, Institute for
Hydrobiology and Fishery Sciences,
University of Hamburg
Hamburg, Germany

R. Liversidge
Alexander MacGregor Memorial
Museum
Kimberley, South Africa

Dr. Konrad Lorenz
Professor and Director, Max Planck
Institute for Behavioral Physiology
Seewiesen/Obb., Germany

Dr. Martin Lühmann
Federal Research Institute for the
Breeding of Small Animals
Celle, Germany

Dr. Johannes Lüttschwager
Oberstudienrat (retired)
Heidelberg, Germany

Dr. Wolfgang Makatsch
Bautzen, Germany

Dr. Hubert Markl
Professor and Director, Zoological
Institute, Technical University of
Darmstadt
Darmstadt, Germany

Basil J. Marlow , BSc (Hons)
Curator, Australian Museum
Sydney, Australia

Dr. Theodor Mebs
Instructor of Biology
Weissenhaus/Ostsee, Germany

Dr. Gerlof Fokko Mees
Curator of Birds, Rijks Museum of
Natural History
Leiden, The Netherlands

Hermann Meinken
Director, Fish Identification Institute,
V.D.A.
Bremen, Germany

Dr. Wilhelm Meise
Chief Curator, Zoological Institute
and Museum, University of Hamburg
Hamburg, Germany

Dr. Joachim Messtorff
Field Station of the Federal Fisheries
Research Institute
Bremerhaven, Germany

Dr. Marian Mlynarski
Professor, Polish Academy of
Sciences, Institute for Systematic and
Experimental Zoology
Cracow, Poland

Dr. Walburga Moeller
Nature Museum
Hamburg, Germany

Dr. H. C. Erna Mohr
Curator (retired), Zoological State
Institute and Museum
Hamburg, Germany

Dr. Karl-Heinz Moll
Waren/Müritz, Germany

Dr. Detlev Müller-Using
Professor, Institute for Game
Management, University of Göttingen
Hannoversch-Münden, Germany

Werner Münster
Instructor of Biology
Ebersbach, Germany

Dr. Joachim Münzing
Altona Museum
Hamburg, Germany

Contributors to the first edition

Dr. Wilbert Neugebauer
Wilhelma Zoo
Stuttgart-Bad Cannstatt, Germany

Dr. Ian Newton
Senior Scientific Officer, The Nature
Conservancy
Edinburgh, Scotland

Dr. Jürgen Nicolai
Max Planck Institute for Behavioral
Physiology
Seewiesen/Obb., Germany

Dr. Günther Niethammer
Professor, Zoological Research
Institute and A. Koenig Museum
Bonn, Germany

Dr. Bernhard Nievergelt
Zoological Museum, University of
Zurich
Zurich, Switzerland

Dr. C. C. Olrog
Institut Miguel Lillo San Miguel de
Tucumán
Tucumán, Argentina

Alwin Pedersen
Mammal Research and Arctic Explorer
Holte, Denmark

Dr. Dieter Stefan Peters
Nature Museum and Senckenberg
Research Institute
Frankfurt a.M., Germany

Dr. Nicolaus Peters
Scientific Councillor and Docent,
Institute of Hydrobiology and
Fisheries, University of Hamburg
Hamburg, Germany

Dr. Hans-Günter Petzold
Assistant Director, Zoological Garden
Berlin, Germany

Dr. Rudolf Piechocki
Docent, Zoological Institute,
University of Halle
Halle a.d.S., Germany

Dr. Ivo Poglayen-Neuwall
Director, Zoological Garden
Louisville, Kentucky, U.S.A.

Dr. Egon Popp
Zoological Collection of the State of
Bavaria
Munich, Germany

Dr. H. C. Adolf Portmann
Professor Emeritus, Zoological
Institute, University of Basel
Basel, Switzerland

Hans Psenner
Professor and Director, Alpine Zoo
Innsbruck, Austria

Dr. Heinz-Siburd Raethel
Oberveterinärrat
Berlin, Germany

Dr. Urs H. Rahm
Professor, Museum of Natural History
Basel, Switzerland

Dr. Werner Rathmayer
Biology Institute, University of
Konstanz
Konstanz, Germany

Walter Reinhard
Biologist
Baden-Baden, Germany

Dr. H. H. Reinsch
Federal Fisheries Research Institute
Bremerhaven, Germany

Dr. Bernhard Rensch
Professor Emeritus, Zoological
Institute, University of Münster
Münster, Germany

Dr. Vernon Reynolds
Docent, Department of Sociology,
University of Bristol
Bristol, England

Dr. Rupert Riedl
Professor, Department of Zoology,
University of North Carolina
Chapel Hill, North Carolina, U.S.A.

Dr. Peter Rietschel
Professor (retired), Zoological
Institute, University of Frankfurt a.M.
Frankfurt a.M., Germany

Dr. Siegfried Rietschel
Docent, University of Frankfurt;
Curator, Nature Museum and
Research Institute Senckenberg
Frankfurt a.M., Germany

Herbert Ringleben
Institute of Ornithology, Heligoland
Ornithological Station
Wilhelmshaven, Germany

Dr. K. Rohde
Institute for General Zoology, Ruhr
University
Bochum, Germany

Dr. Peter Röben
Academic Councillor, Zoological
Institute, Heidelberg University
Heidelberg, Germany

Dr. Anton E. M. De Roo
Royal Museum of Central Africa
Tervuren, South Africa

Dr. Hubert Saint Girons
Research Director, Center for
National Scientific Research
Brunoy (Essonne), France

Dr. Luitfried Von Salvini-Plawen
First Zoological Institute, University
of Vienna
Vienna, Austria

Dr. Kurt Sanft
Oberstudienrat, Diesterweg-
Gymnasium
Berlin, Germany

Dr. E. G. Franz Sauer
Professor, Zoological Research
Institute and A. Koenig Museum,
University of Bonn
Bonn, Germany

Dr. Eleonore M. Sauer
Zoological Research Institute and A.
Koenig Museum, University of Bonn
Bonn, Germany

Dr. Ernst Schäfer
Curator, State Museum of Lower
Saxony
Hannover, Germany

Rhinopomatidae [Family]
  Rhinopoma [Genus]
    R. hardwickei [Species]
    R. microphyllum
    R. muscatellum

Emballonuridae [Family]
  Balantiopteryx [Genus]
    B. infusca [Species]
    B. io
    B. plicata
  Centronycteris [Genus]
    C. maximiliani [Species]
  Coleura [Genus]
    C. afra [Species]
    C. seychellensis
  Cormura [Genus]
    C. brevirostris [Species]
  Cyttarops [Genus]
    C. alecto [Species]
  Diclidurus [Genus]
    D. albus [Species]
    D. ingens
    D. isabellus
    D. scutatus
  Emballonura [Genus]
    E. alecto [Species]
    E. atrata
    E. beccarii
    E. dianae
    E. furax
    E. monticola
    E. raffrayana
    E. semicaudata
  Mosia [Genus]
    M. nigrescens [Species]
  Peropteryx [Genus]
    P. kappleri [Species]
    P. leucoptera
    P. macrotis
  Rhynchonycteris [Genus]
    R. naso [Species]
  Saccolaimus [Genus]
    S. flaviventris [Species]
    S. mixtus
    S. peli
    S. pluto
    S. saccolaimus
  Saccopteryx [Genus]
    S. bilineata [Species]
    S. canescens
    S. gymnura
    S. leptura
  Taphozous [Genus]
    T. australis [Species]
    T. georgianus
    T. hamiltoni
    T. hildegardeae
    T. hilli
    T. kapalgensis
    T. longimanus

T. mauritianus
T. melanopogon
T. nudiventris
T. perforatus
T. philippinensis
T. theobaldi

Craseonycteridae [Family]
  Craseonycteris [Genus]
    C. thonglongyai [Species]

Nycteridae [Family]
  Nycteris [Genus]
    N. arge [Species]
    N. gambiensis
    N. grandis
    N. hispida
    N. intermedia
    N. javanica
    N. macrotis
    N. major
    N. nana
    N. thebaica
    N. tragata
    N. woodi

Megadermatidae [Family]
  Cardioderma [Genus]
    C. cor [Species]
  Lavia [Genus]
    L. frons [Species]
  Macroderma [Genus]
    M. gigas [Species]
  Megaderma [Genus]
    M. lyra [Species]
    M. spasma

Rhinolophidae [Family]
  Rhinolophus [Genus]
    R. acuminatus [Species]
    R. adami
    R. affinis
    R. alcyone
    R. anderseni
    R. arcuatus
    R. blasii
    R. borneensis
    R. canuti
    R. capensis
    R. celebensis
    R. clivosus
    R. coelophyllus
    R. cognatus
    R. cornutus
    R. creaghi
    R. darlingi
    R. deckenii
    R. denti
    R. eloquens
    R. euryale
    R. euryotis

R. ferrumequinum
R. fumigatus
R. guineensis
R. hildebrandti
R. hipposideros
R. imaizumii
R. inops
R. keyensis
R. landeri
R. lepidus
R. luctus
R. maclaudi
R. macrotis
R. malayanus
R. marshalli
R. megaphyllus
R. mehelyi
R. mitratus
R. monoceros
R. nereis
R. osgoodi
R. paradoxolophus
R. pearsoni
R. philippinensis
R. pusillus
R. rex
R. robinsoni
R. rouxi
R. rufus
R. sedulus
R. shameli
R. silvestris
R. simplex
R. simulator
R. stheno
R. subbadius
R. subrufus
R. swinnyi
R. thomasi
R. trifoliatus
R. virgo
R. yunanensis

Hipposideridae [Family]
  Anthops [Genus]
    A. ornatus [Species]
  Asellia [Genus]
    A. patrizii [Species]
    A. tridens
  Aselliscus [Genus]
    A. stoliczkanus [Species]
    A. tricuspidatus
  Cloeotis [Genus]
    C. percivali [Species]
  Coelops [Genus]
    C. frithi [Species]
    C. hirsutus
    C. robinsoni
  Hipposideros [Genus]
    H. abae [Species]
    H. armiger

MAMMALS SPECIES LIST

H. ater
H. beatus
H. bicolor
H. breviceps
H. caffer
H. calcaratus
H. camerunensis
H. cervinus
H. cineraceus
H. commersoni
H. coronatus
H. corynophyllus
H. coxi
H. crumeniferus
H. curtus
H. cyclops
H. diadema
H. dinops
H. doriae
H. dyacorum
H. fuliginosus
H. fulvus
H. galeritus
H. halophyllus
H. inexpectatus
H. jonesi
H. lamottei
H. lankadiva
H. larvatus
H. lekaguli
H. lylei
H. macrobullatus
H. maggietaylorae
H. marisae
H. megalotis
H. muscinus
H. nequam
H. obscurus
H. papua
H. pomona
H. pratti
H. pygmaeus
H. ridleyi
H. ruber
H. sabanus
H. schistaceus
H. semoni
H. speoris
H. stenotis
H. turpis
H. wollastoni
Paracoelops [Genus]
P. megalotis [Species]
Rhinonicteris [Genus]
R. aurantia [Species]
Triaenops [Genus]
T. furculus [Species]
T. persicus

Phyllostomidae [Family]
Ametrida [Genus]

A. centurio [Species]
Anoura [Genus]
A. caudifer [Species]
A. cultrata
A. geoffroyi
A. latidens
Ardops [Genus]
A. nichollsi [Species]
Ariteus [Genus]
A. flavescens [Species]
Artibeus [Genus]
A. amplus [Species]
A. anderseni
A. aztecus
A. cinereus
A. concolor
A. fimbriatus
A. fraterculus
A. glaucus
A. hartii
A. hirsutus
A. inopinatus
A. jamaicensis
A. lituratus
A. obscurus
A. phaeotis
A. planirostris
A. toltecus
Brachyphylla [Genus]
B. cavernarum [Species]
B. nana
Carollia [Genus]
C. brevicauda [Species]
C. castanea
C. perspicillata
C. subrufa
Centurio [Genus]
C. senex [Species]
Chiroderma [Genus]
C. doriae [Species]
C. improvisum
C. salvini
C. trinitatum
C. villosum
Choeroniscus [Genus]
C. godmani [Species]
C. intermedius
C. minor
C. periosus
Choeronycteris [Genus]
C. mexicana [Species]
Chrotopterus [Genus]
C. auritus [Species]
Desmodus [Genus]
D. rotundus [Species]
Diaemus [Genus]
D. youngi [Species]
Diphylla [Genus]
D. ecaudata [Species]
Ectophylla [Genus]

E. alba [Species]
Erophylla [Genus]
E. sezekorni [Species]
Glossophaga [Genus]
G. commissarisi [Species]
G. leachii
G. longirostris
G. morenoi
G. soricina
Hylonycteris [Genus]
H. underwoodi [Species]
Leptonycteris [Genus]
L. curasoae [Species]
L. nivalis
Lichonycteris [Genus]
L. obscura [Species]
Lionycteris [Genus]
L. spurrelli [Species]
Lonchophylla [Genus]
L. bokermanni [Species]
L. dekeyseri
L. handleyi
L. hesperia
L. mordax
L. robusta
L. thomasi
Lonchorhina [Genus]
L. aurita [Species]
L. fernandezi
L. marinkellei
L. orinocensis
Macrophyllum [Genus]
M. macrophyllum [Species]
Macrotus [Genus]
M. californicus [Species]
M. waterhousii
Mesophylla [Genus]
M. macconnelli [Species]
Micronycteris [Genus]
M. behnii [Species]
M. brachyotis
M. daviesi
M. hirsuta
M. megalotis
M. minuta
M. nicefori
M. pusilla
M. schmidtorum
M. sylvestris
Mimon [Genus]
M. bennettii [Species]
M. crenulatum
Monophyllus [Genus]
M. plethodon [Species]
M. redmani
Musonycteris [Genus]
M. harrisoni [Species]
Phylloderma [Genus]
P. stenops [Species]
Phyllonycteris [Genus]

B. major
Berylmys [Genus]
  B. berdmorei [Species]
  B. bowersi
  B. mackenziei
  B. manipulus
Bibimys [Genus]
  B. chacoensis [Species]
  B. labiosus
  B. torresi
Blanfordimys [Genus]
  B. afghanus [Species]
  B. bucharicus
Blarinomys [Genus]
  B. breviceps [Species]
Bolomys [Genus]
  B. amoenus [Species]
  B. lactens
  B. lasiurus
  B. obscurus
  B. punctulatus
  B. temchuki
Brachiones [Genus]
  B. przewalskii [Species]
Brachytarsomys [Genus]
  B. albicauda [Species]
Brachyuromys [Genus]
  B. betsileoensis [Species]
  B. ramirohitra
Bullimus [Genus]
  B. bagobus [Species]
  B. luzonicus
Bunomys [Genus]
  B. andrewsi [Species]
  B. chrysocomus
  B. coelestis
  B. fratrorum
  B. heinrichi
  B. penitus
  B. prolatus
Calomys [Genus]
  C. boliviae [Species]
  C. callidus
  C. callosus
  C. hummelincki
  C. laucha
  C. lepidus
  C. musculinus
  C. sorellus
  C. tener
Calomyscus [Genus]
  C. bailwardi [Species]
  C. baluchi
  C. hotsoni
  C. mystax
  C. tsolovi
  C. urartensis
Canariomys [Genus]
  C. tamarani [Species]
Cannomys [Genus]

C. badius [Species]
Cansumys [Genus]
  C. canus [Species]
Carpomys [Genus]
  C. melanurus [Species]
  C. phaeurus
Celaenomys [Genus]
  C. silaceus [Species]
Chelemys [Genus]
  C. macronyx [Species]
  C. megalonyx
Chibchanomys [Genus]
  C. trichotis [Species]
Chilomys [Genus]
  C. instans [Species]
Chiromyscus [Genus]
  C. chiropus [Species]
Chinchillula [Genus]
  C. sahamae [Species]
Chionomys [Genus]
  C. gud [Species]
  C. nivalis
  C. roberti
Chiropodomys [Genus]
  C. calamianensis [Species]
  C. gliroides
  C. karlkoopmani
  C. major
  C. muroides
  C. pusillus
Chiruromys [Genus]
  C. forbesi [Species]
  C. lamia
  C. vates
Chroeomys [Genus]
  C. andinus [Species]
  C. jelskii
Chrotomys [Genus]
  C. gonzalesi [Species]
  C. mindorensis
  C. whiteheadi
Clethrionomys [Genus]
  C. californicus [Species]
  C. centralis
  C. gapperi
  C. glareolus
  C. rufocanus
  C. rutilus
  C. sikotanensis
Coccymys [Genus]
  C. albidens [Species]
  C. ruemmleri
Colomys [Genus]
  C. goslingi [Species]
Conilurus [Genus]
  C. albipes [Species]
  C. penicillatus
Coryphomys [Genus]
  C. buhleri [Species]
Crateromys [Genus]

C. australis [Species]
  C. paulus
  C. schadenbergi
Cremnomys [Genus]
  C. blanfordi [Species]
  C. cutchicus
  C. elvira
Cricetomys [Genus]
  C. emini [Species]
  C. gambianus
Cricetulus [Genus]
  C. alticola [Species]
  C. barabensis
  C. kamensis
  C. longicaudatus
  C. migratorius
  C. sokolovi
Cricetus [Genus]
  C. cricetus [Species]
Crossomys [Genus]
  C. moncktoni [Species]
Crunomys [Genus]
  C. celebensis [Species]
  C. fallax
  C. melanius
  C. rabori
Dacnomys [Genus]
  D. millardi [Species]
Dasymys [Genus]
  D. foxi [Species]
  D. incomtus
  D. montanus
  D. nudipes
  D. rufulus
Delanymys [Genus]
  D. brooksi [Species]
Delomys [Genus]
  D. dorsalis [Species]
  D. sublineatus
Dendromus [Genus]
  D. insignis [Species]
  D. kahuziensis
  D. kivu
  D. lovati
  D. melanotis
  D. mesomelas
  D. messorius
  D. mystacalis
  D. nyikae
  D. oreas
  D. vernayi
Dendroprionomys [Genus]
  D. rousseloti [Species]
Deomys [Genus]
  D. ferrugineus [Species]
Dephomys [Genus]
  D. defua [Species]
  D. eburnea
Desmodilliscus [Genus]
  D. braueri [Species]

Desmodillus [Genus]
   D. auricularis [Species]
Dicrostonyx [Genus]
   D. exsul [Species]
   D. groenlandicus
   D. hudsonius
   D. kilangmiutak
   D. nelsoni
   D. nunatakensis
   D. richardsoni
   D. rubricatus
   D. torquatus
   D. unalascensis
   D. vinogradovi
Desmomys [Genus]
   D. harringtoni [Species]
Dinaromys [Genus]
   D. bogdanovi [Species]
Diomys [Genus]
   D. crumpi [Species]
Diplothrix [Genus]
   D. legatus [Species]
Echiothrix [Genus]
   E. leucura [Species]
Eropeplus [Genus]
   E. canus [Species]
Eligmodontia [Genus]
   E. moreni [Species]
   E. morgani
   E. puerulus
   E. typus
Eliurus [Genus]
   E. majori [Species]
   E. minor
   E. myoxinus
   E. penicillatus
   E. tanala
   E. webbi
Ellobius [Genus]
   E. alaicus [Species]
   E. fuscocapillus
   E. lutescens
   E. talpinus
   E. tancrei
Eolagurus [Genus]
   E. luteus [Species]
   E. przewalskii
Eothenomys [Genus]
   E. chinensis [Species]
   E. custos
   E. eva
   E. inez
   E. melanogaster
   E. olitor
   E. proditor
   E. regulus
   E. shanseius
Euneomys [Genus]
   E. chinchilloides [Species]
   E. fossor

E. mordax
E. petersoni
Galenomys [Genus]
   G. garleppi [Species]
Geoxus [Genus]
   G. valdivianus [Species]
Gerbillurus [Genus]
   G. paeba [Species]
   G. setzeri
   G. tytonis
   G. vallinus
Gerbillus [Genus]
   G. acticola [Species]
   G. allenbyi
   G. andersoni
   G. bilensis
   G. bottai
   G. burtoni
   G. cheesmani
   G. dalloni
   G. diminutus
   G. dunni
   G. floweri
   G. gerbillus
   G. grobbeni
   G. henleyi
   G. hoogstraali
   G. juliani
   G. lowei
   G. maghrebi
   G. mesopotamiae
   G. nancillus
   G. nigeriae
   G. percivali
   G. poecilops
   G. pulvinatus
   G. pyramidum
   G. riggenbachi
   G. ruberrimus
   G. somalicus
   G. syrticus
   G. vivax
Golunda [Genus]
   G. ellioti [Species]
Grammomys [Genus]
   G. aridulus [Species]
   G. caniceps
   G. dolichurus
   G. gigas
   G. macmillani
   G. rutilans
Graomys [Genus]
   G. domorum [Species]
   G. griseoflavus
Gymnuromys [Genus]
   G. roberti [Species]
Habromys [Genus]
   H. chinanteco [Species]
   H. lepturus
   H. lophurus

H. simulatus
Hadromys [Genus]
   H. humei [Species]
Haeromys [Genus]
   H. margarettae [Species]
   H. minahassae
   H. pusillus
Hapalomys [Genus]
   H. delacouri [Species]
   H. longicaudatus
Heimyscus [Genus]
   H. fumosus [Species]
Hodomys [Genus]
   H. alleni [Species]
Holochilus [Genus]
   H. brasiliensis [Species]
   H. chacarius
   H. magnus
   H. sciureus
Hybomys [Genus]
   H. basilii [Species]
   H. eisentrauti
   H. lunaris
   H. planifrons
   H. trivirgatus
   H. univittatus
Hydromys [Genus]
   H. chrysogaster [Species]
   H. habbema
   H. hussoni
   H. neobrittanicus
   H. shawmayeri
Hylomyscus [Genus]
   H. aeta [Species]
   H. alleni
   H. baeri
   H. carillus
   H. denniae
   H. parvus
   H. stella
Hyomys [Genus]
   H. dammermani [Species]
   H. goliath
Hyperacrius [Genus]
   H. fertilis [Species]
   H. wynnei
Hypogeomys [Genus]
   H. antimena [Species]
Ichthyomys [Genus]
   I. hydrobates [Species]
   I. pittieri
   I. stolzmanni
   I. tweedii
Irenomys [Genus]
   I. tarsalis [Species]
Isthmomys [Genus]
   I. flavidus [Species]
   I. pirrensis
Juscelinomys [Genus]
   J. candango [Species]

INDEX

INDEX

INDEX

INDEX

INDEX

United States Fish and Wildlife Service
Aplodontia rufa nigra, 16:133
Arizona gray squirrels, 16:172
babirusas, 15:280
black-tailed prairie dogs, 16:155
Caprinae, 16:95
gray whales, 15:99
Heteromyidae, 16:203
Idaho ground squirrels, 16:158
Indian rhinoceroses, 12:224
peninsular pronghorns, 15:415
pygmy hogs, 15:280, 15:287
Sonoran pronghorns, 15:415–416
Vespertilioninae, 13:502
Unstriped ground squirrels, 16:159*t*
Urartsk mouse-like hamsters, 16:243
Urea, 12:122–123
Ureters, 12:90
See also Physical characteristics
Urials, 12:178, 16:6, 16:87, 16:89, 16:91,
16:92–93, 16:104*t*
Urinary tract, 12:90–91
See also Physical characteristics
Urocyon cinereoargenteus. See Gray foxes
Urocyon littoralis. See Island foxes
Uroderma bilobatum. See Tent-making bats
Urogale spp., 13:292
Urogale everetti. See Philippine tree shrews
Urogenital sinus, 12:90, 12:111
See also Physical characteristics
Uropsilinae, 13:279, 13:280
Uropsilus spp. *See* Shrew moles
Uropsilus investigator. See Yunnan shrew-
moles
Uropsilus soricipes. See Chinese shrew-moles
Urotrichus talpoides. See Greater Japanese
shrew moles
Ursidae. *See* Bears
Ursinae, 14:295
Ursus americanus. See American black bears
Ursus americanus floridanus, 14:300
Ursus arctos. See Brown bears
Ursus arctos middendorffi. See Kodiak bears
Ursus arctus. See Brown bears
Ursus maritimus. See Polar bears
Ursus thibetanus. See Asiatic black bears
USDI. *See* United States Department of the
Interior
Uta Hick's bearded sakis, 14:148, 14:151
Uterine glands, 12:94
See also Physical characteristics
Uterus, 12:90–91, 12:*102*
See also Physical characteristics
Uzungwa bushbabies, 14:33*t*

V

Vaginas, 12:91
See also Physical characteristics;
Reproduction
Valais goats, 15:*268*
Valley pocket gophers, 12:74, 12:*74,* 16:*188,*
16:*189,* 16:*190,* 16:*191,* 16:*192*
Val's gundis. *See* Desert gundis
Vampire bats, 13:*308,* 13:*311,* 13:*422*
behavior, 13:419
conservation status, 13:420
distribution, 13:415

feeding ecology, 13:314, 13:418, 13:*420*
hairy-legged, 13:314, 13:*434t*
humans and, 13:316, 13:317, 13:420
infrared energy, 12:83–84
physical characteristics, 13:308, 13:414
reproduction, 13:315, 13:419, 13:420
species of, 13:*425–426*
See also False vampire bats
Vampyrum spectrum. See Spectral bats
Van Zyl's golden moles, 13:222*t*
Vancouver Island marmots, 16:147, 16:*150,*
16:*152,* 16:156
Vaquitas, 15:9, 15:33, 15:34, 15:35–36,
15:*39t*
Varecia spp., 14:51
Varecia variegata. See Variegated lemurs
Variable flying foxes. *See* Island flying foxes
Variable pocket gophers, 16:*196t*
Variegated lemurs, 14:*53,* 14:*57,* 14:*58–59*
Variegated spider monkeys, 14:*156*
Variegated squirrels, 16:*168,* 16:*170,* 16:172
Varying hares. *See* Mountain hares
Vas deferens, 12:91
Veldkamp's dwarf epauletted bats. *See* Little
flying cows
Veldkamp's dwarf fruit bats. *See* Little flying
cows
Velvet (antlers), 12:40
See also Physical characteristics
Velvety free-tailed bats. *See* Greater house
bats
Venezuelan hemorrhagic fever, 16:270
Venezuelan red howler monkeys, 14:*157,*
14:*158,* 14:*161,* 14:*162–163*
Venezuelan spiny rats, 16:*450*
Venom, platypus, 12:232, 12:244
Venous hearts, bats, 12:60
Vermilingua, 13:171
Verraux's sifakas, 14:66, 14:67
Vertebral column, 12:8
See also Physical characteristics
Vertebrates, flight, 12:52–61
See also specific vertebrates
Vervet monkeys. *See* Grivets
Vesper mice, 16:267, 16:269, 16:270
Vespertilio murinus. See Parti-colored bats
Vespertilionidae, **13:497–526,** 13:*504,*
13:*505,* 13:*523*
behavior, 13:498–500, 13:521
conservation status, 13:501–502, 13:522
distribution, 13:*497,* 13:498, 13:*520,*
13:520–521
evolution, 13:497, 13:519
feeding ecology, 13:313, 13:500, 13:521
habitats, 13:498, 13:521
humans and, 13:502, 13:522
physical characteristics, 13:310, 13:497–498,
13:519–520
reproduction, 13:500–501, 13:522
species of, 13:*506–514,* 13:*515t–516t,*
13:524–525, 13:*526t*
taxonomy, 13:497, 13:519
Vespertilioninae, **13:497–517,** 13:*504,*
13:*505,* 13:*515t–516t*
Veterinary medicine, in zoos, 12:210–211
Vibrations, 12:83, 12:*83*
See also Physical characteristics
Vibrissae, 12:38–39, 12:*43,* 12:72
See also Physical characteristics
Victoriapithecus spp., 14:172, 14:188–189

Vicugna spp. *See* Vicuñas
Vicugna vicugna. See Vicuñas
Vicuñas, 12:132, 15:142, 15:151, **15:313–323,**
15:*314,* 15:*317,* 15:*319,* 15:*321*
Vietnam warty pigs, 15:263, 15:280, 15:289*t*
Vietnamese pot-bellied pigs, 15:*278*
Viruses, in rabbit control, 12:187–188
Visayan spotted deer, 15:360, 15:*371t*
Visayan warty pigs. *See* Cebu bearded pigs
Viscacha rats, 16:436, 16:*437,* 16:*438,*
16:*439–440*
Viscachas, **16:377–384,** 16:*380,* 16:*381,*
16:*383t*
Vision, 12:5–6, 12:50, 12:76–77, 12:79–80
See also Eyes; Physical characteristics
Visual social signals, 12:160
See also Behavior
Vitamin requirements, 12:120, 12:123
See also Feeding ecology
Vitelline sac. *See* Yolk sac
Viverra spp., 14:335, 14:338
Viverra civettina. See Malabar civets
Viverra megaspila. See Oriental civets
Viverra zibetha. See Indian civets
Viverricula spp., 14:335, 14:338
Viverricula indica, 14:336
Viverridae, **14:335–345,** 14:*339*
behavior, 14:258, 14:337
conservation status, 14:262, 14:338
distribution, 12:135, 12:136, 14:*335,* 14:336
evolution, 14:335
feeding ecology, 14:260, 14:337
habitats, 14:337
humans and, 14:338
physical characteristics, 14:256, 14:335–336
reproduction, 14:338
species of, 14:*340–343,* 14:344*t–*345*t*
taxonomy, 14:256, 14:335
Viverrinae, 14:335
Viviparous reproduction, 12:5, 12:7, 12:106
See also Reproduction
Vizcachas. *See* Plains viscachas
Vlei rats. *See* Angoni vlei rats
Vocalizations, 12:85
See also Behavior
Voice organs, 12:8
See also Physical characteristics
Volant mammals, 12:200–201
Volcano rabbits, 16:482, 16:483, 16:487,
16:505, 16:509, 16:*510,* 16:513–*514*
Volcano shrews, 13:277*t*
Voles, **16:225–238,** 16:*231,* 16:*232*
behavior, 12:141–142, 12:148, 16:226–227
conservation status, 16:229–230
distribution, 16:225, 16:226
evolution, 16:225
feeding ecology, 16:227–228
habitats, 16:124, 16:226
humans and, 16:127
parasitic infestations, 12:78
physical characteristics, 16:225–226
predation and, 12:115
reproduction, 16:228–229
species of, 16:233–237*t,* 16:238*t*
taxonomy, 16:225
Vombatidae. *See* Wombats
Vombatiformes, 13:31, 13:32
Vombatus ursinus. See Common wombats

INDEX

INDEX